# Theory of
# Computational Complexity

# Theory of Computational Complexity

**Ding-Zhu Du**

**Ker-I Ko**

A Wiley-Interscience Publication

**JOHN WILEY & SONS, INC.**

New York • Chichester • Weinheim • Brisbane • Singapore • Toronto

This book is printed on acid-free paper. ☺

Copyright © 2000 by John Wiley & Sons, Inc. All rights reserved.

Published simultaneously in Canada.

No part of this publication may be reproduced, stored in a retrieval system or transmitted in any form or by any means, electronic, mechanical, photocopying, recording, scanning or otherwise, except as permitted under Sections 107 or 108 of the 1976 United States Copyright Act, without either the prior written permission of the Publisher, or authorization through payment of the appropriate per-copy fee to the Copyright Clearance Center, 222 Rosewood Drive, Danvers, MA 01923, (978) 750-8400, fax (978) 750-4744. Requests to the Publisher for permission should be addressed to the Permissions Department, John Wiley & Sons, Inc., 605 Third Avenue, New York, NY 10158-0012, (212) 850-6011, fax (212) 850-6008, E-Mail: PERMREQ@WILEY.COM.

For ordering and customer service, call 1-800-CALL WILEY.

Library of Congress Cataloging in Publication Data is available.

ISBN 0-471-34506-7

Printed in the United States of America

10 9 8 7 6 5 4 3

# Contents

# *Preface*

Computational complexity theory has been a central area of theoretical computer science since its early development in the mid-1960s. Its subsequent rapid development in the next three decades has not only established itself as a rich, exciting theory but also shown strong influence on many other related areas in computer science, mathematics, and operations research. We may roughly divide its development into three periods. The works from the mid-1960s to the early 1970s paved a solid foundation for the theory of feasible computation. The Turing machine-based complexity theory and the axiomatic complexity theory established computational complexity as an independent mathematics discipline. The identification of polynomial-time computability with feasible computability and the discovery of the *NP*-complete problems consolidated the *P* versus *NP* question as the central issue of the complexity theory.

From the early 1970s to the mid-1980s, research in computational complexity expanded at an exponential rate both in depth and in breadth. Various new computational models were proposed as alternatives to the traditional deterministic models. Finer complexity hierarchies and complexity classes were identified from these new models and more accurate classifications have been obtained for the complexity of practical algorithmic problems. Parallel computational models, such as alternating Turing machines and parallel random access machines, together with the *NC* hierarchy, provide a tool for the classification of the complexity of feasible problems. Probabilistic Turing machines are a model for the complexity theory of distribution-independent randomized algorithms. Interactive proof systems, an extension of probabilistic Turing machines, and communication complexity study the complexity aspect of distributed or interactive computing. The study of one-way functions led to a breakthrough in cryptography. A theory of average-case completeness, based on the notion of distribution-faithful reductions, aims at establishing the foundation for the distribution-dependent average-case complexity. Boolean and threshold circuits are models for nonuniform complexity in which algebraic and topological methods have found interesting applications. Oracle Turing machines are the model for the theory of relativization in which combinatorial techniques meet recursion-theoretic techniques. Program-size complexity (or, Kolmogorov complexity) formalizes the notion of descriptive complexity and has strong connections with computational complexity. Although the central questions remain open, these developments demonstrate that computational complexity is a rich discipline with both a deep mathematical theory and a diverse area of applications.

Beginning in the mid-1980s, we have seen a number of deep, surprising results using diverse, sophisticated proof techniques. In addition, seemingly independent subareas have found interesting connections. The exponential lower bounds for monotone circuits and constant-depth circuits have been found using probabilistic and algebraic methods. The connection between constant-depth circuits and relativization led to the relativized separation of the polynomial-time hierarchy. The technique of nondeterministic iterative counting has been used to collapse the nondeterministic space hierarchy. The study of probabilistic reductions gave us the surprising result about the power of the counting class $\#P$ versus the polynomial-time hierarchy. Arithmetization of Boolean computation in interactive proof systems collapses the class of polynomial space to the class of sets with interactive proof systems. Further development of this research, together with techniques of coding theory, have led to strong negative results in combinatorial approximation.

As outlined above, complexity theory has grown fast both in breadth and in depth. With so many new computational models, new proof techniques and applications in different areas, it is simply not possible to cover all important topics of the theory in a single book. The goal of this book is therefore not to provide a comprehensive treatment of complexity theory. Instead, we only select some of the fundamental areas which we believe represent the most important recent advances in complexity theory, in particular, on the $P$ versus $NP$ problem, and present the complete treatment of these subjects. The presentation follows the approach of traditional mathematics textbooks. With a small number of exceptions, all theorems are presented with rigorous mathematical proofs.

We divide the subjects of this book into three parts, each representing a different approach to computational complexity. In Part I, we develop the theory of uniform computational complexity, which is based on the worst-case complexity measures on traditional models of Turing machines, both deterministic ones and nondeterministic ones. The central issue here is the $P$ versus $NP$ question, and we apply the notions of reducibility and completeness to develop a complexity hierarchy. We first develop in Chapter 1 the notion of time and space complexity and complexity classes. Two basic proof techniques, simulation and diagonalization, including Immerman and Szelepcsényi's iterative counting technique, are presented. The knowledge of recursion theory is useful here but is not required. Chapter 2 presents the notion of *NP*-completeness, including Cook's Theorem and a few well-known *NP*-complete problems. The relations between decision problems versus search problems are carefully discussed through the notion of polynomial-time Turing reducibility. Chapter 3 extends the theory of *NP*-completeness to the polynomial-time hierarchy and polynomial space. In addition to complete problems for these complexity classes, we also present the natural characterizations of these complexity classes by alternating Turing machines and alternating quantifiers. In Chapter 4, the structure of the class *NP* is analyzed in several different views. We present both the abstract proof that there exist problems in *NP* that are neither *NP*-complete nor in *P*, assuming *NP* does not collapse to *P*, and some natural problems as the candidates of such problems, as well as their applications in public-key cryptography. The controversial theory of relativization and their interpretations are also introduced and discussed in this chapter.

In Part II, we study the theory of nonuniform computational complexity, including the computational models of decision trees and Boolean circuits, and the notion of sparse sets. The nonuniform computational models grew out of our inability to solve the major open questions in the uniform complexity theory. It is hoped that the simpler structure of these nonuniform models will allow better lower bound results. Although the efforts so far are not strong enough to settle the major open questions in the area of uniform complexity theory, a number of nontrivial lower bound results have been obtained through new proof techniques. The emphasis of this part is thus not on the subjects themselves but on the proof techniques. In Chapter 5, we present both the algebraic and the topological techniques to prove the lower bounds for decision trees of Boolean functions, particularly Boolean functions representing monotone graph properties. In Chapter 6, we present two exponential lower bound results on circuits using the approximation circuit technique and the probabilistic method. The notion of sparse sets links the study of nonuniform complexity with uniform complexity theory. This interesting interconnection between uniform and nonuniform complexity theory, such as the question of *NC* versus *P*, is also studied in Chapter 6. Then, we present, in Chapter 7, the works on the Hartmanis-Berman conjecture about the polynomial-time isomorphism of *NP*-complete problems, which provide further insight into the structure of the complexity class *NP*.

Part III is about the theory of probabilistic complexity, which studies the complexity issues related to randomized computation. Randomization in algorithms started in the late 1970s, and has become increasingly popular. The computational model for randomized algorithms, the probabilistic Turing machine, and the corresponding probabilistic complexity classes are introduced in Chapter 8. The notion of probabilistic quantifiers is used to provide characterizations of these complexity classes, and their relations with deterministic and nondeterministic complexity classes are discussed. The counting problems and the complexity class #*P* may be viewed as an extension of probablistic computation, and they are the main subjects of Chapter 9. Valiant's proof completeness of the permanent problem, as well as Toda's theorem that all problems in the polynomial-time hierarchy are reducible to problems in $\#P$, are presented. The exponential lower bound of constant-depth circuits developed in Chapter 6 has an interesting application to the relativized separation of the complexity class $\#P$ from the polynomial-time hierarchy. This result is also presented in Chapter 9. Chapter 10 studies the notion of interactive proof systems, which is another extension of probabilistic computation. The collapse of the interactive proof systems hierarchy is presented, and the relations between the interactive proof systems and the Arthur-Merlin proof systems are discussed. Shamir's characterization of polynomial space by interactive proof systems is also presented as a prelude to the recent breakthrough on probabilistically checkable proofs. This celebrated result of Arora et al., that probabilistically checkable proofs with a constant number of queries characterize precisely the class $NP$, is presented in Chapter 11. We also present, in this chapter, the application of this result to various combinatorial approximation problems.

Although we have tried hard to include, within this framework, as many subjects in complexity theory as possible, many interesting topics inevitably have to be omitted. Two of the most important topics that we are not able to include here are program-size complexity (or, Kolmogorov complexity) and average-case completeness. Program-size complexity is a central theory which would provide a unified view of the other nonuniform models of Part II. However, this topic has grown into an independent discipline in recent years and has become too big to be included here. Interested readers are referred to the comprehensive book of Li and Vitányi [1997]. Average-case completeness provides a different view toward the notion of distribution-independent average-case complexity, and would complement the works studied in Part III about distribution-independent probabilistic complexity. This theory, however, seems to be still in the early development stage. Much research is needed before we can better understand its proof techniques and its relation to the worst-case complexity theory, and we reluctantly omit it here. We refer interested readers to Wang [1997] for a thorough review of this topic. Exercises at the end of each chapter often include additional topics that are worth studying but are omitted in the main text due to space limitations.

This book is grown out of authors' lecture notes developed in the past ten years at the University of Minnesota and the State University of New York at Stony Brook. We have taught from these notes in several different ways. For a one-semester graduate course in which the students have had limited exposure to theory, we typically cover most of Part I plus a couple of chapters from either Part II or Part III. For better prepared students, a course emphasizing the recent advances can be taught based mainly on either Part II or Part III. Seminars based on some advanced materials in Parts II and III, plus recent journal papers, have also been conducted for Ph.D. students.

We are grateful to all our colleagues and students who have made precious suggestions, corrections, and criticism on the earlier drafts of this book. We are also grateful to the following institutions for their financial support in preparing this book: the National Science Foundation of the United States, National Science Council of Taiwan, National Natural Science Foundation of China, National 973 Fundamental Research Program of China, City University of Hong Kong, and National Chiao-Tung University of Taiwan.

DING-ZHU DU
KER-I KO

# Part I

# Uniform Complexity

*In P or not in P,*
*That is the question.*
— William Shakespeare (?)

# 1

# *Models of Computation and Complexity Classes*

The notions of algorithms and complexity are meaningful only when they are defined in terms of formal computational models. In this chapter, we introduce our basic computational models: deterministic Turing machines and nondeterministic Turing machines. Based on these models, we define the notion of time and space complexity and the fundamental complexity classes including $P$ and $NP$. In the last two sections, we study two best known proof techniques, diagonalization and simulation, that are used to separate and collapse complexity classes, respectively.

## 1.1 Strings, Coding, and Boolean Functions

Our basic data structure is a string. All other data structures are to be encoded and represented by strings. A *string* is a finite sequence of symbols. For instance, the word *"string"* is a string over the symbols of English letters; the arithmetic expression "$3 + 4 - 5$" is a string over symbols 3, 4, 5, $+$, and $-$. Thus, to describe a string, we must specify the set of symbols to occur in that string. We call a finite set of symbols to be used to define strings an

3

*alphabet.* Note that not every finite set can be an alphabet. A finite set $S$ can be an alphabet if and only if the following condition holds:

**Property 1.1** *Two finite sequences of elements in $S$ are identical if and only if the elements in the two sequences are identical respectively in ordering.*

For example, $\{0, 1\}$ and $\{00, 01\}$ are alphabets, but $\{1, 11\}$ is not an alphabet since 11 can be formed by either 11 or (1 and 1).

Assume that $\Sigma$ is an alphabet. A set of strings over the alphabet $\Sigma$ is called a *language*. A collection of languages is called a *language class*, or simply a *class*.

The length of a string $x$ is the number of symbols in the string $x$, denoted by $|x|$. For example, $|string| = 6$ and $|3 + 4 - 5| = 5$. For convenience, we allow a string to contain no symbol. Such a string is called the *empty string*, which is denoted by $\lambda$. So, $|\lambda| = 0$. (The notation $|\cdot|$ is also used on sets. If $S$ is a finite set, we write $|S|$ to denote its cardinality.)

There is a fundamental operation on strings. The *concatenation* of two strings $x$ and $y$ is the string $xy$. The concatenation follows associative law, that is, $x(yz) = (xy)z$. Moreover, $\lambda x = x\lambda = x$. Thus, all strings over an alphabet form a monoid under concatenation.[1] We denote $x^0 = \lambda$ and $x^n = xx^{n-1}$ for $n \geq 1$.

The concatenation operation on strings can be extended to languages. The concatenation of two languages $A$ and $B$ is the language $AB = \{ab : a \in A, b \in B\}$. We also denote $A^0 = \{\lambda\}$ and $A^n = AA^{n-1}$ for $n \geq 1$. In addition, we define $A^* = \bigcup_{i=0}^{\infty} A^i$. The language $A^*$ is called the *Kleene closure* of $A$. The Kleene closure of an alphabet is the set of all strings over the alphabet.

For convenience, we will often work only on strings over the alphabet $\{0, 1\}$. To show that this does not impose a serious restriction on the theory, we note that there exists a simple way of encoding strings over any finite alphabet into the strings over $\{0, 1\}$. Let $X$ be a finite set. A one-one mapping $f$ from $X$ to $\Sigma^*$ is called a *coding* (of $X$ in $\Sigma^*$). If both $X$ and $\{f(x) : x \in X\}$ are alphabets, then, by Property 1.1, $f$ induces a coding from $X^*$ to $\Sigma^*$. Suppose that $X$ is an alphabet of $n$ elements. Choose $k = \lceil \log n \rceil$ and choose a one-one mapping $f$ from $X$ to $\{0, 1\}^k$.[2] Note that any subset of $\{0, 1\}^k$ is an alphabet, and hence $f$ is a coding from $X$ to $\{0, 1\}^*$ and $f$ induces a coding from $X^*$ to $\{0, 1\}^*$.

Given a linear ordering for an alphabet $\Sigma = \{a_1, \cdots, a_n\}$, the *lexicographic ordering* $<$ on $\Sigma^*$ is defined as follows: $x = a_{i_1} a_{i_2} \cdots a_{i_m} < y = a_{j_1} a_{j_2} \cdots a_{j_k}$ if and only if either $[m < k]$ or $[m = k$ and for some $\ell < m$, $i_1 = j_1, \cdots, i_\ell = j_\ell$ and $i_{\ell+1} < j_{\ell+1}]$. The lexicographic ordering is a coding from natural numbers to all strings over an alphabet.

---

[1] A set with an associative multiplication operation and an identity element is a *monoid*. A monoid is a *group* if every element in it has an inverse.

[2] Throughout this book, unless otherwise stated, log denotes the logarithm function with base 2.

A coding from $\Sigma^* \times \Sigma^*$ to $\Sigma^*$ is also called a *pairing function* on $\Sigma^*$. As an example, for $x$, $y \in \{0,1\}^*$ define $\langle x,y \rangle = 0^{|x|}1xy$ and $x\#y = x0y1x^R$, where $x^R$ is the reverse of $x$. Then $\langle \cdot, \cdot \rangle$ and "$\#$" are pairing functions on $\{0,1\}^*$. A pairing function induces a coding from $\underbrace{\Sigma^* \times \cdots \times \Sigma^*}_{n}$ to $\Sigma^*$ by defining

$$\langle x_1, x_2, x_3, \ldots, x_n \rangle = \langle \cdots \langle \langle x_1, x_2 \rangle, x_3 \rangle, \ldots, x_n \rangle.$$

Pairing functions can also be defined on natural numbers. For instance, let $\iota : \{0,1\}^* \to \mathbf{N}$ be the lexicographic ordering function; that is, $\iota(x) = n$ if $x$ is the $n$th string in $\{0,1\}^*$ under the lexicographic ordering (starting with 0). Then, we can define a pairing function on natural numbers from a pairing function on binary strings: $\langle n, m \rangle = \iota(\langle \iota^{-1}(n), \iota^{-1}(m) \rangle)$.

In the above, we have seen some specific simple codings. In general, if $A$ is a finite set of strings over some alphabet, when can $A$ be an alphabet? Clearly, $A$ cannot contain the empty string $\lambda$ because $\lambda x = x\lambda$. The following theorem gives another necessary condition.

**Theorem 1.2** (McMillan's Theorem) *Let $s_1, \ldots, s_q$ be $q$ nonempty strings over an alphabet of $r$ symbols. If $\{s_1, \ldots, s_q\}$ is an alphabet, then*

$$\sum_{i=1}^{q} r^{-|s_i|} \leq 1.$$

*Proof.* For any natural number $n$, consider

$$\left( \sum_{i=1}^{q} r^{-|s_i|} \right)^n = \sum_{k=n}^{n\ell} m_k r^{-k},$$

where $\ell = \max\{|s_1|, \ldots, |s_q|\}$ and $m_k$ is the number of elements in the following set:

$$A_k = \{(i_1, \cdots, i_n) : 1 \leq i_1 \leq q, \ldots, 1 \leq i_n \leq q, k = |s_{i_1}| + \cdots + |s_{i_n}|\}.$$

Since $\{s_1, \ldots, s_q\}$ is an alphabet, different vectors $(i_1, \ldots, i_n)$ correspond to different strings $s_{i_1} \ldots s_{i_n}$. The strings corresponding to vectors in $A_k$ all have length $k$. Note that there are at most $r^k$ strings of length $k$. Therefore, $m_k \leq r^k$. It implies

$$\left( \sum_{i=1}^{q} r^{-|s_i|} \right)^n \leq \sum_{k=n}^{n\ell} r^k r^{-k} = n\ell - (n-1) \leq n\ell. \qquad (1.1)$$

Now, suppose $\sum_{i=1}^{q} r^{-|s_i|} > 1$. Then for sufficiently large $n$, $(\sum_{i=1}^{q} r^{-|s_i|})^n > n\ell$, contradicting (1.1). ∎

A *Boolean function* is a function whose variable values and function value are all in $\{\text{TRUE}, \text{FALSE}\}$. We often denote TRUE by 1 and FALSE by 0. In the following table, we show two Boolean functions of two variables, *conjunction* $\wedge$ and *disjunction* $\vee$, and a Boolean function of a variable, *negation* $\neg$.

| $x$ | $y$ | $x \wedge y$ | $x \vee y$ | $\neg x$ |
|---|---|---|---|---|
| 0 | 0 | 0 | 0 | 1 |
| 0 | 1 | 0 | 1 | 1 |
| 1 | 0 | 0 | 1 | 0 |
| 1 | 1 | 1 | 1 | 0 |

All Boolean functions can be defined in terms of these three functions. For instance, the two-variable function *exclusive-or* $\oplus$ can be defined by

$$x \oplus y = ((\neg x) \wedge y) \vee (x \wedge (\neg y)).$$

For simplicity, we also write $xy$ for $x \wedge y$, $x + y$ for $x \vee y$ and $\bar{x}$ for $\neg x$. A table like the above, in which the value of a Boolean function for each possible input is given explicitly, is called a *truth-table* for the Boolean function. For each Boolean function $f$ over variables $x_1, x_2, \ldots, x_n$, a function $\tau : \{x_1, x_2, \ldots, x_n\} \rightarrow \{0, 1\}$ is called a *Boolean assignment* (or, simply, an *assignment*) for $f$. An assignment on $n$ variables can be seen as a *binary* string of length $n$, that is, a string in $\{0, 1\}^n$. A function $\tau : Y \rightarrow \{0, 1\}$, where $Y = \{x_{i_1}, x_{i_2}, \ldots, x_{i_k}\}$ is a subset of $X = \{x_1, x_2, \ldots, x_n\}$, is called a *partial assignment* on $X$. A partial assignment $\tau$ on $n$ variables can be seen as a string of length $n$ over $\{0, 1, *\}$, with $*$ denoting "unchanged." If $\tau : Y \rightarrow \{0, 1\}$ is a partial assignment for $f$, we write $f|_\tau$ or $f|_{x_{i_1} = \tau(x_{i_1}), \ldots, x_{i_k} = \tau(x_{i_k})}$ to denote the function obtained by substituting $\tau(x_{i_j})$ for $x_{i_j}$, $1 \leq j \leq k$, into $f$. This function $f|_\tau$ is a Boolean function on $X - Y$, and is called the *restriction* of $f$ by $\tau$. We say a partial assignment $\tau$ *satisfies* $f$, or $\tau$ is a *truth assignment* for $f$, if $f|_\tau = 1$.[3]

The functions conjunction, disjunction, and exclusive-or all follow the commutative and associative laws. The distributive law holds for conjunction to disjunction, disjunction to conjunction, and conjunction to exclusive-or, that is, $(x + y)z = xz + yz$, $xy + z = (x + z)(y + z)$, and $(x \oplus y)z = xz \oplus yz$. An interesting and important law about negation is de Morgan's law, that is, $\overline{xy} = \bar{x} + \bar{y}$ and $\overline{x + y} = \bar{x}\bar{y}$. A *Boolean formula* is a formula over Boolean variables using operators $\vee$, $\wedge$, and $\neg$.

A *literal* is either a Boolean variable or the negation of a Boolean variable. An *elementary product* is a product of several literals. Consider an elementary product $p$ and a Boolean function $f$. If $p = 1$ implies $f = 1$, then $p$ is called an *implicant* of $f$. An implicant $p$ is *prime* if no product of any proper subset of the literals defining $p$ is an implicant of $f$. A prime implicant is also called a *minterm*. For example, function $f(x_1, x_2, x_3) = (x_1 + x_2)(\bar{x}_2 + x_3)$ has minterms $x_1 \bar{x}_2$, $x_1 x_3$, and $x_2 x_3$. $x_1 \bar{x}_2 x_3$ is an implicant of $f$ but not a minterm. The *size* of an implicant is the number of variables in the implicant. We let $D_1(f)$ denote the maximum size of minterms of $f$. A DNF (*disjunctive*

---

[3]In the literature, the term *truth assignment* sometimes simply means a Boolean assignment.

*normal form*) is a sum of elementary products. Every Boolean function is equal to the sum of all its minterms. So, every Boolean function can be represented by a DNF with terms of size at most $D_1(f)$. For a constant function $f \equiv 0$ or $f \equiv 1$, we define $D_1(f) = 0$. For a nonconstant function $f$, we always have $D_1(f) \geq 1$.

Similarly, an *elementary sum* is a sum of several literals. Consider an elementary sum $c$ and a Boolean function $f$. If $c = 0$ implies $f = 0$, then $c$ is called a *clause* of $f$. A minimal clause is also called a *prime* clause. The size of a clause is the number of literals in it. We let $D_0(f)$ denote the maximum size of prime clauses of $f$. A CNF (*conjunctive normal form*) is a product of elementary sums. Every Boolean function is equal to the product of all its prime clauses, which is a CNF with clauses of size at most $D_0(f)$. For a constant function $f \equiv 0$ or $f \equiv 1$, we define $D_0(f) = 0$. For a nonconstant function $f$, we always have $D_0(f) \geq 1$.

The following states a relation between implicants and clauses.

**Proposition 1.3** *Any implicant and any clause of a Boolean function $f$ have at least one variable in common.*

*Proof.* Let $p$ and $c$ be an implicant and a clause of $f$, respectively. Suppose that $p$ and $c$ have no variable in common. Then we can assign values to all variables in $p$ to make $p = 1$ and to all variables in $c$ to make $c = 0$ simultaneously. However, $p = 1$ implies $f = 1$ and $c = 0$ implies $f = 0$, which is a contradiction. ∎

## 1.2 Deterministic Turing Machines

Turing machines are a simple and yet powerful enough computational model. Almost all reasonable, general-purpose computational models have been known to be equivalent to Turing machines, in the sense that they define the same class of computable functions. There are many variations of Turing machines studied in literature. We are going to introduce, in this section, the simplest model of Turing machines, namely, the *deterministic Turing machine*. Another model, the *nondeterministic Turing machine*, is to be defined in the next section. Other generalized Turing machine models, such as deterministic and nondeterministic oracle Turing machines, will be defined in later chapters. In addition, we will introduce in Part II other *nonuniform* computational models which are not equivalent to Turing machines.

A deterministic (one-tape) Turing machine (TM, DTM) consists of two basic units: the *control unit* and the *memory unit*. The control unit contains a finite number of states. The memory unit is a tape that extends infinitely to both ends. The tape is divided into an infinite number of tape squares (or, tape cells). Each tape square stores one of a finite number of tape symbols. The communication between the control unit and the tape is through a *read/write tape head* that scans a tape square at a time. (See Figure 1.1.)

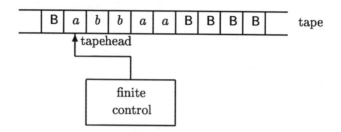

Figure 1.1: A Turing machine.

A normal *move* of a Turing machine consists of the following actions:

(1) Reading the tape symbol from the tape square currently scanned by the tape head;

(2) Writing a new tape symbol on the tape square currently scanned by the tape head;

(3) Moving the tape head to the right or to the left of the current square; and

(4) Changing to a new control state.

The exact actions of (2), (3), and (4) above depend on the current control state and the tape symbol read in (1). This relation between the current state and the current tape symbol and actions (2)–(4) is predefined by a *program.*

Formally, a TM $M$ is defined by the following information:

(1) A finite set $Q$ of states;

(2) An initial state $q_0 \in Q$;

(3) A subset $F \subseteq Q$ of accepting states;

(4) A finite set $\Sigma$ of input symbols;

(5) A finite set $\Gamma \supset \Sigma$ of tape symbols, including a special blank symbol $B \in \Gamma - \Sigma$; and

(6) A *partial* transition function $\delta$ that maps $(Q - F) \times \Gamma$ to $Q \times \Gamma \times \{L,R\}$ (the *program*).

In the above, the transition function $\delta$ is a partial function, meaning that the function $\delta$ may be undefined on some pairs $(q, s) \in (Q - F) \times \Gamma$. The use of the initial state, accepting states, and the blank symbol is explained below.

In order to discuss the notions of *accepting a language* and *computing a function* by a TM, we add some convention to the computation of a TM. First, we assume that initially an *input* string $w$ is stored in the consecutive squares of the tape of $M$, and the other squares contain the blank symbol B. The tape head of $M$ is initially scanning the leftmost square of the input $w$, and the machine starts at the initial state $q_0$. (Figure 1.1 shows the initial setting for a Turing machine with input *abbaa*.) Starting from this initial configuration, the machine $M$ operates move by move according to the transition function $\delta$. The machine may either operate forever, or it may halt when it enters a control state $q$ and reads a tape symbol $s$ for which $\delta(q, s)$ is undefined. If a TM $M$ eventually halts at an accepting state $q \in F$ on input $w$, then we say $M$ *accepts* $w$. If $M$ halts at a nonaccepting state $q \notin F$ on input $w$, then we say $M$ *rejects* $w$.

To formally define the notion of accepting an input string, we need to define the concept of *configurations*. A configuration $\alpha$ of a TM $M$ is a record of all the information of the computation of $M$ at a specific moment, which includes the current state, the current symbols in the tape, and the current position of the tape head. From this information, one can determine what the future computation is. Formally, a configuration of a TM $M$ is an element $(q, x, y)$ of $Q \times \Gamma^* \times \Gamma^*$ such that the leftmost symbol of $x$ and the rightmost symbol of $y$ are not B. A configuration $(q, x, y)$ denotes that the current state is $q$, that the current nonblank symbols in the tape are the string $xy$, and the tape head is scanning the leftmost symbol of $y$ (when $y$ is empty, the tape head is scanning the blank that is immediately to the right of the rightmost symbol of $x$).[4] Assuming $Q \cap \Gamma = \emptyset$, we also write $xqy$ to stand for $(q, x, y)$.

We now generalize the transition function $\delta$ of a TM $M$ to the *next configuration function* $\vdash_M$ (or, simply $\vdash$ if $M$ is understood) defined on configurations of $M$. Intuitively, the function $\vdash$ maps each configuration to the next configuration after one move of $M$. To handle the special nonblank requirement in the definition of configurations, we define two simple functions: $\ell(x) =$ the string $x$ with the leading blanks removed, and $r(x) =$ the string $x$ with the trailing blanks removed. Assume that $(q_1, x_1, y_1)$ is a configuration of $M$. If $y_1$ is not the empty string, then let $y_1 = s_1 y_2$ for some $s_1 \in \Gamma$ and $y_2 \in \Gamma^*$; if $y_1 = \lambda$, then let $s_1 = B$ and $y_2 = \lambda$. Then, we can formally define the function $\vdash$ as follows (we write $\alpha \vdash \beta$ for $\vdash(\alpha) = \beta$):

*Case 1.* $\delta(q_1, s_1) = (q_2, s_2, L)$ for some $q_2 \in Q$ and $s_2 \in \Gamma$. If $x_1 = \lambda$ then let $s_3 = B$ and $x_2 = \lambda$; otherwise, let $x_1 = x_2 s_3$ for some $x_2 \in \Gamma^*$ and $s_3 \in \Gamma$. Then, $(q_1, x_1, y_1) \vdash (q_2, \ell(x_2), r(s_3 s_2 y_2))$.

*Case 2.* $\delta(q_1, s_1) = (q_2, s_2, R)$ for some $q_2 \in Q$ and $s_2 \in \Gamma$. Then, $(q_1, x_1, y_1) \vdash (q_2, \ell(x_1 s_2), r(y_2))$.

*Case 3.* $\delta(q_1, s_1)$ is undefined. Then, $\vdash$ is undefined on $(q_1, x_1, y_1)$.

Now we define the notion of the computation of a TM. A TM $M$ *halts* on an input string $w$ if there exists a finite sequence of configurations $\alpha_0, \alpha_1, \ldots, \alpha_n$ such that

(1) $\alpha_0 = (q_0, \lambda, w)$ (this is called the *initial configuration* for input $w$);

(2) $\alpha_i \vdash \alpha_{i+1}$ for all $i = 0, 1, \ldots, n - 1$; and

(3) $\vdash(\alpha_n)$ is undefined.

---

[4]The nonblank requirement for the leftmost symbol of $x$ and for the rightmost symbol of $y$ is added so that each configuration has a unique finite representation.

A TM $M$ *accepts* an input string $w$ if $M$ halts on $w$ and, in addition, the halting state is in $F$; that is, in (3) above, $\alpha_n = (q, x, y)$ for some $q \in F$ and $x, y \in \Gamma^*$. A TM $M$ *outputs* $y \in \Sigma^*$ on input $w$ if $M$ halts on $w$ and, in addition, the final configuration $\alpha_n$ is of the form $\alpha_n = (q, \lambda, y)$ for some $q \in F$.

**Example 1.4** We describe a TM $M$ that accepts the strings in $L = \{a^i b a^j \mid 0 \le i \le j\}$. The machine $M$ has states $Q = \{q_0, q_1, \ldots, q_5\}$, with the initial state $q_0$ and accepting state $q_5$ (i.e., $F = \{q_5\}$). It accepts input symbols from $\Sigma = \{a, b\}$ and uses tape symbols in $\Gamma = \{a, b, c, B\}$. Figure 1.2 is the transition function $\delta$ of $M$.

| $\delta$ | $a$ | $b$ | $c$ | B |
|---|---|---|---|---|
| $q_0$ | $q_1, c, R$ | $q_4, B, R$ | $q_0, B, R$ | |
| $q_1$ | $q_1, a, R$ | $q_2, b, R$ | | |
| $q_2$ | $q_3, c, L$ | | $q_2, c, R$ | $q_2, B, R$ |
| $q_3$ | $q_3, a, L$ | $q_3, b, L$ | $q_3, c, L$ | $q_0, B, R$ |
| $q_4$ | $q_4, a, R$ | | $q_4, B, R$ | $q_5, B, R$ |

Figure 1.2: The transition function of machine $M$.

It is not hard to check that $M$ halts at state $q_5$ on all strings in $L$, that it halts at a state $q_i$, $0 \le i \le 4$, on strings having zero or more than one $b$, and that it does not halt on strings $a^i b a^j$ with $i > j \ge 0$. In the following, we show the computation paths of machine $M$ on some inputs (we write $x q_i y$ to denote the configuration $(q_i, x, y)$):

On input $abaa$: $q_0 abaa \vdash c q_1 baa \vdash cb q_2 aa \vdash c q_3 bca \vdash q_3 cbca \vdash q_3 Bcbca \vdash q_0 cbca \vdash q_0 bca \vdash q_4 ca \vdash q_4 a \vdash a q_4 \vdash aB q_5$.

On input $aaba$: $q_0 aaba \vdash c q_1 aba \vdash ca q_1 ba \vdash cab q_2 a \vdash ca q_3 bc \vdash c q_3 abc \vdash q_3 cabc \vdash q_3 Bcabc \vdash q_0 cabc \vdash q_0 abc \vdash c q_1 bc \vdash cb q_2 c \vdash cbc q_2 \vdash cbcB q_2 \vdash cbcBB q_2 \vdash \cdots$.

On input $abab$: $q_0 abab \vdash c q_1 bab \vdash cb q_2 ab \vdash c q_3 bcb \vdash q_3 cbcb \vdash q_3 Bcbcb \vdash q_0 cbcb \vdash q_0 bcb \vdash q_4 cb \vdash q_4 b$.                                                        □

The notion of computable languages and computable functions can now be formally defined. In the following, we say $f$ is a *partial function* defined on $\Sigma^*$ if the domain of $f$ is a subset of $\Sigma^*$, and $f$ is a *total function* defined on $\Sigma^*$ if the domain of $f$ is $\Sigma^*$.

**Definition 1.5** *(a) A language $A$ over a finite alphabet $\Sigma$ is* recursively enumerable (r.e.) *if there exists a TM $M$ that halts on all strings $w$ in $A$ and does not halt on any string $w$ in $\Sigma^* - A$.*

*(b) A language $A$ over a finite alphabet $\Sigma$ is* computable *(or, recursive) if there exists a TM $M$ that halts on all strings $w$ in $\Sigma^*$, accepts all strings $w$ in $A$ and does not accept any string $w$ in $\Sigma^* - A$.*

*(c) A partial function f defined from $\Sigma^*$ to $\Sigma^*$ is* partial computable *(or,* partial recursive*) if there exists a TM M that outputs $f(w)$ on all w in the domain of f and does not halt on any w not in the domain of f.*

*(d) A (total) function $f : \Sigma^* \to \Sigma^*$ is* computable *(or,* recursive*) if it is partial computable (i.e., the TM M that computes it halts on all $w \in \Sigma^*$).*

For each Turing machine $M$ with the input alphabet $\Sigma$, we let $L(M)$ denote the set of all strings $w \in \Sigma^*$ that are accepted by $M$. Thus, a language $A$ is recursively enumerable if and only if $A = L(M)$ for some Turing machine $M$. Also, a language $A$ is recursive if and only if $A = L(M)$ for some Turing machine $M$ that halts on all inputs $w$.

Recursive sets, recursively enumerable sets, partial recursive functions, and recursive functions are the main objects studied in *recursive function theory*, or, *recursion theory*. See, for instance, Rogers [1967] for a complete treatment.

The above classes of recursive sets and recursively enumerable sets are defined based on the model of deterministic, one-tape Turing machines. Since Turing machines look very primitive, the question arises whether Turing machines are as powerful as other machine models. In other words, do the classes of recursive sets and recursively enumerable sets remain the same if they are defined based on different computational models? The answer is yes, according to the famous Church-Turing Thesis:

**Church-Turing Thesis.** *A function computable in any reasonable computational model is computable by a Turing machine.*

What is a *reasonable computational model*? Intuitively, it is a model in which the following conditions hold:

(1) The computation of a function is given by a set of finite instructions.

(2) Each instruction can be carried out in this model in a finite number of steps, or in a finite amount of time.

(3) Each instruction can be carried out in this model in a deterministic manner.[5]

Since the notion of *reasonable computational models* in the Church-Turing Thesis is not well defined mathematically, we cannot prove the Church-Turing Thesis as a mathematical statement, but can only collect mathematical proofs as evidence to support it. So far, many different computational models have been proposed and compared with the Turing machine model, and all *reasonable* ones are proven to be equivalent to Turing machines. The Church-Turing Thesis thus remains trustworthy.

---

[5]By condition (1), we exclude the nonuniform models; by condition (2), we exclude the models with infinite amount of resources; and by condition (3), we exclude the nondeterministic models and probabilistic models. Although they are considered unreasonable, meaning probably not realizable by reliable physical devices, these nonstandard models remain as interesting mathematical models, and will be studied extensively in the rest of the book. In fact, we will see that the Church-Turing Thesis still holds even if we allow nondeterministic or probabilistic instructions in the computational model.

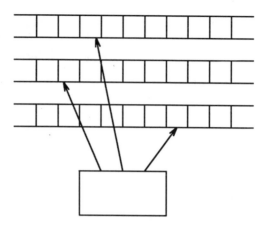

Figure 1.3: A multi-tape TM.

In the following, we show that multi-tape Turing machines compute the same class of functions as one-tape Turing machines. A *multi-tape TM* is similar to a one-tape TM with the following exceptions. First, it has a finite number of tapes that extends infinitely to the both ends. Each tape is equipped with its own tape head. All tape heads are controlled by a common finite control. There are two special tapes: an *input tape* and an *output tape*. The input tape is used to hold the input strings only; it is a read-only tape that prohibits erasing and writing. The output tape is used to hold the output string when the computation of a function is concerned; it is a write-only tape. The other tapes are called the *storage tapes* or the *work tapes*. All work tapes are allowed to read, erase, and write. (See Figure 1.3.)

Next, we allow each tape head in a multi-tape TM, during a move, to stay at the same square without moving to the right or the left. Thus, each move of a $k$-tape TM is defined by a partial transition function $\delta$ that maps $(Q - F) \times \Gamma^k$ to $Q \times \Gamma^k \times \{L, R, S\}^k$ (where S stands for *stay*). The initial setting of the input tape of the multi-tape TM is the same as that of the one-tape TM, and other tapes of the multi-tape TM initially contain only blanks. The formal definition of the computation of a multi-tape TM on an input string $x$ and the concepts of accepting a language and computing a function by a multi-tape TM can be defined similar to that of a one-tape TM. We leave it as an exercise.

**Theorem 1.6** *For any multi-tape TM $M$, there exists a one-tape TM $M_1$ computing the same function as $M$.*

*Proof.* Suppose that $M$ has $k$ tapes and the tape symbol set is $\Gamma$. Then we use the tape symbol set $\Gamma_1 = (\Gamma \times \{X, B\})^k$ for $M_1$, where $X$ is a symbol not in $\Gamma$. This means that we divide the one tape of $M_1$ into $2k$ sections which form $k$ groups. Each group contains two sections: one uses the tape symbol set $\Gamma$, the other uses the tape symbol set $\{X, B\}$. Thus, the blank symbol in

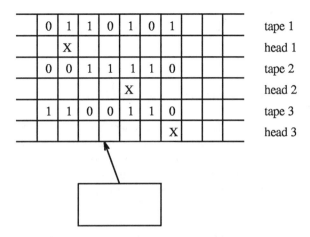

| 0 | 1 | 1 | 0 | 1 | 0 | 1 |   |   |   | tape 1 |
|   | X |   |   |   |   |   |   |   |   | head 1 |
| 0 | 0 | 1 | 1 | 1 | 1 | 0 |   |   |   | tape 2 |
|   |   |   |   | X |   |   |   |   |   | head 2 |
| 1 | 1 | 0 | 0 | 1 | 1 | 0 |   |   |   | tape 3 |
|   |   |   |   |   |   | X |   |   |   | head 3 |

Figure 1.4: The TM $M_1$.

$\Gamma_1$ is $(B, \ldots, B)$ (with $2k$ B's). Each group records the information about a tape of $M$, with the symbol $X$ in the second section indicating the position of the tape head, and the first section containing the corresponding symbol used by $M$. For instance, Figure 1.4 shows the machine $M_1$ that simulates a three-tape TM $M$. The nonblank symbols in the three tapes of $M$ are 0110101, 0011110, and 1100110, and the tape heads of the three tapes are scanning the second, the fifth, and the last symbol of the nonblank symbols, respectively.

For each move of $M$, $M_1$ does the following to simulate it. First, we assume that after each simulation step, $M_1$ is scanning a square such that all symbols $X$ appear to the right of that square. To begin the simulation, $M_1$ moves from left to right scanning all groups to look for the $X$ symbols and the symbols in $\Gamma$ that appear in the same groups as $X$'s. After it finds all $X$ symbols, it has also collected all the tape symbols that are currently scanned by the tape heads of $M$ (cf. Exercise 1.10). Next, $M_1$ moves back from right to left and look for each $X$ symbol again. This time, for each $X$ symbol, $M_1$ properly simulates the action of $M$ on that tape. Namely, it may write over the symbol of the first section of the square where the second section has an $X$ symbol, or it may move the $X$ symbol to the right or to the left. The simulation finishes when actions on all $k$ tapes are taken care of. Note that, by then, all $X$ symbols appear to the right of the tape head of $M_1$.

To complete the description of the machine $M_1$, we only need to add the initialization and the ending parts of the machine program for $M_1$. That is, we first initialize the tape of $M_1$ so that it contains the input in the first group and all blanks in the other groups. At the end, when $M$ reaches a halting configuration, $M_1$ erases all symbols in all groups except the output group. We omit the details.                                                                                ∎

It is obvious that a one-tape TM can be simulated by a multi-tape TM. So, an immediate corollary is that the set of functions computable by one-tape

TMs is the same as that by multi-tape TMs. Similarly, we conclude that the set of recursive sets as well as the set of r.e. sets defined by one-tape TMs are the same as those defined by multi-tape TMs.

## 1.3   Nondeterministic Turing Machines

The Turing machines we defined in the last section are *deterministic*, because from each configuration of a machine there is at most one move to make, and hence there is at most one next configuration. If we allow more than one moves for some configurations, and hence those configurations have more than one next configurations, then the machine is called a *nondeterministic Turing machine* (NTM).

Formally, an NTM $M$ is defined by the following information: states $Q$, initial state $q_0$, accepting states $F$, input symbols $\Sigma$, tape symbols $\Gamma$, including the blank symbol B, and the transition relation $\Delta$. All information except the transition relation $\Delta$ is defined in the same form as a DTM. The transition relation $\Delta$ is a subset of $(Q - F) \times \Gamma \times Q \times \Gamma \times \{L, R\}$. Each quintuple $(q_1, s_1, q_2, s_2, D)$ in $\Delta$ indicates that one of the possible moves of $M$, when it is in state $q_1$ and scanning symbol $s_1$, is to change the current state to $q_2$, to overwrite symbol $s_1$ by $s_2$, and to move the tape head to the direction $D$.

The computation of an NTM can be defined similar to that of a DTM. First, we consider a way of restricting an NTM to a DTM. Let $M$ be an NTM defined by $(Q, q_0, F, \Sigma, \Gamma, \Delta)$ as above. We say $M_1$ is a *restricted* DTM of $M$ if $M_1$ has the same components $Q, q_0, F, \Sigma, \Gamma$ as $M$ and it has a transition function $\delta_1$ that is a subrelation of $\Delta$ satisfying the property that for each $q_1 \in Q$ and $s_1 \in \Gamma$, there is at most one triple $(q_2, s_2, D)$, $D \in \{L, R\}$, such that $(q_1, s_1, q_2, s_2, D) \in \delta_1$. Now we can define the notion of the next configurations of an NTM easily: For each configuration $\alpha = (q_1, x_1, y_1)$ of $M$, we let $\vdash_M(\alpha)$ be the set of all configurations $\beta$ such that $\alpha \vdash_{M_1} \beta$ for some restricted DTM $M_1$ of $M$. We write $\alpha \vdash_M \beta$ if $\beta \in \vdash_M(\alpha)$. Since each configuration of $M$ may have more than one next configurations, the computation of an NTM on an input $w$ is, in general, a *computation tree* rather than a single computation path (as it is in the case of DTMs). In the computation tree, each node is a configuration $\alpha$ and all its next configurations are its children. The root of the tree is the initial configuration.

We say an NTM $M$ *halts* on an input string $w \in \Sigma^*$ if there exists a finite sequence of configurations $\alpha_0, \alpha_1, \ldots, \alpha_n$ such that

(1)  $\alpha_0 = (q_0, \lambda, w)$;

(2)  $\alpha_i \vdash_M \alpha_{i+1}$ for all $i = 0, 1, \ldots, n - 1$; and

(3)  $\vdash_M(\alpha_n)$ is undefined (i.e., it is an empty set).

The notion of an NTM $M$ halting on input $w$ can be rephrased in terms of its computation tree as follows: $M$ halts on $w$ if the computation tree of $M(w)$ contains a finite path. This finite path (i.e., the sequence $\alpha_0, \alpha_1, \ldots, \alpha_n$

of configurations satisfying the above conditions (1)–(3)) is called a *halting path* for $M(w)$.

A halting path is called an *accepting path* if the state of the last configuration is in $F$. An NTM $M$ *accepts* an input string $w$ if there exists an accepting path in the computation tree of $M$ on $w$. (Note that this computation tree may contain some halting paths that are not accepting paths and may contain some nonhalting paths. As long as there exists at least one accepting path, we will say that $M$ accepts $w$.) We say an NTM $M$ *accepts a language* $A \subseteq \Sigma^*$ if $M$ accepts all $w \in A$ and does not accept any $w \in \Sigma^* - A$. For each NTM $M$, we write $L(M)$ to denote the language accepted by $M$.

**Example 1.7** Let $L = \{a^{i_1} b a^{i_2} b \cdots b a^{i_k} b b a^j : i_1, \ldots, i_k, j > 0, \sum_{r \in A} i_r = j$ for some $A \subseteq \{1, 2, \ldots, k\}\}$. We define an NTM $M = (Q = \{q_0, \ldots, q_9\}, q_0, F = \{q_9\}, \Sigma = \{a, b\}, \Gamma = \{a, b, c, \mathrm{B}\}, \Delta)$ that accepts the set $L$. We show $\Delta$ in Figure 1.5.

| $\Delta$ | $a$ | $b$ | $c$ | B |
|---|---|---|---|---|
| $q_0$ | $q_1, \mathrm{B}, \mathrm{R}$ <br> $q_2, c, \mathrm{R}$ | $q_7, \mathrm{B}, \mathrm{R}$ | | |
| $q_1$ | $q_1, \mathrm{B}, \mathrm{R}$ | $q_0, \mathrm{B}, \mathrm{R}$ | | |
| $q_2$ | $q_2, a, \mathrm{R}$ | $q_3, b, \mathrm{R}$ | | |
| $q_3$ | $q_2, a, \mathrm{R}$ | $q_4, b, \mathrm{R}$ | | |
| $q_4$ | $q_5, c, \mathrm{L}$ | | $q_4, c, \mathrm{R}$ | $q_4, \mathrm{B}, \mathrm{R}$ |
| $q_5$ | $q_5, a, \mathrm{L}$ | $q_5, b, \mathrm{L}$ | $q_5, c, \mathrm{L}$ | $q_6, \mathrm{B}, \mathrm{R}$ |
| $q_6$ | $q_2, c, \mathrm{R}$ | $q_0, \mathrm{B}, \mathrm{R}$ | $q_6, \mathrm{B}, R$ | |
| $q_7$ | | | $q_8, \mathrm{B}, \mathrm{R}$ | |
| $q_8$ | | | $q_8, \mathrm{B}, \mathrm{R}$ | $q_9, \mathrm{B}, \mathrm{R}$ |

Figure 1.5: The transition function of machine $M$.

The main idea of the machine $M$ is that, for each block of $a$'s, except for the last one, it nondeterministically chooses (at state $q_0$) to either erase the whole block or, like Example 1.4, to erase the block and the same number of $a$'s from the last block. Thus, if all blocks, including the last one, are erased, then accept. States $q_2$ to $q_6$ are devoted to the task of the second choice. We show the computation tree of $M$ on input $aababbaa$ in Figure 1.6. □

The notion of an NTM $M$ computing a function is potentially ambiguous since for each input $w$, the computation tree of $M(w)$ may contain more than one halting path and each halting path may output a different value. We impose a strong requirement to eliminate the ambiguity.

**Definition 1.8** *(a) We say that an NTM $M$ outputs $y$ on input $x$ if (i) there exists at least one computation path of $M(x)$ that halts in an accepting state with output $y$, and (ii) whenever a computation path of $M(x)$ halts in an accepting state, its output is $y$.*

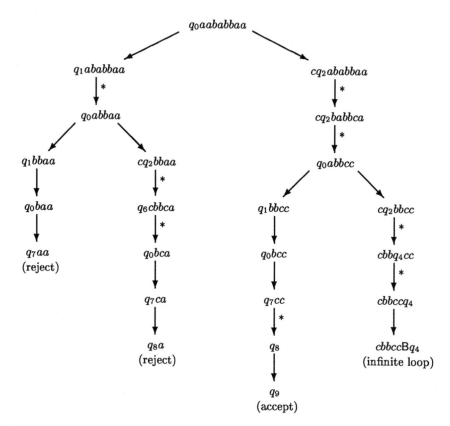

Figure 1.6: The computation tree of machine $M$ on input $aababbaa$. An edge with $*$ means that the transition between configurations takes more than one deterministic moves.

*(b) We say that an NTM $M$ computes a partial function $f$ from $\Sigma^*$ to $\Sigma^*$ if for each input $x \in \Sigma^*$ that is in the domain of $f$, $M$ outputs $f(x)$, and for each input $x \in \Sigma^*$ that is not in the domain of $f$, $M$ does not accept $x$.*

The Church-Turing Thesis claims that deterministic Turing machines are powerful enough to simulate other types of machine models as long as the machine model is a reasonable implementation of the intuitive notion of algorithms. Nondeterministic Turing machines are, however, not a very reasonable machine model, since the nondeterministic moves are obviously beyond the capability of currently existing and foreseeable physical computing devices. Nevertheless, DTMs are still powerful enough to be able to simulate NTMs (though this simulation may require much more resources for DTMs than the corresponding NTMs, as we will see in later sections).

**Theorem 1.9**

   *(a) All languages accepted by NTMs are recursively enumerable.*
   *(b) All functions computed by NTMs are partial recursive.*

*Proof.* Let $M$ be a one-tape NTM defined by $(Q, q_0, F, \Sigma, \Gamma, \Delta)$. We may regard $\Delta$ as a multi-valued function from $Q \times \Gamma$ to $Q \times \Gamma \times \{L, R\}$. Let $k$ be the maximum number of values that $\Delta$ can assume on some $(q, s) \in Q \times \Gamma$.

We are going to design a DTM $M_1$ to simulate $M$. Our DTM $M_1$ is a three-tape DTM which uses $k$ additional symbols $\eta_1, \ldots, \eta_k$ that are not in $\Gamma$. The DTM $M_1$ uses the third tape to simulate the tape of $M$ and uses the second tape to store the current simulation information. More precisely, $M_1$ operates on input $x$ in stages. At stage $r > 0$, $M_1$ performs the following actions:

   (1) $M_1$ erases anything in tape 3 that may have been left over from stage $r - 1$, and copies the input $x$ from tape 1 to tape 3.

   (2) $M_1$ generates the $r$th string $\eta_{i_1} \ldots \eta_{i_m}$ in $\{\eta_1, \ldots, \eta_k\}^*$, in the lexicographic ordering, on tape 2. (This string overwrites the $(r - 1)$st string generated in stage $r - 1$.)

   (3) $M_1$ simulates $M$ on input $x$ on tape 3 for at most $m$ moves. At the $j$th move, $1 \leq j \leq m$, $M_1$ examines the $j$th symbol $\eta_{i_j}$ on the second tape to determine which transition of the relation $\Delta$ is to be simulated. More precisely, if the current state is $q$ and the symbol currently scanned on tape 3 is $s$, and $\Delta(q, s)$ contains at least $i_j$ values, then $M_1$ follows the $i_j$th move of $\Delta(q, s)$; if $\Delta(q, s)$ has less than $i_j$ values, then $M_1$ goes to stage $r + 1$.

   (4) If the simulation halts within $m$ moves in an accepting state, then $M_1$ enters a state in $F$ and halts; otherwise, it goes to stage $r + 1$.

It is clear that if $M$ does not accept $x$ then $M_1$ never accepts $x$ either. Conversely, if $M$ accepts $x$, then there is a finite string $\eta_{i_1} \ldots \eta_{i_m}$ with respect to which $M_1$ will simulate the accepting path of $M(x)$ and hence will accept $x$.

The above proves part (a). In addition, the above simulation of $M_1$ also yields the same output as $M$, and so part (b) also follows. ∎

Informally we often describe a nondeterministic algorithm as a *guess-and-verify* algorithm. That is, the nondeterministic moves of the algorithm are to guess a computation path and, for each computation path, a deterministic subroutine verifies that it is indeed an accepting path. The critical guess that determines an accepting path is called a *witness* to the input. For instance, the NTM of Example 1.7 guesses a subset $A \subseteq \{1, \ldots, k\}$ and then verifies that $\sum_{r \in A} i_r = j$. In the above simulation of NTM $M$ by a DTM $M_1$, $M_1$ generates all strings $w = \eta_{i_1} \cdots \eta_{i_m}$ in tape 2 one by one and then verifies that this string $w$ is a witness to the input.

## 1.4   Complexity Classes

Computational complexity of a machine is the measure of the resources used
by the machine in the computation. Computational complexity of a problem
is the measure of the minimum resources required by any machine that solves
the problem. For Turing machines, time and space are the two most important
types of resources of concern. Let $M$ be a one-tape DTM. For an input string
$x$, the *running time* of $M$ on $x$ is the number of moves made by $M$ from
the beginning until it halts, denoted by $time_M(x)$. (We allow $time_M(x)$ to
assume the value $\infty$.) The *working space* of $M$ on input $x$ is the number of
squares which the tape head of $M$ visited at least once during the computation,
denoted by $space_M(x)$. (Again, $space_M(x)$ may be equal to $\infty$. Note that
$space_M(x)$ may be finite even if $M$ does not halt on $x$.) For a multi-tape
DTM $M$, the time complexity $time_M(x)$ is defined in a similar way, and the
space complexity $space_M(x)$ is defined to count only the squares visited by
the heads of working tapes, excluding the input and output tapes. This allows
the possibility of having $space_M(x) < |x|$.

The functions $time_M$ and $space_M$ are defined on each input string. This
makes it difficult to compare the complexity of two different machines. The
common practice in complexity theory is to compare the complexity of two different machines based on their growth rates with respect to the input length.
Thus, we define the time complexity of a DTM $M$ to be the function $t_M : \mathbf{N} \to
\mathbf{N}$ with $t_M(n) = \max\{time_M(x) : |x| = n\}$. The space complexity function
$s_M$ of $M$ is similarly defined to be $s_M(n) = \max\{space_M(x) : |x| = n\}$.

We would like to define the (deterministic) time complexity of a function
$f$ to be the minimum $t_M$ with respect to DTMs $M$ which compute $f$. Unfortunately, for some recursive function $f$, there is no *best* DTM. In other
words, for any machine $M_1$ which computes $f$, there is another machine $M_2$
also computing $f$ such that $t_{M_2}(n) < t_{M_1}(n)$ for infinitely many $n$ (*Blum's
Speed-up Theorem*). So, formally, we can only talk about the upper bound
and lower bound of the time complexity of a function. We say that the time
(space) complexity of a recursive function $f$ is bounded (above) by the function $\phi : N \to N$ if there exists a TM $M$ which computes $f$ such that for almost
all $n \in N$, $t_M(n) \leq \phi(n)$ ($s_M(n) \leq \phi(n)$, respectively). The time and space
complexity of a recursive language $L$ is defined to be, respectively, the time
and space complexity of its characteristic function $\chi_L$. (The *characteristic
function* $\chi_L$ of a language $L$ is defined by $\chi_L(x) = 1$ if $x \in L$ and $\chi_L(x) = 0$
if $x \notin L$.)

Now we can define complexity classes of languages as follows: Let $t : \mathbf{N} \to
\mathbf{N}$ be a nondecreasing function from integers to integers, and $C$ be a collection
of such functions.

**Definition 1.10** *(a) We define DTIME(t) to be the class of languages L that
are accepted by DTMs M with $t_M(n) \leq t(n)$ for almost all $n \geq 0$. We let
$DTIME(C) = \bigcup_{t \in C} DTIME(t)$.*

*(b) Similarly, we define DSPACE(s) to be the class of languages L that*

*are accepted by DTMs M with $s_M(n) \leq s(n)$ for almost all $n \geq 0$. We let $DSPACE(C) = \bigcup_{s \in C} DSPACE(s)$.*

For NTMs, the notion of time and space complexity is a little more complex. Let $M$ be an NTM. On each input $x$, $M(x)$ defines a computation tree. For each finite computation path of this tree, we define the time of the path to be the number of moves in the path and the space of the path to be the total number of squares visited in the path by the tape head of $M$. We define the time complexity $time_M(x)$ of $M$ on an input $x$ to be the minimum time of the computation paths among all *accepting* paths of the computation tree of $M(x)$. The space complexity $space_M(x)$ is the minimum space of the computation paths among all *accepting* paths of the computation tree of $M(x)$. (When $M$ is a multi-tape NTM, $space_M(x)$ only counts the space in the work tapes.) If $M$ does not accept $x$ (and, hence, there is no accepting path in the computation tree of $M(x)$), then $time_M(x)$ and $space_M(x)$ are undefined. (Note that the minimum time computation path and the minimum space computation path in the computation tree of $M(x)$ are not necessarily the same path.)

We now define the time complexity function $t_M : \mathbf{N} \to \mathbf{N}$ of an NTM $M$ to be $t_M(n) = \max(\{n + 1\} \cup \{time_M(x) : |x| = n, M \text{ accepts } x\})$.[6] Similarly, the space complexity function $s_M$ of an NTM $M$ is defined to be $s_M(n) = \max(\{1\} \cup \{space_M(x) : |x| = n, M \text{ accepts } x\})$.

**Definition 1.11** *(a)* $NTIME(t) = \{L : L$ *is accepted by an NTM M with* $t_M(n) \leq t(n)$ *for almost all* $n > 0\}$; $NTIME(C) = \bigcup_{t \in C} NTIME(t)$.
*(b)* $NSPACE(s) = \{L : L$ *is accepted by an NTM M with* $s_M(n) \leq s(n)$ *for almost all* $n > 0\}$; $NSPACE(C) = \bigcup_{s \in C} NSPACE(s)$.

It is interesting to point out that for these complexity classes, an increase of a constant factor on the time or the space bounds does not change the complexity classes. The reason is that a TM with a larger alphabet set and a larger set of states can simulate, in one move, a constant number of moves of a TM with a smaller alphabet set and a smaller number of states. We only show these properties for deterministic complexity classes.

**Proposition 1.12** (Tape Compression Theorem) *Assume that $c > 0$. Then,* $DSPACE(s(n)) = DSPACE(c \cdot s(n))$.

*Proof.* Let $M$ be a TM that runs in space bound $s(n)$. We will construct a new TM $M'$ which simulates $M$ within space bound $c \cdot s(n)$, using the same number of tapes. Assume that $M$ uses the alphabet $\Gamma$ of size $k$, has $r$ states $Q = \{q_1, \ldots, q_r\}$, and uses $t$ tapes. Let $m$ be a positive integer such that $m > 1/c$. Divide the squares in each work tape of $M$ into groups; each group contains exactly $m$ squares. The machine $M'$ uses an alphabet

---

[6]The extra value $n + 1$ is added so that when $M$ does not accept any $x$ of length $n$, the time complexity is $n + 1$, the time to read the input $x$.

of size $k^m$ so that each square of $M'$ can simulate a group of squares of $M$. In addition, $M'$ has $rm^t$ states, each state represented by $\langle q_i, j_1, \ldots, j_t \rangle$, for some $q_i \in Q$, $1 \leq j_1 \leq m$, $\ldots$, $1 \leq j_t \leq m$. Each state $\langle q_i, j_1, \ldots, j_t \rangle$ encodes the information that the machine $M$ is in state $q_i$, and the local positions of its $t$ tape heads within a group of squares. So, together with the position of its own tape heads, it is easy for $M'$ to simulate each move of $M$ using a smaller size of space. For instance, if a tape head of $M$ moves left, and if its current local position within the group is $j$ and $j > 1$, then the corresponding tape head of $M'$ does not move but the state of $M$ is modified so that its current local position within the group becomes $j - 1$; if $j = 1$, then the corresponding tape head of $M'$ moves left and the local position is changed to $m$. By this simulation, $L(M) = L(M')$ and $M'$ has a space bound $s(n)/m \leq c \cdot s(n)$. ∎

**Proposition 1.13** (Linear Speed-Up Theorem) *Suppose* $\lim_{n \to \infty} t(n)/n = \infty$. *Then for any* $c > 0$, $DTIME(t(n)) = DTIME(c \cdot t(n))$.

*Proof.* Let $M$ be a TM that runs in time bound $t(n)$. We will construct a new TM $M'$ that simulates $M$ within time $c \cdot t(n)$, using an additional tape. Let $m$ be a large positive integer whose exact value is to be determined later. Divide the squares in each tape of $M$ into groups; each group contains $m$ squares. That is, if $M$ uses alphabet $\Gamma$, then each group $g$ is an element of $\Gamma^m$. For each group $g$, call its left neighboring group $g_\ell$ and the right neighboring group $g_r$. Let $H(g, g_\ell, g_r)$ be a history of the moves of $M$ around $g$; that is, the collection of the moves of $M$ starting from entering the group $g$ until halting, entering a loop, or leaving the three groups $g$, $g_\ell$, and $g_r$. (Note that the number of possible histories $H(g, g_\ell, g_r)$ is bounded by a function of $m$, independent of the input size $n$.)

The new machine $M'$ encodes each group of $M$ into a symbol. Initially, it encodes the input and copies it to tape 2. Then, it simulates $M$ using the same number of work tapes (in addition to tape 2). It simulates each history $H(g, g_\ell, g_r)$ of $M$ by a constant number of moves as follows: $M'$ visits the groups $g$, $g_\ell$, and $g_r$. Then, according to the contents of the three squares, it finds the history $H(g, g_\ell, g_r)$ of $M$. From the history $H(g, g_\ell, g_r)$, it overwrites new symbols on $g$, $g_\ell$, and $g_r$, and sets up the new configuration for the simulation of the next step. In the above simulation step, the contents of the three groups are stored in the states of $M'$, and the history $H(g, g_\ell, g_r)$ is stored in the form of the transition function of $M'$. For instance, after $M'$ gets the contents $a$, $a_\ell$, and $a_r$ of the three groups $g$, $g_\ell$, and $g_r$, respectively, the state of $M'$ is of the form $\langle q_i, j, a, a_\ell, a_r \rangle$, where $q_i$ is a state of $M$ and $j$ is the starting position of the tape head of $M$ within the group $g$ ($j = 1, \ldots, m$). (Here, we assume that $M'$ uses only one tape.)

Note that each history $H(g, g_\ell, g_r)$ encodes at least $m$ moves of $M$, and its simulation takes $c_1$ moves of $M'$ where $c_1$ is an absolute constant, independent of $m$. Since the initial setup of $M'$, including the encoding of the input string and returning the tape head to the leftmost square, takes $n + \lceil n/m \rceil$ moves,

$M'$ has the time bound

$$n + \lceil n/m \rceil + c_1 \lceil t(n)/m \rceil \le n + n/m + c_1 t(n)/m + c_1 + 1.$$

Choose $m > c_1/c$. Because $\lim_{n \to \infty} t(n)/n = \infty$, we have that for sufficiently large $n$, $n + n/m + c_1 \cdot t(n)/m + c_1 + 1 \le c \cdot t(n)$. Thus, $A \in DTIME(c \cdot t(n))$. ∎

The above two theorems allow us to write simply, for instance, $DSPACE(\log n)$ to mean $\bigcup_{c>0} DSPACE(c \cdot \log n)$, and $NTIME(n^2)$ to mean $\bigcup_{c>0} NTIME(c \cdot n^2)$.

Next, let us introduce the notion of polynomial-time computability. Let *poly* be the collection of all integer polynomial functions with nonnegative coefficients. Define

$$P = DTIME(poly),$$
$$NP = NTIME(poly),$$
$$PSPACE = DSPACE(poly),$$
$$NPSPACE = NSPACE(poly).$$

We say a language $L$ is polynomial-time (polynomial-space) computable if $L \in P$ ($L \in PSPACE$, respectively).

In addition to polynomial time/space-bounded complexity classes, some other important complexity classes are

$$LOGSPACE = DSPACE(\log n),$$
$$NLOGSPACE = NSPACE(\log n),$$
$$EXP = \bigcup_{c>0} DTIME(2^{cn}),$$
$$NEXP = \bigcup_{c>0} NTIME(2^{cn}),$$
$$EXPSPACE = \bigcup_{c>0} DSPACE(2^{cn}).$$

For the classes of functions, we only define the classes of (deterministically) polynomial-time and polynomial-space computable functions. Other complexity classes of functions will be defined later. We let *FP* denote the class of all recursive functions $f$ which are computable by a DTM $M$ of time complexity $t_M(n) \le p(n)$ for some $p \in poly$, and let *FPSPACE* denote the class of all recursive functions $f$ which are computable by a DTM $M$ of space complexity $s_M(n) \le p(n)$ for some $p \in poly$.

The classes $P$ and $FP$ are often referred as the mathematical equivalence of the classes of *feasibly computable problems*. That is, a problem is considered to be feasibly solvable if it has a solution whose time complexity grows in a polynomial rate; on the other hand, if a solution has a super-polynomial growth rate on its running time then the solution is not feasible. Although this formulation is not totally accepted by practitioners, so far it remains the best one, since these complexity classes have many nice mathematical properties. For instance, if two functions $f, g$ are feasibly computable, then we expect their composition $h(x) = f(g(x))$ also to be feasibly computable. The class *FP* is indeed closed under composition; as a matter of fact, it is the smallest class of functions which contains all quadratic-time computable functions and is closed under composition.

Another important property is that the classes $P$ and *FP* are *independent of computational models*. We note that we have defined the classes $P$ and *FP* based on the Turing machine model. We have argued, based on the Church-Turing Thesis, that the Turing machine model is as general as any other models, as far as the notion of computability is concerned. Here, when the notion of time and space complexity is concerned, the Church-Turing Thesis is no longer applicable. It is obvious that different machine models define different complexity classes of languages. However, as far as the notion of polynomial-time computability is concerned, it can be verified that most familiar computational models define the same classes $P$ and *FP*. The idea that the classes $P$ and *FP* are machine independent can be formulated as follows:

**Extended Church-Turing Thesis**. A function computable in polynomial time in any *reasonable* computational model using a *reasonable* time complexity measure is computable by a deterministic Turing machine in polynomial time.

The intuitive notion of *reasonable computational models* has been discussed in Section 1.2. They exclude any machine models which are not realizable by physical devices. In particular, we do not consider an NTM to be a reasonable model. In addition, we also exclude any time complexity measures that do not reflect the physical time requirements. For instance, in a model like the random access machine (see Exercises 1.14 and 1.15), the *uniform time complexity measure* for which an arithmetic operation on integers counts only one unit of time, no matter how large the integers are, is not considered as a reasonable time complexity measure.[7]

The Extended Church-Turing Thesis is a much stronger statement than the Church-Turing Thesis, because it implies that TMs can simulate each instruction of any other reasonable models within time polynomial in the length of the input. It is not as widely accepted as the Church-Turing Thesis. To disprove it, however, one needs to demonstrate in a formal manner a reasonable model and a problem $A$ and prove formally that $A$ is computable in the new model in polynomial time and $A$ is not solvable by deterministic TMs in

---

[7]The reader should keep in mind that we are here developing a general theory of computational complexity for all computational problems. Depending on the nature of the problems, we may sometime want to apply more specific complexity measures to specific problems. For instance, when we compare different algorithms for the integer matrix multiplication problem, the uniform time complexity measure appears to be more convenient than the bit-operation (or, logarithmic) time complexity measure.

polynomial time. Although there are some candidates for the counterexample $A$, such a formal proof has not yet been found.[8]

To partially support the Extended Church-Turing Thesis, we verify that the simulation of a multi-tape TM by a one-tape TM described in Theorem 1.6 can be implemented in polynomial time.

**Corollary 1.14** *For any multi-tape TM $M$, there exists a one-tape TM $M_1$ that computes the same function as $M$ in time $t_{M_1}(n) = O((t_M(n))^2)$.*

*Proof.* In the simulation of Theorem 1.6, for each move of $M$, the machine $M_1$ makes one pass from left to right and then one pass from right to left. In the first pass, the machine $M_1$ looks for all $X$ symbols and then stores the information of tape symbols currently scanned by $M$ in its states, and it takes at most $time_M(x)$ moves to find all $X$ symbols. (Note that at any point of computation, the tape heads of $M$ can move at most $time_M(x)$ squares away from the starting square.) In the second pass, $M_1$ needs to adjust the positions of the $X$ symbols, and to erase and write new symbols in the first section of each group. For each group, these actions take two additional moves, and the total time for the second pass takes only $time_M(x) + 2k$, where $k$ is the number of tapes of $M$. Thus, the total running time of $M_1$ on $x$ is at most $(time_M(x))^2 + (2k+2)time_M(x)$ (where the extra $2time_M(x)$ is for the initialization and ending stages). ∎

The above quadratic simulation time can be reduced to quasi-linear time (i.e., $t_M(n) \cdot \log(t_M(n))$) if we allow $M_1$ to have two work tapes (in addition to the input tape).

**Theorem 1.15** *For any multi-tape TM $M$, there exists a two-worktape TM $M_1$ that computes the same function as $M$ in time $t_{M_1}(n) = O(t_M(n) \cdot \log(t_M(n)))$.*

*Proof.* See Exercise 1.19. ∎

For the space complexity, we observe that the machine $M_1$ in Theorem 1.6 does not use any extra space than $M_1$.

**Corollary 1.16** *For any multi-tape TM $M$, there exists a one-worktape TM $M_1$ that computes the same function as $M$ in space $s_{M_1}(n) = O(s_M(n))$.*

The relationship between the time- and space-bounded complexity classes and between the deterministic and nondeterministic complexity classes is one

---

[8]One of the new computational models that challenges the validity of the Extended Church's Thesis is the quantum Turing machine model. It has been proved recently that some number theoretic problems which are believed not solvable in polynomial time by deterministic TMs (or, even by probabilistic TMs) are solvable in polynomial time by quantum TMs. See, for instance, Shor [1997] and Bernstein and Vazirani [1997] for more discussions.

of the main topics in complexity theory. In particular, the relationship between the polynomial bounded classes, such as whether $P = NP$ and whether $P = PSPACE$, presents the ultimate challenge to the researchers. We discuss these questions in the following sections.

## 1.5   Universal Turing Machine

One of the most important properties of a computation system like Turing machines is that there exists a universal machine that can simulate each machine from its code.

Let us first consider one-tape DTMs with the input alphabet $\{0,1\}$, the working alphabet $\{0,1,B\}$, the initial state $q_0$, and the final state set $\{q_1\}$, that is, DTMs defined by $(Q, q_0, \{q_1\}, \{0,1\}, \{0,1,B\}, \delta)$. Such a TM can be determined by the definition of the transition function $\delta$ only, for $Q$ is assumed to be the set of states appearing in the definition of $\delta$. Let us use the notation $q_i$, $0 \le i \le |Q| - 1$, for a state in $Q$, $X_j$, $j = 0,1,2$, for a tape symbol where $X_0 = 0, X_1 = 1$ and $X_2 = B$, and $D_k$, $k = 0,1$, for a moving direction for the tape head where $D_0 = L$ and $D_1 = R$. For each equation $\delta(q_i, X_j) = (q_k, X_\ell, D_h)$, we encode it by the following string in $\{0,1\}^*$:

$$0^{i+1}10^{j+1}10^{k+1}10^{\ell+1}10^{h+1}. \qquad (1.2)$$

Assume that there are $m$ equations in the definition of $\delta$. Let $code_i$ be the code of the $i$th equation. Then we combine the codes for equations together to get the following code for the TM:

$$code_1 11 code_2 11 \cdots 11 code_m. \qquad (1.3)$$

Note that since different orderings of the equations give different codes, there are $m!$ equivalent codes for a TM of $m$ equations.

The above coding system is a one-to-many mapping $\phi$ from TMs to $\{0,1\}^*$. Each string $x$ in $\{0,1\}^*$ encodes at most one TM $\phi^{-1}(x)$. Let us extend $\phi^{-1}$ into a function mapping each string $x \in \{0,1\}^*$ to a TM $M$ by mapping each $x$ not encoding a TM to a fixed empty TM $M_0$ whose code is $\lambda$ and which rejects all strings. Call this mapping $\iota$. Observe that $\iota$ is a mapping from $\{0,1\}^*$ to TMs with the following properties:

(i) For every $x \in \Sigma^*$, $\iota(x)$ represents a TM;

(ii) Every TM is represented by at least one $\iota(x)$; and

(iii) The transition function $\delta$ of the TM $\iota(x)$ can be easily decoded from the string $x$.

We say a coding system $\iota$ is an *enumeration* of one-tape DTMs if $\iota$ satisfies properties (i) and (ii). In addition, property (iii) means that this enumeration admits a *universal Turing machine*. In the following, we write $M_x$ to mean the TM $\iota(x)$. We assume that $\langle \cdot, \cdot \rangle$ is a pairing function on $\{0,1\}^*$ such that both the function and its inverse are computable in linear time, for instance, $\langle x, y \rangle = 0^{|x|}1xy$.

**Proposition 1.17** *There exists a TM $M_u$, which, on input $\langle x, y \rangle$, simulates the machine $M_x$ on input $y$ so that $L(M_u) = \{\langle x, y \rangle : y \in L(M_x)\}$. Furthermore, for each $x$, there is a constant $c$ such that $time_{M_u}(\langle x, y \rangle) \le c \cdot (time_{M_x}(y))^2$.*

*Proof.* We first construct a four-tape machine $M_u$. The machine $M_u$ first decodes $\langle x, y \rangle$ and copies the string $x$ to the second tape and copies string $y$ to the third tape. After this, the machine $M_u$ checks that the first string $x$ is a legal code for a TM; that is, it verifies that it is of the form (1.3) such that each $code_i$ is of the form (1.2) with $0 \le j \le 2$, $0 \le \ell \le 2$ and $0 \le h \le 1$. Furthermore, it verifies that it encodes a *deterministic* TM by verifying that no two $code_p$ and $code_q$ begin with the same initial segment $0^{i+1}10^{j+1}1$. If $x$ does not legally encode a DTM, then $M_u$ halts and rejects the input. Otherwise, $M_u$ begins to simulate $M_x$ on input $y$. During the simulation, the machine $M_u$ stores the current state $q_i$ (in the form $0^{i+1}$) of $M_x$ in tape 4 and, for each move, it looks on tape 2 for the equation of the transition function of $M_x$ that applies to the current configuration on tape 3, and then simulate the move of $M_x$ on tapes 3 and 4.

To analyze the running time of $M_u$, we note that decoding of $\langle x, y \rangle$ into $x$ and $y$ takes only time $c_1(|x| + |y|)$ for some constant $c_1$, and the verification of $x$ being a legal code for a DTM takes only $O(|x|^2)$ moves. In addition, the simulation of each move takes only time linearly proportional to the length of $x$. Thus, the total simulation time of $M_u$ on $\langle x, y \rangle$ is only $c_2 \cdot time_{M_x}(y)$ for some constant $c_2$ that depends on $x$ only. The proposition now follows from Theorem 1.14. ∎

Although the above coding system and the universal Turing machine are defined for a special class of one-tape TMs, it is not hard to see how to extend it to the class of multi-tape TMs. For instance, we note that a universal TM exists for the enumeration of one-worktape TMs and its simulation can be done in linear space.

**Proposition 1.18** *Assume that $\{M_x\}$ is an enumeration of one-worktape TMs over the alphabet $\{0, 1, B\}$. Then, there exists a one-worktape TM $M_u$ such that $L(M_u) = \{\langle x, y \rangle : y \in L(M_x)\}$. Furthermore, for each $x$, there is a constant $c$ such that $space_{M_u}(\langle x, y \rangle) \le c \cdot space_{M_x}(y)$.*

*Proof.* Follows from Corollary 1.16. ∎

The above results can be extended to other types of TMs. For instance, for each $k > 1$ and any fixed alphabet $\Sigma$, we may easily extend our coding system to get a one-to-one mapping from $\{0, 1\}^*$ to $k$-tape TMs over $\Sigma$, and obtain an enumeration $\{M_x\}$ for such TMs. Based on Theorem 1.15 and Corollary 1.16, it can be easily seen that for this enumeration system, a universal TM $M_u$ exists that simulates each machine with $time_{M_u}(\langle x, y \rangle) \le c \cdot time_{M_x}(y) \cdot \log(time_{M_x}(y))$, and $space_{M_u}(\langle x, y \rangle) \le c \cdot space_{M_x}(y)$ for some constant $c$ (depending on $k$, $\Sigma$, and $x$). In addition, this enumeration can be extended to nondeterministic Turing machines.

**Proposition 1.19** *There exist a mapping from each string $x \in \{0,1\}^*$ to a multi-tape NTM $M_x$, and a universal NTM $M_u$, such that*

(i) $L(M_u) = \{\langle x, y \rangle : y \in L(M_x)\}$; and

(ii) *For each $x$, there is a polynomial function $p$ such that $time_{M_u}(\langle x, y \rangle) \leq p(time_{M_x}(y))$ for all $y$.*

*Proof.* The proof is essentially the same as Proposition 1.17, together with the polynomial-time simulation of multi-tape NTMs by two-tape NTMs. We omit the details.                                                                    ∎

When we consider the complexity classes like $P$ and $NP$, we are concerned only with a subclass of DTMs or NTMs. It is convenient to enumerate only these machines. We present an enumeration of polynomial-time "clocked" machines. First we need some notations. A function $t : \mathbf{N} \to \mathbf{N}$ is called a *fully time-constructible* function if there is a multi-tape DTM that on any input $x$ of length $n$ halts in exactly $t(n)$ moves. Most common integer-valued functions such as $cn$, $n \cdot \lceil \log n \rceil$, $n^c$, $n^{\lceil \log n \rceil}$, $2^n$ and $n!$ are known to be fully time-constructible (see Exercise 1.20). A DTM that halts in exactly $t(n)$ moves on all inputs of length $n$ is called a $t(n)$-*clock machine*. For any $k_1$-tape DTM $M_1$ and any fully time-constructible function $t$ for which there exists a $k_2$-tape clock machine $M_2$, we can construct a $t(n)$-clocked $(k_1 + k_2)$-tape DTM $M_3$ that simulates $M_1$ for exactly $t(n)$ moves on inputs of length $n$. The machine $M_3$ first copies the input to the $(k_1 + 1)$st tape, and then simultaneously simulates on the first $k_1$ tapes machine $M_1$ and on the last $k_2$ tapes the $t(n)$-clock machine $M_2$. $M_3$ halts when either $M_1$ or $M_2$ halts, and it accepts only if $M_1$ accepts. The two moves of $M_1$ and $M_2$ can be combined into one move by using a state set that is the product of the state sets of $M_1$ and $M_2$. For instance, assume that $M_1$ is a one-tape DTM defined by $(Q_1, q_0^1, F_1, \Sigma_1, \Gamma_1, \delta_1)$ and $M_2$ is a one-tape machine defined by $(Q_2, q_0^2, F_2, \Sigma_2, \Gamma_2, \delta_2)$. Then, $M_3$ is defined to be $(Q_1 \times Q_2, \langle q_0^1, q_0^2 \rangle, F_1 \times Q_2, \Sigma_1 \times \Sigma_2, \Gamma_1 \times \Gamma_2, \delta_3)$, where

$$\delta_3(\langle q_i^1, q_{i'}^2 \rangle, \langle s_j^1, s_{j'}^2 \rangle) = (\langle q_k^1, q_{k'}^2 \rangle, \langle s_\ell^1, s_{\ell'}^2 \rangle, \langle D_h, D_{h'} \rangle)$$

if $\delta_1(q_i^1, s_j^1) = (q_k^1, s_\ell^1, D_h)$ and $\delta_2(q_{i'}^2, s_{j'}^2) = (q_{k'}^2, s_{\ell'}^2, D_{h'})$. It is clear that machine $M_3$ always halts in time $t(n) + 2n$. Furthermore, if $t_{M_1}(n) \leq t(n)$ for all $n \geq 0$, then $L(M_2) = L(M_1)$.

Now, consider the functions $q_i(n) = n^i + i$, for $i \geq 1$. It is easy to see that for any polynomial function $p$, there exists an integer $i \geq 1$ such that $p(n) \leq q_i(n)$ for all $n \geq 0$. Assume that $\{M_x\}$ is an enumeration of DTMs as described above. For each string $x \in \{0,1\}^*$, and each integer $i$, let $M_{\langle x, i \rangle}$ denote the machine $M_x$ attached with the $q_i(n)$-clock machine. Then, $\{M_{\langle x, i \rangle}\}$ is an enumeration of all polynomial-time clocked machines. This enumeration has the following properties and is called an effective enumeration of languages in $P$:

(i) Every machine $M_{\langle x, i \rangle}$ in the enumeration accepts a set in $P$.

(ii) Every set $A$ in $P$ is accepted by at least one machine in the enumeration (in fact, an infinite number of machines).

(iii) There exists a universal machine $M_u$ such that for given input $\langle x, i, y \rangle$, $M_u$ simulates $M_{\langle x,i \rangle}$ on $y$ in time $p(q_i(|y|))$, where $p$ is a polynomial function depending on $x$ only.

Note that the above statements claim that the class $P$ of polynomial-time computable languages are enumerable through the enumeration of polynomial-time clocked machines. They do not claim that we can enumerate *all* polynomial time-bounded machines, since the question of determining whether a given machine always halts in polynomial time is in fact not decidable (Exercise 1.23).

By a similar setting, we obtain an enumeration of clocked NTMs that accept exactly the languages in *NP*.

We can also enumerate all languages in *PSPACE* through the enumeration of polynomial *space-marking* machines. A function $s : \mathbf{N} \to \mathbf{N}$ is called a *fully space-constructible* function if there is a two-tape DTM that on any input $x$ of length $n$ halts visiting exactly $s(n)$ squares of the work tape. Again, most common functions, including $\lceil \log n \rceil, \lceil \log^2 n \rceil, cn, n^c$, are fully space-constructible. For any fully space-constructible function $s$, we can construct a two-tape, $s(n)$-space marking machine $M$ that, on any input of length $n$, places the symbol $\#$ on exactly $s(n)$ squares on the work tape. For any $k$-tape DTM $M_1$, we can attach this $s(n)$-space marking machine to it to form a new machine $M_2$ that does the following:

(1) It first reads the input and simulates the marking machine $M_1$ to mark each tape with $s(n)$ $\#$ symbols.

(2) It then simulates $M_1$ using these marked tapes, treating each $\#$ symbol as the blank symbol, such that whenever the machine $M_1$ sees a real blank symbol $\mathsf{B}$, the machine $M_2$ halts and rejects. (We assume that $M_2$ begins step (2) with each tape head scanning the square of the leftmost $\#$ mark. It can be proved that for any machine $M_1$ that operates within a space bound $s(n)$, there is a machine $M_1'$ that computes the same function and operates in space bound $s(n)$ starting with this initial setting (Exercise 1.24).)

We can then enumerate all polynomial space-marked machines as we did for polynomial clocked machines. This enumeration effectively enumerates all languages in *PSPACE*.

## 1.6 Diagonalization

Diagonalization is an important proof technique widely used in recursive function theory and complexity theory. One of the earliest applications of diagonalization is Cantor's proof for the fact that the set of real numbers is not countable. We give a similar proof for the set of functions on $\{0,1\}^*$. A set

$S$ is *countable* (or, *enumerable*) if there exists a one-one, onto mapping from the set of natural numbers to $S$.

**Proposition 1.20** *The set of functions from $\{0,1\}^*$ to $\{0,1\}$ is not countable.*

*Proof.* Suppose, by way of contradiction, that such a set is countable, that is, it can be represented as $\{f_0, f_1, f_2, \cdots\}$. Let $a_i$ denote the $i$th string in $\{0,1\}^*$ under the lexicographic ordering. Then we can define a function $f$ as follows: For each $i \geq 0$, $f(a_i) = 1$ if $f_i(a_i) = 0$, and $f(a_i) = 0$ if $f_i(a_i) = 1$. Clearly, $f$ is a function from $\{0,1\}^*$ to $\{0,1\}$. However, it is not in the list $f_0, f_1, f_2, \cdots$, since it differs from each $f_i$ on at least one input string $a_i$. This establishes a contradiction.                                                                               ∎

An immediate consequence of Proposition 1.20 is that there exists a non-computable function from $\{0,1\}^*$ to $\{0,1\}$, since we have just shown that the set of all TMs, and hence the set of all computable functions, is countable. In the following, we use diagonalization to construct directly an undecidable (i.e., nonrecursive) problem: the *halting problem*. The halting problem is the set $K = \{x \in \{0,1\}^* : M_x \text{ halts on } x\}$, where $\{M_x\}$ is an enumeration of TMs.

**Theorem 1.21** *$K$ is r.e. but not recursive.*

*Proof.* The fact that $K$ is r.e. follows immediately from the existence of the universal TM $M_u$ (Proposition 1.17). To see that $K$ is not recursive, we note that the complement of a recursive set is also recursive and, hence, r.e. Thus, if $K$ were recursive, then $\overline{K}$ would be r.e., and there would be an integer $y$ such that $M_y$ halts on all $x \in \overline{K}$ and does not halt on any $x \in K$. Then, a contradiction could be found when we consider whether or not $y$ itself is in $K$: if $y \in K$ then $M_y$ must not halt on $y$ and it follows from the definition of $K$ that $y \notin K$; and if $y \notin K$ then $M_y$ must halt on $y$ and it follows from the definition of $K$ that $y \in K$.                                                                     ∎

Now we apply the diagonalization technique to separate complexity classes. We first consider deterministic space-bounded classes. From the Tape Compression Theorem, we know that $DSPACE(s_1(n)) = DSPACE(s_2(n))$ if $c_1 \cdot s_1(n) \leq s_2(n) \leq c_2 \cdot s_1(n)$ for some constants $c_1$ and $c_2$. The following theorem shows that if $s_2(n)$ asymptotically diverges faster than $c \cdot s_1(n)$ for all constants $c$ then $DSPACE(s_1(n)) \neq DSPACE(s_2(n))$.

**Theorem 1.22** (Space Hierarchy Theorem) *Let $s_2(n)$ be a fully space-constructible function. Suppose that $\liminf_{n \to \infty} s_1(n)/s_2(n) = 0$ and $s_2(n) \geq s_1(n) \geq \log_2 n$. Then, $DSPACE(s_1(n)) \subsetneq DSPACE(s_2(n))$.*

*Proof.* We are going to construct a TM $M^*$ that works within space $s_2(n)$ to diagonalize against all TMs $M_x$ that use space $s_1(n)$. To set up for the diagonalization process, we modify the coding of the TMs a little. Each set $A$ in $DSPACE(s_1(n))$ is computed by a multi-tape TM within space bound $s_1(n)$. By Corollary 1.16, together with a simple coding of any alphabet $\Gamma$ by the fixed alphabet $\{0,1\}$, there is a one-worktape TM over the alphabet $\{0,1,B\}$ that accepts $A$ in space $c_1 \cdot s_1(n)$ for some constant $c_1 > 0$. So, it is sufficient to consider the enumeration $\{M_x\}$ of one-worktape TMs over the alphabet $\{0,1,B\}$ as defined in Section 1.5. Recall that each TM was encoded by a string of the form (1.3). We now let all strings of the form $1^k x$, $k \geq 0$, encode $M_x$ if $x$ is of the form (1.3); we still let all other illegal strings $y$ encode the fixed empty TM $M_0$. Thus, for each TM $M$, there now are infinitely many strings that encode it.

Our machine $M^*$ is a two-worktape TM that uses the tape alphabet $\{0,1,B,\#\}$. On each input $x$, $M^*$ does the following:

(1) $M^*$ simulates a $s_2(n)$-space marking TM to place $s_2(n)$ $\#$ symbols on each work tape.

(2) Let $t_2(n) = 2^{s_2(n)}$. $M^*$ writes the integer $t_2(|x|)$, in the binary form, on tape 2. That is, $M^*$ writes 1 to the left of the leftmost $\#$ symbol and writes 0 over all $\#$'s.

(3) $M^*$ simulates $M_x$ on input $x$ on tape 1, one move at a time. For each move of $M_x$, if $M_x$ attempts to move off the $\#$ symbols then $M^*$ halts and rejects; otherwise, $M^*$ subtracts one from the number on tape 2. If tape 2 contains the number 0, then $M^*$ halts and accepts.

(4) If $M_x$ halts on $x$ within space $s_2(|x|)$ and time $t_2(|x|)$, then $M^*$ halts, and $M^*$ accepts $x$ if and only if $M_x$ rejects $x$.

It is clear that $M^*$ works within space $2s_2(n) + 1$, and so, by the tape compression theorem, $L(M^*) \in DSPACE(s_2(n))$. Now we claim that $L(M^*) \notin DSPACE(s_1(n))$. To see this, assume by way of contradiction that $L(M^*)$ is accepted by a one-worktape TM $M$ that works in space $c_1 s_1(n)$ and uses the tape alphabet $\{0,1,B\}$ for some constant $c_1 > 0$. Then, there is a sufficiently large $x$ such that $L(M_x) = L(M^*)$, $c_1 s_1(|x|) \leq s_2(|x|)$ and $t_1(|x|) \leq t_2(|x|)$, where $t_1(n) = c_1 \cdot n^2 \cdot s_1(n) \cdot 3^{c_1 s_1(n)}$. (Note that $s_1(n) \geq \log n$ and $\liminf_{n \to \infty} s_1(n)/s_2(n) = 0$, and so such an $x$ exists.) Consider the simulation of $M_x$ on $x$ by $M^*$. Since $c_1 \cdot s_1(|x|) \leq s_2(|x|)$, $M^*$ on $x$ always works within the $\#$ marks.

*Case 1.* $M_x(x)$ halts in $t_2(|x|)$ moves. Then, $M^*$ accepts $x$ if and only if $M_x$ rejects $x$. Thus, $x \in L(M^*)$ if and only if $x \notin L(M_x)$. This is a contradiction.

*Case 2.* $M_x(x)$ does not halt in $t_2(|x|)$ moves. Since $M_x(x)$ uses only space $c_1 s_1(|x|)$, it may have at most $r \cdot c_1 \cdot s_1(|x|) \cdot |x| \cdot 3^{c_1 s_1(|x|)}$ different configurations, where $r$ is the number of states of $M_x$. (The value $s_1(|x|)$ is the number of possible tape head positions for the worktape, $|x|$ is the number of possible tape head position for the input tape, and $3^{c_1 s_1(|x|)}$ is the number of possible tape configurations.) Since $r \leq |x|$, we know that this value is bounded by $t_2(|x|)$. So, if $M_x$ does not halt in $t_2(|x|)$ moves, then it must have reached a configuration twice already. For a deterministic TM, this implies that it loops forever, and so $x \notin L(M_x)$. But $M^*$ accepts $x$ in this case, and again this is a contradiction.                                                                        ∎

For the time-bounded classes, we show a weaker result.

**Theorem 1.23** (Time Hierarchy Theorem) *If $t_2$ is a fully time-constructible function, $t_2(n) \geq t_1(n) \geq n$ and*

$$\liminf_{n \to \infty} \frac{t_1(n) \log(t_1(n))}{t_2(n)} = 0,$$

*then $DTIME(t_1(n)) \subsetneq DTIME(t_2(n))$.*

*Proof.* The proof is similar to the Space Hierarchy Theorem. The only thing that needs extra attention is that the machines $M_x$ may have an arbitrarily large number of tapes over an arbitrarily large alphabet, while the new machine $M^*$ has only a fixed number of tapes and a fixed-size alphabet. Thus, the simulation of $M_x$ on $x$ by $M^*$ may require extra time. This problem is resolved by considering the enumeration $\{M_x\}$ of two-worktape TMs over a fixed alphabet $\{0, 1, \mathsf{B}\}$ and relaxing the time bound to $O(t_1(n) \log(t_1(n)))$ (Theorem 1.15).                                                                        ∎

A nondecreasing function $f(n)$ is called *super-polynomial* if

$$\liminf_{n \to \infty} \frac{n^i}{f(n)} = 0$$

for all $i \geq 1$. For instance, the exponential function $2^n$ is super-polynomial, and the functions $2^{\log^k n}$, for $k > 1$, are also super-polynomial.

**Corollary 1.24** $P \subsetneq DTIME(f(n))$ *for all super-polynomial functions $f$.*

*Proof.* If $f(n)$ is super-polynomial, then so is $g(n) = (f(n))^{1/2}$. By the Time Hierarchy Theorem, $P \subseteq DTIME(g(n)) \subsetneq DTIME(f(n))$.                                      ∎

**Corollary 1.25** $PSPACE \subsetneq DSPACE(f(n))$ *for all super-polynomial functions $f$.*

It follows that $P \subsetneq EXP$, and $LOGSPACE \subsetneq PSPACE \subsetneq EXPSPACE$.

The above time and space hierarchy theorems can also be extended to nondeterministic time- and space-bounded complexity classes. The proofs, however, are more involved, because acceptance and rejection in NTMs are not symmetric. We only list the simpler results on nondeterministic space-bounded complexity classes which can be proved by a straightforward diagonalization. For the nondeterministic time hierarchy, see Exercise 1.28.

**Theorem 1.26** *(a) $NLOGSPACE \subsetneq NPSPACE$.*
*(b) For $k \geq 1$, $NSPACE(n^k) \subsetneq NSPACE(n^{k+1})$.*

*Proof.* See Exercise 1.27. ∎

In addition to the diagonalization technique, some other proof techniques for separating complexity classes are known. For instance, the following result separating the classes *EXP* and *PSPACE* is based on a closure property of *PSPACE* that does not hold for *EXP*. Unfortunately, this type of indirect proof techniques is not able to resolve the question of whether $PSPACE \subseteq EXP$ or $EXP \subseteq PSPACE$.

**Theorem 1.27** $EXP \neq PSPACE$.

*Proof.* From the Time Hierarchy Theorem, we get $EXP \subseteq DTIME(2^{n^{3/2}}) \subsetneq DTIME(2^{n^2})$. Thus, it suffices to show that if $PSPACE = EXP$ then $DTIME(2^{n^2}) \subseteq PSPACE$.

Assume that $L \in DTIME(2^{n^2})$. Let \$ be a symbol not used in $L$, and let $L' = \{x\$^t : x \in L, |x| + t = |x|^2\}$. Clearly, $L' \in DTIME(2^n)$. So, by the assumption that $PSPACE = EXP$, we have $L' \in PSPACE$; that is, there exists an integer $k > 0$ such that $L' \in DSPACE(n^k)$. Let $M$ be a DTM accepting $L'$ with the space bound $n^k$. We can construct a new DTM $M'$ that operates as follows.

> On input $x$, $M'$ copies $x$ into a tape and then adds $|x|^2 - |x|$ \$'s. Then, $M'$ simulates $M$ on $x\$^{|x|^2-|x|}$.

Clearly, $L(M') = L$. Note that $M'$ uses space $n^{2k}$, and so $L \in PSPACE$. Therefore, $DTIME(2^{n^2}) \subseteq PSPACE$, and the theorem is proven. ∎

## 1.7 Simulation

We study, in this section, the relationship between deterministic and nondeterministic complexity classes, as well as the relationship between time and space bounded complexity classes. We show several different simulations of nondeterministic machines by deterministic ones.

**Theorem 1.28** *(a) For any fully space-constructible function $f(n) \geq n$,*

$$DTIME(f(n)) \subseteq NTIME(f(n)) \subseteq DSPACE(f(n)).$$

*(b) For any fully space-constructible function $f(n) \geq \log n$,*

$$DSPACE(f(n)) \subseteq NSPACE(f(n)) \subseteq \bigcup_{c>0} DTIME(2^{cf(n)}).$$

*Proof.* (a) The relation $DTIME(f(n)) \subseteq NTIME(f(n))$ follows immediately from the fact that DTMs are just a subclass of NTMs. For the relation $NTIME(f(n)) \subseteq DSPACE(f(n))$, we recall the simulation of an NTM $M$ by a DTM $M_1$ as described in Theorem 1.9. Suppose that $M$ has time complexity bounded by $f(n)$; then $M_1$ needs to simulate $M$ for at most $f(n)$ moves. That is, we restrict $M_1$ to only execute the first $\sum_{i=1}^{f(n)} k^i$ stages such that the strings written in tape 2 are at most $f(n)$ symbols long. Since $f(n)$ is fully space-constructible, this restriction can be done by first marking off $f(n)$ squares on tape 2. It is clear that such a restricted simulation works within space $f(n)$.

(b) Again, $DSPACE(f(n)) \subseteq NSPACE(f(n))$ is obvious. To show that $NSPACE(f(n)) \subseteq \bigcup_{c>0} DTIME(2^{cf(n)})$, assume that $M$ is an NTM with the space bound $f(n)$. We are going to construct a DTM $M_1$ to simulate $M$ in time $2^{cf(n)}$ for some $c > 0$. Since $M$ uses only space $f(n)$, there is a constant $c_1 > 0$ such that the shortest accepting computation for each $x \in L(M)$ is of length $\leq 2^{c_1 f(|x|)}$. Thus, the machine $M_1$ needs only to simulate $M(x)$ for, at most, $2^{c_1 f(n)}$ moves. However, $M$ is a nondeterministic machine and so its computation tree of depth $2^{c_1 f(n)}$ could have $2^{2^{O(f(n))}}$ leaves, and the naive simulation as (a) above takes too much time.

To reduce the deterministic simulation time, we notice that this computation tree, although of size $2^{2^{O(f(n))}}$, has at most $2^{O(f(n))}$ different configurations: Each configuration is determined by at most $f(n)$ tape symbols on the work tape, one of $f(n)$ positions for the work tape head, one of $n$ positions for the input tape head, and one of $r$ states, where $r$ is a constant. Thus, the total number of possible configurations of $M(x)$ is $2^{O(f(n))} \cdot f(n) \cdot n \cdot r = 2^{O(f(n))}$. (Note that $f(n) \geq \log n$ implies $n \leq 2^{f(n)}$.)

Let $T$ be the computation tree of $M(x)$ with depth $2^{c_1 f(n)}$, with each node labeled by its configuration. We define a breadth-first ordering $\prec$ on the nodes of $T$ and prune the subtree of $T$ rooted at a node $v$ as long as there is a node $u \prec v$ that has the same configuration as $v$. Then, the resulting pruned tree $T'$ has at most $2^{O(f(n))}$ internal nodes and hence is of size $2^{O(f(n))}$. In addition, the tree $T'$ contains an accepting configuration if and only if $x \in L(M)$, since all deleted subtrees occur somewhere else in $T'$. In other words, our DTM $M_1$ works as follows: it simulates $M(x)$ by making a breadth-first traversal over the tree $T$, keeping a record of the configurations encountered so far. When it visits a new node $v$ of the tree, it checks the history record to see if the configuration has occurred before, and prunes the subtree rooted at $v$ if this is the case. In this way, $M_1$ only visits the nodes in tree $T'$ and works within time $2^{cf(n)}$ for some constant $c > 0$. ∎

**Corollary 1.29** $LOGSPACE \subseteq NLOGSPACE \subseteq P \subseteq NP \subseteq PSPACE$.

Next we consider the space complexity of the deterministic simulation of nondeterministic space-bounded machines.

**Theorem 1.30** (Savitch's Theorem) *For any fully space-constructible function $s(n) \geq \log n$, $NSPACE(s(n)) \subseteq DSPACE((s(n))^2)$.*

*Proof.* Let $M$ be a one-worktape $s(n)$-space-bounded NTM. We are going to construct a DTM $M_1$ to simulate $M$ using space $(s(n))^2$. Without loss of generality, we may assume that $M$ has a unique accepting configuration for all inputs $x$; that is, we require that $M$ cleans up the worktape and enters a fixed accepting state when it accepts an input $x$. As pointed out in the proof of Theorem 1.28, the shortest accepting computation of $M$ on an $x \in L(M)$ is at most $2^{O(s(n))}$, and each configuration is of length $s(n)$. (We say a configuration is of length $\ell$ if its work tape has at most $\ell$ nonblank symbols. Note that there are only $2^{O(\ell)}$ configurations on input $x$ that has length $\ell$.) Thus, for each input $x$, the goal of the DTM $M_1$ is to search for a computation path of length at most $2^{O(s(n))}$ from the initial configuration $\alpha_0$ to the accepting configuration $\alpha_f$.

Define a predicate $reach(\beta, \gamma, k)$ to mean that both $\beta$ and $\gamma$ are configurations of $M$ such that $\gamma$ is *reachable* from $\beta$ in at most $k$ moves, that is, that there exists a sequence $\beta = \beta_0, \beta_1, \ldots, \beta_k = \gamma$ such that $\beta_i \vdash_M \beta_{i+1}$ or $\beta_i = \beta_{i+1}$, for each $i = 0, \ldots, k-1$. Using this definition, we see that $M$ accepts $x$ if and only if $reach(\alpha_0, \alpha_f, 2^{cs(n)})$, where $\alpha_0$ is the unique initial configuration of $M$ on input $x$, $\alpha_f$ is the unique accepting configuration of $M$ on $x$, and $2^{cs(n)}$ is the number of possible configurations of $M$ of length $s(n)$. The following is a recursive algorithm computing the predicate *reach*. The main observation here is that

$$reach(\alpha_1, \alpha_2, j+k) \iff (\exists \alpha_3) \, [reach(\alpha_1, \alpha_3, j) \text{ and } reach(\alpha_3, \alpha_2, k)].$$

*Algorithm for* $reach(\alpha_1, \alpha_2, i)$:
First, if $i \leq 1$, then return TRUE if and only if $\alpha_1 = \alpha_2$ or $\alpha_1 \vdash_M \alpha_2$.
If $i \geq 2$, then for all possible configurations $\alpha_3$ of $M$ of length $s(n)$, recursively compute whether it is true that $reach(\alpha_1, \alpha_3, \lceil i/2 \rceil)$ and $reach(\alpha_3, \alpha_2, \lfloor i/2 \rfloor)$; return TRUE if and only if there exists such an $\alpha_3$.

It is clear that the algorithm is correct. This recursive algorithm can be implemented by a standard nonrecursive simulation using a stack of depth $O(s(n))$, with each level of the stack using space $O(s(n))$ to keep track of the current configuration $\alpha_3$. Thus the total space used is $O((s(n))^2)$. (See Exercise 1.34 for the detail.) ∎

**Corollary 1.31** *PSPACE = NPSPACE.*

For any complexity class $\mathcal{C}$, let $co\mathcal{C}$ (or, $co$-$\mathcal{C}$) denote the class of complements of sets in $\mathcal{C}$; that is, $co\mathcal{C} = \{S : \overline{S} \in \mathcal{C}\}$, where $\overline{S} = \Sigma^* - S$, and $\Sigma$ is the smallest alphabet such that $S \subseteq \Sigma^*$. One of the differences between deterministic and nondeterministic complexity classes is that deterministic classes are closed under complementation, but this is not known to be true for nondeterministic classes. For instance, it is not known whether $NP = coNP$. The following theorem shows that for most interesting nondeterministic space-bounded classes $\mathcal{C}$, $co\mathcal{C} = \mathcal{C}$.

**Theorem 1.32** *For any fully space-constructible function* $s(n) \geq \log n$, $NSPACE(s(n)) = coNSPACE(s(n))$.

*Proof.* Let $M$ be a one-worktape NTM with the space bound $s(n)$. Recall the predicate $reach(\alpha, \beta, k)$ defined in the proof of Theorem 1.30. In this proof, we further explore the concept of reachable configurations from a given configuration of length $s(n)$. Consider a fixed input $x$ of length $n$. Let $C_x$ be the class of all configurations of $M$ on $x$ that is of length $s(n)$. Let $\prec$ be a fixed ordering of configurations in $C_x$ such that an $O(s(n))$ space-bounded DTM can generate these configurations in the increasing order. First we observe that the predicate $reach(\alpha, \beta, k)$ is acceptable by an NTM $M_1$ in space $O(s(n) + \log k)$ if $\alpha$ and $\beta$ are from $C_x$. The NTM $M_1$ operates as follows:

> *Machine $M_1$.* Let $\alpha_0 = \alpha$. For each $i = 0, \ldots, k-2$, $M_1$ guesses a configuration $\alpha_{i+1} \in C_x$ and verifies that $\alpha_i = \alpha_{i+1}$ or $\alpha_i \vdash_M \alpha_{i+1}$. (If neither $\alpha_i \neq \alpha_{i+1}$ nor $\alpha_i \vdash_M \alpha_{i+1}$ holds, then $M_1$ rejects on this computation path.) Finally, $M_1$ verifies that $\alpha_{k-1} = \beta$ or $\alpha_{k-1} \vdash_M \beta$, and accepts if this holds.

Apparently, this NTM $M_1$ uses space $O(s(n) + \log k)$ and accepts $(\alpha, \beta, k)$ if and only if $reach(\alpha, \beta, k)$.

Next, we apply this machine $M_1$ to construct another NTM $M_2$ which, on any given configuration $\beta$, computes the exact number $N$ of configurations in $C_x$ that are reachable from $\beta$, using space $O(s(n))$. (This is an NTM computing a function as defined in Definition 1.8.) Let $m$ be the maximum length of an accepting path on input $x$. Then, $m = 2^{O(s(n))}$. The machine $M_2$ uses the following algorithm to compute iteratively the number $N_k$ of configurations in $C_x$ that are reachable from $\beta$ in at most $k$ moves, for $k = 0, \ldots, m+1$. Then, when for some $k$ it is found that $N_k = N_{k+1}$, it halts and outputs $N = N_k$.

> *Algorithm for computing $N_k$:*
>
> For $k = 0$, just let $N_0 = 1$ ($\beta$ is the only reachable configuration). For each $k > 0$, assume that $N_k$ has been found. To compute $N_{k+1}$, machine $M_2$ maintains a counter $N_{k+1}$ which is initialized to 0 and, then, for each configuration $\alpha \in C_x$, $M_2$ does the following:
>
> (1) First, let $r_\alpha$ be FALSE. For each $i = 1, \ldots, N_k$, $M_2$ guesses a configuration $\gamma_i$ in $C_x$, and verifies that (i) $\gamma_{i-1} \prec \gamma_i$ if $i > 1$, and (ii) $reach(\beta, \gamma_i, k)$ (this can be done by machine $M_1$). It rejects this computation path if (i) or (ii) does not hold. Next, it deterministically checks whether $reach(\gamma_i, \alpha, 1)$. If it is true that $reach(\gamma_i, \alpha, 1)$, then it sets the flag $r_\alpha$ to TRUE. In either case, it continues to the next $i$.

(2) When the above is done for all $i = 1, \ldots, N_k$ and if the computation does not reject, then $M_2$ adds one to $N_{k+1}$ if and only if $r_\alpha = \text{TRUE}$, and goes to the next configuration.

Note that $M_2$ uses only space $O(s(n))$, because at each step corresponding to configuration $\alpha$ and integer $i$, it only needs to keep the following information: $i$, $k$, $N_k$, the current $N_{k+1}$, $r_\alpha$, $\alpha$, $\gamma_{i-1}$ and $\gamma_i$. Furthermore, it can be checked that this algorithm indeed computes the function that maps $\beta$ to the number $N$ of reachable configurations: At stage $k + 1$, assume that $N_k$ has been correctly computed. Then, for each $\alpha$, there exists one nonrejecting path in stage $(k + 1)$—the path that guesses the $N_k$ configurations $\gamma_i$ that are reachable from $\beta$ in $k$ moves, in the increasing order. All other paths are rejecting paths. For each $\alpha$, this unique path must determine whether $reach(\beta, \alpha, k + 1)$ correctly. (See Exercise 1.35 for more discussions.)

Next, we construct a third NTM $M_3$ for $\overline{L(M)}$ as follows:

> *Machine $M_3$.* First, $M_3$ simulates $M_2$ to compute the number $N$ of reachable configurations from the initial configuration $\alpha_0$. Then, it guesses $N$ configurations $\gamma_1, \ldots, \gamma_N$, one by one and in the increasing order as in $M_2$ above, and checks that each is reachable from $\alpha_0$ (by machine $M_1$) and none of them is an accepting configuration. It accepts if the above are checked; otherwise, it rejects this computation path.

We claim that this machine $M_3$ accepts $\overline{L(M)}$. First, it is easy to see that if $x \notin L(M)$, then all reachable configurations from $\alpha_0$ are nonaccepting configurations. So, the computation path of $M_3$ that guesses correctly all $N$ reachable configurations of $\alpha_0$ will accept $x$. Conversely, if $x \in L(M)$, then one of the reachable configuration from $\alpha_0$ must be an accepting configuration. So, a computation path of $M_3$ must guess either all reachable configurations that include one accepting configuration or guess at least one nonreachable configuration. In either case, this computation path must reject. Thus, $M_3$ accepts exactly those $x \notin L(M)$.

Finally, the same argument for $M_2$ verifies that $M_3$ uses space $O(s(n))$. The theorem then follows from the tape compression theorem for NTMs. ∎

Recall that in the formal language theory, a language $L$ is called *context-sensitive* if there exists a grammar $G$ generating the language $L$ with the right-hand side of each grammar rule in $G$ being at least as long as its left-hand side. A well-known characterization for the context-sensitive languages is that the context-sensitive languages are exactly the class $NSPACE(n)$.

**Corollary 1.33** *(a) $NLOGSPACE = coNLOGSPACE$.*

*(b) The class of context-sensitive languages is closed under complementation.*

The above results, together with the ones proved in the last section, are the best we know about the relationship among time/space-bounded determinis-

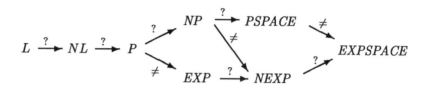

Figure 1.7: Inclusive relations among the complexity classes. We write $L$ to denote $LOGSPACE$ and $NL$ to denote $NLOGSPACE$. The label $\neq$ means the proper inclusion. The label ? means the inclusion is not known to be proper.

tic/nondeterministic complexity classes. Many other important relations are not known. We summarize in Figure 1.7 the known relations among the most familiar complexity classes. More complexity classes between $P$ and $PSPACE$ will be introduced in Chapters 3, 8, 9, and 10. Complexity classes between $LOGSPACE$ and $P$ will be introduced in Chapter 6.

## Exercises

**1.1** Let $\ell_1, \cdots, \ell_q$ be $q$ natural numbers, and $\Sigma$ be an alphabet of $r$ symbols. Show that there exist $q$ strings $s_1, \cdots, s_q$ over $\Sigma$, of lengths $\ell_1, \cdots, \ell_q$, respectively, which form an alphabet if and only if $\sum_{i=1}^{q} r^{-\ell_i} \leq 1$.

**1.2** (a) Let $c \geq 0$ be a constant. Prove that no pairing function $\langle \cdot, \cdot \rangle$ exists such that for all $x, y \in \Sigma^*$, $|\langle x, y \rangle| \leq |x| + |y| + c$.

(b) Prove that there exists a pairing function $\pi : \Sigma^* \times \Sigma^* \to \Sigma^*$ such that (i) for any $x, y \in \Sigma^*$, $|\pi(x, y)| \leq |x| + |y| + O(\log |x|)$, (ii) it is polynomial-time computable, and (iii) its inverse function is polynomial-time computable (if $z \notin range(\pi)$, then $\pi^{-1}(z) = (\lambda, \lambda)$).

(c) Let $f : \mathbf{N} \times \mathbf{N} \to \mathbf{N}$ be defined by $f(n, m) = (n + m)(n + m + 1)/2 + n$. Prove that $f$ is one-one, onto, polynomial-time computable (with respect to the binary representations of natural numbers) and that $f^{-1}$ is polynomial-time computable. (Therefore, $f$ is an efficient pairing function on natural numbers.)

**1.3** There are two cities $T$ and $F$. The residents of city $T$ always tell the truth and the residents of city $F$ always lie. The two cities are very close. Their residents visit each other very often. When you walk in city $T$ you may meet a person who came from city $F$, and vice versa. Now, suppose that you are in one of the two cities and you want to find out which city you are in. Also suppose that you are allowed to ask a person on the street for only one YES/NO question. What question should you ask? [*Hint:* You may first design a Boolean function $f$ of two variables, the resident and the city, for your need, then find a question corresponding to $f$.]

**1.4** How many Boolean functions of $n$ variables are there?

**1.5** Assume that $f$ is a Boolean function of more than two variables. Prove that for any minterm $p$ of $f|_{x_1=0, x_2=1}$, there exists a minterm of $f$ containing $p$ as a subterm.

**1.6** Design a multi-tape DTM $M$ to accept the language $L = \{a^i b a^j b a^k : i, j, k \geq 0, i + j = k\}$. Show the computation paths of $M$ on inputs $a^2 b a^3 b a^5$ and $a^2 b a b a^4$.

**1.7** Design a multi-tape NTM $M$ to accept the language $L = \{a^{i_1} b a^{i_2} b \cdots b a^{i_k} : i_1, i_2, \ldots, i_k \geq 0, i_r = i_s \text{ for some } 1 \leq r < s \leq k\}$. Show the computation tree of $M$ on input $a^3 b a^2 b a^2 b a^4 b a^2$. What is the time complexity of $M$?

**1.8** Show that the problem of Example 1.7 can be solved by a DTM in polynomial time.

**1.9** Give formal definitions of the configurations and the computation of a $k$-tape DTM.

**1.10** Assume that $M$ is a three-tape DTM using tape alphabet $\Sigma = \{a, b, B\}$. Consider the one-tape DTM $M_1$ that simulates $M$ as described in Theorem 1.6. Describe explicitly the part of the transition function of $M_1$ that moves the tape head from left to right to collect the information of the current tape symbols scanned by $M$.

**1.11** Let $C = \{\langle G, u, v \rangle : G = (V, E) \text{ is an undirected graph, } u, v \in V \text{ are connected}\}$. Design an NTM $M$ that accepts set $C$ in space $O(\log |V|)$. Can you find a DTM accepting $C$ in space $O(\log |V|)$? (Assume that $V = \{v_1, \ldots, v_n\}$. Then, the input to the machine $M$ is $\langle 0^n 1 e_{11} e_{12} \cdots e_{1n} e_{21} \cdots e_{nn}, o^i, o^j \rangle$, where $e_{ij} = 1$ if $\{v_i, v_j\} \in E$ and $e_{ij} = 0$ otherwise.)

**1.12** Prove that any finite set of strings belongs to $DTIME(n)$.

**1.13** Estimate an upper bound for the number of possible computation histories $H(g, g_\ell, g_r)$ in the proof of Proposition 1.13.

**1.14** A *random access machine* (RAM) is a machine model for computing integer functions. The memories of a RAM $M$ consists of a one-way read-only input tape, a one-way write-only output tape, and an infinite number of registers named $R_0, R_1, R_2, \ldots$. Each square of the input and output tapes and each register can store an integer of an arbitrary size. Let $c(R_i)$ denote the content of the register $R_i$. An instruction of a RAM may access a register

$R_i$ to get its content $c(R_i)$ or it may access the register $R_{c(R_i)}$ by the indirect addressing scheme. Each instruction has an integer label, beginning from 1. A RAM begins the computation on instruction with label 1, and halts when it reaches an empty instruction. The following table lists the instruction types of a RAM:

| Instruction | Meaning |
|---|---|
| READ($R_i$) | read the next input integer into $R_i$ |
| WRITE($R_i$) | write $c(R_i)$ on the output tape |
| COPY($R_i, R_j$) | write $c(R_i)$ to $R_j$ |
| ADD($R_i, R_j, R_k$) | write $c(R_i) + c(R_j)$ to $R_k$ |
| SUB($R_i, R_j, R_k$) | write $c(R_i) - c(R_j)$ to $R_k$ |
| MULT($R_i, R_j, R_k$) | write $c(R_i) \cdot c(R_j)$ to $R_k$ |
| DIV($R_i, R_j, R_k$) | write $\lfloor c(R_i)/c(R_j) \rfloor$ to $R_k$ (write 0 if $c(R_j) = 0$) |
| GOTO($i$) | go to the instruction with label $i$ |
| IF-THEN($R_i, j$) | if $c(R_i) \geq 0$ then go to the instruction with label $j$ |

In the above table, we only listed the arguments in the direct addressing scheme. It can be changed to indirect addressing scheme by changing the argument $R_i$ to $R_i^*$. For instance, ADD($R_i, R_j^*, R_k$) means to write $c(R_i) + c(R_{c(R_j)})$ to $R_k$. Each instruction can also use a constant argument $i$ instead of $R_j$. For instance, COPY($i, R_j^*$) means to write integer $i$ to the register $R_{c(R_j)}$.

(a) Design a RAM that reads inputs $i_1, i_2, \ldots, i_k, j$ and outputs 1 if $\sum_{r \in A} i_r = j$ for some $A \subseteq \{1, 2, \ldots, k\}$ and outputs 0 otherwise (cf. Example 1.7).

(b) Show that for any TM $M$ there is a RAM $M'$ that computes the same function as $M$. (You need to first specify how to encode tape symbols of the TM $M$ by integers.)

(c) Show that for any RAM $M$ there is a TM $M'$ that computes the same function as $M$ (with the integer $n$ encoded by the string $a^n$).

**1.15** In this exercise, we consider the complexity of RAMs. There are two possible ways of defining the computational time of a RAM $M$ on input $x$. The *uniform* time measure counts each instruction as taking one unit of time and so the total runtime of $M$ on $x$ is the number of times $M$ executes an instruction. The *logarithmic* time measure counts, for each instruction, the number of bits of the arguments involved. For instance, the total time to execute the instruction MULT($i, R_j^*, R_k$) is $\lceil \log i \rceil + \lceil \log j \rceil + \lceil \log c(R_j) \rceil + \lceil \log c(R_{c(R_j)}) \rceil + \lceil \log k \rceil$. The total runtime of $M$ on $x$, in the logarithmic time measure, is the sum of the instruction time over all instructions executed by $M$ on input $x$.

Use both the uniform and the logarithmic time measures to analyze the simulations of parts (b) and (c) of Exercise 1.14. In particular, show that the notion of polynomial-time computability is equivalent between TMs and RAMs with respect to the logarithmic time measure.

**1.16** Suppose that a TM is allowed to have infinitely many tapes. Does this increase its computation power as far as the class of computable sets is concerned?

**1.17** Prove Blum's Speed-up Theorem. That is, find a function $f : \{0,1\}^* \to \{0,1\}^*$ such that for any DTM $M_1$ computing $f$ there exists another DTM $M_2$ computing $f$ with $time_{M_2}(x) < time_{M_1}(x)$ for infinitely many $x \in \{0,1\}^*$.

**1.18** Prove that if $c > 1$, then for any $\epsilon > 0$,

$$DTIME(cn) = DTIME((1+\epsilon)n).$$

**1.19** Prove Theorem 1.15.

**1.20** Prove that if $f(n)$ is an integer function such that (i) $f(n) \geq 2n$ and (ii) $f$, regarded as a function mapping $a^n$ to $a^{f(n)}$, is computable in deterministic time $O(f(n))$, then $f$ is fully time-constructible. [*Hint*: Use the linear speed-up of Proposition 1.13 to compute the function $f$ in *less than* $f(n)$ *moves*, with the output $a^{f(n)}$ compressed.]

**1.21** Prove that if $f$ is fully space-constructible and $f$ is not a constant function, then $f(n) = \Omega(\log\log n)$. (We say $g(n) = \Omega(h(n))$ if there exists a constant $c > 0$ such that $g(n) \geq c \cdot h(n)$ for almost all $n \geq 0$.)

**1.22** Prove that $f(n) = \lceil \log n \rceil$ and $g(n) = \lceil \sqrt{n} \rceil$ are fully space-constructible.

**1.23** Prove that the question of determining whether a given TM halts in time $p(n)$ for some polynomial function $p$ is undecidable.

**1.24** Prove that for any one-worktape DTM $M$ that uses space $s(n)$, there is a one-worktape DTM $M'$ with $L(M') = L(M)$ such that it uses space $s(n)$ and its tape head never moves to the left of its starting square.

**1.25** Recall that $\log^* n = \min\{k : k \geq 1, \log\log\cdots\log n \ (k \text{ iterations of } \log \text{ on } n) \leq 1\}$. Let $k > 0$ and $t_2$ be a fully time-constructible function. Assume that

$$\liminf_{n\to\infty} \frac{t_1(n)\log^*(t_1(n))}{t_2(n)} = 0.$$

Prove that there exists a language $L$ that is accepted by a $k$-tape DTM in time $t_2(n)$ but not by any $k$-tape DTM in time $t_1(n)$.

**1.26** Prove that $DTIME(n^2) \subsetneq DTIME(n^2 \log n)$.

**1.27** (a) Prove Theorem 1.26 (cf. Theorems 1.30 and 1.32).

(b) Prove that for any integer $k \geq 1$ and any rational $\epsilon > 0$, $NSPACE(n^k) \subsetneq NSPACE(n^{k+\epsilon})$.

**1.28** (a) Prove that $NP \subsetneq NEXP$.

(b) Prove that $NTIME(n^k) \subsetneq NTIME(n^{k+1})$ for any $k \geq 1$.

**1.29** Prove that $P \neq DSPACE(n)$ and $NP \neq EXP$.

**1.30** A set $A$ is called a *tally set* if $A \subseteq \{a\}^*$ for some symbol $a$. A set $A$ is called a *sparse set* if there exists a polynomial function $p$ such that $|\{x \in A : |x| \leq n\}| \leq p(n)$ for all $n \geq 1$. Prove that the following are equivalent:

(a) $EXP = NEXP$.

(b) All tally sets in $NP$ are actually in $P$.

(c) All sparse sets in $NP$ are actually in $P$.

(*Note*: The above implies that if $EXP \neq NEXP$ then $P \neq NP$.)

**1.31** (Busy Beaver) In this exercise, we show the existence of an r.e., nonrecursive set without using the diagonalization technique. For each TM $M$, we let $|M|$ denote the length of the codes for $M$. We consider only one-tape TMs with a fixed alphabet $\Gamma = \{0, 1, B\}$. Let $f(x) = \min\{|M| : M(\lambda) = x\}$ and $g(m) = \min\{x \in \{0,1\}^* : |M| \leq m \Rightarrow M(\lambda) \neq x\}$. That is, $f(x)$ is the size of the minimum TM that outputs $x$ on the input $\lambda$, and $g(m)$ is the least $x$ that is not printable by any TM $M$ of size $|M| \leq m$ on the input $\lambda$. It is clear that both $f$ and $g$ are total functions.

(a) Prove that neither $f$ nor $g$ is a recursive function.

(b) Define $A = \{\langle x, m \rangle : (\exists M, |M| \leq m)\ M(\lambda) = x\}$. Apply part (a) above to show that $A$ is r.e. but is not recursive.

**1.32** ( Busy Beaver, *time-bounded version*) In this exercise, we apply the busy beaver technique to show the existence of a set computable in exponential time but not in polynomial time.

(a) Let $g(0^m, 0^k)$ be the least $x \in \{0,1\}^*$ that is not printable by any TM $M$ of size $|M| \leq m$ on input $\lambda$ in time $(m + k)^{\log k}$. It is easy to see that $|g(0^m, 0^k)| \leq m + 1$, and hence $g$ is polynomial length-bounded. Prove that $g(0^m, 0^k)$ is computable in time $2^{O(m+k)}$ but is not computable in time polynomial in $m + k$.

(b) Can you modify part (a) above to prove the existence of a function that is computable in subexponential time (e.g., $n^{O(\log n)}$) but is not polynomial-time computable?

**1.33** Assume that an NTM computes the characteristic function $\chi_A$ of a set $A$ in polynomial time, in the sense of Definition 1.8. What can you infer about the complexity of set $A$?

**1.34** In the proof of Theorem 1.30, the predicate *reach* was solved by a deterministic recursive algorithm. Convert it to a nonrecursive algorithm for the predicate *reach* that only uses space $O((s(n))^2)$.

**1.35** What is wrong if we use the following simpler algorithm to compute $N_k$ in the proof of Theorem 1.32?

> For each $k$, to compute $N_k$, we generate each configuration $\alpha \in C_x$ one by one and, for each one, nondeterministically verify whether $reach(\beta, \alpha, k)$ (by machine $M_1$), and increments the counter for $N_k$ by one if $reach(\beta, \alpha, k)$ holds.

**1.36** In this exercise, we study the notion of computability of real numbers. First, for each $n \in \mathbf{N}$, we let $bin(n)$ denote its binary expansion, and for each $x \in \mathbf{R}$, let $sgn(x) = \lambda$ if $x \geq 0$, and $sgn(x) = -$ if $x < 0$. An integer $m \in \mathbf{Z}$ is represented as $sgn(m)bin(|m|)$. A rational number $r = \pm a/b$ with $a, b \in \mathbf{N}$, $b \neq 0$, and $gcd(a, b) = 1$, has a unique representation over alphabet $\{-, 0, 1, /\}$: $sgn(r)bin(a)/bin(b)$. We write $\mathbf{N}$ ($\mathbf{Z}$ and $\mathbf{Q}$) to denote both the set of natural numbers (integers, and rational numbers, respectively) and the set of their representations.

We say a real number $x$ is *Cauchy computable* if there exist two computable functions $f : \mathbf{N} \to \mathbf{Q}$ and $m : \mathbf{N} \to \mathbf{N}$ satisfying the property $C_{f,m}$: $n \geq m(k) \Rightarrow |f(n) - x| \leq 2^{-k}$. A real number $x$ is *Dedekind computable* if the set $L_x = \{r \in \mathbf{Q} : r < x\}$ is computable. A real number $x$ is *binary computable* if there is a computable function $b_x : \mathbf{N} \to \mathbf{N}$, with the property $b_x(n) \in \{0, 1\}$ for all $n > 0$, such that

$$x = sgn_x \cdot \sum_{n \geq 0} b_x(n) 2^{-n},$$

where $sgn_x = 1$ if $x \geq 0$ and $sgn_x = -1$ if $x < 0$. (The function $b_x$ is not unique for some $x$, but notice that for such $x$, both functions $b_x$ are computable.) A real number $x$ is *Turing computable* if there exists a DTM $M$ that on the empty input prints a binary expansion of $x$ (i.e., it prints the string $b_x(0)$ followed by the binary point, and then followed by the infinite string $b_x(1)b_x(2) \cdots$).

Prove that the above four notions of computability of real numbers are equivalent. That is, prove that a real number $x$ is Cauchy computable if and only if it is Dedekind computable if and only if it is binary computable if and only if it is Turing computable.

**1.37** In this exercise, we study the computational complexity of a real number. We define a real number $x$ to be *polynomial-time Cauchy computable* if there exist a polynomial-time computable function $f : \mathbf{N} \to \mathbf{Q}$, and a polynomial function $m : \mathbf{N} \to \mathbf{N}$, satisfying $C_{f,m}$, where $f$ being polynomial-time computable means that $f(n)$ is computable by a DTM in time $p(n)$ for some polynomial $p$ (i.e., the input $n$ is written in the unary form). We say $x$ is *polynomial-time Dedekind computable* if $L_x$ is in $P$. We say $x$ is *polynomial-time binary computable* if $b_x$, restricted to inputs $n > 0$, is polynomial-time computable, assuming that the inputs $n$ are written in the unary form.

Prove that the above three notions of polynomial-time computability of real numbers are not equivalent. That is, let $P_C$ ($P_D$, and $P_b$) denote the class of polynomial-time Cauchy (Dedekind and binary, respectively) computable real numbers. Prove that $P_D \underset{\neq}{\subseteq} P_b \underset{\neq}{\subseteq} P_C$.

## Historical Notes

McMillan's Theorem is well known in coding theory; see, for example, Roman [1992]. Turing machines were first defined by Turing [1936, 1937]. The equivalent variations of Turing machines, the notion of computability, and the Church-Turing Thesis are the main topics of recursive function theory; see, for example, Rogers [1967] and Soare [1987]. Kleene [1979] contains an interesting personal account of the history. The complexity theory based on Turing machines was developed by Hartmanis and Stearns [1965], Stearns, Hartmanis, and Lewis [1965] and Lewis, Stearns, and Hartmanis [1965]. The Tape Compression Theorem, the Linear Speed-up Theorem, the tape-reduction simulations (Corollaries 1.14–1.16) and the time and space hierarchy theorems are from these works. A machine-independent complexity theory has been established by Blum [1967]. Blum's Speed-up Theorem is from there. Cook [1973a] first established the nondeterministic time hierarchy. Seiferas [1977a, 1977b] contain further studies on the hierarchy theorems for nondeterministic machines. The indirect separation result, Theorem 1.27, is from Book [1974a]. The identification of $P$ as the class of feasible problems was first suggested by Cobham [1964] and Edmonds [1965]. Theorem 1.30 is from Savitch [1970]. Theorem 1.32 was independently proved by Immerman [1988] and Szelepcsényi [1988]. Hopcroft et al. [1977] and Paul et al. [1983] contain separation results between $DSPACE(n)$, $NTIME(O(n))$ and $DTIME(O(n))$. Context-sensitive languages and grammars are major topics in formal language theory; see, for example, Hopcroft and Ullman [1979].

RAMs were first studied in Shepherdson and Sturgis [1963] and Elgot and Robinson [1964]. The improvements over the Time Hierarchy Theorem, including Exercises 1.25 and 1.26, can be found in Paul [1979] and Fürer [1984]. Exercise 1.30 is an application of the Translation Lemma of Book [1974b]; see Hartmanis et al. [1983]. The busy beaver problems (Exercises 1.31 and 1.32) are related to the notion of Kolmogorov complexity of finite strings. The arguments based on Kolmogorov complexity avoids the direct use of diagonalization. See Daley [1980] for discussions on the busy beaver proof technique, and Li and Vitányi [1997] for a complete treatment of Kolmogorov complexity. Computable real numbers were first studied by Turing [1936] and Rice [1954]. Polynomial-time computable real numbers were studied in Ko and Friedman [1982] and Ko [1991a].

# 2

# NP-Completeness

The notions of reducibility and completeness play an important role in the classification of the complexity of problems from diverse areas of applications. In this chapter, we formally define two types of reducibilities: the polynomial-time many-one reducibility and the polynomial-time Turing reducibility. Then we prove several basic *NP*-complete problems and discuss the complexity of some combinatorial optimization problems.

## 2.1  *NP*

The complexity class *NP* has a nice characterization in terms of class *P* and polynomial length-bounded existential quantifiers.

**Theorem 2.1** *A language A is in NP if and only if there exist a language B in P, and a polynomial function p, such that for each instance x,*

$$x \in A \iff (\exists y, |y| \le p(|x|)) \langle x, y \rangle \in B. \tag{2.1}$$

*Proof.* Let $A = L(M)$ where $M = (Q, q_0, F, \Sigma, \Gamma, \Delta)$ is a one-tape NTM with a polynomial time bound $p(n)$. Similar to Theorem 1.9, we regard $\Delta$ as a multi-valued function from $Q \times \Gamma$ to $Q \times \Gamma \times \{L, R\}$, and let $k$ be the maximum number of values that $\Delta$ can assume on some $(q, s) \in Q \times \Gamma$. Let $\eta_1, \ldots, \eta_k$ be $k$ new symbols not in $\Gamma$. We define a DTM $M^*$ over the strings $\langle x, y \rangle$, where

$x \in \Sigma$ and $y \in \{\eta_1, \cdots, \eta_k\}^*$. The DTM $M^*$ is similar to the third step of the machine $M_1$ of Theorem 1.9. Namely, on input $\langle x, y \rangle$, where $y = \eta_{i_1} \cdots \eta_{i_m}$, $M^*$ simulates $M$ on input $x$ for at most $m$ moves. At the $j$th move, $M^*$ reads the $j$th symbol $\eta_{i_j}$ of $y$, and follows the $i_j$th value of the multi-valued function $\Delta$. $M^*$ accepts the input $\langle x, y \rangle$ if and only if this simulation accepts in $m$ moves. It is clear that $x \in L(M)$ if and only if $\langle x, y \rangle \in L(M^*)$ for some $y \in \{\eta_1, \ldots, \eta_k\}^*$ of length $\leq p(|x|)$.

Conversely, if $A$ can be represented by set $B$ and function $p$ as in (2.1), then we can design a 3-tape NTM $M$ for $A$ as follows: on input $x$, $M$ first makes $p(|x|)$ nondeterministic moves to write down a string $y$ of length $p(|x|)$ on tape 2. Then, it copies $\langle x, y \rangle$ to tape 3, and simulates the DTM $M_B$ for $B$ on tape 3. It accepts if $M_B$ accepts $\langle x, y \rangle$. It is clear that $M$ accepts the set $A$.                                                                                                      ■

Based on the above characterization, we often say a nondeterministic computation is a *guess-and-verify* computation. That is, if set $A$ has the property (2.1), then for any input $x$, an NTM $M$ can determine whether $x \in A$ by first *guessing* a string $y$ of length $p(|x|)$, and then deterministically *verifying* that $\langle x, y \rangle \in B$, in time polynomial in $|x|$. For each $x \in A$, a string $y$ of length $p(|x|)$ such that $\langle x, y \rangle \in B$ is called a *witness* or a *certificate* for $x$. This notion of guess-and-check computation is a simple way of presenting a nondeterministic algorithm.

In the following, we introduce some well-known problems in *NP*. We define each problem $\Pi$ in the form of a decision problem: "Given an input instance $x$, determine whether it satisfies the property $Q_\Pi(x)$." Each problem $\Pi$ has a corresponding language $L_\Pi = \{x : Q_\Pi(x)\}$. We say that $\Pi$ is in *NP* to mean that the corresponding language $L_\Pi$ is in *NP*.

**Example 2.2** (SATISFIABILITY) The problem SATISFIABILITY (SAT) is defined as follows: Given a Boolean formula $\phi$, determine whether there is an assignment that satisfies it (i.e., more formally, SAT is the set of all satisfiable Boolean formulas). Note that for any Boolean formula $\phi$ of length $n$ and any assignment $\tau$ on its variables, it can be checked deterministically in time $O(n^2)$ whether $\tau$ satisfies $\phi$. Thus, SAT is in *NP*, since we can first guess a Boolean assignment $\tau$ for each variable that occurs in $\phi$ ($\tau$ is of length $O(n)$) and then check that $\tau$ satisfies $\phi$ in time $O(n^2)$.                                                                       □

**Example 2.3** (HAMILTONIAN CIRCUIT). We first review the terminology of graphs. A *graph* is an ordered pair of disjoint sets $(V, E)$ such that $V \neq \emptyset$ and $E$ is a set of pairs of elements in $V$. The set $V$ is the set of *vertices* and the set $E$ is the set of *edges*. Two vertices are *adjacent* if there is an edge between them. A *path* is a sequence of vertices of which any two consecutive vertices are adjacent. A path is *simple* if all its vertices are distinct. A *cycle* is a path that is simple except that the first and the last vertices are identical.

Let $\{v_1, \cdots, v_n\}$ be the vertex set of a graph $G = (V, E)$. Then its *adjacency matrix* is an $n \times n$ matrix $[x_{ij}]$ defined by

$$x_{ij} = \begin{cases} 1 & \text{if } \{v_i, v_j\} \in E, \\ 0 & \text{otherwise.} \end{cases}$$

Note that $x_{ij} = x_{ji}$ and $x_{ii} = 0$. So, we have only $n(n-1)/2$ independent variables in an $n \times n$ adjacency matrix, for example, $x_{ij}$, $1 \le i < j \le n$. We use the string $x_{12}x_{13} \cdots x_{1n}x_{23} \cdots x_{n-1,n}$ to encode the graph $G$.

The problem HAMILTONIAN CIRCUIT (HC) asks, for a given graph $G$, whether it has a *Hamiltonian circuit*, that is, a cycle that passes through every vertex in $G$ exactly once. Assume that a graph $G = (V, E)$, with $|V| = n$, is given by its adjacency matrix (so the input is of length $n(n-1)/2$). Then, HC is in *NP* by the following nondeterministic algorithm: we first guess a string $y = y_1 \# y_2 \# \cdots \# y_n$ where each $y_i$ is an integer between 1 and $n$, and then verify that (i) $y$ forms a permutation of $\{1, \ldots, n\}$, and (ii) for each $i$, $1 \le i \le n-1$, $\{v_{y_i}, v_{y_{i+1}}\} \in E$, and $\{v_{y_n}, v_{y_1}\} \in E$. Note that $y$ is of length $O(n \log n)$ and the checking of conditions (i) and (ii) can be done in time $O(n^2)$. □

**Example 2.4** (INTEGER PROGRAMMING). The problem INTEGER PROGRAMMING (IP) asks, for a given pair $(A, b)$, where $A$ is an $n \times m$ integer matrix and $b$ is an $n$-dimensional integer vector, whether there exists an $m$-dimensional integer vector $x$ such that $Ax \ge b$.[1] Like the above examples, in order to demonstrate that $(A, b) \in$ IP, we may guess an $m$-dimensional integer vector $x$ and check whether $x$ satisfies $Ax \ge b$. However, for the purpose of proving IP $\in$ *NP*, we need to make sure that if the problem has a solution, then there is a solution of polynomial size. That is, we need to show that if $Ay \ge b$ for some $y$, then there exists a vector $x$ satisfying $Ax \ge b$ whose elements are integers of length at most $p(\log_2 \alpha + q)$ for some polynomial $p$, where $\alpha$ is the maximum absolute value of elements in $A$ and $b$, and $q = \max\{m, n\}$. (The length of an integer $k > 0$ written in the binary form is $\lceil \log_2(k+1) \rceil$.) Note that the inputs $A$ and $b$ have the total length $\ge \log_2 \alpha + q$, and hence the length of the above solution $x$ is polynomially bounded by the input length. We establish this result in the following three lemmas. For any square matrix $X$, we let det$X$ denote its determinant. (In the following three lemmas, $|x|$ denotes the absolute value of a real number $x$, not the length of the codes for $x$.)

**Lemma 2.5** *If $B$ is a square submatrix of $A$, then $|\det B| \le (\alpha q)^q$.*

*Proof.* Let $k$ be the order of $B$ (i.e., $B$ is a $k \times k$ matrix). Then $|\det B| \le k! \alpha^k \le k^k \alpha^k \le q^q \alpha^q = (\alpha q)^q$. ∎

---

[1] Let $x$ and $y$ be two $n$-dimensional integer vectors. We write $x \ge y$ to mean that each element of $x$ is greater than or equal to the corresponding element of $y$.

**Lemma 2.6** *If $rank(A) = r < m$, then there exists a nonzero vector $z$ such that $Az = 0$ and the absolute value of each component of $z$ is at most $(\alpha q)^q$.*

*Proof.* Without loss of generality, assume that the upper-left $r \times r$ submatrix $B$ of $A$ is nonsingular. Apply Cramer's rule to the system of equations

$$B(x_1, \cdots, x_r)^T = (a_{1m}, \cdots, a_{rm})^T,$$

where $a_{ij}$ is the element of $A$ on the $i$th row and the $j$th column. Then we obtain, for each $i = 1, \ldots, r$, $x_i = \det B_i / \det B$, where $B_i$ is the matrix obtained from matrix $B$ by replacing its $i$th column by $(a_{1m}, \ldots, a_{rm})^T$, and hence is a submatrix of $A$. By Lemma 2.5, $|\det B_i| \leq (\alpha q)^q$. Now, set $z_1 = \det B_1, \cdots, z_r = \det B_r, z_{r+1} = \cdots = z_{m-1} = 0$, and $z_m = -\det B$. Then, $Az = 0$.  ∎

**Lemma 2.7** *If $Ax \geq b$ has an integer solution, then it must have an integer solution of which the absolute value of each component is at most $2(\alpha q)^{2q+2}$.*

*Proof.* Let $a_i$ denote the $i$th row of $A$ and $b_i$ the $i$th component of $b$. Suppose that $Ax \geq b$ has an integer solution. Then we choose a solution $x$ that maximizes the the number of elements in the following set:

$$B_x = \{a_i : b_i \leq a_i x \leq b_i + (\alpha q)^{q+1}, 1 \leq i \leq n\}$$
$$\cup \{e_j : |x_j| \leq (\alpha q)^q, 1 \leq j \leq m\},$$

where $e_j$ is the $j$th $m$-dimensional unit vector, that is, $e_j = (0, \cdots, 0, 1, 0, \cdots, 0)$, with the $j$th component equal to 1 and the rest equal to 0. We first prove that the rank of $B_x$ is $m$. For the sake of contradiction, suppose that the rank of $B_x$ is less than $m$. Then, by Lemma 2.6, we can find an $m$-dimensional nonzero integer vector $z$ such that for any $d \in B_x$, $dz = 0$ and each component of $z$ does not exceed $(\alpha q)^q$. Note that for each $1 \leq j \leq m$, $e_j \in B_x$ implies that the $j$th component $z_j$ of $z$ is 0 since $0 = e_j z = z_j$. Since $z$ is nonzero, $z_k \neq 0$ for some $k$, $1 \leq k \leq m$, and for this $k$, $e_k \notin B_x$, and so $|x_k| > (\alpha q)^q$. Set $y = x + z$ or $x - z$ such that $|y_k| < |x_k|$. (This is possible since $|z_k| \leq (\alpha q)^q < |x_k|$.) Note that for every $e_j \in B_x$, we have $y_j = x_j$, and so $e_j \in B_y$, and for every $a_i \in B_x$, $a_i y = a_i x \pm a_i z = a_i x$ and, hence, $a_i \in B_y$. Thus, $B_y$ contains $B_x$. Moreover, for $a_i \notin B_x$, $a_i y \geq a_i x - |a_i z| \geq b_i + (\alpha q)^{q+1} - m\alpha(\alpha q)^q \geq b_i$. Thus, $y$ is an integer solution of $Ax \geq b$. By the maximality of $B_x$, $B_y = B_x$. So, we have shown that for each solution $x$ that maximizes $B_x$ there is a solution $y$ with $B_y = B_x$ of which the absolute value of the $k$th component is less than that of $x$. Applying the same procedure to $y$, we can decrease the absolute value of the $k$th component again. However, it cannot be decreased forever, and eventually a contradiction would occur. The above argument establishes that $B_x$ must have rank $m$.

Now, choose $m$ linearly independent vectors $d_1, \cdots, d_m$ from $B_x$. Denote $c_i = d_i x$. Then, by the definition of $B_x$, $|c_i| \leq \alpha + (\alpha q)^{q+1}$ for all $1 \leq i \leq m$.

Applying Cramer's rule to the system of equations $d_i x = c_i$, $i = 1, 2, \cdots, m$, we obtain a representation of $x$ in terms of $c_i$'s: $x_i = \det D_i / \det D$, where $D$ is a square submatrix of $(A^T, I)^T$ and $D_i$ is a square matrix obtained from $D$ by replacing the $i$th column by vector $(c_1, \cdots, c_m)^T$. Note that the determinant of any submatrix of $(A^T, I)^T$ is just the determinant of a submatrix of $A$. By the Laplace expansion and Lemma 2.5, it is easy to see that $|x_i| \leq |\det D_i| \leq (\alpha q)^q(|c_1| + \cdots + |c_m|) \leq (\alpha q)^q m(\alpha + (\alpha q)^{q+1}) \leq 2(\alpha q)^{2q+2}$. ∎

## 2.2 Cook's Theorem

The notion of reducibilities was first developed in recursion theory. In general, a reducibility $\leq_r$ is a binary relation on languages that satisfies the reflexivity and transitivity properties and, hence, it defines a partial ordering on the class of all languages. In this section, we introduce the notion of polynomial-time many-one reducibility. Let $A \subseteq \Sigma^*$ and $B \subseteq \Gamma^*$ be two languages. We say that $A$ is *many-one reducible* to $B$, denoted by $A \leq_m B$, if there exists a computable function $f : \Sigma^* \to \Gamma^*$ such that for each $x \in \Sigma^*$, $x \in A$ if and only if $f(x) \in B$. If the reduction function $f$ is further known to be computable in polynomial time, then we say that $A$ is *polynomial-time many-one reducible* to $B$, and write $A \leq_m^P B$. It is easy to see that polynomial-time many-one reducibility does satisfy the reflexivity and transitivity properties and, hence, indeed is a reducibility.

**Proposition 2.8** *The following hold for all sets $A$, $B$, and $C$:*
 *(a) $A \leq_m^P A$.*
 *(b) $A \leq_m^P B$, $B \leq_m^P C \Rightarrow A \leq_m^P C$.*

Note that if $A \leq_m^P B$ and $B \in P$ then $A \in P$. In general, we say a complexity class $\mathcal{C}$ is *closed under the reducibility* $\leq_r$ if $A \leq_r B$ and $B \in \mathcal{C}$ imply $A \in \mathcal{C}$.

**Proposition 2.9** *The complexity classes P, NP, PSPACE, and EXP are all closed under $\leq_m^P$.*

For any complexity class $\mathcal{C}$ that is closed under a reducibility $\leq_r$, we say a set $B$ is $\leq_r$-*hard* for class $\mathcal{C}$ if $A \leq_r B$ for all $A \in \mathcal{C}$, and we say a set $B$ is $\leq_r$-*complete* for class $\mathcal{C}$ if $B \in \mathcal{C}$ and $B$ is $\leq_r$-hard for $\mathcal{C}$. For convenience, we say a set $B$ is $\mathcal{C}$-complete if $B$ is $\leq_m^P$-complete for the class $\mathcal{C}$.[2] Thus, a set $B$ is *NP-complete* if $B \in NP$ and $A \leq_m^P B$ for all $A \in NP$. An NP-complete set $B$ is a maximal element in $NP$ under the partial ordering $\leq_m^P$. Thus, it is not in $P$ if and only if $P \neq NP$.

**Proposition 2.10** *Let $B$ be an NP-complete set. Then, $B \in P$ if and only if $P = NP$.*

---

[2]The term $\mathcal{C}$-hard is reserved for $\leq_T^P$-hard sets for $\mathcal{C}$; see Section 2.4.

In recursion theory, the halting problem is the first problem proved to be complete for the class of recursively enumerable (r.e.) sets under the reducibility $\leq_m$. Since the class of r.e. sets properly contains the class of recursive sets, the halting problem, and other complete sets for the class of r.e. sets, are not recursive. The analogous problem for the class $NP$, called the *time-bounded halting problem* for NTMs, is our first $NP$-complete problem. (Again, for a natural problem $\Pi$ that asks whether the given input $x$ satisfies $Q_\Pi(x)$, we say it is $NP$-complete to mean that the set $L_\Pi = \{x : Q_\Pi(x)\}$ is $NP$-complete.)

> BOUNDED HALTING PROBLEM (BHP): Given the code $x$ of an NTM $M_x$, an input $y$ and a string $0^t$, determine whether machine $M$ accepts $y$ within $t$ moves.

**Theorem 2.11** BHP *is NP-complete.*

*Proof.* It is clear that BHP is in $NP$, since a universal NTM $M_u$ can simulate $M_x$ on $y$ for $t$ moves to see whether $M_x$ accepts $y$. The time needed by $M_u$ on input $\langle x, y, 0^t \rangle$ is $p(t)$ for some polynomial $p$ (see Proposition 1.17).

Next, assume that $A \in NP$. Then, there exist an NTM $M$ and a polynomial $p$ such that $M$ accepts $y$ in time $p(|y|)$ if and only if $y \in A$. The reduction function for $A \leq_m^P$ BHP is the simple padding function $g(y) = \langle x, y, 0^{p(|y|)} \rangle$, where $x$ is the code of $M$.                                                                          ∎

The above theorem demonstrates that $NP$ contains at least one complete problem. By the transitivity property of $\leq_m^P$, we would like to prove other $NP$-complete problems by reducing a known $NP$-complete problem to them, which is presumably easier than reducing all problems in $NP$ to them. However, the problem BHP, being a problem concerning with the properties of NTMs, is hardly a *natural* problem. To prove the $NP$-completeness of natural problems, we first need a natural $NP$-complete problem. Our first natural $NP$-complete problem is the satisfiability problem SAT. It was first proved to be $NP$-complete by Stephen Cook in 1970 [Cook, 1971].

**Theorem 2.12** (Cook's Theorem) SAT *is NP-complete.*

*Proof.* The fact that SAT is in $NP$ has been proved in Example 2.2. We need to prove that for each $A \in NP$, $A \leq_m^P$ SAT. Let $A = L(M)$, where $M = (Q, q_0, F, \Sigma, \Gamma, \Delta)$ is a one-tape NTM with time bound $p(n)$ for some polynomial $p$. Then, we need to find a Boolean formula $F_x$ for each input string $x$ such that $x \in A$ if and only if $F_x$ is satisfiable. To do so, let an input string $x$ of length $n$ be given. Since $M$ halts on $x$ in $p(n)$ moves, the computation of $M(x)$ has at most $p(n) + 1$ configurations. We may assume that there are exactly $p(n) + 1$ configurations by adding dummy moves from a halting configuration; that is, we allow $\alpha \vdash \alpha$ if $\alpha$ is a halting configuration. Furthermore, we may assume that each configuration is of length $p(n)$, where each symbol $a$ in a configuration is either a symbol in $\Gamma$ or a symbol in $Q \times \Gamma$. (In the latter case, symbol $a = \langle q, s \rangle$ indicates that the tape head is scanning this square and the current state is $q$.) Now, for each triple $(i, j, a)$, with $0 \leq i \leq p(n)$, $1 \leq j \leq p(n)$ and $a \in \Gamma' = \Gamma \cup (Q \times \Gamma)$, we define a Boolean variable $y_{i,j,a}$ to denote the property that the $j$th square of the $i$th configuration of the computation of $M(x)$ has the symbol $a$.

We will construct a Boolean formula $F_x$ of these variables $y_{i,j,a}$, such that $F_x$ is satisfiable if and only if the following conditions hold under the above interpretation of the variables $y_{i,j,a}$:

(1) For each pair $(i, j)$, with $0 \leq i \leq p(n)$ and $1 \leq j \leq p(n)$, there is exactly one $a \in \Gamma'$ such that $y_{i,j,a} = 1$. (Based on the above interpretation of variables $y_{i,j,a}$, this means that each square of the $i$th configuration holds exactly one symbol.)

(2) For each pair $(j, a)$, with $1 \leq j \leq p(n)$ and $a \in \Gamma'$, $y_{0,j,a} = 1$ if and only if the $j$th symbol in the initial configuration of $M(x)$ is $a$.

(3) $y_{p(n),j,a} = 1$ for some $j$, $1 \leq j \leq p(n)$, and some $a \in F \times \Gamma$. (This means that the $p(n)$th configuration of $M(x)$ contains a final state.)

(4) For each $i$ such that $0 \leq i < p(n)$, the $(i+1)$st configuration is one of the next configurations of the $i$th configuration.

Using the above interpretation of the variables $y_{i,j,a}$, it is easy to see that the above four conditions are equivalent to the condition that the configurations defined by variables $y_{i,j,a}$ is an accepting computation of $M(x)$.

The first three conditions are easy to translate into Boolean formulas:

$$\phi_1 = \prod_{i=0}^{p(n)} \prod_{j=1}^{p(n)} \left( \sum_{a \in \Gamma'} y_{i,j,a} \right) \prod_{a,b \in \Gamma', a \neq b} (\bar{y}_{i,j,a} + \bar{y}_{i,j,b}),$$

$$\phi_2 = y_{0,1,\langle q_0, x_1 \rangle} \prod_{j=2}^{n} y_{0,j,x_j} \prod_{j=n+1}^{p(n)} y_{0,j,\mathrm{B}},$$

$$\phi_3 = \sum_{j=1}^{p(n)} \sum_{a \in F \times \Gamma} y_{p(n),j,a},$$

where for each $j$, $1 \leq j \leq n$, $x_j$ denotes the $j$th symbol of $x$.

For the fourth condition, we first observe a simple relation between consecutive configurations. For each configuration $\alpha$, and each $k$, $1 \leq k \leq p(n)$, let $\alpha_k$ denote the $k$th symbol of $\alpha$. Assume that the $i$th configuration of $M(x)$ is $\alpha$, and the $(i+1)$st is $\beta$. Then, we can verify this fact locally on the symbols $\alpha_{k-1}$, $\alpha_k$, $\alpha_{k+1}$, $\beta_{k-1}$, $\beta_k$ and $\beta_{k+1}$ for each $k$ from 1 to $p(n)$. (To avoid problems around the two ends, we may add blank symbols at positions 0 and $p(n) + 1$ to each configuration.) More precisely, the following conditions are equivalent to $\alpha \vdash_M \beta$:

(4.1) If $\alpha_{k-1}, \alpha_k, \alpha_{k+1} \notin Q \times \Gamma$ then $\beta_k = \alpha_k$; and

(4.2) If $\alpha_k \in Q \times \Gamma$ then $\alpha_{k-1}\alpha_k\alpha_{k+1} \vdash'_M \beta_{k-1}\beta_k\beta_{k+1}$, where $\vdash'_M$ is the subrelation of $\vdash_M$ restricted to configurations of length 3.

To translate the above condition (4.2) into a Boolean formula, we define, for each pair $(q,a) \in Q \times \Gamma$ and integers $i, j, 0 \leq i \leq p(n) - 1, 1 \leq j \leq p(n)$, a Boolean formula $\phi_{q,a,i,j}$ that encodes the relation $\vdash'_M$, with $\alpha_k = \langle q,a \rangle$. For instance, if $(q,a,r,b,R)$ and $(q,a,r',b',L)$ are the only two quintuples in $\Delta$ that begin with $(q,a)$, then

$$\phi_{q,a,i,j} = \prod_{c,d \in \Gamma} (\bar{y}_{i,j-1,c} + \bar{y}_{i,j,\langle q,a \rangle} + \bar{y}_{i,j+1,d}$$
$$+ y_{i+1,j-1,c}\, y_{i+1,j,b}\, y_{i+1,j+1,\langle r,d \rangle}$$
$$+ y_{i+1,j-1,\langle r',c \rangle}\, y_{i+1,j,b'}\, y_{i+1,j+1,d}).$$

Now, we can define $\phi_4$ as

$$\phi_4 = \prod_{i=0}^{p(n)-1} \prod_{j=1}^{p(n)} \prod_{a,b,c \in \Gamma} (\bar{y}_{i,j-1,a} + \bar{y}_{i,j,b} + \bar{y}_{i,j+1,c} + y_{i+1,j,b}) \prod_{(q,a) \in Q \times \Gamma} \phi_{q,a,i,j}.$$

Define $F_x = \phi_1 \phi_2 \phi_3 \phi_4$. It is clear now that $F_x$ is satisfiable if and only if $M$ accepts $x$.

It is left to verity that $F_x$ is of length polynomial in $|x|$, and can be uniformly generated from $x$. Let $|\Gamma| = m$ and $|\Delta| = r$. It is easy to see that $\phi_1$ is of length $O(m^2(p(n))^2)$, $\phi_2$ is of length $O(p(n))$, $\phi_3$ is of length $O(mp(n))$, and $\phi_4$ is of length $O(rm^3(p(n))^2)$. Together, formula $F_x$ is of length $O(rm^3(p(n))^2)$. Furthermore, these formulas, except for $\phi_2$, do not depend on input $x$ (but only on $n = |x|$), and $\phi_2$ is easily derived from $x$. Therefore, $F_x$ is polynomial-time computable from $x$, and the theorem is proven. ∎

To apply SAT to prove other natural *NP*-complete problems, it would be easier to work with Boolean formulas of simpler forms, such as CNF. Recall that a CNF formula is a product of elementary sums. Each factor of a CNF formula $F$ is a clause of $F$. A CNF formula $F$ is called a 3-CNF formula if each clause (implicant, respectively) of $F$ contains exactly three literals of three distinct variables. The following variation of SAT is very useful for proving new *NP*-complete problems:

3-SAT: Given a 3-CNF formula, determine whether it is satisfiable.

**Corollary 2.13** 3-SAT *is NP-complete.*

*Proof.* We first observe that, in Theorem 2.12, the Boolean formula $F_x$ can be rewritten as a CNF formula of length $O((p(n))^2)$. Indeed, $\phi_1$ is a product of clauses, $\phi_2$ is a product of literals, and $\phi_3$ is a single clause. The only subformulas of $F_x$ that are not in CNF are $\phi_{q,a,i,j}$, each of which is a product of DNF formulas. These DNF formulas can be transformed, by the distributive law, into CNF formulas of at most $3^k$ clauses, where $k$ is the maximum number of nondeterministic moves allowed from any configuration. Thus, $\phi_4$ can be transformed into a CNF formula of length $O((p(n))^2)$. This shows that the problem SAT restricted to CNF formulas remains *NP*-complete.

Next, we reduce SAT restricted to CNF formulas to SAT restricted to 3-CNF formulas. Let $F$ be a CNF formula. First, we add three variables $w_1, w_2, w_3$, and seven clauses: $(w_1 + w_2 + w_3)$, $(w_1 + w_2 + \bar{w}_3)$, $(w_1 + \bar{w}_2 + w_3)$, $(\bar{w}_1 + w_2 + w_3)$, $(w_1 + \bar{w}_2 + \bar{w}_3)$, $(\bar{w}_1 + \bar{w}_2 + w_3)$, and $(\bar{w}_1 + w_2 + \bar{w}_3)$. Note that there is a unique assignment $\tau_0$ that can satisfy all these seven clauses: $\tau_0(w_1) = \tau(w_2) = \tau(w_3) = 1$.

Now, we consider each clause of $F$. A clause of $F$ is of the form $\phi = z_1 + z_2 + \cdots + z_k$ for some $k \geq 1$, where each $z_j$, $1 \leq j \leq k$, is a literal. We will replace $\phi$ by a formula $\psi$ as follows:

*Case 1.* $k = 1$. We let $\psi = (z_1 + \bar{w}_1 + \bar{w}_2)$.

*Case 2.* $k = 2$. We let $\psi = (z_1 + z_2 + \bar{w}_1)$.

*Case 3.* $k = 3$. We let $\psi = \phi$.

*Case 4.* $k = 4$. We introduce a new variable $u = u_\phi$. Then, we define $\psi = (z_1 + z_2 + u)(z_3 + z_4 + \bar{u})(\bar{z}_3 + u + \bar{w}_1)(\bar{z}_4 + u + \bar{w}_1)$, where $w_1$ is the new variable defined above. Note that the last three clauses of $\psi$ means $(z_3 + z_4)$ is equivalent to $u$. Thus, an assignment $\tau$ on $z_i$, $1 \leq i \leq 4$, satisfies $\phi$ if and only if the assignment $\tau'$, defined by $\tau'(z_i) = \tau(z_i)$ for $1 \leq i \leq 4$ and $\tau'(u) = \tau(z_3) + \tau(z_4)$, satisfies $\psi$.

*Case 5.* $k > 4$. We inductively apply the procedure of Case 4 until we have only clauses of 3 literals left.

We let $G$ be the product of all new clauses $\psi$. It is easy to check that $F$ is satisfiable if and only if $G$ is satisfiable. The above transformation from $\phi$ to $\psi$ can be easily done in time polynomial in $k$, and so it is a polynomial-time reduction from SAT to 3-SAT. ∎

## 2.3 More *NP*-Complete Problems

The importance of the notion of *NP*-completeness is witnessed by thousands of *NP*-complete problems from a variety of areas in computer science, discrete mathematics, and operations research. Theoretically, all these problems can be proved to be *NP*-complete by reducing SAT to them. It is practically much easier to prove new *NP*-complete problems from some other known *NP*-complete problems that have similar structures as the new problems. In this section, we study some best-known *NP*-complete problems that may be useful to obtain new *NP*-completeness results.

> VERTEX COVER (VC): Given a graph $G = (V, E)$ and an integer $K \geq 0$, determine whether $G$ has a vertex cover of size at most $K$; that is, determine whether $V$ has a subset $V'$ of size $\leq K$ such that each $e \in E$ has at least one endpoint in $V'$.

**Theorem 2.14** VC *is NP-complete.*

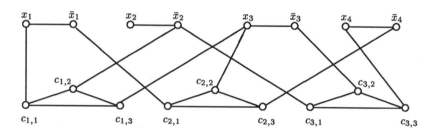

Figure 2.1: The graph $G_F$.

*Proof.* It is easy to see that VC is in *NP*. To show that VC is complete for *NP*, we reduce 3-SAT to it.

Let $F$ be a 3-CNF formula with $m$ clauses $C_1, C_2, \ldots, C_m$, over $n$ variables $x_1, x_2, \ldots, x_n$. We construct a graph $G_F$ of $2n + 3m$ vertices as follows. The vertices are named $x_i$, $\bar{x}_i$, for $1 \leq i \leq n$, and $c_{j,k}$ for $1 \leq j \leq m$, $1 \leq k \leq 3$. The vertices are connected by the following edges: for each $i$, $1 \leq i \leq n$, there is an edge connecting $x_i$ and $\bar{x}_i$; for each $j$, $1 \leq j \leq m$, there are three edges connecting $c_{j,1}, c_{j,2}, c_{j,3}$ into a triangle and, in addition, if $C_j = \ell_1 + \ell_2 + \ell_3$, then there are three edges connecting each $c_{j,k}$ to the vertex named $\ell_k$, $1 \leq k \leq 3$. Figure 2.1 shows the graph $G_F$ for $F = (x_1 + \bar{x}_2 + x_3)\,(\bar{x}_1 + x_3 + \bar{x}_4)\,(\bar{x}_2 + \bar{x}_3 + x_4)$.

We claim that $F$ is satisfiable if and only if $G_F$ has a vertex cover of size $n + 2m$. First, suppose that $F$ is satisfiable by a truth assignment $\tau$. Let $S_1 = \{x_i : \tau(x_i) = 1, 1 \leq i \leq n\} \cup \{\bar{x}_i : \tau(x_i) = 0, 1 \leq i \leq n\}$. Next for each $j$, $1 \leq j \leq m$, let $c_{j,j_k}$ be the vertex of the least index $j_k$ such that $c_{j,j_k}$ is adjacent to a vertex in $S_1$. (By the assumption that $\tau$ satisfies $F$, such an index $j_k$ always exists.) Then, let $S_2 = \{c_{j,r} : 1 \leq r \leq 3, r \neq j_k, 1 \leq j \leq m\}$, and $S = S_1 \cup S_2$. It is clear that $S$ is a vertex cover for $G_F$ of size $n + 2m$.

Conversely, suppose that $G_F$ has a vertex cover $S$ of size at most $n + 2m$. Since each triangle over $c_{j_1}, c_{j_2}, c_{j_3}$ must have at least two vertices in $S$ and each edge $\{x_i, \bar{x}_i\}$ has at least one vertex in $S$, $S$ is of size exactly $n + 2m$ with exactly two vertices from each triangle $c_{j_1}, c_{j_2}, c_{j_3}$, and exactly one vertex from each edge $\{x_i, \bar{x}_i\}$. Define $\tau(x_i) = 1$ if $x_i \in S$ and $\tau(x_i) = 0$ if $\bar{x}_i \in S$. Then, each clause $C_j$ must have a true literal which is the one adjacent to the vertex $c_{j,k}$ that is not in $S$. Thus, $F$ is satisfied by $\tau$.

The above construction is clearly polynomial-time computable. Hence, we have proved 3-SAT $\leq_m^P$ VC.                                           ∎

CLIQUE: Given a graph $G = (V, E)$ and an integer $K \geq 0$, determine whether $G$ has a clique of size $K$, that is, whether $G$ has a complete subgraph of size $K$.

INDEPENDENT SET (IS): Given a graph $G = (V, E)$ and an integer $K \geq 0$, determine whether $G$ has an independent set of size $K$, that is, whether there is a subset $V' \subseteq V$ of size $K$ such that for any two vertices $u, v \in V'$, $\{u, v\} \notin E$.

**Theorem 2.15** CLIQUE *and* IS *are NP-complete.*

*Proof.* It is easy to verify that IS is in *NP*. To see that IS is *NP*-complete, we reduce VC to it. For any given graph $G = (V, E)$ and given integer $K \geq 0$, we observe that a subset $V' \subseteq V$ is a vertex cover of $G$ if and only if $V - V'$ is an independent set of $G$. In other words, $G$ has a vertex cover of size $K$ if and only if $G$ has an independent set of size $|V| - K$, and so the function mapping $(G, K)$ to $(G, |V| - K)$ reduces VC to IS. Thus, IS is *NP*-complete.

For any graph $G = (V, E)$, let $G^c$ be its complement graph; that is, let $G^c = (V, E')$, where $E' = \{\{u, v\} : u, v \in V, \{u, v\} \notin E\}$. Then, a subset $V' \subseteq V$ is an independent set of $G$ if and only if $V'$ is a clique of $G^c$. This observation gives a simple reduction from IS to CLIQUE, and shows that CLIQUE is also *NP*-complete. ∎

**Theorem 2.16** HC *is NP-complete.*

*Proof.* We have shown in Example 2.3 that HC is in NP. We prove that HC is *NP*-complete by reducing 3-SAT to it. For each 3-CNF formula $F$ of $m$ clauses $C_1, C_2, \ldots, C_m$, over $n$ variables $x_1, x_2, \ldots, x_n$, we construct a graph $G_F$ as follows: For each variable $x_i$, construct a ladder $H_i$ of $12m + 7$ vertices as shown in Figure 2.2(a) (the vertex $y_{i+1}$ is part of $H_{i+1}$). Connect $n$ subgraphs $H_i$, $1 \leq i \leq n$, into a graph $H$ as shown in Figure 2.2(b). Next, for each clause $C_j$, $1 \leq j \leq m$, we add a new vertex $z_j$ and connect $z_j$ to $u_{i,4j-1}$ and $u_{i,4j-2}$ if $C_j$ contains literal $x_i$ and to $w_{i,4j-1}$ and $w_{i,4j-2}$ if $C_j$ contains literal $\bar{x}_i$ (so that each $z_j$ is of degree 6). We let $G_F$ be the resultant graph.

We now prove that $F$ is satisfiable if and only if $G_F$ has a Hamiltonian circuit. First, we assume that $F$ is satisfiable by a truth assignment $\tau$ on variables $x_i$, $1 \leq i \leq n$. We observe that for each subgraph $H_i$, $1 \leq i \leq n$, there are exactly two Hamiltonian paths from $y_i$ to $y_{i+1}$:

$$P_{i,1} = (y_i, u_{i,1}, v_{i,1}, w_{i,1}, w_{i,2}, v_{i,2}, u_{i,2}, u_{i,3}, \cdots, v_{i,4m+2}, u_{i,4m+2}, y_{i+1}),$$

$$P_{i,0} = (y_i, w_{i,1}, v_{i,1}, u_{i,1}, u_{i,2}, v_{i,2}, w_{i,2}, w_{i,3}, \cdots, v_{i,4m+2}, w_{i,4m+2}, y_{i+1}).$$

Note that for all $j$, $1 \leq j \leq m$, the edges $\{u_{i,4j-1}, u_{i,4j-2}\}$ are in path $P_{i,1}$, and the edges $\{w_{i,4j-1}, w_{i,4j-2}\}$ are in path $P_{i,0}$. We associate $P_{i,j}$, $j = 0, 1$, to the assignment that assigns $x_i$ with the value $j$. Thus, there are $2^n$ Hamiltonian paths from $y_1$ to $y_{n+1}$ in the subgraph $H$, which correspond to $2^n$ assignments on variables $x_i$, $1 \leq i \leq n$. Choose the path $Q$ that corresponds to the assignment $\tau$. For each $j$, $1 \leq j \leq m$, fix a literal in $C_j$ which is assigned by $\tau$ with value 1. If the literal is $x_i$ ($\bar{x}_i$), then $Q$ must pass the edge $\{u_{i,4j-1}, u_{i,4j-2}\}$ (or, respectively, the edge $\{w_{i,4j-1}, w_{i,4j-2}\}$). Now,

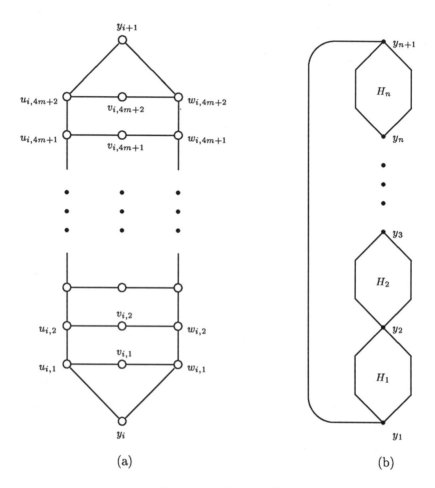

Figure 2.2: $H_i$ and $H$

replace this edge by two edges $\{u_{i,4j-1}, z_j\}$ and $\{z_j, u_{i,4j-2}\}$ (or, respectively, the edges $\{w_{i,4j-1}, z_j\}$ and $\{z_j, w_{i,4j-2}\}$). Clearly, after the modification, the path $Q$ plus the edge $\{y_{n+1}, y_1\}$ become a circuit passing through each vertex of $G_F$ exactly once, i.e., a Hamiltonian circuit of $G_F$.

For the converse direction, assume that graph $G_F$ has a Hamiltonian circuit $Q$. Let us observe some properties of the graph $G_F$ and the Hamiltonian circuit $Q$.

(a) The Hamiltonian circuit $Q$ must contain edges $\{u_{i,k}, v_{i,k}\}$ and $\{v_{i,k}, w_{i,k}\}$ for all $i$, $1 \leq i \leq n$, and all $k$, $1 \leq k \leq 4m + 2$, for otherwise $v_{i,k}$ is not in $Q$.

(b) From observation (a), it easily follows that to avoid a small cycle around vertex $y_i$, $Q$ must contain exactly one of the edges $\{y_i, u_{i,1}\}$ and $\{y_i, w_{i,1}\}$ for each $i$, $1 \leq i \leq n$.

(c) Similarly, the Hamiltonian circuit $Q$ can have at most one of the edges $\{u_{i,k}, u_{i,k+1}\}$ and $\{w_{i,k}, w_{i,k+1}\}$ for any $i$, $1 \leq i \leq n$, and any $k$, $1 \leq k \leq 4m + 1$.

Next, for each $j$, $1 \leq j \leq m$, consider the vertex $z_j$. Without loss of generality, assume that $Q$ contains an edge $\{z_j, u_{i,4j-1}\}$. We claim that

(d) $Q$ must also contain the edge $\{z_j, u_{i,4j-2}\}$.

To see this, we check that, from the observation (a), $Q$ contains the following path: $z_j, u_{i,4j-1}, v_{i,4j-1}, w_{i,4j-1}$, and it must continue to either $w_{i,4j-2}$ or to $w_{i,4j}$ (since $C_j$ cannot contain both $x_i$ and $\bar{x}_i$). In the former case, none of the edges $\{u_{i,4j}, u_{i,4j-1}\}$ and $\{w_{i,4j}, w_{i,4j-1}\}$ is in $Q$ and so both edges $\{u_{i,4j}, u_{i,4j+1}\}$ and $\{w_{i,4j}, w_{i,4j+1}\}$ are in $Q$, violating observation (c) above. Therefore, the path of $Q$ must go from $z_j$ to $u_{i,4j-1}, v_{i,4j-1}, w_{i,4j-1}$ to $w_{i,4j}$. This means $\{w_{i,4j-1}, w_{i,4j-2}\} \notin Q$ and so $\{w_{i,4j-2}, w_{i,4j-3}\} \in Q$. It follows by property (c) then $\{u_{i,4j-2}, u_{i,4j-3}\} \notin Q$, and so $\{u_{i,4j-2}, z_j\}$ must be in $Q$.

Now, from observations (b) and (d), we see that, for each $i$, $1 \leq i \leq n$, the path of $Q$ between $y_i$ and $y_{i+1}$ is simply one of the paths $P_{i,1}$ or $P_{i,0}$ with some edges $\{u_{i,4j-1}, u_{i,4j-2}\}$ replaced by the edges $\{u_{i,4j-1}, z_j\}$ and $\{z_j, u_{i,4j-2}\}$ (or, some edges $\{w_{i,4j-1}, w_{i,4j-2}\}$ replaced by the edges $\{w_{i,4j-1}, z_j\}$ and $\{z_j, w_{i,4j-2}\}$). Thus $Q$ defines an assignment $\tau$ such that $\tau(x_i) = 1$ if and only if $\{y_i, u_{i,1}\}$ is in $Q$. Now, if $z_j$ is in the path of $Q$ from $y_i$ to $y_{i+1}$ such that edges $\{u_{i,4j-1}, z_j\}$, $\{z_j, u_{i,4j-2}\}$ are in $Q$, then $Q$ must also contain edge $\{y_i, u_{i,1}\}$ and hence $\tau(x_i) = 1$. Therefore, clause $C_j$ is satisfied by $\tau$. Similarly, if $\{w_{i,4j-1}, z_j\}$, $\{z_j, w_{i,4j-2}\} \in Q$, then $Q$ contains the edge $\{y_i, w_{i,1}\}$ and $\tau(x_i) = 0$. Again, clause $C_j$ is satisfied by $\tau$. This shows that $\tau$ satisfies the formula $F$. ∎

**Theorem 2.17** IP *is NP-complete.*

*Proof.* We have seen in Example 2.4 that IP is in *NP*. To show that IP is *NP*-complete, we reduce 3-SAT to IP. Let $F$ be a 3-CNF formula with $m$ clauses $C_1, \ldots, C_m$, over $n$ variables $u_1, \ldots, u_n$. For each $i$, $1 \leq i \leq m$, let $r_i$ be the number of $j$, $1 \leq j \leq n$, such that $\bar{u}_j$ occurs in $C_i$. We define an instance $(A, b)$ of IP as follows: $A$ is an $m \times n$ integer matrix, such that for each $i$, $1 \leq i \leq m$, and each $j$, $1 \leq j \leq n$,

$$A[i,j] = \begin{cases} 1 & \text{if } u_j \text{ occurs in } C_i, \\ -1 & \text{if } \bar{u}_j \text{ occurs in } C_i, \\ 0 & \text{otherwise,} \end{cases}$$

and $b$ is an $m$-dimensional integer vector such that for each $i$, $1 \leq i \leq m$, $b[i] = 1 - r_i$. We claim that $F \in$ 3-SAT if and only if $(A, b) \in$ IP.

First assume that $F$ is satisfiable. We consider 0-1-valued, $n$-dimensional vectors $x$. Notice that for such vectors, the minimum of $\sum_{j=1}^{n} A[i,j]x[j]$ for each $1 \leq i \leq m$ is $-r_i$, achieved by $x_0$ with $x_0[j] = 1$ if and only if $\bar{u}_j$ occurs in $C_i$. Furthermore, if for some $1 \leq j \leq n$, either $u_j$ occurs in $C_i$ and $x[j] = 1$ or $\bar{u}_j$ occurs in $C_i$ and $x[j] = 0$, then $\sum_{j=1}^n A[i,j]x[j] \geq 1 - r_i = b[i]$. Since $F$ is satisfiable, there exists a Boolean assignment $\tau$ on $\{u_1, \ldots, u_n\}$ satisfying each clause $C_i$. Define, for each $1 \leq j \leq n$, $x[j] = \tau(u_j)$. Then, for each $i$, $1 \leq i \leq m$, either there is a $\bar{u}_j$ in $C_i$ such that $x[j] = 0$ or there is a $u_k$ in $C_i$ such that $x[k] = 1$. Either way, as observed above, $\sum_{j=1}^n A[i,j]x[j] \geq b[i]$.

Conversely, assume that $x$ is an $n$-dimensional integer vector such that $Ax \geq b$. Define a Boolean assignment $\tau$ on $\{u_1, \ldots, u_n\}$ by $\tau(u_j) = 1$ if and only if $x[j] \geq 1$. We claim that $\tau$ satisfies $F$. Suppose otherwise that there is a clause $C_i$, $1 \leq i \leq m$, such that $\tau(u_j) = 1$ for all $\bar{u}_j$ occurring in $C_i$ and $\tau(u_k) = 0$ for all $u_k$ occurring in $C_i$. That means $x[j] \geq 1$ for all $\bar{u}_j$ in $C_i$ and $x[j] < 1$ (and hence $x[j] \leq 0$) for all $u_k$ in $C_i$. By the definition of the matrix $A$, we know that

$$\sum_{u_k \in C_i} A[i,k]x[k] \leq 0,$$

$$\sum_{\bar{u}_j \in C_i} A[i,j]x[j] \leq \sum_{\bar{u}_j \in C_i} (-1) = -r_i,$$

and, hence, $\sum_{j=1}^n A[i,j]x[j] \leq -r_i < b[i]$, which contradicts our assumption. ∎

SUBSET SUM (SS): Given a list of integers $L = (a_1, a_2, \ldots, a_n)$ and an integer $S$, determine whether there exists a set $I \subseteq \{1, 2, \ldots, n\}$ such that $\sum_{i \in I} a_i = S$.

**Theorem 2.18** SS *is NP-complete.*

*Proof.* Again, the membership in *NP* is easy for the problem SS. We show that SS is *NP*-complete by a reduction from 3-SAT to it.

Let $F$ be a 3-CNF formula with $m$ clauses $C_1, \ldots, C_m$, over $n$ variables $x_1, \ldots, x_n$. We are going to define a list $L$ of $2n + 2m$ integers named $b_{j,k}$, $1 \leq j \leq n$, $0 \leq k \leq 1$, and $c_{i,k}$, $1 \leq i \leq m$, $0 \leq k \leq 1$. All integers in the list and the integer $S$ are of value between 0 and $10^{n+m}$. Thus, each integer has a unique decimal representation of *exactly* $n + m$ digits (possibly with leading zeroes). We let $r[k]$ denote the $k$th most significant digit of the $(n+m)$-digit decimal representation of integer $r$. First, we define $S$ by

$$S[k] = \begin{cases} 1 & \text{if } 1 \leq k \leq n, \\ 3 & \text{if } n+1 \leq k \leq n+m. \end{cases}$$

Next, for each $1 \leq j \leq n$ and $1 \leq k \leq n$, define

$$b_{j,0}[k] = b_{j,1}[k] = \begin{cases} 1 & \text{if } k = j, \\ 0 & \text{if } 1 \leq k \leq n \text{ and } k \neq j. \end{cases}$$

For each $1 \leq j \leq n$ and $1 \leq k \leq m$, define

$$b_{j,0}[n+k] = \begin{cases} 1 & \text{if } \bar{x}_j \text{ occurs in } C_k, \\ 0 & \text{otherwise}, \end{cases}$$

and

$$b_{j,1}[n+k] = \begin{cases} 1 & \text{if } x_j \text{ occurs in } C_k, \\ 0 & \text{otherwise}. \end{cases}$$

The above defines the first $2n$ integers in the list $L$. For the other $2m$ integers, we define, for each $1 \leq i \leq m$, $1 \leq k \leq n+m$,

$$c_{i,0}[k] = c_{i,1}[k] = \begin{cases} 1 & \text{if } k = n+i, \\ 0 & \text{otherwise}. \end{cases}$$

First, we observe that for any $k$, $1 \leq k \leq n+m$, there are at most five integers in the list $L$ having a nonzero digit (which must be 1) at the $k$th digit. Thus, to get the sum $S$, we must choose a sublist in which exactly one integer has value 1 at digit $j$, for $1 \leq j \leq n$, and exactly three integers have value 1 at digit $n+i$, for $1 \leq i \leq m$. For the first $n$ digits, this implies that we must choose exactly one of $b_{j,0}$ and $b_{j,1}$, for each $1 \leq j \leq n$.

Now assume that $F$ is satisfiable by a Boolean assignment $\tau$ on variables $x_1, \ldots, x_n$. We define a sublist $L_1$ of $L$ as follows:

(1) For each $j$, $1 \leq j \leq n$,

$$L_1 \cap \{b_{j,0}, b_{j,1}\} = \begin{cases} \{b_{j,0}\} & \text{if } \tau(x_j) = 0, \\ \{b_{j,1}\} & \text{if } \tau(x_j) = 1. \end{cases}$$

(2) For each $i$, $1 \leq i \leq m$,

$$L_1 \cap \{c_{j,0}, c_{j,1}\} = \begin{cases} \emptyset & \text{if } \tau \text{ assigns 1 to all 3 literals in } C_i, \\ \{c_{j,0}\} & \text{if } \tau \text{ assigns 1 to 2 of 3 literals in } C_i, \\ \{c_{j,0}, c_{j,1}\} & \text{if } \tau \text{ assigns 1 to 1 of 3 literals in } C_i. \end{cases}$$

Clearly, the sum of the sublist $L_1$ is equal to $S$ digit by digit and, hence, $(L, S) \in SS$.

Next, assume that $L_1$ is a sublist of $L$ whose sum is equal to $S$. Then, from our earlier observation, $L_1 \cap \{b_{j,0}, b_{j,1}\}$ must be a singleton for all $1 \leq j \leq n$. Define $\tau(x_j) = 1$ if and only if $L_1 \cap \{b_{j,0}, b_{j,1}\} = b_{j,1}$. We verify that $\tau$ satisfies each clause $C_i$. Since the $(n+i)$th digit of the sum of $L_1$ is 3, and since there are at most two integers among the last $2m$ integers (i.e., $c_{p,q}$'s) that have value 1 at the $(n+i)$th digit, there must be a $b_{j,k}$, $k = 0, 1$, in $L_1$ that has value 1 at the $(n+i)$th digit. It implies that $\tau(x_j) = 0$ and $\bar{x}_j$ occurs in $C_i$ (in the case of $k = 0$), or that $\tau(x_j) = 1$ and $x_j$ occurs in $C_i$ (in the case of $k = 1$). In either case, the clause $C_i$ is satisfied by $\tau$. ∎

A special form of the problem SS is when the sum of the whole list is equal to $2S$. This subproblem of SS is called PARTITION: given a list of integers $L = (a_1, \ldots, a_n)$, determine whether there is a set $I \subseteq \{1, \ldots, n\}$ such that $\sum_{i \in I} a_i = \sum_{i \notin I} a_i$.

**Corollary 2.19** PARTITION *is NP-complete.*

*Proof.* For any given instance $(L, S)$ for SS, just append the list $L$ with an extra integer $b$ which is equal to $|2S - sum(L)|$, where $sum(L)$ is the sum of the list $L$.                                                                                  ∎

## 2.4   Polynomial-Time Turing Reducibility

Polynomial-time many-one reducibility is a strong type of reducibility on decision problems (i.e., languages) that preserves the membership in the class $P$. In this section, we extend this notion to a weaker type of reducibility called polynomial-time Turing reducibility that also preserves the membership in $P$. Moreover, this weak reducibility can also be applied to search problems (i.e., functions).

Intuitively, a problem $A$ is *Turing reducible* to a problem $B$, denoted by $A \leq_T B$, if there is an algorithm $M$ for $A$ which can ask, during its computation, some membership questions about set $B$. If the total amount of time used by $M$ on an input $x$, excluding the querying time, is bounded by $p(|x|)$ for some polynomial $p$, and furthermore if the length of each query asked by $M$ on input $x$ is also bounded by $p(|x|)$, then we say $A$ is *polynomial-time Turing reducible* to $B$, and denote it by $A \leq_T^P B$. Let us look at some examples.

**Example 2.20** (a) For any set $A$, $\overline{A} \leq_T^P A$, where $\overline{A}$ is the complement of $A$. This is achieved easily by asking the oracle $A$ whether the input $x$ is in $A$ or not and then reversing the answer. Note that if $NP \neq coNP$ then $\overline{\text{SAT}}$ is not polynomial-time many-one reducible to SAT. So, this demonstrates that the $\leq_T^P$-reducibility is potentially weaker than the $\leq_m^P$-reducibility (cf. Exercise 2.14).

(b) Recall the $NP$-complete problem CLIQUE. We define a variation of the problem CLIQUE as follows:

> EXACT-CLIQUE: Given a graph $G = (V, E)$ and an integer $K \geq 0$, determine whether it is true that the maximum-size clique of $G$ is of size $K$.

It is not clear whether EXACT-CLIQUE is in $NP$. We can guess a subset $V'$ of $V$ of size $K$ and verify in polynomial time that the subgraph of $G$ induced by $V'$ is a clique. However, there does not seem to be a nondeterministic algorithm that can check that there is no clique of size greater than $K$ in polynomial time. Therefore, this problem may seem even harder than the $NP$-complete problem CLIQUE. In the following, however, we show that this problem is actually polynomial-time equivalent to CLIQUE in the sense that they are polynomial-time Turing reducible to each other. Thus, either they are both in $P$ or they are both not in $P$.

First, let us describe an algorithm for the problem CLIQUE which can ask queries to the problem EXACT-CLIQUE. Assume that $G = (V, E)$ is a graph and $K$ is a given integer. We want to know whether there is a clique in $G$ that is of size $K$. We ask whether $(G, k)$ is in EXACT-CLIQUE for each $k = 1, 2, \ldots, |V|$. Then, we will get the maximum size $k^*$ of the cliques of $G$. We answer YES to the original problem CLIQUE if and only if $K \leq k^*$.

Conversely, let $G = (V, E)$ be a graph and $K$ a given integer. Note that the maximum clique size of $G$ is $K$ if and only if $(G, K) \in$ CLIQUE and $(G, K + 1) \notin$ CLIQUE. Thus, the question of whether $(G, K) \in$ EXACT-CLIQUE can be solved by asking two queries to the problem CLIQUE. (See Exercise 3.3(b) for more studies on EXACT-CLIQUE.)

(c) The notion of polynomial-time Turing reducibility also applies to search problems. Consider the following optimization version of the problem CLIQUE:

MAX-CLIQUE: Given a graph $G = (V, E)$, find a maximum-size clique $Q$ of $G$.

It is clear that CLIQUE is polynomial time Turing reducible to MAX-CLIQUE: For any input $(G, K)$, simply ask MAX-CLIQUE to obtain a maximum clique $Q$, and then verify that $K \leq |Q|$. Conversely, the following algorithm shows that MAX-CLIQUE is also polynomial time Turing reducible to CLIQUE:

*Algorithm for* MAX-CLIQUE.

Let $G = (V, E)$ be an input graph with $|V| = n$.

(1) For each $k = 1, \ldots, n$, ask whether $(G, k) \in$ CLIQUE. Let $k^*$ be the maximum $k$ such that $(G, k) \in$ CLIQUE.

(2) Let $H := G$, $U := V$, $Q := \emptyset$ and $K := k^*$. Then repeat the following steps until it halts:

  (2.1) If $U$ is empty, then output $Q$ and halt.

  (2.2) Choose a vertex $v \in U$. Let $H_v$ be the subgraph of $G$ induced by the vertex set $U - \{v\}$. Ask whether $(H_v, K) \in$ CLIQUE.

  (2.3) If the answer to (2.2) is YES, then let $H := H_v$ and $U := U - \{v\}$.

  (2.4) If the answer to (2.2) is NO, then add $v$ to the set $Q$ and let $K := K - 1$. Also, let $U := \{u \in U - \{v\} : u$ is adjacent to $v\}$, and let $H$ be the subgraph of $G$ induced by the vertex set $U$.

To see that the above algorithm does find a maximum-size clique, we observe that the following properties hold after each iteration of steps (2.1)–(2.4):

(a) $Q$ is a clique of size $k^* - K$.

(b) All vertices $u \in U$ are adjacent to each $q \in Q$.

(c) $H$ contains a clique of size $K$.

We give an induction proof for these properties. First, it is clear that the above are satisfied before the first iteration. Now assume that (a), (b), and (c) hold at the end of the $m$th iteration of step (2). If, in the $(m+1)$st iteration, the answer to (2.2) is YES, then $Q$ and $K$ are not changed and $U$ becomes smaller. Thus, (a) and (b) still hold. In addition, property (c) also holds, since $(H_v, K) \in$ CLIQUE. If the answer is NO, then $Q$ is increased by a vertex $v$ that was adjacent to all $q \in Q$, and $K$ is decreased by 1. It follows that (a) holds after (2.4). Property (b) also holds since all vertices in $U$ (the new $U$ after step (2.4)) are adjacent to $v$. For property (c), we observe that the answer NO to the query "$(H_v, K) \in$ CLIQUE?" implies that all cliques of $H$ of size $K$ contain the vertex $v$. In particular, all cliques of $H$ of size $K$ are contained in the set $\{u \in U : u = v$ or $u$ is adjacent to $v\}$, and property (c) follows.                                                                                    □

The above example (c) illustrates a typical relation between a decision problem and the corresponding optimization search problem. That is, the optimization search problem is polynomial-time solvable if it is allowed to make one or more queries to the corresponding decision problem.

In the following, we define oracle Turing machines as the formal model for the notion of polynomial-time Turing reducibility. A *function-oracle DTM* is an ordinary DTM equipped with an extra tape, called the *query* tape, and two extra states, called the *query* state and the *answer* state. The oracle machine $M$ works as follows: First, on input $x$ and with oracle function $f$ (which is a total function on $\Sigma^*$), it begins the computation at the initial state and behaves exactly like the ordinary TM when it is not in any of the special states. The machine is allowed to enter the query state to make queries to the oracle, but it is not allowed to enter the answer state from any ordinary state. Before it enters the query state, machine $M$ needs to prepare the query string $y$ by writing the string $y$ on the query tape and leaving the tape head of the query tape scanning the square to the right of the rightmost square of $y$. After the oracle machine $M$ enters the query state, the computation is taken over by the "oracle" $f$, which will do the following for the machine: it reads the string $y$ on the query tape; it replaces $y$ by the string $f(y)$; it puts the tape head of the query tape back scanning the leftmost square of $f(y)$; and it puts the machine into the answer state. Then the machine continues from the answer state as usual. The actions taken by the oracle count as only one unit of time.

Let $M$ be an oracle Turing machine. Then, $M$ defines an operator that maps each total function $f$ to a partial function $g$, the function computed by $M$ with oracle $f$. We write $M^f$ to denote the computation of $M$ using oracle $f$, and let $L(M, f)$ denote the set of strings accepted by $M$ with oracle $f$ (i.e., the domain of function $g$).

A special type of oracle machines are those which only use 0-1 functions as oracles. Each 0-1 function $f$ may be regarded as the characteristic function of a set $A = \{x : f(x) = 1\}$. In this case, we write $M^A$ and $L(M, A)$ for $M^f$ and $L(M, f)$, respectively. This type of oracle machine has a simpler but equivalent model: the *set-oracle Turing machine*. A set-oracle Turing machine $M$ has an extra query tape and three extra states: the *query* state, the *yes* state, and the *no* state. The machine behaves like an ordinary DTM when it is not in the query state. When it enters the query state, the oracle (set $A$) reads the query $y$ and puts the machine into either the yes state or the no state, depending upon whether $y \in A$ or $y \notin A$, respectively. The oracle also erases the content $y$ of the query tape. Afterwards, the machine $M$ continues the computation from the yes or the no state as usual.

Since we will be mostly working with set-oracle DTMs, we only give a formal definition of the computation of a set-oracle DTM. We leave the formal definition of the computation of a function-oracle TM as an exercise (Exercise 2.12). Formally, a two-tape, (set-)oracle DTM $M$ can be defined by nine components $(Q, q_0, q_?, q_Y, q_N, F, \Sigma, \Gamma, \delta)$. Similar to the ordinary DTMs, $Q$ is a finite set of states, $q_0 \in Q$ is the initial state, $F \subseteq Q - \{q_?, q_Y, q_N\}$ is the set of final states, $\Sigma$ is the set of input alphabet, $\Gamma \supset \Sigma$ is the set of tape symbols, including the blank symbol $B \in \Gamma - \Sigma$. In addition, $q_? \in Q$ is the query state, $q_Y$ is the yes state, $q_N$ is the no state, and $\delta$ is the transition function from $(Q - F - \{q_?\}) \times \Gamma^2$ to $Q \times \Gamma^2 \times \{L, R, S\}^2$.

A configuration of a two-tape oracle DTM is formally defined to be an element in $Q \times (\Gamma^*)^4$. A configuration $\alpha = (q, x_1, y_1, x_2, y_2)$ denotes that the machine is currently in state $q$, the regular tape contains the string $B^\infty x_1 y_1 B^\infty$, the tape head is scanning the leftmost symbol of $y_1$, the query tape contains the string $B^\infty x_2 y_2 B^\infty$, and the tape head of the query tape is scanning the leftmost symbol of $y_2$; the string $x_2$ is called the *query string* of $\alpha$. We extend the transition function $\delta$ to configurations to define a move of the machine $M$. Let $\alpha$ and $\beta$ be two configurations of $M$. We write $\alpha \vdash_M \beta$ to mean that $\alpha$ is not in state $q_?$, and $\beta$ is the next configuration obtained from $\alpha$ by applying the function $\delta$ to it (see Section 1.2). If $\alpha = (q_?, x_1, y_1, x_2, y_2)$ then we write $\alpha \vdash^{yes} \beta$ if $\beta = (q_Y, x_1, y_1, \lambda, \lambda)$, and write $\alpha \vdash^{no} \beta$ if $\beta = (q_N, x_1, y_1, \lambda, \lambda)$. The computation of an oracle DTM $M = (Q, q_0, q_?, q_Y, q_N, F, \Sigma, \Gamma, \delta)$ on an input $x$ is a sequence of configurations $\alpha_0, \alpha_1, \dots, \alpha_m$, satsifying the following properties:

(i) $\alpha_0 = (q_0, \lambda, x, \lambda, \lambda)$;

(ii) $\alpha_m$ is not in the query state and it does not have a next configuration; and

(iii) For each $i$, $1 \le i < m$, either $\alpha_i \vdash \alpha_{i+1}$, or $\alpha_i \vdash^{yes} \alpha_{i+1}$ or $\alpha_i \vdash^{no} \alpha_{i+1}$.

A computation $(\alpha_0, \alpha_1, \dots, \alpha_m)$ of $M$ on input $x$ is *consistent with an oracle set* $A$, if for each $i$, $1 \le i < m$, such that $\alpha_i$ is in state $q_?$, $\alpha_i \vdash^{yes} \alpha_{i+1}$ if and only if the query string in $\alpha_i$ is in $A$, and $\alpha_i \vdash^{no} \alpha_{i+1}$ if and only if the query string in $\alpha_i$ is not in $A$. We say $M$ accepts $x$, relative to oracle set $A$, if there exists a computation of $M$ on $x$ that is consistent with $A$ and whose halting state is in $F$. We define $L(M, A) = \{x \in \Sigma^* : M \text{ accepts } x \text{ relative to } A\}$.

We now formally define the notion of Turing reducibility based on the notion of oracle Turing machines.

**Definition 2.21** *(a) A partial function $f$ is* Turing reducible *to a total function $g$, denoted by $f \leq_T g$, if there is an oracle TM $M$ such that $M^g$ computes the function $f$.*

*(b) A partial function $f$ is* Turing reducible *to a set $B$ if $f \leq_T \chi_B$, where $\chi_B$ is the characteristic function of $B$, that is, $\chi_B(x) = 1$ if $x \in B$ and $\chi_B(x) = 0$ if $x \notin B$.*

*(c) A set $A$ is* Turing reducible *to a total function $f$ if $\chi_A \leq_T f$.*

*(d) A set $A$ is* Turing reducible *to a set $B$ if $\chi_A \leq_T \chi_B$.*

The time complexity of a set-oracle Turing machine can be defined in a similar way as that of an ordinary Turing machine. The single extra rule is that the transition from the query state to either the *yes* state or the *no* state (performed by the oracle) counts only one step. Thus, for any set-oracle machine $M$, oracle $A$ and any input $x$, $time_M^A(x)$ is defined to be the length of the halting computation of $M$ on $x$ relative to $A$. We then define $t_M^A(n) = \max\{time_M^A(x) : |x| = n\}$. We say that $t_M(n) \leq g(n)$ for some function $g$ if for any oracle $A$ and any $n \geq 0$, $t_M^A(n) \leq g(n)$. Note that the measure $t_M(n)$ is taken over the maximum of $t_M^A(n)$ over all possible oracle sets $A$. In the applications, we often find this measure too severe and consider only $t_M^A(n)$ for some specific set $A$ or for some specific group of sets $A$. An oracle Turing machine $M$ is called a *polynomial-time oracle Turing machine* if $t_M(n) \leq p(n)$ for some polynomial $p$.

The time complexity of a function-oracle Turing machine can be defined similar to that of a set-oracle Turing machine. A potential problem may occur though if the oracle function $f$ is length-increasing. For instance, if $f$ can replace a query string $y$ of length $n$ by the answer string $z = f(y)$ of length $2^n$, then it is possible that the oracle machine $M$ can use $z$ as the new query string to obtain, in one move, the answer $w = f(z) = f(f(y))$ which would have taken $2^n$ moves if $M$ were to write down the query string $z$ on the query tape by itself. This problem cannot be avoided by requiring, for instance, that the length $|f(y)|$ of the answer relative to the length $|y|$ of the query be bounded by a polynomial function, since repeated queries using the previous answers could create large queries in a small number of moves. Note, however, that we have carefully required that the oracle $f$ must, after writing down the answer $z$, move the tape head of the query tape to the leftmost square of $z$, and yet $M$ needs to move it to the rightmost square of $z$ to query for $f(z)$. Thus, this problem is avoided by this seemingly arbitrary requirement.

**Definition 2.22** *(a) A function $f$ is* polynomial-time Turing reducible *to a function $g$, denoted by $f \leq_T^P g$ or $f \in FP^g$, if $f \leq_T g$ via a polynomial-time oracle Turing machine.*

*(b) A set $A$ is* polynomial-time Turing reducible *to a set $B$, denoted by $A \leq_T^P B$ or $A \in P^B$, if $\chi_A \leq_T^P \chi_B$. We sometimes say that $A$ is polynomial-time computable relative to set $B$ if $A \leq_T^P B$.*

In addition to the notations $FP^g$ and $P^B$, we also write $P^C$ to denote the class $\bigcup_{A \in C} P^A$ and $FP^C$ to denote the class $\bigcup_{A \in C} FP^A$, if $C$ is a class of sets.

The reader is expected to verify that the reductions described in Example 2.20 can indeed be implemented by polynomial-time oracle machines.

It is easy to see that $\leq_T^P$ is both reflexive and transitive and, hence, is indeed a reducibility. Furthermore, it is easy to see that it is weaker than $\leq_m^P$.

**Proposition 2.23** *The following hold for all sets $A$, $B$, and $C$:*
  (a) $A \leq_T^P A$.
  (b) $A \leq_T^P B$, $B \leq_T^P C \Rightarrow A \leq_T^P C$.
  (c) $A \leq_m^P B \Rightarrow A \leq_T^P B$.
  (d) $A \leq_T^P B \Rightarrow A \leq_T^P \overline{B}$.

**Proposition 2.24** *(a) $P$ and $PSPACE$ are closed under $\leq_T^P$.*
  *(b) $NP = coNP$ if and only if $NP$ is closed under $\leq_T^P$.*

*Proof.* The only nontrivial part is the forward direction of part (b). Assume that $NP = coNP$, and that $A \in P^B$ for some $B \in NP$. Let $M_1$ be an oracle DTM such that $M_1^B$ accepts $A$ in polynomial time $p_1$, $M_2$ an NTM accepting $B$ in polynomial time $p_2$, and $M_3$ an NTM accepting $\overline{B}$ in polynomial time $p_3$. We describe an NTM $M$ for $A$ as follows: On input $x$, $M$ begins the computation by simulating $M_1$ on $x$. In the simulation, if $M_1$ asks a query "$y \in ?B$" then $M$ nondeterministically simulates both $M_2$ and $M_3$ on the input $y$ for at most $p_2(|y|) + p_3(|y|)$ moves. If a computation path of $M_2(y)$ accepts then it continues the simulation of $M_1(x)$ with the answer YES; if a computation path of $M_3(y)$ accepts then it continues it with the answer NO; and if a computation path of either $M_2(y)$ or $M_3(y)$ does not accept then this path of $M(x)$ rejects. The machine $M$ halts and accepts if this simulation of $M_1$ accepts.

Since $B = L(M_2)$ and $\overline{B} = L(M_3)$, for each query "$y \in ?B$" made by $M_1$, one of the simulations must accept $y$, and so the simulation of $M$ must be correct. In addition, each query $y$ of $M_1$ must be of length at most $p_1(|x|)$, the answer to each query can be obtained in the simulation in time $q(|x|) = p_2(p_1(|x|)) + p_3(p_1(|x|))$. Therefore, the above simulation works correctly in time $q(n) \cdot p_1(n)$. ∎

Let $\mathcal{C}$ be a complexity class. Recall that a set $A$ is $\leq_T^P$-hard for $\mathcal{C}$ if $B \leq_T^P A$ for all $B \in \mathcal{C}$. We say a set is $\mathcal{C}$-*hard* to mean that it is $\leq_T^P$-hard for class $\mathcal{C}$.

The notion of the space complexity of an oracle DTM $M$ is not as easy to define as the time complexity. The straightforward approach is to define $space_M^A(x)$ to be the number of squares the machine $M$ visits in the computation of $M^A(x)$, excluding the input and output tapes. A problem with this definition is that the space complexity of accepting a set $A$, using itself as the oracle, would require a linear space, since we need to copy the input to the query tape to make the query. This seems too much space necessary to compute the identity operator. When we consider the set-oracle TMs, note that each query is written to the query tape and then, after the query is answered, is erased by the oracle. Thus, the query tape serves a function similar to the output tape, except that this "output" is written to be read by the oracle instead of the user, and it seems more reasonable to exclude the query tape from the space measure. To prevent the machine from using the space of query tape "free" for other purposes, we can require that the query tape be a write-only tape, just like the output tape. These observations suggest us to use the alternative approach for the space complexity of oracle TMs in which the query tape is a write-only tape and is not counted in the space measure. It should be warned, however, that this space complexity measure still leads to many unsatisfactory properties. For instance, the polynomial-space Turing reducibility defined by polynomial space-bounded oracle TMs is not really a reducibility relation since it does not have the transitivity property (see Exercise 2.16). Furthermore, if we extend this notion to more general types of oracle machines such as nondeterministic and alternating oracle TMs (to be introduced in the next chapter), there are even more undesirable properties about polynomial space-bounded oracle machines. Therefore, we leave this issue open here, and refer the interested reader to, for instance, Buss [1986] for more discussions.

## 2.5  *NP*-Complete Optimization Problems

Based on the notion of polynomial-time Turing reducibility, we can see that many important combinatorial optimization problems are *NP*-hard search problems. We prove these results by first showing that the corresponding decision problems are $\leq_m^P$-complete for *NP*, and then proving that the problems of searching for the optimum solutions are $\leq_T^P$-equivalent to the corresponding decision problems. In practice, however, we often do not need the optimum solution. A *nearly* optimum solution is sufficient for most applications. In general, the *NP*-hardness of the optimization problem does not necessarily imply the *NP*-hardness of the approximation to the optimization problem. In this section, we demonstrate that for some *NP*-complete optimization problems, their approximation versions are also *NP*-hard and, yet, for some problems, polynomial-time approximation is achievable. These types of results are often more difficult to prove than other *NP*-completeness results. We only present some easier results, and delay the more involved results until Chapter 11.

We first introduce a general framework to deal with the approximation problems. Very often, an optimization problem $\Pi$ has the following general structure: for each input instance $x$ to the problem $\Pi$, there are a number of *solutions* $y$ to $x$. For each solution $y$, we associate a *value* $v_\Pi(y)$ (or, simply, $v(y)$, if $\Pi$ is known from the context) to it. The problem $\Pi$ is to find, for the given input $x$, a solution $y$ to $x$ such that its value $v(y)$ is maximized (or, minimized). For instance, we can fit the problem MAX-CLIQUE into this framework as follows: an input to the problem is a graph $G$; a solution to $G$ is a clique $C$ in $G$; the value $v(C)$ of a solution $C$ is the number of its vertices; and the problem is to find, for a given graph $G$, a clique of the maximum size.

Let $r$ be a real number with $r > 1$. For a maximization problem $\Pi$ with the above structure, we define its approximation version, with the approximation ratio $r$, as follows:

> $r$-APPROX-$\Pi$: For a given input $x$, find a solution $y$ to $x$ such that $v(y) \geq v^*(x)/r$, where $v^*(x) = \max\{v(z) : z$ is a solution to $x\}$.

Similarly, for a minimization problem $\Pi$, its approximation version with the approximation ratio $r$ is as follows:

> $r$-APPROX-$\Pi$: For a given input $x$, find a solution $y$ to $x$ such that $v(y) \leq r \cdot v^*(x)$, where $v^*(x) = \min\{v(z) : z$ is a solution to $x\}$.

In general, the approximation ratio $r$ could be a function $r(n)$ of the input size. In this section, we consider only approximation problems with a constant ratio $r$.

We first consider a famous optimization problem: the traveling salesman problem. The traveling salesman problem is an optimization problem which asks, from a given map of $n$ cities, for a shortest tour of all these cities. The following is the standard formulation of its corresponding decision problem:

> TRAVELING SALESMAN PROBLEM (TSP): Given a complete graph $G = (V, E)$ with a cost function $c : E \rightarrow \mathbf{N}$, and an integer $K > 0$, determine whether there is a tour (i.e., a Hamiltonian circuit) of $G$ whose total cost is less than or equal to $K$.

**Theorem 2.25** TSP *is NP-complete.*

*Proof.* There is a simple reduction from HC to TSP: For any graph $G = (V, E)$, we map it to a complete graph $G' = (V, E')$ with the costs $c(e) = 1$ if $e \in E$, and $c(e) = 2$ if $e \notin E$, and let $K = |V|$. Then, it is clear that $G$ has a Hamiltonian circuit if and only if $G'$ has a tour of total cost $K$. ∎

Using the notion of polynomial-time Turing reducibility, it is not hard to show that the search problem of TSP that asks for the minimum-cost tour is *equivalent* to the above decision problem (see Exercise 2.18). Therefore, the search problem of TSP is also *NP*-hard. In the following, we show that the approximation to TSP is also *NP*-hard.

For each graph $G$ with a cost function $c$ on its edges, let $K^*(G, c)$ denote the cost of a minimum-cost tour of $G$. We formulate the problem of approximation to TSP as the following search problem. Let $r$ be a rational number with $r \geq 1$.

> $r$-APPROX-TSP: Given a complete graph $G = (V, E)$ with a cost function $c : E \rightarrow \mathbf{N}$, find a tour of $G$ with the total cost $\leq r \cdot K^*(G, c)$.

**Theorem 2.26** *For any $r > 1$, the problem $r$-APPROX-TSP is NP-hard.*

*Proof.* We modify the reduction from HC to TSP as follows: For any graph $G = (V, E)$, we define a complete graph $G' = (V, E')$ with the costs $c(\{u, v\}) = 1$ if $\{u, v\} \in E$, and $c(\{u, v\}) = r|V|$ if $\{u, v\} \notin E$.

Assume that $|V| = n$. If $G$ has a Hamiltonian circuit, then that tour in $G'$ is of cost $n$; otherwise, a minimum-cost tour in $G'$ is of cost $\geq (n-1) + rn$. Thus, it is easy to see how to solve the question of whether $G \in HC$ using $r$-APPROX-TSP as an oracle: We ask the oracle to find a tour $T$ of $G'$ with the cost at most $r \cdot K^*(G', c)$, and decide that $G \in HC$ if and only if the cost of $T$ is $\leq rn$. Notice that if $G$ has a Hamiltonian circuit, then $K^*(G', c) = n$, and so the cost of $T$ is $\leq rn$. On the other hand, if $G$ does not have a Hamiltonian circuit, then the cost of $T$ is $\geq K^*(G', c) \geq rn + (n-1)$. Thus, the above reduction is correct.                                    ∎

An important subproblem of TSP is the traveling salesman problem satisfying the *triangle inequality*, that is, the problem whose cost function $c$ satisfies

$$c(\{u, v\}) \leq c(\{u, w\}) + c(\{w, v\})$$

for any three vertices $u, v, w$ of the graph $G$. For instance, if the input graph $G$ is a Euclidean graph in the sense that its vertices are points in an Euclidean space $E^k$, and the cost of an edge is the distance of the two endpoints in this space, then this cost function satisfies the triangle inequality.

Note that the graph $G'$ and its cost function $c$ constructed in the reduction from HC to TSP of Theorem 2.25 do not satisfy the triangle inequality. Yet, it is easy to modify that proof to show that TSP with the triangle inequality restriction on the cost function is still *NP*-complete (see Exercise 2.17). On the other hand, the story is quite different when we consider the approximation to this restricted version of TSP. Namely, there exists a polynomial-time algorithm that solves the problem $r$-APPROX-TSP, for all $r \geq 3/2$, on graphs having the triangle inequality. It is an open question whether this bound $r = 3/2$ can be improved. Furthermore, if the graph $G$ is a Euclidean graph on the two-dimensional plane $E^2$, then there exist polynomial-time algorithm that solves the problem $r$-APPROX-TSP, for all $r > 1$. In the following, we present a polynomial-time approximation algorithm for $(3/2)$-APPROX-TSP on graphs satisfying the triangle inequality. We leave the problem $r$-APPROX-TSP for graphs on the Euclidean plane as an exercise (Exercise 2.19).

To establish this result, we first review three well-known problems that all are known to have polynomial-time solutions. Let $G = (V, E)$ be a connected graph. A *spanning tree* of the graph $G$ is a subgraph $T = (V, E')$ with the same vertex set $V$ such that $T$ is a tree, that is, $T$ is connected and contains no cycle. For a graph $G$ with a cost function $c$ on edges, the *minimum spanning tree problem* is to find a spanning tree $T$ of $G$ with the minimum total cost. The existence of a polynomial-time algorithm for the minimum spanning tree problem is well known (see, e.g., Aho, Hopcroft and Ullman [1974]).

Let $G = (V, E)$ be a complete graph of an even number of vertices, and $c$ a cost function on the edges. A *matching* of the graph $G$ is a partition of the set $V$ into $|V|/2$ subsets each of two vertices. The cost of a matching is the total cost of the edges connecting the vertices in the same subsets of the matching. The *minimum matching problem* is to find a minimum-cost matching of a given graph $G$ with a cost function $c$. The minimum matching problem is polynomial-time solvable [Lawler, 1976].

A graph $G = (V, E)$ is an *Eulerian graph* if each vertex $v \in V$ has a positive, even degree. An *Eulerian circuit* of a graph $G = (V, E)$ is a path that passes through each edge $e \in E$ exactly once. The *Eulerian circuit problem* is to find an Eulerian circuit for a given graph. This problem, although similar to the problem HAMILTONIAN CIRCUIT, has a simple solution in polynomial time. That is, a graph has an Eulerian circuit if and only if it is an Eulerian graph. In addition, there is a polynomial-time algorithm finding an Eulerian tour of any given Eulerian graph [Liu, 1968].

Now we are ready for the following result:

**Theorem 2.27** *The problem* $(3/2)$-APPROX-TSP *is polynomial-time solvable, when the input cost functions satisfy the triangle inequality.*

*Proof.* Let $G = (V, E)$ be a complete graph, and $c : E \to \mathbf{N}$ be a cost function satisfying the triangle inequality. Let $T = (V, E_T)$ be an minimum spanning tree of the graph $G$. Assume that the total cost of $E_T$ is $K_T$. Then, obviously $K_T \leq K^*(G, c)$, since we can remove one edge from the minimum-cost tour of $G$ to obtain a spanning tree. We claim that there is a tour $Q$ of $G$ of cost at most $3K_T/2$ and, hence, at most $(3/2) \cdot K^*(G, c)$.

Let $V_1$ be the set of all vertices in $T$ that have an odd degree. Then, $|V_1|$ must be even, since the sum of the degrees of all vertices must be even. Let $G_1 = (V_1, E_1)$ be the complete subgraph of $G$ induced by the vertex set $V_1$, and let $c_1(e) = c(e)$ for all $e \in E_1$. Consider the minimum matching $M$ of the graph $G_1$. That is, $M$ is the minimum-cost set of edges such that each vertex belongs to exactly one of the edges in $M$. We observe that the cost $K_M$ of $M$ does not exceed $(1/2)K^*(G_1, c_1)$. To see this, let $T_1$ be a minimum-cost tour of $G_1$ that passes through each vertex exactly once. Then, by taking the alternating edges of $T_1$, we obtain two matchings of $G_1$. (*Note:* $G_1$ has an even number of vertices.) The one with the lower cost must have the cost less than or equal to one half of the cost of $T_1$, and so $K_M \leq (1/2)K^*(G_1, c_1) \leq (1/2)K^*(G, c)$.

Now we define a new graph $G_2 = (V, E_2)$ on the vertex set $V$ with $E_2 = E_T \uplus M$, where $\uplus$ means the exclusive union; that is, we allow more than one edge between two vertices. For each edge $e$ in $E_2$, we assign the same cost $c(e)$ as before. Then, this new graph $G_2$ is an Eulerian graph, and has an Eulerian circuit $R$. Note that the Eulerian circuit $R$ has total cost $K_T + K_M \leq (3/2)K^*(G, c)$. Furthermore, this cycle $R$ passes through every vertex in $V$ at least once. It, however, may pass through some vertices more than once. To avoid this problem, we can convert this cycle $R$ to a Hamiltonian circuit of

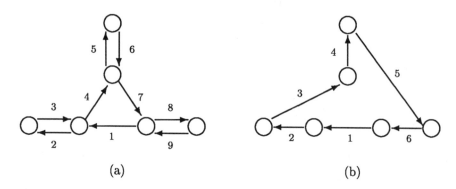

Figure 2.3: (a) An Eulerian circuit; (b) A shortcut.

$G$ by taking a shortcut whenever the path $R$ visits the second time a vertex other than the starting vertex. To be more precise, assume that the cycle $R$ is defined by the following sequence of vertices: $(v_{i_1}, v_{i_2}, \ldots, v_{i_m})$, where $v_{i_m} = v_{i_1}$. Then, we let $Q$ be the path defined by the subsequence of $R$ that contains, for each vertex, only the first occurrence of that vertex (see Figure 2.3). By the triangle inequality property, it is clear that the cost of $Q$ is less than or equal to the cost of $R$, and so $Q$ is a Hamiltonian tour of cost $\leq (3/2)K^*(G, c)$.                                                                     ∎

In the above, we have considered three variations of the problem TSP, namely, TSP on arbitrary graphs, TSP on graphs that satisfy the triangle inequality, and TSP on graphs on the Euclidean plane. We have seen that the optimization versions of these three variations of TSP are all *NP*-complete, but their approximation versions belong to three different complexity classes. Let us review these three complexity classes of approximation problems.

The first class consists of approximation problems that are *NP*-hard for all approximation ratios $r > 1$. The problem APPROX-TSP on arbitrary graphs belongs to this class.

The second class consists of approximation problems for each of which there exists a constant $r_0$ such that the problem is *NP*-hard for all approximation ratios $r < r_0$ and is polynomial-time solvable for $r \geq r_0$. The problem APPROX-TSP on graphs satisfying the triangle inequality is a candidate of problems in the second class. We will see in Chapter 11 more approximation problems that belong to the first and second classes.

The third class consists of all approximation problems that are polynomial-time solvable for all ratios $r > 1$. Problems in the third class can be further classified into two subclasses. A problem may have, for each $k \geq 1$, a polynomial-time algorithm $A_k$ for the approximation ratio $r = 1 + 1/k$, such that the runtime of the algorithm $A_k$ is polynomial in the input size but super-polynomial in the parameter $k$. We say such a problem has a *polynomial-time approximation scheme*. The problem APPROX-TSP on graphs on the Euclidean plane belongs to this subclass. A problem with a polynomial-time approximation scheme, though considered as polynomial-time approximable, is not feasibly solvable when we need good approximation with a small ratio $r \approx 1$. Instead, we would like to have a *fully polynomial-time approximation scheme A* that takes a problem instance and an approximation parameter $k$ as inputs and outputs an approximate solution with the approximation ratio $r = 1 + 1/k$ in time polynomial in the instance size and the parameter $k$. Such a scheme would be the best we can expect, short of proving $P = NP$. In the following, we demonstrate that the approximation version of the knapsack problem has a fully polynomial-time approximation scheme.

> KNAPSACK (KS): Given a list of positive integer pairs $L = (\langle c_1, v_1 \rangle, \langle c_2, v_2 \rangle, \ldots, \langle c_n, v_n \rangle)$ and an integer $B \geq \max\{c_i\}_{i=1}^n$, find the set $I \subseteq \{1, 2, \ldots, n\}$ that maximizes $V_I = \sum_{i \in I} v_i$, subject to the condition of $C_I = \sum_{i \in I} c_i \leq B$.

Intuitively, we say each $c_i$ is the cost of item $i$ and $v_i$ is the value of item $i$, and the problem KS is to maximize the total value while keeping the total cost below a given bound $B$. The approximation version of the optimization problem KNAPSACK, called APPROX-KS, is to find, for a given list $L$, a given integer $B$ and a given rational $r > 1$, an index set $I \subseteq \{1, 2, \ldots, n\}$, such that $V_I \geq V^*(L, B)/r$ and $C_I \leq B$, where $V^*(L, B)$ is the maximum solution $V_I$ with respect to input $(L, B)$. It is clear that the sum of subset problem SS is a subproblem of KS and hence KS is $NP$-hard.

**Theorem 2.28** *There is a fully polynomial-time approximation scheme for the problem* APPROX-KS.

*Proof.* Our approximation algorithm for APPROX-KS is based on a simple *pseudo* polynomial-time algorithm $D$ for the optimization version KS. This algorithm is called pseudo polynomial-time because it runs in time polynomial in $n$ and $M$, where $M = \max\{v_i : 1 \leq i \leq n\}$. Since the input length is of the order $\Theta(n \log M)$, this is not really a polynomial-time algorithm. It is, however, very simple and we will modify it later to run in real polynomial time.

Assume that $(L, B)$, with $L = (\langle c_1, v_1 \rangle, \ldots, \langle c_n, v_n \rangle)$, is a given instance of the problem KS. Let $S = \sum_{i=1}^n v_i$, and for each $I \subseteq \{1, \ldots, n\}$, let $C_I = \sum_{i \in I} c_i$ and $V_I = \sum_{i \in I} v_i$. For each $1 \leq i \leq n$ and $0 \leq j \leq S$, let

$$C_{i,j}^* = \min\{C_I : I \subseteq \{1, \ldots, i\}, V_I = j\},$$

if the underlying set is nonempty. Our algorithm $D$ will build a table $T(i, j)$, for $1 \leq i \leq n$, $0 \leq j \leq S$, with the following property:

> *Property $T_1$:* If there is no subset $I \subseteq \{1, \ldots, i\}$ having $V_I = j$ and $C_I \leq B$, then $T(i, j)$ is empty (denoted by $\lambda$); otherwise, $T(i, j)$ is a subset $I^* \subseteq \{1, \ldots, i\}$ such that $V_{I^*} = j$ and $C_{I^*} = C_{i,j}^*$.

(Notice that an empty entry $T(i,j) = \lambda$ is different from an entry $T(i,j)$ whose value is the empty set $\emptyset$; $T(i,0) = \emptyset$ means that $C_\emptyset = 0 \leq B$ and $V_\emptyset = 0$.) Suppose that $T$ satisfies the above property. Let $j^*$ be the maximum $j \leq S$ such that $T(n,j) \neq \lambda$. Then, $T(n,j^*)$ is a maximum solution. Thus, the problem KS is solved when the table $T$ is built.

The table $T(i,j)$ is easy to build by the dynamic programming technique below. We recall that $B \geq \max\{c_i : 1 \leq i \leq n\}$.

*Algorithm D.*

(1) Initialize $T(i,j) := \lambda$ for all $1 \leq i \leq n$, $1 \leq j \leq S$;

(2) Let $T(i,0) := \emptyset$ for all $1 \leq i \leq n$, and $T(1,v_1) := \{1\}$;

(3) For $i := 2$ to $n$ do

    (3.1) For each $j$, $1 \leq j \leq v_i - 1$, let $T(i,j) := T(i-1,j)$;

    (3.2) For each $j$, $v_i \leq j \leq S$, if $[T(i-1,j-v_i) = I \neq \lambda]$ and $[C_I + c_i \leq B]$ and $[T(i-1,j) = J \neq \lambda \Rightarrow C_I + c_i \leq C_J]$, then we let $T(i,j) := I \cup \{i\}$, else we let $T(i,j) := T(i-1,j)$.

We prove by induction that the table $T(i,j)$ built by algorithm $D$ indeed satisfies property $T_1$. Consider $T(i,j)$ for some $2 \leq i \leq n$ and $v_i \leq j \leq S$. In Step (3.2), $T(i,j)$ is defined to be either equal to $I \cup \{i\}$ or $J = T(i-1,j)$, where $I = T(i-1,j-v_i)$. The former case happens only when $V_{I\cup\{i\}} = j$, $C_{I\cup\{i\}} \leq B$ and $C_I + c_i \leq C_J$ (when $T(i-1,j) = J \neq \lambda$). If there is no subset $I \subseteq \{1,\ldots,i\}$ with $V_I = j$ and $C_I \leq B$ then, by the inductive hypothesis, $T(i-1,j) = \lambda$ and step (3.2) defines $T(i,j) = T(i-1,j)$ correctly. Next, assume that there is at least one subset $K \subseteq \{1,\ldots,i\}$ satisfying $V_K = j$ and $C_K \leq B$. Then, no matter whether $T(i,j)$ is defined to be $T(i-1,j-v_i) \cup \{i\}$ or $T(i-1,j)$, it holds that $V_{T(i,j)} = j$ and $C_{T(i,j)} \leq B$. We only need to check that $C_{T(i,j)} = C_{i,j}^*$. By the inductive hypothesis, if $T(i-1,j) = J \neq \lambda$ exists, then $C_J = C_{i-1,j}^*$. Similarly, if $T(i-1,j-v_i) = I \neq \lambda$ exists, then $C_I = C_{i-1,j-v_i}^*$. Now, if $T(i,j)$ is defined to be $I \cup \{i\}$, then it must be true that $C_{I\cup\{i\}}$ is the minimum among all $C_K$'s with $K \subseteq \{1,\ldots,i\}$, $i \in K$, and $V_K = j$. By the condition that $C_I + c_i \leq C_J$, we know that $C_{I\cup\{i\}}$ is also the minimum among all $C_K$'s with $K \subseteq \{1,\ldots,i-1\}$ and $V_K = j$. Thus, $C_{T(i,j)} = C_{i,j}^*$. Otherwise, if $T(i,j) = T(i-1,j)$, then the condition of step (3.2) implies that there is no $I \subseteq \{1,\ldots,i-1\}$ having the properties $V_{I\cup\{i\}} = j$, and $C_{I\cup\{i\}} \leq C_{i-1,j}^*$. Thus, $C_{i,j}^* = C_{i-1,j}^*$ and $C_{T(i,j)} = C_{i,j}^*$.

It is apparent that the above algorithm builds the table $T$ in time $O(n^3 M \log(MB))$ and, hence, solves the problem KS in pseudo polynomial time. Next, we modify the above algorithm to make it run in time polynomial in $n$ and $\log(MB)$. For any integer $k > 0$, we let $u_i := \lfloor v_i n(k+1)/M \rfloor$ for each $1 \leq i \leq n$. Then, we run the above algorithm $D$ on the instance $(L',B)$ of KS, where $L' = (\langle c_1,u_1 \rangle, \langle c_2,u_2 \rangle, \ldots, \langle c_n,u_n \rangle)$.

It is clear that the algorithm $D$ runs on this instance $(L',B)$ in time $O(n^4 k \log(nkB))$, since now the table $T$ has size $n$ by $(k+1)n^2$ and the calculation for each entry $T(i,j)$ involves only integers of length $\leq \log(nkB)$. Thus, the time complexity of this new algorithm is polynomial in both the size of the input $(L,B)$ and in the parameter $k$.

Let $T$ be the table built by the algorithm $D$ on input $(L',B)$. Let $j^*$ be the maximum $j \leq (k+1)n^2$ such that $T(n,j) \neq \lambda$. We claim that the output $T(n,j^*) = J^*$ has the property that $C_{J^*} \leq B$ and $V_{J^*} \geq (1-1/k)V^*(L,B)$, where $V^*(L,B)$ is the maximum solution for the instance $(L,B)$. We first observe that all nonempty entries $T(i,j) = I$ of the table must have the cost $C_I \leq B$, since we did not change the costs $c_i$ and the upper bound $B$. Next, suppose that $I^* \subseteq \{1,\ldots,n\}$ is the optimum solution for the instance $(L,B)$ (so that $V_{I^*} = V^*(L,B)$), and that $J^* \subseteq \{1,\ldots,n\}$ is the optimum solution for the instance $(L',B)$. Then, we must have

$$\sum_{i \in J^*} v_i \geq \sum_{i \in J^*} \left\lfloor \frac{v_i n(k+1)}{M} \right\rfloor \cdot \frac{M}{n(k+1)} = \frac{M}{n(k+1)} \sum_{i \in J^*} u_i$$

$$\geq \frac{M}{n(k+1)} \sum_{i \in I^*} u_i$$

$$\geq \frac{M}{n(k+1)} \sum_{i \in I^*} \left( \frac{v_i n(k+1)}{M} - 1 \right)$$

$$\geq \frac{M}{n(k+1)} \left( \frac{n(k+1)}{M} V^*(L,B) - n \right)$$

$$= V^*(L,B) - \frac{M}{k+1} \geq V^*(L,B)\left(1 - \frac{1}{k}\right).$$

(The last inequality follows from the assumption that $B \geq \max\{c_i\}_{i=1}^n$ and, hence, $V^*(L,B) \geq M$.) Thus, the output is within the factor $1 - 1/k$ of the optimum solution. ∎

*Remark.* The above algorithm is not the most efficient approximation algorithm for APPROX-KS. Our purpose is only to demonstrate the fact that the problem APPROX-KS has a fully polynomial-time approximation scheme. For more sophisticated fine tuning of the algorithm, see Ibarra and Kim [1975] and Lawler [1976].

## Exercises

**2.1** A binary relation $R \subseteq \Sigma^* \times \Sigma^*$ is called *polynomially honest* if there exists a polynomial function $p$ such that $\langle x,y \rangle \in R$ only if $|x| \leq p(|y|)$ and $|y| \leq p(|x|)$. A function $f : \Sigma^* \to \Sigma^*$ is *polynomially honest* if the relation $\{\langle x, f(x)\rangle : x \in \Sigma^*\}$ is polynomially honest. Prove that $A \subseteq \Sigma^*$ is in *NP* if and only if $A = Range(f)$ for some polynomially honest function $f \in FP$.

**2.2** For any set $B$, define $prefix(B) = \{\langle x,u \rangle : (\exists v) \langle x, uv \rangle \in B\}$. Assume that $A \in NP$ such that for each $x$, $x \in A \iff (\exists y) \langle x,y \rangle \in B$ for some polynomially honest relation $B \in P$. Prove that $prefix(B) \in NP$, and $A \leq_T^P$ $prefix(B)$. Under what conditions does the relation $prefix(B) \leq_T^P A$ hold?

**2.3** (a) Prove that if $A, B \in P$ then $A \cup B, A \cap B, AB, A^* \in P$.
(b) Prove that if $A, B \in NP$, then $A \cup B, A \cap B, AB, A^* \in NP$.

**2.4** Assume that $A$ is $NP$-complete and $B \in P$. Prove that if $A \cap B = \emptyset$, then $A \cup B$ is $NP$-complete. What can you say about the complexity of $A \cup B$ if $A$ and $B$ are not known to be disjoint?

**2.5** Prove that a nonempty finite set is $NP$-complete if and only if $P = NP$.

**2.6** Assume that set $A$ has the following property: There exist a set $B \in P$ and two polynomial functions $p, q$, such that for any string $x \in \Sigma^*$, $x \in A$ if and only if there are at least $p(|x|)$ strings $y$ of length $\leq q(|x|)$ satisfying $\langle x, y \rangle \in B$. Prove that $A \in NP$.

**2.7** Let $F_x$ be the Boolean formula constructed from an input $x$ of length $2^n$ in the proof of Cook's theorem, and $F_x'$ be the 3-CNF formula obtained from $F_x$ by the reduction of Corollary 2.13. Show that the clauses of the formula $F_x'$ can be labeled as $C_1, C_2, \ldots, C_k$ so that the function $g(i) = C_i$ is computable in time polynomial in $n$.

**2.8** Prove that a DNF formula can be switched in polynomial time to a CNF formula, with possibly more variables, preserving satisfiability.

**2.9** Prove that if $P \neq NP$, then there is no polynomial-time algorithm that switches a CNF formula to a DNF formula preserving satisfiability.

**2.10** Define 2-SAT to be the problem of determining whether a given 2-CNF formula, which is a CNF formula with each clause having at most two literals, is satisfiable. Prove that 2-SAT is in $P$.

**2.11** For each of the following problems, prove that it is $NP$-complete.
(a) 1-IN-3-3SAT: Given a 3-CNF formula, determine whether there is a Boolean assignment that assigns the value TRUE to exactly one literal of each clause.
(b) NOT-ALL-EQUAL-3SAT: Given a 3-CNF formula, determine whether there is a Boolean assignment that assigns the value TRUE to at least one and at most two literals of each clause.
(c) 3-DIMENSIONAL MATCHING (3DM): Given three disjoint sets $A, B, C$, each of $n$ elements and a set $S \subseteq A \times B \times C$, determine whether there is a subset $T \subseteq S$ of size $n$ such that no two elements of $T$ agree in any of three coordinates.
(d) SUBGRAPH ISOMORPHISM: Given two graphs $G$ and $H$, determine whether $G$ has a subgraph that is isomorphic to graph $H$.
(e) SET SPLITTING: Given a finite set $S$ and a collection $C$ of subsets of $S$, determine whether there is a partition of $S$ into two subsets $S_1$ and $S_2$, such that no subset $T$ in $C$ is contained in either $S_1$ or $S_2$.

**2.12** Formally define the notion of the computation of a function-oracle DTM, and the notion of a function-oracle DTM computing a partial function $g$ relative to a total function $f$.

**2.13** Prove that if $A$ is computable by an oracle Turing machine $M$ using oracle $B$, such that $t_M^B(n)$ is bounded by a polynomial function $p$, then $A \leq_T^P B$. (That is, $A \leq_T^P B$ via some other oracle machine $M_1$ that halts in polynomial time for all oracles.)

**2.14** The polynomial-time Turing reducibility is often called an *adaptive* reducibility since the $(k+1)$st query made by an oracle machine $M$ on an input $x$ may depend on the answers to the first $k$ queries. A nonadaptive version of the polynomial-time Turing reducibility works as follows: The oracle machine must determine all the queries it wants to ask before the first query is made, and then it presents all queries to the oracle together. Such a reducibility is called a *polynomial-time truth-table reducibility,* and denoted by $\leq_{tt}^P$. Formally we define $A \leq_{tt}^P B$ if there exists a function $f \in FP$ and a set $C \in P$ such that for any $x$, $x \in A$ if and only if $\langle x, \chi_B(y_1), \chi_B(y_2), \ldots, \chi_B(y_m) \rangle \in C$, where $f(x) = \langle y_1, y_2, \ldots, y_m \rangle$ and $\chi_B$ is the characteristic function of set $B$.
   (a) Prove that $A \leq_m^P B \Rightarrow A \leq_{tt}^P B \Rightarrow A \leq_T^P B$.
   (b) Prove that there exist sets $A, B$, such that $A \leq_{tt}^P B$ but $A \not\leq_m^P B$.
   (c) Prove that there exist sets $C, D$, such that $C \leq_T^P D$ but $C \not\leq_{tt}^P D$.

**2.15** An oracle Turing machine $M$ is called a *positive* oracle machine if for all oracles $X, Y$, $X \subseteq Y$ implies $L(M, X) \subseteq L(M, Y)$. We say that $A$ is *polynomial-time positive Turing reducible* to $B$, denoted by $A \leq_{pos\text{-}T}^P B$, if $A \leq_T^P B$ via a positive oracle machine $M$.
   (a) Prove that $NP$ is closed under the $\leq_{pos\text{-}T}^P$-reducibility; that is, $A \leq_{pos\text{-}T}^P B$ and $B \in NP$ imply $A \in NP$.
   (b) Show that there exist sets $A, B$, such that $A \leq_T^P B$ but $A \not\leq_{pos\text{-}T}^P B$.

**2.16** (a) Using the first type of space measure for oracle TMs (i.e., the space measure that includes the space in the query tape), show that there exists a set $A$ that is not log-space Turing reducible to itself.
   (b) Using the second type of space measure for oracle TMs (i.e., the space measure that excludes the query tape while requiring the query tape to be write-only), show that there exist sets $A, B$, such that $B \in PSPACE$, $A \notin PSPACE$, but $A$ is computable by a polynomial space-bounded oracle machine with respect to the oracle $B$; that is, $PSPACE$ is not closed under the polynomial-space Turing reducibility.
   (c) Using the second type of space measure for oracle TMs, show that the polynomial-space Turing reducibility does not satisfy the transitivity property (and, hence, should not be called a reducibility).

**2.17** Prove that the problem TSP on graphs that satisfy the triangle inequality is still $NP$-complete.

**2.18** In this exercise, we consider the decision problem EXACT-TSP that asks, for a given weighted graph $G$ and an integer $K$, whether the cost of a minimum-cost tour of $G$ is exactly $K$, and the search problem MIN-TSP that asks for a minimum-cost tour of a given weighted graph $G$.

(a) Show that EXACT-TSP is $\leq_T^P$-equivalent to TSP.

(b) Show that MIN-TSP is $\leq_T^P$-equivalent to TSP.

(c) The notion of polynomial-time many-one reducibility can be extended to functions. Namely, we say function $\phi$ is *polynomial-time many-one reducible* to function $\psi$ if there exist two polynomial-time computable functions $f$ and $g$ such that for each $x$, $\phi(x) = g(\psi(f(x)))$. Let MAX-3SAT be the problem that asks, for a given 3-CNF formula and a given weight function on clauses, to find the Boolean assignment that maximizes the weight of satisfied clauses. Show that MAX-3SAT is $\leq_m^P$-complete for the class $FP^{NP}$.

(d) Prove that MIN-TSP is $\leq_m^P$-complete for the class $FP^{NP}$ by reducing MAX-3SAT to MIN-TSP.

**2.19** Prove that $r$-APPROX-TSP is polynomial-time solvable for all $r > 1$ when it is restricted to graphs on the two-dimensional Euclidean plane.

**2.20** For each of the following optimization problems, first define a decision version of the problem, then prove that the search version is $\leq_T^P$-equivalent to the decision version and that the decision version is *NP*-complete.

(a) GRAPH COLORING (GCOLOR): Given a graph $G = (V, E)$, find a (coloring) function $c : V \to \{1, 2, \dots, k\}$ that uses the minimum number $k$ of colors such that $c(u) \neq c(v)$ for any two adjacent vertices $u, v \in V$.

(b) BIN PACKING (BP): Given a list of rational numbers $L = (r_1, r_2, \dots, r_n)$, with $0 < r_i < 1$, for each $1 \leq i \leq n$, find a partition $L$ into $k$ parts $L_1, \dots, L_k$ such that the sum of each part $L_i$, $1 \leq i \leq k$, is at most 1 and the number $k$ of parts is minimized.

(c) SHORTEST COMMON SUPERSTRING (SCS): Given a list of strings $L = (s_1, s_2, \dots, s_n)$ over a finite alphabet $\Sigma$, find the shortest string $t$ such that each $s_i$ is a substring of $t$.

**2.21** (a) Prove that $(4/3)$-APPROX-GCOLOR is NP-hard.

(b) Prove that APPROX-BP has a polynomial-time approximation scheme.

(c) Prove that 3-APPROX-SCS is solvable in polynomial time.

**2.22** In this problem, we explore the similarity between the computation of an NTM and the computation of an oracle DTM. Assume that $M_1$ is an NTM in which each nondeterministic move has exactly two choices. Then, we can construct an oracle DTM $M_2$ to simulate $M_1$ as follows: On any input $x$, $M_2(x)$ simulates $M_1(x)$ deterministically; when $M_1$ makes a nondeterministic move, $M_2$ asks the oracle $H$ whether $y$ is in $H$, where $y$ is a string encoding the computation of $M_2^H(x)$ up to this point. It follows the first choice if and only if the answer to the query is YES.

(a) Assume that $M_1$ is a polynomial-time NTM. Prove that there exists an oracle $H \in NP$ such that for each $x$, $x \in L(M_1)$ if and only if $M_2^H$ accepts $x$ in polynomial time.

(b) We say $M_2$ is a *robust* oracle machine if for any oracle $H$, $L(M_2, H) = L(M_2, \emptyset)$; that is, the set of strings accepted by $M_2$ does not depend on the oracle. Show that a set $A$ is computable by a robust oracle machine $M$ with the property that relative to some oracle $H$, $M^H(x)$ always halts in polynomial time, if and only if $A \in NP \cap coNP$.

**2.23** Determine the complexity of the following problem: Given a set of points on the first quadrant of the rectilinear plane (the plane with the rectilinear metric: $d(\langle x_1, y_1 \rangle, \langle x_2, y_2 \rangle) = |x_1 - x_2| + |y_1 - y_2|$), find a minimum total-edge-length directed tree rooted at the origin that connect the given points with edges in direction either to the right or upward.

> *When I feel like exercising,*
> *I just lie down until the feeling goes away.*
> — Robert M. Hutchins

## Historical Notes

Many-one and Turing reducibilities are major tools in recursion theory. Rogers [1967] contains a complete treatment of these topics. The first natural *NP*-complete problem, SAT, was proved by Cook [1971], who used the polynomial-time Turing reducibility. Karp [1972] proved more natural *NP*-complete problems, including the problems VC, CLIQUE, IS, HC, IP, SS, PARTITION, KS, 3DM, and GCOLOR, under the polynomial-time many-one reducibility. Levin [1973] independently proved the *NP*-completeness of the tiling problem. Garey and Johnson [1979] collected several hundred *NP*-complete problems. They also included a discussion on the proof techniques to establish *NP*-completeness results. The problem TSP has been studied widely in literature. The approximation algorithm for TSP satisfying the triangle inequality is from Christofides [1976]. The polynomial-time approximation scheme for TSP on two-dimensional Euclidean plane (Exercise 2.19) is from Arora [1996]. The fully polynomial-time approximation scheme for KS is from Ibarra and Kim [1975].

The fact that 2-SAT is polynomial-time solvable was proved in Cook [1971]. A number of variations of 3-SAT, including 1-IN-3-SAT and NOT-ALL-EQUAL-3SAT, are shown to be *NP*-complete in Schaefer [1978a]. The problem SET SPLITTING was proved to be *NP*-complete by Lovasz [1973]. Different notions of polynomial-time reducibilities have been compared in Ladner et al. [1975]. Exercise 2.14 is from there. The two models of space-bounded oracle machines discussed in Section 2.4 and other modifications have been studied by Ladner and Lynch [1976], Orponen [1983], Ruzzo et al. [1984], Buss [1986], and Wilson [1988]. The nonapproximability results on GCOLOR can be found in Garey and Johnson [1976], Lund and Yannakakis [1994], and Khanna et al. [1993]. Karmarkar and Karp [1982] gave a polynomial-time approximation scheme for BIN PACKING. See also Coffman et al. [1985] for a survey on BIN PACKING. The problem SHORTEST COMMON SUPERSTRING was first proved to be NP-complete by Maier and Storer [1977]. The ratio-3 approximation algorithm was given by Blum et al. [1991], who also showed, by the results of Chapter 11, that $r$-APPROX-SCS is NP-hard for some $r > 1$. Exercise 2.22 is from Schöning [1985] and Ko [1987].

# 3

# The Polynomial-Time Hierarchy and Polynomial Space

We study complexity classes beyond the class *NP*. First, the polynomial-time hierarchy of complexity classes is introduced based on nondeterministic oracle machines. This hierarchy lies between the class *P* and the class *PSPACE*, and the class *NP* is the first level of the hierarchy. Characterizations of these complexity classes in terms of alternating quantifiers and alternating Turing machines are proven. Finally, we present some natural complete problems for this hierarchy and for complexity classes *PSPACE* and *EXP*.

## 3.1   Nondeterministic Oracle Turing Machines

We have defined in the last chapter the notions of polynomial-time Turing reducibility and oracle Turing machines, and have seen that many optimization problems, when formulated in the search problem form, are solvable in polynomial time relative to a set in *NP*. We now extend this notion to nondeterministic oracle Turing machines, and study problems that are solvable in nondeterministic polynomial time relative to sets in *NP*.

A *nondeterministic (function-)oracle Turing machine* (*oracle NTM*) is a nondeterministic Turing machine equipped with an additional *query* tape and two additional states: the *query* state and the *answer* state. The computation of an oracle NTM is similar to that of an oracle DTM, except that at each nonquery state an oracle NTM can make a nondeterministic move. We require that the query step of the computation be a deterministic move determined by the oracle. Let $M$ be an oracle NTM, and $f$ an oracle function. We write $M^f(x)$ to denote the computation of $M$ on input $x$, using $f$ as the oracle function (note that this is a computation tree). If the oracle function is a characteristic function of a set $A$, we say $M$ is a set-oracle NTM and write $M^A$ to denote $M^f$, and write $L(M, A)$ to denote the set of strings accepted by $M^A$.

The time complexity of a set-oracle NTM is also defined similar to that of a set-oracle DTM. In particular, the actions from the query state to the answer state count as only one step. For any fixed oracle set $A$, we let $time_M^A(x)$ be the length of the shortest accepting computation path of $M^A(x)$, and $t_M^A(n) = \max(\{n+1\} \cup \{time_M^A(x) : |x| = n, M^A \text{ accepts } x\})$. For a set-oracle NTM $M$, we say $t_M(n)$ is bounded by a function $g(n)$, if for all oracle sets $A$, $t_M^A(n) \le g(n)$. An oracle NTM $M$ is a *polynomial-time oracle NTM* if $t_M(n)$ is bounded by a polynomial function $p$. Let $A$ be a set, and $C$ be a complexity class. We let $NP^A$ denote the class of sets accepted by polynomial-time oracle NTMs relative to the oracle $A$, and let $NP^C$ (or, $NP(C)$) denote the class of sets accepted by polynomial-time oracle NTMs using an oracle $B \in C$ (i.e., $NP^C = \bigcup_{B \in C} NP^B$).

**Proposition 3.1** *The following hold for all sets $A, B,$ and $C$:*
    *(a)* $A \in NP^A$.
    *(b)* $A \in P^B \Rightarrow A \in NP^B$.
    *(c)* $A \in NP^B \Rightarrow A \in NP^{\overline{B}}$.
    *(d)* $A \in NP^B, B \in P^C \Rightarrow A \in NP^C$.

Let $A \le_T^{NP} B$ denote that $A \in NP^B$. It is provable that $\le_T^{NP}$ does not have the transitivity property and hence is not really a reducibility (see Exercise 4.14). Thus, property (d) above does not fully extend to the case when $B \in NP^C$. In Exercise 3.1, we give a partial extension of this property.

**Proposition 3.2** *(a)* $NP(P) = NP$.
    *(b)* $NP(PSPACE) = PSPACE$.

*Proof.* Part (a) is a special case of Proposition 3.1(d). Part (b) can be proved as a *relativization* of the proof of Theorem 1.28(a), where we showed how an $f(n)$-space bounded DTM $M_1$ can simulate an $f(n)$-time bounded NTM $M$. Now, suppose that $M$ is a $p(n)$-time bounded oracle NTM with an oracle $A \in PSPACE$, where $p(n)$ is a polynomial function. It is easy to verify that we can modify machine $M_1$ into a $p(n)$-space bounded oracle DTM such that it, when given the oracle $A$, accepts the same set as $L(M, A)$. (We say that the proof of Theorem 1.28(a) *relativizes*.) Furthermore, on any input $x$ of length $n$, each query $y$ asked by machine $M_1$ is of length at most $p(n)$. Thus, instead of asking the query of whether $y$ is in $A$, we can further modify machine $M_1$ to simulate a polynomial-space DTM $M_A$ that accepts $A$ on input $y$ to get the correct answer. The total space used for the simulation is only $q(p(n))$ for some polynomial $q$ and so $L(M, A) \in PSPACE$. ∎

## 3.2 Polynomial-Time Hierarchy

The polynomial-time hierarchy is the polynomial analog of the arithmetic hierarchy in recursion theory [Rogers, 1967]. It can be defined inductively by oracle NTMs.

**Definition 3.3** *For integers $n \in \mathbf{N}$, complexity classes $\Delta_n^P$, $\Sigma_n^P$, and $\Pi_n^P$ are defined as follows:*

$$\Sigma_0^P = \Pi_0^P = \Delta_0^P = P,$$
$$\Sigma_{n+1}^P = NP(\Sigma_n^P),$$
$$\Pi_{n+1}^P = co\text{-}\Sigma_{n+1}^P,$$
$$\Delta_{n+1}^P = P(\Sigma_n^P), \quad n \geq 0.$$

*The class PH is defined to be the union of $\Sigma_n^P$ over all $n \geq 0$.*

Thus, $\Sigma_1^P = NP$, $\Sigma_2^P = NP^{NP}$, and $\Sigma_3^P = NP(NP^{NP})$, and so on. It is easy to verify that these classes form a hierarchy.

**Proposition 3.4** *For all $k > 0$,*

$$\Sigma_k^P \cup \Pi_k^P \subseteq \Delta_{k+1}^P \subseteq \Sigma_{k+1}^P \cap \Pi_{k+1}^P \subseteq PSPACE.$$

*Proof.* Note that $P^A = P^{\overline{A}}$, and so $\Pi_k^P \subseteq P(\Pi_k^P) = P(\Sigma_k^P) = \Delta_{k+1}^P$. Other inclusive relations among classes in *PH* follow easily from the definition. Finally, the whole hierarchy *PH* is included in *PSPACE* following from Proposition 3.2(b). ∎

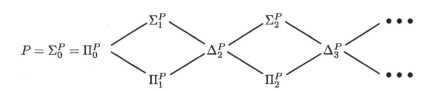

Figure 3.1: The polynomial-time hierarchy

Based on the above proposition, we show in Figure 3.1 the basic structure of the polynomial-time hierarchy. To further understand the structure of the polynomial-time hierarchy, we first extend Theorem 2.1 to a characterization of the polynomial-time hierarchy in terms of the polynomial length-bounded quantifiers.

First, we observe some closure properties of the polynomial-time hierarchy under the Boolean operations.

**Lemma 3.5** *Let $k \geq 0$.*
*(a) If $A, B \in \Sigma_k^P$, then $A \cup B$, $A \cap B \in \Sigma_k^P$, and $\overline{A} \in \Pi_k^P$.*

*(b) If $A, B \in \Pi_k^P$, then $A \cup B$, $A \cap B \in \Pi_k^P$, and $\overline{A} \in \Sigma_k^P$.*
*(c) If $A, B \in \Delta_k^P$, then $A \cup B$, $A \cap B, \overline{A} \in \Delta_k^P$.*

*Proof.* We prove by induction that $A, B \in \Sigma_k^P$ implies $A \cup B \in \Sigma_k^P$. The rest can be proved similarly. First for the case $k = 0$, $\Sigma_0^P = \Pi_0^P = P$, and so the statement is true (Exercise 2.3(a)). Next, assume that the statement is true for integer $k$, and let $A, B \in \Sigma_{k+1}^P$. Then, there are two polynomial-time oracle NTMs $M_A, M_B$, and two sets $C, D \in \Sigma_k^P$, such that $A = L(M_A, C)$ and $B = L(M_B, D)$. Let $p_A$ and $p_B$ be two polynomial functions that bound the runtime $t_{M_A}(n)$ and $t_{M_B}(n)$, respectively. Let $E = \{0x : x \in C\} \cup \{1y : y \in D\}$. (The set $E$ is called the *join* of sets $C$ and $D$.) By the inductive hypothesis, it is clear that $E$ is in $\Sigma_k^P$. Now, it is easy to see that the following oracle NTM $M$ accepts set $A \cup B$ in polynomial time, when set $E$ is used as the oracle:

> *Oracle NTM M*: On input $x$ and oracle $X$, $M$ first simulates $M_A$ on $x$ for $p_A(|x|)$ moves, using the oracle $X_0 = \{z : 0z \in X\}$. If the simulation accepts, then $M$ halts and accepts. Otherwise, it simulates $M_B$ on $x$ for $p_B(|x|)$ moves, using the oracle $X_1 = \{z : 1z \in X\}$. $M$ then accepts $x$ if and only if the simulation accepts.  ∎

The following lemma can be proved in a similar way.

**Lemma 3.6** *For any set $A$ and any $m > 0$, define $A_m = \{\langle x_1, x_2, \ldots, x_m \rangle : (\exists i, 1 \le i \le m)\ x_i \in A\}$ and $A_m' = \{\langle x_1, x_2, \ldots, x_m \rangle : (\forall i, 1 \le i \le m)\ x_i \in A\}$. For any $k, m > 0$, if $A \in \Sigma_k^P$ then $A_m, A_m' \in \Sigma_k^P$ and if $A \in \Pi_k^P$ then $A_m, A_m' \in \Pi_k^P$.*

In the following, we say a predicate $Q(x_1, \ldots, x_n)$ of $n$ variables $x_1, \ldots, x_n$ is a $\Sigma_k^P$-*predicate* (or, a $\Pi_k^P$-*predicate*), where $k \ge 0$, if the set $A = \{\langle x_1, \ldots, x_n \rangle : Q(x_1, \ldots, x_n)\}$ is in $\Sigma_k^P$ (or, respectively, in $\Pi_k^P$). Lemma 3.5 then states that if $Q_1$ and $Q_2$ are two $\Sigma_k^P$-predicates (or, $\Pi_k^P$-predicates), then $Q_1 \wedge Q_2$ and $Q_1 \vee Q_2$ are also $\Sigma_k^P$-predicates (or, $\Pi_k^P$-predicates), and $\neg Q_1$ is a $\Pi_k^P$-predicate (or, respectively, $\Sigma_k^P$-predicate).

**Lemma 3.7** *Let $k \ge 1$. A set $A$ is in $\Sigma_k^P$ if and only if there exist a set $B \in \Pi_{k-1}^P$ and a polynomial function $p$ such that for every $x \in \Sigma^*$,*

$$x \in A \iff (\exists y, |y| \le p(n))\langle x, y \rangle \in B. \tag{3.1}$$

*Proof.* We prove the lemma by induction. For the case $k = 1$, we note that $\Sigma_1^P = NP$ and $\Pi_0^P = P$, and so this is just Theorem 2.1.

For the inductive step, we assume that the lemma holds for all sets $A$ in $\Sigma_k^P$. First, for the backward direction, assume that there exist a set $B \in \Pi_k^P$ and a polynomial function $p$ satisfying (3.1). We first observe that $NP^B \subseteq NP(\Sigma_k^P) = \Sigma_{k+1}^P$ (Proposition 3.1(c)). Now, we design an oracle NTM $M$ that behaves as follows:

On input $x$, and with oracle $C$, the machine $M$ nondetermin-
istically guesses a string $y$ of length $\leq p(|x|)$, and checks that
$\langle x, y \rangle \in C$. $M$ accepts $x$ if and only if $\langle x, y \rangle \in C$.

It is clear that when $M$ uses set $B$ as the oracle, it accepts exactly the set
$A$. Furthermore, $M$ operates in polynomial time, and so $A$ is in $NP^B \subseteq \Sigma_{k+1}^P$.

Next, for the forward direction of the inductive step, we assume that $M = (Q, q_0, q_?, q_Y, q_N, F, \Sigma, \Gamma, \Delta)$ is a polynomial-time oracle NTM that accepts $A$
relative to a set $D \in \Sigma_k^P$. Also assume that $M$ has a polynomial time bound
$q$. Let us consider the computation of $M^D(x)$. For each $x \in \Sigma^*$ of length
$n$, $x \in A$ if and only if there is an accepting computation path of $M^D(x)$
of length $\leq q(n)$. This computation path can be encoded as a sequence of
configurations $z = \langle \alpha_0, \alpha_1, \ldots, \alpha_m \rangle$, with $m \leq q(n)$, and $|\alpha_i| \leq q(n)$ for each
$i$, $0 \leq i \leq m$, such that

(a) $\alpha_0$ is the initial configuration of $M^D(x)$;

(b) $\alpha_m$ is in a final state $q_f \in F$;

(c) For each $i$, $0 \leq i < m$, such that $\alpha_i$ is not in the query state $q_?$, $\alpha_i \vdash \alpha_{i+1}$;
and

(d) For each $i$, $0 \leq i < m$, if $\alpha_i$ is in the query state $q_?$, and $query(\alpha_i) \in D$
then $\alpha_i \vdash^{yes} \alpha_{i+1}$; if $\alpha_i$ is in the state $q_?$, and $query(\alpha_i) \notin D$ then
$\alpha_i \vdash^{no} \alpha_{i+1}$, where $query(\alpha_i)$ denotes the string on the query tape of
$\alpha_i$ (recall the notations $\alpha_i \vdash^{yes} \alpha_{i+1}$ and $\alpha_i \vdash^{no} \alpha_{i+1}$ defined in Section
2.4).

For a given sequence of configurations of $M$, how do we check whether it
satisfies the above conditions? Conditions (a), (b), and (c) can be checked
by a *deterministic* TM in time polynomial in $n$. Condition (d) involves the
oracle set $D$, and it is harder to verify by an NTM without using the oracle $D$.
We can simplify our task by encoding the query strings in the computation of
$M^D(x)$ into two finite lists $u = \langle u_1, u_2, \ldots, u_I \rangle$ and $v = \langle v_1, v_2, \ldots, v_J \rangle$ and
replacing the condition (d) by the following four new conditions:

(e) For each $i$, $1 \leq i < m$, if $\alpha_i$ is in the query state, then $query(\alpha_i)$ is
either in the list $u$ or in the list $v$;

(f) For each $i$, $1 \leq i < m$, such that $\alpha_i$ is in the query state $q_?$, if $query(\alpha_i)$
is in the list $u$ then $\alpha_i \vdash^{yes} \alpha_{i+1}$, and if $query(\alpha_i)$ is in the list $v$ then
$\alpha_i \vdash^{no} \alpha_{i+1}$;

(g) All strings $u_i$, $1 \leq i \leq I$, in the list $u$ are in set $D$; and

(h) None of the strings $v_j$, $1 \leq j \leq J$, in the list $v$ is in set $D$.

Now, conditions (e) and (f) can also be verified by a DTM in time polynomial in $n$. That is, the set

$$B_1 = \{\langle x, z, u, v \rangle : x, z, u, v \text{ satisfy conditions (a), (b), (c), (e), and (f)}\}$$

is in $P$. In addition, since $D \in \Sigma_k^P$, the predicate $[v_i \in \overline{D}]$ is a $\Pi_k^P$-predicate, and so it follows from Lemma 3.6 that set

$$B_2 = \{v : v \text{ satisfies condition (h)}\}$$

is in $\Pi_k^P$.

For the condition (g), we note from the inductive hypothesis that there exist a set $E \in \Pi_{k-1}^P$ and a polynomial $r$ such that for each string $s$, $s \in D$ if and only if $(\exists t, |t| \leq r(|s|)) \langle s, t \rangle \in E$. Thus, condition (g) is equivalent to

(g′) There exists a string $w = \langle w_1, w_2, \ldots, w_I \rangle$, such that for each $i$, $1 \leq i \leq I$, $|w_i| \leq r(|u_i|)$, and $\langle u_i, w_i \rangle \in E$.

From this condition (g′), we define

$$B_3 = \{\langle u, w \rangle : (\forall i, 1 \leq i \leq I) \, |w_i| \leq r(|u_i|), \langle u_i, w_i \rangle \in E\}.$$

By Lemma 3.6, $B_3 \in \Pi_{k-1}^P \subseteq \Pi_k^P$.

We are now ready to see that $A$ satisfies (3.1):

$$
\begin{aligned}
x \in A \iff & (\exists z) \langle x, z \rangle \text{ satisfies (a), (b), (c), (d)} \\
\iff & (\exists \langle z, u, v \rangle) \langle x, z, u, v \rangle \text{ satisfies (a), (b), (c), (e), (f), (g), (h)} \\
\iff & (\exists y = \langle z, u, v, w \rangle) [\langle x, z, u, v \rangle \in B_1, \, v \in B_2, \, \langle u, w \rangle \in B_3].
\end{aligned}
$$

From Lemma 3.5, the predicate $[\langle x, z, u, v \rangle \in B_1] \wedge [v \in B_2] \wedge [\langle u, w \rangle \in B_3]$ is a $\Pi_k^P$-predicate. Furthermore, if $x \in A$, then there must be a witness string $y$ of length bounded by $p(n)$ for some polynomial $p$, since each of $z, u, v, w$ encodes a list of at most $q(n)$ strings, each of length at most $q(n)$. Thus, we can add a bound $p(n)$ to $y$, and obtain

$$x \in A \iff (\exists y, |y| \leq p(|x|)) \langle x, y \rangle \in B,$$

where $B = \{\langle x, y \rangle : y = \langle z, u, v, w \rangle, \langle x, z, u, v \rangle \in B_1, v \in B_2, \text{ and } \langle u, w \rangle \in B_3\} \in \Pi_k^P$. This completes the proof of the inductive step and, hence, the lemma. ∎

**Theorem 3.8** Let $k \geq 1$. A set $A$ is in $\Sigma_k^P$ if and only if there exist a set $B \in P$ and a polynomial function $p$ such that for every $x \in \Sigma^*$ of length $n$,

$$
\begin{aligned}
x \in A \iff & (\exists y_1, |y_1| \leq p(n))(\forall y_2, |y_2| \leq p(n)) \cdots \\
& (Q_k y_k, |y_k| \leq p(n)) \langle x, y_1, y_2, \ldots, y_k \rangle \in B,
\end{aligned}
\tag{3.2}
$$

where $Q_k$ is the quantifier $\exists$ if $k$ is odd, and it is $\forall$ if $k$ is even.

*Proof.* The theorem can be proved by induction, using Lemma 3.7. The case $k = 1$ is simply Theorem 2.1. For the inductive step, assume, for some $k \geq 1$, that the theorem holds for all sets in $\Sigma_k^P$. Also assume that $A \in \Sigma_{k+1}^P$. Then, by Lemma 3.7, there exists a set $C \in \Pi_k^P$ and a polynomial function $q$ such that $x \in A$ if and only if $(\exists y_1, |y_1| \le q(|x|)) \langle x, y_1 \rangle \in C$. Now, from the inductive hypothesis, there exist a set $D \in P$ and a polynomial $r$ such that the following is true for all $x$ and $y_1$, with $|y_1| \le q(|x|)$:

$$\langle x, y_1 \rangle \in \overline{C} \iff (\exists y_2, |y_2| \le r(|\langle x, y_1 \rangle|)) \cdots (Q_k y_{k+1}, |y_{k+1}| \le r(|\langle x, y_1 \rangle|))$$
$$\langle \langle x, y_1 \rangle, y_2, \dots, y_{k+1} \rangle \notin D.$$

Equivalently,

$$\langle x, y_1 \rangle \in C \iff (\forall y_2, |y_2| \le r(|\langle x, y_1 \rangle|)) \cdots (Q_{k+1} y_{k+1}, |y_{k+1}| \le r(|\langle x, y_1 \rangle|))$$
$$\langle \langle x, y_1 \rangle, y_2, \dots, y_{k+1} \rangle \in D.$$

Assume that our pairing function on strings satisfies $|\langle x, y \rangle| \le p_1(|x| + |y|)$. Let $B = \{\langle x, y_1, \dots, y_{k+1} \rangle : \langle \langle x, y_1 \rangle, y_2, \dots, y_{k+1} \rangle \in D\}$, and $p(n) = r(p_1(n + q(n)))$. Then, sets $A, B$, and polynomial $p$ satisfy (3.2).

The converse direction is easy and is left as an exercise. ∎

Using the above characterization, we can prove that if any two levels of the polynomial-time hierarchy collapses then the whole hierarchy collapses to that level.

**Theorem 3.9** *For every $k \ge 0$, $NP(\Sigma_k^P \cap \Pi_k^P) = \Sigma_k^P$.*

*Proof.* Assume that $D \in \Sigma_k^P \cap \Pi_k^P$, and $M$ is an oracle NTM with a polynomial time bound $p$. We need to show that $L(M, D)$ is in $\Sigma_k^P$. First, from the proof of Lemma 3.7, we know that for any string $x$ of length $n$, $M^D$ accepts $x$ if and only if there exist a sequence of at most $p(n)$ configurations $z = \langle \alpha_0, \alpha_1, \dots, \alpha_m \rangle$ and two lists $u, v$ of strings such that conditions (a), (b), (c), (e), (f), (g), (h) in the proof of Lemma 3.7 are satisfied. Now, $D \in \Sigma_k^P$ implies that condition (g) is a $\Sigma_k^P$-predicate and $D \in \Pi_k^P$ implies that condition (h) also is a $\Sigma_k^P$-predicate. By Lemma 3.5, these two $\Sigma_k^P$-predicates and the other five $P$-predicates can be combined into a single $\Sigma_k^P$-predicate $R(x, z, u, v)$. Now from Lemma 3.7, there exist a set $B \in \Pi_{k-1}^P$ and a polynomial $q$ such that

$$R(x, z, u, v) \iff (\exists w, |w| \le q(|\langle x, z, u, v \rangle|)) \langle x, z, u, v, w \rangle \in B.$$

It follows that

$$x \in L(M, D) \iff (\exists \langle z, u, v, w \rangle, |\langle z, u, v, w \rangle| \le r(|x|)) \langle x, z, u, v, w \rangle \in B,$$

for some polynomial function $r$. The theorem now follows from Lemma 3.7. ∎

**Corollary 3.10** *For every $k \ge 1$, $\Sigma_k^P = \Pi_k^P$ implies $PH = \Sigma_k^P$ and $\Sigma_k^P = \Delta_k^P$ implies $PH = \Delta_k^P$.*

*Proof.* We prove this by induction. Assume that $\Sigma_k^P = \Pi_k^P$, and $j \geq k$. If $\Sigma_j^P = \Sigma_k^P$, then, by Theorem 3.9, $\Sigma_{j+1}^P = NP(\Sigma_j^P) = NP(\Sigma_k^P) = NP(\Sigma_k^P \cap \Pi_k^P) = \Sigma_k^P$.

For the second part, if $\Sigma_k^P = \Delta_k^P$, then $\Sigma_k^P = \Pi_k^P$ since $\Delta_k^P$ is closed under complementation. It follows then $PH = \Sigma_k^P = \Delta_k^P$.                ∎

The above corollary reveals an interesting property of the polynomial-time hierarchy; namely, if any two levels of the hierarchy collapses into one, then the whole hierarchy collapses to that level. Still, the main question of whether the hierarchy is properly infinite is not known. Many questions about relations between complexity classes are often reduced to the question of whether the polynomial-time hierarchy collapses. Notice that the assumption of the polynomial-time hierarchy not collapsing is stronger than the assumption of $P \neq NP$. Thus, when the assumption of $P \neq NP$ is not sufficient to establish the intractability of a problem $A$, this stronger assumption might be useful. We will present some applications in Chapters 6 and 10.

## 3.3  Complete Problems in $PH$

We have proved in Example 2.20 that the problem EXACT-CLIQUE is in $P(NP) = \Delta_2^P$, and it is $\leq_T^P$-hard for $NP$. Therefore, it is $\leq_T^P$-complete for $\Delta_2^P$, since all problems in $\Delta_2^P$ are $\leq_T^P$-reducible to a problem in $NP$. For most $NP$-complete optimization problems, it can be easily shown that the corresponding version of the decision problem of determining whether a given integer $K$ is the size of the optimum solutions is $\leq_T^P$-complete for $\Delta_2^P$.

In addition to complete problems in $\Delta_2^P$, there are also natural problems complete for the classes in the higher levels of the polynomial-time hierarchy. First, we show that the generic $NP$-complete problem BOUNDED HALTING PROBLEM, or BHP, has relativized versions that are complete for classes $\Sigma_k^P$ for $k > 1$. Let $A$ be an arbitrary set.

> BHP relative to set $A$ (BHP$^A$): Given an oracle NTM $M$, an input $w$, and a time bound $t$, written in the unary form $0^t$, determine whether $M^A$ accepts $w$ in $t$ moves.

First, a simple modification of the universal NTM yields a universal oracle NTM $M_u$ that can simulate any oracle NTM in polynomial time. Thus, BHP$^A \in NP^A$. Next, we observe that the reduction in Theorem 2.11 from any set $B \in NP$ to BHP relativizes. So, for any set $A$, BHP$^A$ is $\leq_m^P$-complete for the class $NP^A$.

Now, define $A_0 = \emptyset$, and for $k \geq 0$, $A_{k+1} = $ BHP$^{A_k}$.

**Lemma 3.11** *If $A$ is $\leq_T^P$-complete for $\Sigma_k^P$ for some $k \geq 1$, then $NP^A = \Sigma_{k+1}^P$.*

*Proof.* $B \in \Sigma_{k+1}^P$ means that $B \in NP^C$ for some $C \in \Sigma_k^P$. So, by Proposition 3.1(d), $B \in NP^C$ and $C \in P^A$ imply $B \in NP^A$.                ∎

**Theorem 3.12** *For each $k \geq 1$, $A_k$ is $\leq_m^P$-complete for $\Sigma_k^P$.*

*Proof.* We show this by induction. First, $A_1 = \text{BHP}^{\emptyset}$, and so by the relativized version of Theorem 2.11, $A_1$ is $\leq_m^P$-complete for $NP^{\emptyset} = NP$.

Next, assume that $A_k$ is $\leq_m^P$-complete for $\Sigma_k^P$. Then, from Lemma 3.11, $A_{k+1} = \text{BHP}^{A_k}$ is $\leq_m^P$-complete for $NP^{A_k} = \Sigma_{k+1}^P$. ∎

Next, for natural complete problems, we show that for each $k \geq 1$, there is a generalization of the problem SAT that is complete for the class $\Sigma_k^P$. In the following, we write $\tau : X \to \{0, 1\}$ to mean a Boolean assignment $\tau$ defined on variables in the set $X$. Let $k \geq 1$.

> SAT$_k$: Given $k$ sets of variables $X_1, X_2, \ldots, X_k$, and a Boolean formula $F$ over variables in $X = \bigcup_{i=1}^k X_i$, determine whether it is true that
>
> $$(\exists \tau_1 : X_1 \to \{0, 1\})(\forall \tau_2 : X_2 \to \{0, 1\}) \cdots$$
> $$(Q_k \tau_k : X_k \to \{0, 1\}) \, F|_{\tau_1, \tau_2, \ldots, \tau_k} = 1, \tag{3.3}$$
>
> where $Q_k$ denotes $\exists$ if $k$ is odd, and it denotes $\forall$ if $k$ is even.

In the above, $F|_{\tau_1, \tau_2, \ldots, \tau_k} = 1$ denotes that the combined assignment $\tau = (\tau_1, \tau_2, \ldots, \tau_k)$ satisfies $F$.

**Theorem 3.13** *For each $k \geq 1$, SAT$_k$ is $\leq_m^P$-complete for $\Sigma_k^P$.*

*Proof.* The proof is a generalization of Cook's Theorem. From the characterization of Theorem 3.8, it is clear that SAT$_k$ is in $\Sigma_k^P$. So, we only need to show that for each set $A \in \Sigma_k^P$, there is a polynomial-time computable function $f$ mapping each instance $w$ of $A$ to a Boolean formula $F_w$ such that $w \in A$ if and only if $F_w$ satisfies (3.3).

First, we consider the case where $k$ is odd. Let $A \subseteq \Sigma^*$ be a set in $\Sigma_k^P$. From Theorem 3.8, there exist a set $B \in P$ and a polynomial $q$ such that for any $w$ of length $n$, $w \in A$ if and only if

$$(\exists u_1, |u_1| = q(n)) \, (\forall u_2, |u_2| = q(n)) \cdots$$
$$(\exists u_k, |u_k| = q(n)) \, \langle w, u_1, \ldots, u_k \rangle \in B, \tag{3.4}$$

where each $u_\ell$, $1 \leq \ell \leq k$, is a string in $\{0, 1\}^*$. Notice that in the above, we have made a minor change from Theorem 3.8. That is, we require that all strings $u_\ell$, $1 \leq \ell \leq k$, are strings from $\{0, 1\}^*$ of length exactly $q(n)$. This is equivalent to the requirement in Theorem 3.8, since we can first encode all strings in $\Sigma^*$ of length $\leq p(n)$ by strings in $\{0, 1\}^*$ of length $\leq c \cdot p(n)$ for some constant $c > 0$, and then add, for each string of length $cp(n) - k$, $k \geq 0$, the string $10^k$ to its right to make the length exactly $q(n) = cp(n) + 1$, with the trailing substring $10^k$ removable easily.

Assume that each input $\langle w, u_1, u_2, \ldots, u_k \rangle$ to $B$ is encoded as $w\#u_1\#u_2\#\cdots\#u_k$, with $w \in \Sigma^*$, each $u_\ell$, $1 \leq \ell \leq k$, in $\{0, 1\}^*$, and $\#$ a new symbol not in $\Sigma$. Also assume that for such an input of length $kq(n)+n+k$, there is a DTM $M$ that halts on such an input in time $p(n)$ and accepts exactly those in $B$. Let us review Cook's Theorem on machine $M$, which showed that for each such input $v = w\#u_1\#\cdots\#u_k$, there is a Boolean formula $G$ over variables in a set $Y$ such that $v \in B$ if and only if there is a truth assignment $\tau$ on variables in $Y$ such that $G|_\tau = 1$. The variables in $Y$ are of the form $y_{i,j,a}$, with $0 \le i \le p(n)$, $1 \le j \le p(n)$, and $a \in \Gamma' = \Gamma \cup (Q \times \Gamma)$, where $\Gamma$ is the worktape alphabet of $M$ and $Q$ is the set of states of $M$. The intended interpretation of $y_{i,j,a}$'s is that $y_{i,j,a} = 1$ if and only if the $j$th symbol of the $i$th configuration $\alpha_i$ of the computation of $M(v)$ is the symbol $a$. The formula $G$ is the product of four subformulas $\phi_1, \phi_2, \phi_3, \phi_4$, where $\phi_1, \phi_3, \phi_4$ are independent of the input $v$, and $\phi_2$ asserts that the input is $v$.

To be more precise, let $m(\ell) = (\ell-1)(q(n)+1)+n+1$ for each $1 \le \ell \le k$, and let $u_{\ell,j}$, with $1 \le \ell \le k$ and $1 \le j \le q(n)$, denote the $j$th symbol of the word $u_\ell$; then we have $\phi_2 = \phi_{2,1}\phi_{2,2}$, where

$$\phi_{2,1} = y_{0,1,\langle q_0,w_1\rangle} \prod_{j=2}^{n} y_{0,j,w_j} \prod_{\ell=1}^{k} y_{0,m(\ell),\#} \prod_{j=kq(n)+n+k+1}^{p(n)} y_{0,j,B},$$

and

$$\phi_{2,2} = \prod_{\ell=1}^{k} \prod_{j=1}^{q(n)} y_{0,m(\ell)+j,u_{\ell,j}}.$$

For our reduction, we define, in addition to variables in $Y$, a set of new variables $z_{\ell,j}$, with $1 \le \ell \le k$ and $1 \le j \le q(n)$. These variables are to be interpreted as follows: $z_{\ell,j} = 1$ if and only if $u_{\ell,j} = 1$. We use these new variables to define a new formula $\phi_{2,3}$ to replace formula $\phi_{2,2}$:

$$\phi_{2,3} = \prod_{\ell=1}^{k} \prod_{j=1}^{q(n)} (\bar{z}_{\ell,j} + y_{0,m(\ell)+j,1}) (z_{\ell,j} + y_{0,m(\ell)+j,0}).$$

In other words, we require that the input string $v$ to $B$ be equal to $w\#u_1\#\cdots\#u_k$, using the above interpretation on $z$-variables. Now we are ready to define the reduction function: For each $w$ of length $n$, we define, for each $\ell$, $1 \le \ell < k$, $X_\ell = \{z_{\ell,j} : 1 \le j \le q(n)\}$, and we let $X_k = \{z_{k,j} : 1 \le j \le q(n)\} \cup Y$. We also define $F_w = \phi_1\phi_3\phi_4\phi_{2,1}\phi_{2,3}$.

Following the above interpretations of the variables, it is easy to see that there is a one-to-one correspondence between the assignments $\tau_\ell$ on all variables $z_{\ell,j}$ and the strings $u_\ell \in \{0,1\}^*$ of length $q(n)$ such that (3.3) holds for $F_w$ if and only if (3.4) holds. Therefore, this is a reduction from $A$ to $SAT_k$.

Next, for the case of even $k \ge 2$, we note that set $A \in \Sigma_k^P$ can be characterized as follows: $w \notin A$ if and only if

$$(\forall u_1, |u_1| = q(n)) \, (\exists u_2, |u_2| = q(n))$$
$$\cdots (\exists u_k, |u_k| = q(n)) \, \langle x, u_1, \ldots, u_k\rangle \notin B, \tag{3.5}$$

where each $u_\ell$, $1 \leq \ell \leq k$, is in $\{0,1\}^*$. Now, from a similar reduction as above, we can find a formula $F_w$ such that $w \notin A$ if and only if

$$
(\forall \tau_1 : X_1 \to \{0,1\}) \, (\exists \tau_2 : X_2 \to \{0,1\}) \cdots \\
(\exists \tau_k : X_k \to \{0,1\}) \, F_w|_{\tau_1,\ldots,\tau_k} = 1. \tag{3.6}
$$

(The formula $F_w$ is defined from a *rejecting* computation of $M$ for $B$.) That is, $w \in A$ if and only if (3.6) does not hold, or if and only if the following holds:

$$
(\exists \tau_1 : X_1 \to \{0,1\}) \, (\forall \tau_2 : X_2 \to \{0,1\}) \cdots \\
(\forall \tau_k : X_k \to \{0,1\}) \, F_w|_{\tau_1,\ldots,\tau_k} = 0.
$$

So, the mapping from $w$ to $\neg F_w$ is a reduction from $A$ to $\text{SAT}_k$.     ∎

Again, for the reductions from $\text{SAT}_k$ to other $\Sigma_k^P$-complete problems, the restricted forms of $\text{SAT}_k$ would be more useful. Recall that a 3-CNF formula is a formula in CNF such that each clause has exactly three literals defined from three distinct variables. A 3-DNF formula is a formula in DNF such that each term has exactly three literals defined from three distinct variables. We let 3-CNF-$\text{SAT}_k$ (3-DNF-$\text{SAT}_k$) denote the problem of determining whether a given 3-CNF (3-DNF, respectively) formula $F$ satisfies (3.3).

**Corollary 3.14** *(a) For each odd $k \geq 1$, 3-CNF-$\text{SAT}_k$ is $\leq_m^P$-complete for $\Sigma_k^P$, and 3-DNF-$\text{SAT}_k$ is $\leq_m^P$-complete for $\Pi_k^P$*
*(b) For each even $k \geq 2$, 3-DNF-$\text{SAT}_k$ is $\leq_m^P$-complete for $\Sigma_k^P$, and 3-CNF-$\text{SAT}_k$ is $\leq_m^P$-complete for $\Pi_k^P$*

*Proof.* We observe that the new subformula $\phi_{2,3}$ in the above proof is in CNF. Therefore, the first half of part (a) follows. For the other three claims, we notice that a formula $F$ is in CNF if and only if $\neg F$ is in DNF.     ∎

In addition to $\text{SAT}_k$, there are a few natural problems known to be complete for the complexity classes in the first three levels of the polynomial-time hierarchy, but very few known to be complete for classes in the higher levels. In the following, we show a generalization of the maximum clique problem to be $\leq_m^P$-complete for $\Pi_2^P$. Some other examples are included in exercises.

We say that a graph $G = (V, E)$ is *k-colored* if it is associated with a function $c : E \to \{1, 2, \ldots, k\}$. The Ramsey number $R_k$, for each integer $k > 0$, is the minimum integer $n$ such that every 2-colored complete graph $G$ of size $n$ contains a monochromatic clique $Q$ of size $k$ (i.e., all edges between two vertices in $Q$ are of the same color). The existence of the number $R_k$ for each $k > 0$ is the famous Ramsey Theorem. A generalized form of the question of computing the Ramsey numbers is the following:

GENERALIZED RAMSEY NUMBER (GRN): Given a complete graph $G = (V, E)$ with its edges partially 2-colored (i.e., it has a function $c : E \to \{0, 1, *\}$), and given an integer $K > 0$, determine whether it is true that for any 2-colored restriction $c'$ of $c$ (i.e., $c'(e) \in \{0,1\}$ for all $e \in E$ and $c'(e) = c(e)$ whenever $c(e) \neq *$), there is a monochromatic clique in $G$ of size $K$.

**Theorem 3.15** *The problem* GRN *is* $\leq_m^P$-*complete for* $\Pi_2^P$.

*Proof.* The fact that GRN is in $\Pi_2^P$ is obvious, since both the restricted coloring function $c'$ and the monochromatic clique $Q$ have size $\leq |E|$. We need to construct a reduction from 3-CNF-SAT$_2$ to GRN.

Let $F$ be a 3-CNF formula over variables $X_1 = \{x_1, \ldots, x_n\}$ and $X_2 = \{y_1, \ldots, y_m\}$. Assume that $F = C_1 C_2 \cdots C_k$, where each $C_i$ is the sum of three literals. We further assume that $n \geq 2$ and $k \geq 3$. Let $K = 2n + k$. The graph $G$ has $N = 6n + 4k - 4$ vertices. We divide them into three groups: $V_X = \{x_{i,j}, \bar{x}_{i,j} : 1 \leq i \leq n, 1 \leq j \leq 2\}$, $V_C = \{c_{i,j} : 1 \leq i \leq k, 1 \leq j \leq 3\}$, and $V_R = \{r_i : 1 \leq i \leq 2n + k - 4\}$. The partial coloring $c$ on the edges of $G$ is defined as follows (we use colors blue and red instead of 0 and 1):

(1) The edges among $x_{i,1}$, $x_{i,2}$, $\bar{x}_{i,1}$, and $\bar{x}_{i,2}$, for each $i$, $1 \leq i \leq n$, are colored by red, except that the edges $e_i = \{x_{i,1}, x_{i,2}\}$ and $\bar{e}_i = \{\bar{x}_{i,1}, \bar{x}_{i,2}\}$ are not colored (i.e., $c(e_i) = c(\bar{e}_i) = *$).

(2) All other edges between two vertices in $V_X$ are colored by blue; that is, $c(\{x_{i,j}, x_{i',j'}\}) = c(\{x_{i,j}, \bar{x}_{i',j'}\}) = c(\{\bar{x}_{i,j}, \bar{x}_{i',j'}\}) = $ blue if $i \neq i'$.

(3) All edges among vertices in $V_R$ are colored by red.

(4) For each $i$, $1 \leq i \leq k$, the three edges among $c_{i,1}, c_{i,2}$, and $c_{i,3}$ are colored by red.

(5) The edge between two vertices $c_{i,j}$ and $c_{i',j'}$, where $i \neq i'$, is colored by red if the $j$th literal of $C_i$ and the $j'$th literal of $C_{i'}$ are complementary (i.e., one is $x_q$ and the other is $\bar{x}_q$, or one is $y_q$ and the other is $\bar{y}_q$ for some $q$). Otherwise, it is colored by blue.

(6) The edge between any vertex in $V_R$ and any vertex in $V_X$ is colored by red, and the edge between any vertex in $V_R$ and any vertex in $V_C$ is colored by blue.

(7) For each vertex $c_{i,j}$ in $V_C$, if the $j$th literal of $C_i$ is $y_q$ or $\bar{y}_q$ for some $q$, then all edges between $c_{i,j}$ and any vertex in $V_X$ are colored by blue. If the $j$th literal of $C_i$ is $x_q$ for some $q$, then all edges between $c_{i,j}$ and any vertex in $V_X$, except $\bar{x}_{q,1}$ and $\bar{x}_{q,2}$, are colored by blue, and $c(\{c_{i,j}, \bar{x}_{q,1}\}) = c(\{c_{i,j}, \bar{x}_{q,2}\}) = $ red. The case where the $j$th literal of $C_i$ is $\bar{x}_q$ for some $q$ is symmetric; that is, all edges between $c_{i,j}$ and any vertex in $V_X$, except $x_{q,1}$ and $x_{q,2}$, are colored by blue, and $c(\{c_{i,j}, x_{q,1}\}) = c(\{c_{i,j}, x_{q,2}\}) = $ red.

The above completes the construction of the graph $G$ and its partial coloring $c$. Notice that the partial coloring $c$ has $c(e) \neq *$ for all edges $e$ except $e_i$ and $\bar{e}_i$, for $1 \leq i \leq n$. Now we prove that this construction is correct. First assume that for each assignment $\tau_1 : X_1 \to \{0,1\}$, there is an assignment $\tau_2 : X_2 \to \{0,1\}$ such that $F|_{\tau_1,\tau_2} = 1$. We verify that for any two-coloring restriction $c'$ of $c$, there must be a size-$K$ monochromatic clique $Q$.

We note that if $c'(e_i) = c'(\bar{e}_i) = $ red for some $i \leq n$, then the vertices $x_{i,1}, x_{i,2}, \bar{x}_{i,1}, \bar{x}_{i,2}$, together with vertices in $V_R$, form a red clique of size

$|V_R| + 4 = K$. Therefore, we may assume that for each $i$, $1 \leq i \leq n$, at least one of $c'(e_i)$ and $c'(\bar{e}_i)$ is blue. Now we define an assignment $\tau_1$ on variables in $X_1$ by $\tau_1(x_i) = 1$ if and only if $c'(e_i) = $ blue. For this assignment $\tau_1$, there is an assignment $\tau_2$ on variables in $X_2$ such that each clause $C_i$ has a true literal. For each $i$, $1 \leq i \leq k$, let $j_i$ be the least $j$, $1 \leq j \leq 3$, such that the $j$th literal of $C_i$ is true to $\tau_1$ and $\tau_2$. Let $Q_C = \{c_{i,j_i} : 1 \leq i \leq k\}$ and $Q_X = \{x_{i,j} : c'(e_i) = \text{blue}, 1 \leq j \leq 2\} \cup \{\bar{x}_{i,j} : c'(e_i) = \text{red}, 1 \leq j \leq 2\}$. Let $Q = Q_C \cup Q_X$. It is clear that $Q$ is of size $2n + k$. Furthermore, $Q$ is a blue clique: (i) every two vertices in $Q_C$ are connected by a blue edge because they both have value true under $\tau_1$ and $\tau_2$ and so are not complementary; (ii) every two vertices in $Q_X$ are connected by a blue edge by the definition of $Q_X$, and (iii) if a vertex $c_{i,j_i}$ in $Q_C$ corresponds to a literal $x_q$, then $\tau_1(x_q) = 1$ and so $\bar{x}_{q,1}, \bar{x}_{q,2} \notin Q_X$ and hence all the edges between $c_{i,j_i}$ and each of $x_{i',j'}$ or $\bar{x}_{i',j'} \in Q_X$ are colored blue. A similar argument works for the case $c_{i,j_i} = \bar{x}_q$.

Conversely, assume that there exists an assignment $\tau_1$ on $X_1$ such that for all assignments $\tau_2$ on $X_2$, $F|_{\tau_1,\tau_2} = 0$. Then, consider the following coloring $c'$ on edges $e_i$ and $\bar{e}_i$: $c'(e_i) = $ blue and $c'(\bar{e}_i) = $ red if $\tau_1(x_i) = 1$, and $c'(e_i) = $ red and $c'(\bar{e}_i) = $ blue if $\tau_1(x_i) = 0$. By the definition of $c'$, the largest red clique in $V_X$ is of size 3. Also, the largest red clique in $V_C$ is of size 3, since every edge connecting two noncomplementary literals in two different clauses is colored by blue. Thus, the largest red clique containing $V_R$ is of size $K - 1$, and the largest red clique containing at least one vertex of $V_C$ is of size $\leq 6 < K$.

Next, assume by way of contradiction that there is a blue clique $Q$ of $G$ of size $K$. From our coloring, it is clear that, for each $i$, $1 \leq i \leq n$, $Q$ contains exactly two vertices in $\{x_{i,1}, x_{i,2}, \bar{x}_{i,1}, \bar{x}_{i,2}\}$, and for each $i$, $1 \leq i \leq k$, $Q$ contains exactly one $c_{i,j_i}$, for some $1 \leq j_i \leq 3$. Define $\tau_2 : X_2 \to \{0,1\}$ by $\tau_2(y_q) = 1$ if and only if the $j_i$th literal of $C_i$ is $y_q$ for some $i$, $1 \leq i \leq k$. Then, there is a clause $C_i$ such that $C_i$ is not satisfied by $\tau_1$ and $\tau_2$. In particular, the $j_i$th literal of $C_i$ is false to $\tau_1$ and $\tau_2$.

*Case 1.* The $j_i$th literal of $C_i$ is $x_q$ for some $q$. Then, $\tau_1(x_q) = 0$, and so $c'(e_q) = $ red, and the edges between $c_{i,j_i}$ and each of $\bar{x}_{q,1}$ and $\bar{x}_{q,2}$ are red. This contradicts the above observation that $Q$ contains two vertices in $\{x_{q,1}, x_{q,2}, \bar{x}_{q,1}, \bar{x}_{q,2}\}$.

*Case 2.* The $j_i$th literal of $C_i$ is $\bar{x}_q$ for some $q$. This is symmetric to Case 1.

*Case 3.* The $j_i$th literal of $C_i$ is $y_q$ for some $q$. This is not possible, because by the definition of $\tau_2$, $\tau_2(y_q) = 1$, but by the property that $C_i$ is not satisfied by $\tau_1$ and $\tau_2$, $\tau_2(y_q) = 0$.

*Case 4.* The $j_i$th literal of $C_i$ is $\bar{y}_q$ for some $q$. Then, $\tau_2(y_q) = 1$ and, hence, by the definition of $\tau_2$, there must be another $i' \leq k$, $i' \neq i$, such that $c_{i',j_{i'}}$ is in $Q$ and the $j_{i'}$th literal of $C_{i'}$ is $y_q$. So, the edge between $c_{i,j_i}$ and $c_{i',j_{i'}}$ is colored by red. This is again a contradiction.

The above case analysis shows that there is no blue clique in $G$ of size $K$ either. So the theorem is proven. ∎

## 3.4   Alternating Turing Machines

The polynomial-time hierarchy was formally defined by oracle Turing machines. Since the oracles play a mysterious role in the computation of an oracle TM, it is relatively more difficult to analyze the computation of such machines. The characterization of Theorem 3.8 provides a different view, and it has been found useful for many applications. In this section, we formalize this characterization as a computational model, called the *alternating Turing machine* (abbreviated as ATM), that can be used to define the complexity classes in the polynomial-time hierarchy without using the notion of oracles.

An ATM $M$ is an ordinary NTM with its states partitioned into two subsets, called the *universal* states and the *existential* states. An ATM operates exactly the same as an NTM, but the notion of its acceptance of an input is defined differently. Thus, the computation of an ATM $M$ is a computation tree of configurations with each pair $(\alpha, \beta)$ of parent and child configurations satisfying $\alpha \vdash_M \beta$. We say a configuration is a *universal configuration* if it is in a universal state, and it is an *existential configuration* if it is in an existential state.

To define the notion of an ATM accepting an input, we assign, inductively, the label ACCEPT to some of the nodes in this computation tree as follows: A leaf is labeled ACCEPT if and only if it is in the accepting state. An internal node in the universal state is labeled with ACCEPT if and only if all of its children are labeled with ACCEPT. An internal node in the existential state is labeled with ACCEPT if and only if at least one of its children is labeled with ACCEPT. We say an ATM $M$ *accepts* an input $x$ if the the root of the computation tree is labeled with ACCEPT using the above labeling system. Thus an NTM is just an ATM in which all states are classified as existential states.

When an NTM $M$ accepts an input $x$, an accepting computation path is a *witness* to this fact. Also, we define $time_M(x)$ to be the length of a shortest accepting path. For an ATM $M$, to demonstrate that it accepts an input $x$, we need to display the *accepting computation subtree* $T_{acc}$ of the computation tree $T$ of $M(x)$ that has the following properties:

(i)   The root of $T$ is in $T_{acc}$.

(ii)  If $u$ is an internal existential node in $T_{acc}$, then exactly one child of $u$ in $T$ is in $T_{acc}$.

(iii) If $u$ is an internal universal node in $T_{acc}$, then all children of $u$ in $T$ are in $T_{acc}$.

(iv)  All nodes in $T_{acc}$ are labeled with ACCEPT.

Based on this notion of accepting computation subtrees, we define $time_M(x)$ to be the height of a shortest accepting subtree. We also define $t_M(n) = \max(\{n+1\} \cup \{time_M(x) : |x| = n, M \text{ accepts } x\})$. It is ready to verify that this definition of time complexity of ATMs is consistent with that of NTMs. We say an ATM is a *polynomial-time ATM* if its time complexity $t_M(n)$ is bounded by a polynomial function $p$.

The space complexity of an ATM can be defined in a similar way. We say that an ATM $M$ accepts $x$ within space $m$ if there exists an accepting subtree for $x$ such that each of its configuration has at most $m$ symbols in the work tapes; we let $space_M(x)$ be the smallest $m$ such that $M$ accepts $x$ within space $m$. Then, we define $s_M(n) = \max(\{1\} \cup \{space_M(x) : |x| = n, M \text{ accepts } x\})$.

To characterize the complexity classes of the polynomial-time hierarchy in terms of the ATM model, we add another complexity measure to ATMs, namely, the number of *quantifier changes*. For any computation path $\alpha_1, \ldots, \alpha_m$, $m \geq 1$, of an ATM $M$, we say it makes $k$ quantifier changes if there are exactly $k - 1$ integers $1 \leq i_1 < i_2 < \cdots < i_{k-1} < m$ such that for each $j$, $1 \leq j \leq k-1$, one of $\alpha_{i_j}$ and $\alpha_{i_j+1}$ is a universal configuration and the other is an existential configuration. For each accepting computation subtree $T$ of an ATM $M$ on an input $x$, we let $qc(T)$ be the maximum number of quantifier changes in any computation path of the tree $T$. We say that an ATM $M$ is a $p(n)$-time, $\exists_{q(n)}$-ATM (or, $\forall_{q(n)}$-ATM), if for each $x \in L(M)$ of length $n$, there exists an accepting computation subtree $T$ of $x$ of height $p(n)$ and $qc(T) \leq q(n)$, whose root is an existential (or, respectively, a universal) configuration.

**Proposition 3.16** *For each $k \geq 1$, $\Sigma_k^P$ is the class of sets accepted by polynomial-time $\exists_k$-ATMs, and $\Pi_k^P$ is the class of sets accepted by polynomial-time $\forall_k$-ATMs.*

In addition to the classes in the polynomial-time hierarchy, ATMs also provide nice characterizations of many other known complexity classes. We present these results through the simulations between ATMs and DTMs. Let $ATIME(t(n))$ denote the class of sets accepted by multi-tape ATMs in time $t(n)$, and $ASPACE(s(n))$ denote the class of sets accepted by multi-tape ATMs in space $s(n)$.

**Theorem 3.17** *For any fully time constructible function $t$, with $t(n) \geq n$, $ATIME(t(n)) \subseteq DSPACE((t(n))^2)$.*

*Proof.* Let $M$ be an ATM with time bound $t(n)$. For any input $x$, we perform a deterministic simulation of its computation tree of height $t(n)$. This simulation follows the depth-first traversal of the computation tree. For each node, after simulating all its child nodes, we attach a label ACCEPT or REJECT to the node, according to the labels of its child nodes and whether it is a universal or an existential node. After the label is attached, all its children can be deleted from the tree. Therefore, at any stage of the simulation, we only need to keep the record of a single computation path, the path between the root and the current node, plus all the children of all nodes in this path. Thus, the simulation always keeps in the worktapes the information of at most $O(t(n))$ nodes, and each node uses space $O(t(n))$. ∎

**Theorem 3.18** *For any fully space constructible function $s$, with $s(n) \geq \log n$, $NSPACE(s(n)) \subseteq ATIME((s(n))^2)$.*

*Proof.* The proof uses the technique of Savitch's Theorem (Theorem 1.29). Let $M$ be a one-worktape NTM that works within space bound $s(n)$. Define $reach(\alpha, \beta, k)$ on two configurations $\alpha$, $\beta$ to mean that there exist configurations $\alpha = \alpha_0, \alpha_1, \ldots, \alpha_k = \beta$ such that for each $i$, $1 \leq i \leq k$, $\alpha_i = \alpha_{i+1}$ or $\alpha_i \vdash \alpha_{i+1}$. Assume that there is a unique accepting configuration $\beta_0$, and assume that $c$ is a constant such that there are at most $c^{s(n)}$ configurations of $M$ having size $s(n)$. Then, the longest halting computation of $M(x)$ is less than or equal to $c^{s(n)}$. Therefore, $M$ accepting $x$ is equivalent to $reach(\alpha_0, \beta_0, c^{s(n)})$, where $\alpha_0$ is the initial configuration of $M(x)$.

In Theorem 1.29, we used the following reduction to construct a DTM that simulates $M$ in space $(s(n))^2$:

$$reach(\alpha, \beta, 2^{k+1}) \equiv (\exists \gamma) \; reach(\alpha, \gamma, 2^k) \text{ and } reach(\gamma, \beta, 2^k).$$

When we use an ATM to simulate $M$, the two conditions $reach(\alpha, \gamma, 2^k)$ and $reach(\gamma, \beta, 2^k)$ can be simulated *in parallel* by adding a universal move. To be more precise, let us call a tree an *alternating tree* if the nodes of the tree are partitioned into existential nodes and universal nodes. In the following, we describe an alternating tree $T$ that can be easily simulated by an ATM $M_1$ on input $x$. We assume that there is a fixed ordering on all possible configurations of $M$ that have at most $s(n)$ symbols in the work tape.

(1) Each node of $T$ has a name consisting of 2 or 3 configurations and an integer $m$, $1 \leq m \leq c^{s(n)}$.

(2) The root of $T$ is an existential node with the name $(\alpha_0, \beta_0, c^{s(n)})$.

(3) Each node with the name $(\alpha_1, \beta_1, 1)$ is a leaf. It is labeled with ACCEPT if and only if $reach(\alpha_1, \beta_1, 1)$.

(4) Each existential node with the name $(\alpha_1, \beta_1, m)$, with $m > 1$, has $c^{s(n)}$ children, with the $j$th child being a universal node with the name $(\alpha_1, \gamma_j, \beta_1, m)$, $1 \leq j \leq c^{s(n)}$, where $\gamma_j$ is the $j$th configuration of $M(x)$.

(5) Each universal node with the name $(\alpha_1, \gamma, \beta_1, m)$, with $m > 1$, has two children, each an existential node. The first child has the name $(\alpha_1, \gamma, \lceil m/2 \rceil)$ and the second one has the name $(\gamma, \beta_1, \lfloor m/2 \rfloor)$.

A simple induction shows that each existential node with the name $(\alpha_1, \beta_1, m)$ is labeled with ACCEPT if and only if $reach(\alpha_1, \beta_1, m)$, and each universal node with the name $(\alpha_1, \gamma, \beta_1, m)$ is labeled with ACCEPT if and only if $reach(\alpha_1, \gamma, \lceil m/2 \rceil)$ and $reach(\gamma, \beta_1, \lfloor m/2 \rfloor)$. Thus, the root of the tree $T$ has label ACCEPT if and only if $M$ accepts $x$.

The above tree is of height $O(s(n))$. It is, however, not easily simulated by an ATM in time $s(n)$. One of the main difficulties is that each existential node $(\alpha_1, \beta_1, m)$ in $T$ has $c^{s(n)}$ children, while a computation tree of an ATM can have at most a constant fan-out. To resolve this problem, we need to insert

$s(n)$ levels of intermediate existential nodes between each node $(\alpha_1, \beta_1, m)$ and its children so that the fan-out of each intermediate node is at most $c$. The second problem is that it takes $O(s(n))$ moves for an ATM to change from configuration with a name of the form $(\alpha_1, \beta_1, m)$ to a name $(\alpha_1, \gamma, \beta_1, m)$ (as well as other name changes). So, altogether, we need to insert $O(s(n))$ nodes between two parent/child nodes to convert it to a computation tree of an ATM. This modification increases the height of the tree by a factor of $O(s(n))$, and so the height of the final ATM is $O((s(n))^2)$. ∎

**Corollary 3.19** $\bigcup_{k>0} ATIME(n^k) = PSPACE.$

The above corollary provides an alternating quantifier characterization of the class *PSPACE*. This is a generalization of Theorem 3.8.

**Corollary 3.20** *For every set $A \in PSPACE$, there exist a polynomial-time computable set $B$ and a polynomial $p$ such that for every $x$ of length $n$, $x \in A$ if and only if*

$$
\begin{aligned}
(Q_1 y_1, |y_1| \le p(n)) \, (Q_2 y_2, |y_2| \le p(n)) \cdots \\
(Q_m y_m, |y_m| \le p(n)) \langle x, y_1, y_2, \ldots, y_m \rangle \in B,
\end{aligned}
\tag{3.7}
$$

*where each $Q_i$ is either $\exists$ or $\forall$, and $m \le p(n)$.*

**Theorem 3.21** *For any fully space constructible function $s(n)$, with $s(n) \ge \log n$, $ASPACE(s(n)) = DTIME(2^{O(s(n))})$.*

*Proof.* The direction $ASPACE(s(n)) \subseteq DTIME(2^{O(s(n))})$ can be proved as in Theorem 1.28(b), where it is proved that $NSPACE(s(n)) \subseteq DTIME(2^{O(s(n))})$. The main idea is that a computation tree of an ATM with space bound $s(n)$ is of height $2^{O(s(n))}$ and has $2^{2^{O(s(n))}}$ nodes. However, since the space bound is $s(n)$, there are only $2^{O(s(n))}$ distinct configurations. Thus, a breadth-first traversal of the tree, pruning along the way the redundant nodes, can simulate the tree deterministically in time $2^{O(s(n))}$.

For the backward direction, let $M = (Q, q_0, F, \Sigma, \Gamma, \delta)$ be a DTM with the time bound $t(n) = 2^{O(s(n))}$. We need to construct an ATM $M_1$ that simulates the computation of $M$ using only space $O(s(n))$. (Formally, we also need the tape compression theorem for ATMs. Its proof is similar to the tape compression theorem for DTMs and NTMs, and is omitted here.) Assume that $x$ is an input of length $n$. If $M$ accepts $x$, then there exists a sequence of configurations $\alpha_0, \alpha_1, \ldots, \alpha_{t(n)}$, with each $\alpha_i$ of length $t(n)$, and satisfying the following conditions:

(1) $\alpha_0$ is the initial configuration of $M(x)$;

(2) $\alpha_{t(n)}$ is a final configuration; and

(3) For each $i$, $0 \le i \le t(n) - 1$, $\alpha_i \vdash \alpha_{i+1}$.

Let $\alpha_{i,j}$ denote the $j$th symbol of $\alpha_i$. As in Cook's Theorem, each symbol of a configuration $\alpha_i$ is a symbol in $\Gamma' = \Gamma \cup (Q \times \Gamma)$. For each pair $(i,j)$, with $1 \le i, j \le t(n)$, $\alpha_{i,j}$ is uniquely determined by $\alpha_{i-1,j-1}$, $\alpha_{i-1,j}$ and $\alpha_{i-1,j+1}$, since $M$ is a DTM. (When $j = 1$ (or, $j = t(n)$), $\alpha_{i-1,j-1}$ (or, respectively, $\alpha_{i-1,j+1}$) is undefined. These cases can be taken care of by letting $\alpha_{i-1,0} = \alpha_{i-1,t(n)+1} = \text{B}$ for all $0 \le i \le t(n)$.) That is, $B = \{\langle a, b, c, d \rangle :$ if $\alpha_{i-1,j-1}\,\alpha_{i-1,j}\,\alpha_{i-1,j+1} = bcd$ then $\alpha_{i,j} = a\}$ is in $P$. We assume, as in Savitch's Theorem, that the final configuration $\alpha_{t(n)}$ is unique: $\langle q_f, \text{B} \rangle \text{B} \cdots \text{B}$. We now describe an alternating tree $T$ as follows:

(1) Each existential node of $T$ has a name of the form $(i, j, a)$, with $0 \le i \le t(n)$, $1 \le j \le t(n)$ and $a \in \Gamma'$.

(2) Each universal node of $T$ has a name of the form $(i, j, a, b, c, d)$, with $1 \le i \le t(n)$, $1 \le j \le t(n)$ and $a, b, c, d \in \Gamma'$.

(3) The root of $T$ is an existential node with the name $(t(n), 1, \langle q_f, \text{B} \rangle)$.

(4) An existential node with the name $(1, j, a)$ is a leaf; it is labeled with ACCEPT if and only if $\alpha_{1,j} = a$.

(5) A universal node with the name $(i, j, a, b, c, d)$, with $i \ge 1$, is a leaf if $\langle a, b, c, d \rangle \notin B$. (It is not labeled with ACCEPT.)

(6) An existential node with the name $(i, j, a)$, with $i \ge 1$, is an internal node with $|\Gamma'|^3$ children, each a universal node with a name $(i, j, a, b, c, d)$, with $(b, c, d)$ ranges over all triples in $(\Gamma')^3$.

(7) A universal node with the name $(i, j, a, b, c, d)$, with $i \ge 1$ and $\langle a, b, c, d \rangle \in B$ is an internal node with three children, each an existential node. Their names are $(i - 1, j - 1, b)$, $(i - 1, j, c)$, and $(i - 1, j + 1, d)$. (If $j = 1$ or $j = t(n)$, it has only two children.)

In other words, at each existential node $(i, j, a)$, an ATM simulating tree $T$ tries to check that $\alpha_{i,j} = a$. It checks this by guessing three symbols $b, c, d$, that are supposed to be $\alpha_{i-1,j-1}$, $\alpha_{i_1,j}$, and $\alpha_{i-1,j+1}$, respectively. It then checks that $\langle a, b, c, d \rangle$ is in $B$; if not, this computation path rejects. If $\langle a, b, c, d \rangle$ is indeed in $B$, then it recursively checks that $\alpha_{i-1,j-1} = b$, $\alpha_{i-1,j} = c$, and $\alpha_{i-1,j+1} = d$.

We can inductively prove that an existential node with the name $(i, j, a)$ has a label ACCEPT if and only if $\alpha_{i,j} = a$. First, this is correct for each leaf node, as property (4) indicates. Next, assume that an existential node $(i - 1, j, a)$ has a label ACCEPT if and only if $\alpha_{i-1,j} = a$, for all $j$, $1 \le j \le t(n)$. Then, we can see that the label of an internal universal node $(i, j, a, b, c, d)$ has a label ACCEPT if and only if (a) $\alpha_{i-1,j-1} = b$, $\alpha_{i-1,j} = c$, $\alpha_{i-1,j+1} = d$, and (b) $\langle a, b, c, d \rangle \in B$. From (a) and (b), we see that $\alpha_{i,j}$ must be equal to $a$ because for any triple $(b, c, d)$ there is at most one $a$ satisfying $\langle a, b, c, d \rangle \in B$. Thus, the universal node $(i, j, a, b, c, d)$ has a label ACCEPT if and only if its parent $(i, j, a)$ has a label ACCEPT if and only if $\alpha_{i,j} = a$.

Finally, we observe that the name of each node is of length $O(s(n))$. Therefore, an ATM $M_1$ can be designed to simulate tree $T$ in space $O(s(n))$.  ∎

**Corollary 3.22** *(a)* $\bigcup_{c>0} ASPACE(c \log n) = P$.
*(b)* $\bigcup_{c>0} ASPACE(cn) = EXP$.

## 3.5 *PSPACE*-Complete Problems

Our first *PSPACE*-complete problem is the space-bounded halting problem.

> SPACE BOUNDED HALTING PROBLEM (SBHP): Given a DTM $M$,
> an input $x$, and an integer $s$, written in the unary form $0^s$, deter-
> mine whether $M$ accepts $x$ within space bound $s$.

**Theorem 3.23** *SBHP is $\leq_m^P$-complete for PSPACE.*

*Proof.* The proof is similar to that of Theorem 2.11. ∎

The existence of a *PSPACE*-complete set implies that if the polynomial-
time hierarchy is properly infinite then *PSPACE* properly contains *PH*.

**Theorem 3.24** *If PH = PSPACE, then the polynomial-time hierarchy collapses to $\Sigma_k^P$ for some $k > 0$.*

*Proof.* If $PH = PSPACE$, then SBHP $\in PH = \bigcup_{k \geq 0} \Sigma_k^P$ and, hence, SBHP $\in$
$\Sigma_k^P$ for some $k \geq 0$. This implies that $PSPACE \subseteq \Sigma_k^P$, since $\Sigma_k^P$ is closed under
the $\leq_m^P$-reducibility. ∎

The first natural *PSPACE*-complete problem is a generalization of SAT$_k$.
The inputs to this problem are Boolean formulas with quantifiers $(\exists x)$ and
$(\forall x)$. An occurrence of a variable $v$ in a Boolean formula $F$ is a *bounded*
variable if there is a quantifier $(\exists v)$ or $(\forall v)$ in $F$ such that this occurrence of
$v$ is in the scope of the quantifier. A Boolean formula $F$ is called a *quantified*
Boolean formula if every occurrence of every variable in $F$ is a bounded vari-
able. For instance, $F = (\forall x) \, [(\forall y) \, [(\exists z) \, [x \bar{y} z + \bar{x} y \bar{z}] \rightarrow (\exists z)[(x \bar{z} + \bar{x} z)(y \bar{z} + \bar{y} z)]]]$
is a quantified Boolean formula. In the above, we used brackets $[\ldots]$ to
denote the scope of a quantifier, and $\rightarrow$ to denote the Boolean operation
$(a \rightarrow b) = (\bar{a} + b)$. Each quantified Boolean formula has a *normal form*
in which all quantifiers occur before any occurrence of a Boolean variable,
and the scope of each quantifier is the rest of the formula to its right.
For instance, the normal form (with renaming) of the above formula $F$ is
$(\forall x) \, (\forall y) \, (\forall z) \, (\exists w) \, [(x \bar{y} z + \bar{x} y \bar{z}) \rightarrow ((x \bar{w} + \bar{x} w)(y \bar{w} + \bar{y} w))]$.

> QUANTIFIED BOOLEAN FORMULA (QBF): Given a quantified
> Boolean formula $F$, determine whether $F$ is true.

**Corollary 3.25** *QBF is $\leq_m^P$-complete for PSPACE.*

*Proof.* The fact that QBF is in *PSPACE* follows immediately from the existence of the normal forms for quantified Boolean formulas and from the characterization of Corollary 3.20. To see that QBF is $\leq_m^P$-complete for *PSPACE*,

we need to show that an arbitrary set $A \in PSPACE$ is $\leq_m^P$-reducible to QBF. The proof is a simple extension of that of Theorem 3.13. First, if $A \in PSPACE$ then, by Corollary 3.20, there exist a set $B \in P$ and a polynomial $p$ such that for each $x$ of length $n$, $x \in A$ if and only if (3.7) holds. Now, as in the proof of Theorem 3.13, the predicate $w = \langle x, y_1, \ldots, y_m \rangle \in B$ can be transformed into a Boolean formula $F_w$ such that $w \in B$ if and only if $F_w$ is satisfiable. In addition, we can encode each string $y_i$, $1 \leq i \leq m$, by Boolean variables $z_{i,j}$, $1 \leq j \leq p(n)$, such that (3.7) holds if and only if

$$
(Q_1 \tau_1 : X_1 \rightarrow \{0,1\}) \cdots (Q_m \tau_m : X_m \rightarrow \{0,1\})
$$
$$
(\exists \tau_{m+1} : X_{m+1} \rightarrow \{0,1\}) \, F_w|_{\tau_1, \ldots, \tau_{m+1}} = 1, \tag{3.8}
$$

where for $1 \leq i \leq m$, $X_i = \{z_{i,j} : 1 \leq j \leq p(n)\}$, and $X_{m+1}$ is the set of extra variables used in $F_w$. Thus, if we quantify each variable in $X_i$ according to the quantifier $Q_i$ in (3.8) above then we obtain a quantified Boolean formula $F$ that is true if and only if (3.8) holds if and only if $x \in A$.  ∎

A quantified Boolean formula $F$ is in 3-CNF if $F$ is a formula with a sequence of quantifiers followed by an (unquantified) 3-CNF formula.

> 3-QBF: Given a 3-CNF quantified Boolean formula $F$, determine whether $F$ is true.

**Corollary 3.26** *The problem* 3-QBF *is PSPACE-complete.*

Another basic *PSPACE*-complete problem is the *regular expression* problem. Recall that a regular expression is an expression of languages using the following operators: concatenation, union, and Kleene (star) closure. For each regular expression $R$, we let $L(R)$ denote the language represented by $R$.

> TOTALITY OF REGULAR EXPRESSIONS (TRE): Given a regular expression $R$, determine whether $L(R) = \Sigma^*$.

**Theorem 3.27** TRE *is PSPACE-complete.*

*Proof.* First, we show that TRE is in *PSPACE*. It is well known in automata theory that there is a polynomial-time algorithm to convert a regular expression $R$ into a nondeterministic finite automaton $M$ (see, e.g., Hopcroft and Ullman [1979]). So the problem of whether $L(R) = \Sigma^*$ is reduced to the problem of whether $L(M) = \Sigma^*$ for a given nondeterministic finite automaton $M$. Suppose we now convert $M$ into an equivalent minimum-sized deterministic finite automaton. Then we can easily determine whether $M$ accepts all strings in $\Sigma^*$. It would take, however, in the worst case, an exponential amount of space to perform the conversion. Instead, we can directly use an NTM $M'$ to simulate the computation of $M$ to determine whether there exists a string $x$ that is not accepted by $M$. The NTM $M'$ works on a nondeterministic finite automaton $M$ as follows:

Set $K := \{q :$ state $q$ is reachable from the starting state using $\lambda$-transitions$\}$.
Repeat the following until $K$ does not contain any accepting state:

(1) Guess a character $a \in \Sigma$;
(2) Let $K'$ be the set of all states reachable from a state in $K$ processing character $a$;
(3) Set $K := K'$.

It is clear that $L(M) = \Sigma^*$ if and only if $M'$ does not halt on input $M$. Furthermore, during the simulation above, the work space required by $M'$ is linearly bounded by the input size $|M|$. Thus, this is a polynomial-space NTM determining whether $L(M) \neq \Sigma^*$, and so TRE is in $NPSACE = PSPACE$.

Next we show that TRE is complete for $PSPACE$. The idea is to, like Cook's Theorem, encode the nonaccepting computations of a polynomial-space DTM $M$ on an input $x$ by a regular expression $R$ so that if $M$ accepts $x$ then at least one string is not in $L(R)$. The difference from Cook's Theorem is that here we can use a regular expression of a polynomial length to encode a computation path of an exponential length, while a Boolean formula of a polynomial length can only encode a computation path of a polynomial length.

Assume that $M = (Q, q_0, F, \Sigma, \Gamma, \delta)$ is a DTM with a polynomial space bound $q(n)$. Also assume that if $M$ does not accept $x$ then $M$ does not halt on $x$. For any input $x$ of length $n$, we know that $M$ accepts $x$ if and only if there exists a sequence of configurations $\alpha_0, \alpha_1, \ldots, \alpha_m$, with $m \leq c^{q(n)}$ and $|\alpha_i| = c \cdot q(n)$ for some constant $c$, such that the following hold:

(1) $\alpha_0$ is the initial configuration of $M(x)$;
(2) $\alpha_m$ has an accepting state; and
(3) For each $i$, $0 \leq i \leq m - 1$, $\alpha_i \vdash \alpha_{i+1}$.

Each configuration $\alpha_i$, as in Cook's Theorem, is encoded by exactly $t = c \cdot q(n)$ symbols from $\Gamma_0 = \Gamma \cup (Q \times \Gamma)$, where $\langle q, a \rangle$ in $\alpha_i$, with $q \in Q$ and $a \in \Gamma$, indicates that the state of $\alpha_i$ is $q$ and the tape head is scanning the square containing $a$. We also encode a sequence of configurations $\alpha_0, \alpha_1, \ldots, \alpha_m$ into a single string $y = \#\alpha_0\#\alpha_1\# \cdots \#\alpha_m\#$, where $\#$ is a new symbol not in $\Gamma_0$. We will design, in the following, a regular expression $R_x$ over the alphabet $\Gamma_1 = \{\#\} \cup \Gamma_0$, such that $y \in L(R_x)$ if and only if $y$ does not encode an accepting computation of $M(x)$. It then follows that $L(R_x) = \Gamma_1^*$ if and only if $M$ does not accept $x$, or that the mapping from $x$ to $R_x$ is a reduction for $\overline{L(M)} \leq_m^P$ TRE. Since $PSPACE$ is closed under complementation, the theorem follows.

We will define $R_x = R_1 \cup R_2 \cup R_3 \cup R_4$, where

(i) $L(R_1)$ is the set of all strings $y$ that do not encode a sequence of configurations of length $t$;
(ii) $L(R_2)$ is the set of all strings $y$ whose first configuration is not the initial configuration of $M(x)$;

(iii) $L(R_3)$ is the set of all strings $y$ whose last configuration is not an accepting configuration; and

(iv) $L(R_4)$ is the set of all strings $y$ that have two consecutive configurations $\alpha$, $\beta$ not satisfying $\alpha \vdash \beta$.

Let $R_{1,1}$ be a regular expression such that $L(R_{1,1})$ is equal to the set of all strings over $\Gamma_0^*$ of length not equal to $t$, and $R_{1,2}$ be a regular expression such that $L(R_{1,2})$ is the set of all strings over $\Gamma_0^*$ that have either no symbol in $Q \times \Gamma$ or have at least two symbols in $Q \times \Gamma$. It is easy to see that there exist regular expressions $R_{1,1}$ and $R_{1,2}$ satisfying the above properties and having length polynomial in $n$. We can then define $R_1$ as follows:

$$R_1 = \lambda \cup (\Gamma_0\Gamma_1^*) \cup (\Gamma_1^*\Gamma_0) \cup (\Gamma_1^*\#(R_{1,1} \cup R_{1,2})\#\Gamma_1^*).$$

Thus, only strings $y$ that encode a sequence of configurations of length $t$ are in $\overline{L(R_1)}$.

Next, assume that the initial configuration $\alpha_0$ is $\alpha_0 = u_1u_2 \ldots u_t$, where each $u_i$ is a single symbol in $\Gamma_0$. Then, we define, for each $1 \leq i \leq t$,

$$R_{2,i} = \#\Gamma_0^{i-1}(\Gamma_0 - u_i)\Gamma_1^*$$

and $R_2 = R_{2,1} \cup \cdots \cup R_{2,t}$. (We write $R^k$ to denote the concatenation of $k$ copies of the regular expression $R$.) Similarly, we define

$$R_3 = \Gamma_1^*((Q - F) \times \Gamma)\Gamma_0^*\#.$$

Finally, for $R_4$, we recall, from the proof of Theorem 2.12, that $\alpha_i \vdash \alpha_{i+1}$ is equivalent to the following two conditions ($\alpha_{i,k}$ denotes the $k$th symbol of $\alpha_i$):

(i) If $\alpha_{i,k-1}, \alpha_{i,k}, \alpha_{i,k+1} \notin Q \times \Gamma$ then $\alpha_{i+1,k} = \alpha_{i,k}$; and

(ii) If $\alpha_{i,k} \in Q \times \Gamma$ then $\alpha_{i,k-1}\alpha_{i,k}\alpha_{i,k+1} \vdash_M' \alpha_{i+1,k-1}\alpha_{i+1,k}\alpha_{i+1,k+1}$, where $\vdash_M'$ is the subrelation of $\vdash_M$ restricted to configurations of length 3.

So, for each triple $(a, b, c)$ with $b \in \Gamma$ and $a, c \in \Gamma \cup \{\#\}$, we can define

$$R_{a,b,c} = \Gamma_1^*abc\Gamma_1^{t-1}(\Gamma_1 - b)\Gamma_1^*.$$

Also, for each pair $(q, a) \in Q \times \Gamma$, and $b, c \in \Gamma \cup \{\#\}$, we can define a regular expression $R_{q,a,b,c}$ to represent strings $y$ in which two consecutive configurations do not satisfy (ii) above. For instance, if $\delta(q, a) = (p, d, R)$, then

$$R_{q,a,b,c} = \Gamma_1^* \, b \, \langle q, a \rangle \, c \, (\Gamma_1^{t-2}(\Gamma_1 - b) \cup \Gamma_1^{t-1}(\Gamma_1 - d) \cup \Gamma_1^t(\Gamma_1 - \langle p, c \rangle))\Gamma_1^*.$$

Together, we define

$$R_4 = \left( \bigcup_{(a,b,c)\in A} R_{a,b,c} \right) \cup \left( \bigcup_{(q,a,b,c)\in B} R_{q,a,b,c} \right),$$

where $A = \{(a, b, c) : b \in \Gamma, a, c \in \Gamma \cup \{\#\}\}$ and $B = \{(q, a, b, c) : q \in Q, a \in \Gamma, b, c \in \Gamma \cup \{\#\}\}$.

From the above discussion, it follows that $R$ represents all strings $y$ that do not encode accepting computations of $M(x)$. It is easy to verify that each of the above regular expressions are of length bounded by $p(n)$ for some polynomial $p$, and so this is indeed a polynomial time reduction from $\overline{L(M)}$ to TRE. ∎

One of the interesting applications of the characterization of Corollary 3.19 is to show the *PSPACE*-completeness of some two-person games. A two-person game played by two players, 0 and 1, is specified by a set $G \subseteq \Sigma^*$, for some fixed alphabet $\Sigma$, of *game configurations* and two binary relations $R_0$ and $R_1$ on $G$. Let $j \in \{0, 1\}$. A sequence of moves, beginning with a move of player $j$, is a sequence of game configurations $x_0, x_1, \ldots, x_m$ such that for each even $i$, $0 \leq i < m$, $(x_i, x_{i+1}) \in R_j$, and for each odd $i$, $0 \leq i < m$, $(x_i, x_{i+1}) \in R_{1-j}$. For each $j = 0, 1$, let the set of *losing configurations* for player $j$ be $H_j = \{x \in G :$ there is no $y \in G$ such that $(x, y) \in R_j\}$. That is, a player $j$ loses if it is his/her move on a configuration $x$ but there is no legal move on $x$ by player $j$. The player $j$ is said to have a *winning strategy* on a game configuration $x_0$, if, for some $m \geq 0$,

$$(\exists x_1, (x_0, x_1) \in R_j) (\forall x_2, (x_1, x_2) \in R_{1-j}) (\exists x_3, (x_2, x_3) \in R_j)$$
$$\cdots (\exists x_{2m+1}, (x_{2m}, x_{2m+1}) \in R_j) [x_{2m+1} \in H_{1-j}]. \quad (3.9)$$

A game $(G, R_0, R_1)$ is called a *polynomially bounded game* if the following conditions hold:

(i) $G, R_0, R_1 \in P$;

(ii) There is a polynomial $p$ such that any sequence of moves starting from a configuration $\alpha \in G$ has length $\leq p(|\alpha|)$; and

(iii) The transitive closures of $R_0$ and $R_1$ are polynomial length-bounded; that is, if $(x_0, \ldots, x_m)$ is a sequence of moves of the game $G$, then $|x_m| \leq q(|x_0|)$ for some polynomial $q$.

It is not hard to see that the *winning strategy* problem for any polynomially bounded game is solvable by an ATM in polynomial time and, hence, is in *PSPACE*.

**Theorem 3.28** *Assume that $(G, R_0, R_1)$ is a polynomially bounded game. There is a polynomial-time ATM $M$ such that on each $x \in G$ and $j \in \{0, 1\}$, $M$ determines whether the player $j$ has a winning strategy on $x$.*

*Proof.* The formula (3.9) defines a search tree of height $p(|x_0|)$ that can be implemented by an ATM $M$ if $(G, R_0, R_1)$ has a polynomial bound $p(n)$ on the length of a sequence of moves starting with a game configuration of length $n$. Since the transitive closures of $R_0$ and $R_1$ are polynomial length-bounded, each node in the search tree is of length $q(|x_0|)$ for some polynomial $q$, and so each node can have at most $c^{q(|x_0|)}$ children for some constant $c > 0$. We can introduce $q(|x_0|)$ levels of new intermediate nodes (of the same quantifier as the parent node) between each node and its children to make the fan-out of $M$ bounded. This will only increase a polynomial factor on the height, and so the resulting tree is still computable by a polynomial-time ATM.     ∎

A number of two-person games have been shown to be *PSPACE*-complete. Indeed, the alternating quantifier characterization of Corollary 3.20 suggests that every problem in *PSPACE* may be viewed as a two-person game. For instance, the problem QBF may be viewed as a game played between two players ∃ and ∀. Without loss of generality, we may assume that a given quantified Boolean formula $F$ has alternating quantifiers in its normal form:

$$F = (\exists x_1) \, (\forall x_2) \cdots (Q_m x_m) \, F_1,$$

where $Q_m = \exists$ if $m$ is odd, $Q_m = \forall$ if $m$ is even, and $F_1$ is an unquantified Boolean formula over variables $x_1, \ldots, x_m$. For any Boolean formula not of this form, we can insert a quantifier over a dummy variable between any two consecutive quantifiers of the same type to convert it to an equivalent formula of this form. On such a formula $F$, player ∃ tries to prove that $F$ is true, and player ∀ works as an adversary trying to prove that $F$ is false (see Exercise 3.12). When the problem QBF is presented as a game, the notion of the player ∃ having a winning strategy is equivalent to the notion of $F$ being true. Thus, the game QBF is *PSPACE*-complete.

In the following, we show a natural two-person game GEOGRAPHY to be *PSPACE*-complete. A game configuration of GEOGRAPHY is a directed graph $H = (V, E)$ with a starting node $v \in V$ explicitly *marked*. There is a move from $\langle H_1, v_1 \rangle$ to $\langle H_2, v_2 \rangle$ (i.e., $(\langle H_1, u_1 \rangle, \langle H_2, v_2 \rangle) \in R_0 = R_1$) if $H_2$ is the subgraph of $H_1$ with the node $v_1$ removed, and $(v_1, v_2)$ is an edge of $H_1$. In other words, the players, starting with an initial graph $H$ and a starting node $v$, take turns to visit one of the unvisited neighbors of the current node until one of the players reaches a node of which all the neighbors have been visited.

> GEOGRAPHY: Given a digraph $H$ with a marked node $v$, determine whether the next player has a winning strategy.

**Theorem 3.29** *The game* GEOGRAPHY *is PSPACE-complete.*

*Proof.* It is easy to see that GEOGRAPHY is a polynomially bounded game, and so, by Theorem 3.28, it is in *PSPACE*. To see that GEOGRAPHY is *PSPACE*-complete, we reduce the problem 3-QBF to it. Let $F$ be a 3-CNF quantified formula of the form

$$F = (\exists x_1) \, (\forall y_1) \, (\exists x_2) \cdots (\exists x_n) \, (\forall y_n) \, C_1 C_2 \cdots C_m,$$

where each $C_i$, $1 \leq i \leq m$, is a clause of three literals. (As we discussed above, we may assume that the quantifiers in $F$ alternate between ∃ and ∀.) For each variable $x_i$ (and $y_i$), $1 \leq i \leq n$, we let $H_i$ (and, respectively, $K_i$) be a four-node digraph shown in Figure 3.2(a). The graph $H$ consists of:

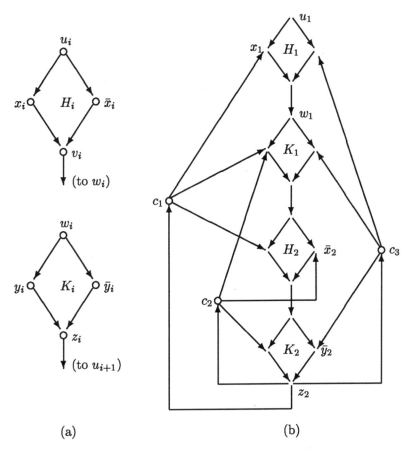

Figure 3.2: (a) The digraphs $H_i$ and $K_i$. (b) An example of digraph $H$.

(i) Graphs $H_1, K_1, H_2, \ldots, K_n$, concatenated with edges $(v_i, w_i)$, for $1 \leq i \leq n$, and $(z_j, u_{j+1})$, for $1 \leq j < n$;

(ii) $m$ new nodes $c_1, c_2, \ldots, c_m$ together with the new edges $(z_n, c_j)$, for $j = 1, 2, \ldots, m$; and

(iii) For each clause $C_j = \ell_{j,1} + \ell_{j,2} + \ell_{j,3}$, $1 \leq j \leq m$, three edges $(c_j, \ell_{j,k})$, $k = 1, 2, 3$.

Figure 3.2(b) shows the graph $H$ for formula $F = (\exists x_1)\ (\forall y_1)\ (\exists x_2)\ (\forall y_2)$ $(x_1 + x_2 + y_1)(\bar{x}_2 + y_1 + y_2)(\bar{x}_1 + \bar{y}_1 + \bar{y}_2)$. Finally, we let $u_1$ be the starting node.

We claim that there is a winning strategy on the game configuration $(H, u_1)$ of GEOGRAPHY if and only if $F$ is true. Let us observe how the game can be played. Assume that the first player is player 0. First, in the first $6n - 1$

moves, the two players are forced to visit from node $u_1$ to node $z_n$, with player 0 at the $(6i - 5)$th move visiting either node $x_i$ or node $\bar{x}_i$, and the player 1 at the $(6i - 2)$nd move visiting either node $y_i$ or $\bar{y}_i$. When these moves are made, the resulting graph $H'$ contains only the nodes $z_n$, $c_1, c_2, \ldots, c_m$ and one node from each pair $\{x_i, \bar{x}_i\}$ and $\{y_i, \bar{y}_i\}$, $1 \le i \le n$. At this point, it is player 1's turn. Player 0 can win if and only if for any move from $z_n$ to any $c_j$, $1 \le j \le m$, played by player 1, player 0 can find a move from $c_j$ to some $\ell_{j,k}$, $1 \le k \le 3$, in $H'$.

Now we may define an assignment $\tau$ on $x_1, y_1, \ldots, x_n, y_n$ according to the first $6n - 1$ moves on $(H, u_1)$ (i.e., let $\tau(x_i) = 1$ if and only if $x_i \in H'$ and $\tau(y_i) = 1$ if and only if $y_i \in H'$), then $C_1 C_2 \cdots C_m$ is true to $\tau$ if and only if player 0 wins after $6n + 1$ moves. Thus, player 0 has a winning strategy on $(H, u_1)$ if and only if

$$(\exists \text{ choice of } x_1 \text{ or } \bar{x}_1 \text{ to } H') \ (\forall \text{ choices of } y_1 \text{ or } \bar{y}_1 \text{ to } H') \cdots$$
$$(\exists \text{ choice of } x_n \text{ or } \bar{x}_n \text{ to } H') \ (\forall \text{ choices of } y_n \text{ or } \bar{y}_n \text{ to } H')$$
$$(\forall j, 1 \le j \le m) \ (\exists k \in \{1, 2, 3\}) \ [\ell_{j,k} \in H']$$

if and only if

$$(\exists \tau(x_1)) \ (\forall \tau(y_1)) \cdots (\exists \tau(x_n)) \ (\forall \tau(y_n))$$
$$(\forall j, 1 \le j \le m) \ (\exists k \in \{1, 2, 3\}) \ [\tau(\ell_{j,k}) = 1]$$

if and only if $F$ is true.                                                                    ∎


## 3.6  EXP-Complete Problems

All complete problems studied so far are candidates for intractable problems, but their intractability still depends on the separation of the classes *NP*, *PSPACE*, and *PH* from the class *P*. Are there natural problems that are *provably* intractable in the sense that they can be proved not belonging to *P*? In this section, we present a few problems that are complete for *EXP* and, hence, not in *P*.

Our first *EXP*-complete problem is the bounded halting problem on deterministic machines with the time bound encoded in binary form.

> EXPONENTIAL-TIME BOUNDED HALTING PROBLEM (EXP-BHP):
> Given a DTM $M$, a string $x$, and an integer $n > 0$, written in the binary form, determine whether $M(x)$ halts in $n$ moves.

**Proposition 3.30** EXP-BHP *is EXP-complete.*

*Proof.* If $L$ is accepted by a DTM $M$ in time $2^{cn}$, then the function $f(x) = \langle M, x, 2^{c|x|} \rangle$ is a polynomial-time reduction from $L$ to EXP-BHP.            ∎

We note that in the above problem, if the time bound $n$ is written in the unary form (as in the problem BHP), then the problem becomes polynomial-time solvable. Indeed, there is a simple translation of most *P-complete* problems[1] to *EXP*-complete problems by more succinct encodings of the inputs. In the following, we demonstrate this idea on the famous *P*-complete problem, CIRCUIT VALUE PROBLEM (CVP).

Let $C$ be a Boolean circuit[2] satisfying the following property: $C$ has $n$ gates numbered from 1 to $n$; we let $C(i)$ denote the gate of $C$ numbered $i$. There are four types of gates in circuit $C$: ZERO gates, ONE gates, AND gates, and OR gates. A ZERO (ONE) gate has no input and one output whose value is 0 (1, respectively). An AND (OR) gate has two inputs and one output whose value is the Boolean product (Boolean sum, respectively) of the two inputs. If the gate $i$ is an AND or OR gate, then its two inputs are the outputs of two gates whose numbers are lower than $i$. Note that this circuit $C$ does not have input gates and so it computes a unique Boolean value (the output of gate $n$). If the circuit is given explicitly, then its output value is computable in polynomial time. (In fact, it is *P*-complete; see Theorem 6.41). In the following, we consider the encoding of the circuit by a DTM. We say that a DTM $M$ *generates a circuit* $C$ of size $n$ in time $m$ if for all $i$, $0 \leq i \leq n$, $M(i)$ outputs a triple $\langle b, j, k \rangle$ in $m$ moves, with $0 \leq b \leq 3$ and $1 \leq j, k < i$ if $b \leq 1$, such that

(i) If $b = 0$ then $C(i) = C(j) \cdot C(k)$;

(ii) If $b = 1$ then $C(i) = C(j) + C(k)$;

(iii) If $b = 2$ then $C(i) = 0$;

(iv) If $b = 3$ then $C(i) = 1$.

> EXPONENTIAL-SIZE CIRCUIT VALUE PROBLEM (EXP-CVP): Given a DTM $M$, an integer $t > 0$, written in the unary form $0^t$, and an integer $s > 0$, written in the binary form, such that $M$ generates a circuit $C$ of size $s$ in time $t$, determine whether circuit $C$ outputs 1.

**Theorem 3.31** EXP-CVP *is EXP-complete.*

*Proof.* It is easy to see that EXP-CVP is in *EXP*. For each input $\langle M, 0^t, s \rangle$, the circuit $C$ generated by $M$ has size $s$ and we can write down the whole circuit $C$ in $O(ts \log s)$ moves. Then, we can simulate the circuit in time polynomial in $s$.

To see that EXP-CVP is complete for *EXP*, assume that $M = (Q, q_0, F, \Sigma, \Gamma, \delta)$ is a DTM with runtime bounded by $2^{cn}$ for some constant $c > 0$. Let $m = 2^{cn}$. Then, for any input $w$ of length $n$, there is a unique sequence of configurations $\alpha_0, \alpha_1, \ldots, \alpha_m$ such that

(i) Each $\alpha_i$, $0 \leq i \leq m$, is of length $\leq m$;

---

[1] A problem $A$ is *P-complete* if $A \in P$ and every set $B \in P$ is $\leq_m$-reducible to $A$ via a logspace computable function. See Chapter 6 for more on P-complete problems.

[2] See Section 6.1 for the definition of Boolean circuits

(ii) $\alpha_i \vdash \alpha_{i+1}$, for each $0 \leq i \leq m - 1$;

(iii) $\alpha_0$ is the initial configuration of $M(w)$; and

(iv) $\alpha_m$ is in an accepting state if and only if $w \in L(M)$.

As in Cook's Theorem, we may define a set of Boolean variables $y_{i,j,a}$ to encode this sequence of configurations. To be more precise, we define a circuit $C$ that contains gates named $y_{i,j,a}$, for all $0 \leq i \leq m$, $1 \leq j \leq m$, and $a \in \Gamma' = \Gamma \cup (Q \times \Gamma)$, plus some auxiliary gates, so that the output of gate $y_{i,j,a}$ is 1 if and only if the $j$th character of $\alpha_i$ is $a$. To satisfy this condition, we let gates $y_{0,j,a}$ be constant gates corresponding to the initial configuration $\alpha_0$. For each $i \geq 1$ and each $1 \leq j \leq m$, $a \in \Gamma'$, we define a subcircuit $C_{i,j,a}$, whose output gate is $y_{i,j,a}$ and input gates are $y_{i-1,k,b}$, $j - 1 \leq k \leq j + 1$, $b \in \Gamma'$. This subcircuit is defined from the transition function $\delta$ of $M$, just as in the Boolean formula $\phi_4$ of Cook's Theorem. We assume that $M$ has a unique accepting configuration $\langle q_f, \text{B} \rangle \text{BB} \cdots$, and we let $y_{m,1,\langle q_f, \text{B} \rangle}$ be the output gate of $C$. Then, it is clear that the output of circuit $C$ is 1 if and only if $M$ accepts $w$. Furthermore, the size of circuit $C$ is polynomial in $m$.

Now, we observe that the circuit $C$ can be encoded into a DTM $M_C$ that generates it in time polynomial in $n = |w|$. This is true because each subcircuit $C_{i,j,a}$, $i \geq 1$, has a constant size and its structure depends only on the transition function $\delta$ of $M$ (e.g., for a fixed $a$, $C_{i,j,a}$ and $C_{i',j',a}$ are identical except for the input gates), and so it can be generated by $M_C$ using a table look-up method, and the runtime on each gate is polynomial in $n$. (Note that the length of the binary expansion of the integer $i$ is linear in $n$.) In addition, the value of each constant gate $y_{i,j,a}$ can be determined from input $w$ only and so can be generated in time linear in $n$. Thus, circuit $C$ can be generated by $M_C$ in time polynomial in $n$, and the codes of $M_C$ itself is easily computed from $M$ and $w$. (See Exercise 3.16 for more discussions.) It follows that the reduction from $L(M)$ to Exp-CVP can be done in polynomial time.  ∎

In Section 3.5, we have shown that a polynomially bounded two-person game is solvable in *PSPACE*, and that some natural games, such as GEOGRAPHY, are *PSPACE*-complete. In the following, we consider a game that is not polynomially bounded, and show that it is complete for the class *EXP*.

Informally, the BOOLEAN FORMULA GAME (BFG) is a two-person game played on a Boolean formula $F$. The variables of the formula $F$ are partitioned into three subsets: $X, Y$, and $\{t\}$. The game begins with an initial assignment $\tau_0$ on variables in $X \cup Y \cup \{t\}$ such that $F$ evaluates to 1 under $\tau_0$. The two players 0 and 1 take turns to change the assignment to variables, subject to the following rules:

(a) Player 0 must assign value 1 to the variable $t$, and must not change the Boolean value of any variable $y \in Y$.

(b) Player 1 must assign value 0 to the variable $t$, and must not change the Boolean value of any variable $x \in X$.

A player loses if the formula $F$ becomes 0 after his/her move.

We notice that a Boolean formula $F$ of $m$ variables has $2^m$ different assignments for its variables and, hence, this game is not polynomially bounded. In fact, the game may last forever, with two players repeating the same assignments. It is easy to see, nevertheless, that these repeated moves can be identified and ignored as far as the winning strategy is concerned.

In the following formal definition of the game BFG, we write $F(A, B, C)$ to denote a Boolean function with its variables partitioned into three sets $A$, $B$, and $C$. Now we define, following the notation of the last section, $\text{BFG} = (G, R_0, R_1)$, where

$$G = \{\langle F, \tau \rangle : F = F(X, Y, \{t\}), \tau : X \cup Y \cup \{t\} \to \{0, 1\}\};$$
$$R_0 = \{(\langle F, \tau_1 \rangle, \langle F, \tau_2 \rangle) : \tau_2(t) = 1, (\forall y \in Y)\, \tau_2(y) = \tau_1(y), F|_{\tau_2} = 1\};\ \text{and}$$
$$R_1 = \{(\langle F, \tau_1 \rangle, \langle F, \tau_2 \rangle) : \tau_2(t) = 0, (\forall x \in X)\, \tau_2(x) = \tau_1(x), F|_{\tau_2} = 1\}.$$

> BOOLEAN FORMULA GAME (BFG): Given a Boolean formula $F(X, Y, \{t\})$ and a truth assignment $\tau : X \cup Y \cup \{t\} \to \{0, 1\}$ such that $F|_\tau = 1$, determine whether the player 0 has a winning strategy on the game configuration $\langle F, \tau \rangle$.

**Theorem 3.32** BFG *is EXP-complete.*

*Proof.* Our proof is based on the ATM characterization of the class *EXP*. Recall that we proved in Theorem 3.21 that $EXP = \bigcup_{c>0} ASPACE(cn)$. To see that BFG $\in EXP$, we describe an alternating tree $T$ for input $\langle F, \tau_0 \rangle$ as follows:

(1) The root of $T$ is an existential node with the name $(F, \tau_0)$.

(2) For each existential node $v$ of $T$, with the name $(F, \tau)$, let $A = \{\tau' : X \cup Y \cup \{t\} \to \{0, 1\} : \tau'(t) = 1, (\forall y \in Y)\,\tau'(y) = \tau(y),\ F|_{\tau'} = 1\}$. If $A = \emptyset$, then $v$ is a leaf. Otherwise, node $v$ has $|A| \leq 2^{|X|}$ children, each a universal node with a unique name $(F, \tau'),\ \tau' \in A$.

(3) For each universal node $v$ of $T$, with the name $(F, \tau)$, let $B = \{\tau' : X \cup Y \cup \{t\} \to \{0, 1\} : \tau'(t) = 0, (\forall x \in X)\,\tau'(x) = \tau(x),\ F|_{\tau'} = 1\}$. If $B = \emptyset$, then $v$ is a leaf with label ACCEPT. Otherwise, node $v$ has $|B| \leq 2^{|Y|}$ children, each an existential node with a unique name $(F, \tau'),\ \tau' \in B$.

It is obvious that this tree can be simulated by an ATM $M$ in linear space. Furthermore, $M$ accepts input $\langle F, \tau_0 \rangle$ if and only if $\langle F, \tau_0 \rangle \in$ BFG.

Next, we show that each set $A \in EXP$ is $\leq_m^P$-reducible to BFG and, hence, BFG is *EXP*-complete. Assume that $A \in EXP$. Then, by Theorem 3.21, there is an ATM $M = (Q, q_0, F, \Sigma, \Gamma, \Delta)$ accepting $A$ in space $cn$ for some constant $c > 0$. That is, on input $w$, each configuration of the computation tree of $M$ on $w$ has length at most $c|w|$. We pad the short configurations with extra blanks and assume that every configuration has a uniform length $n = c|w|$. For our proof below, we further assume the following settings:

(i) The initial configuration of $M(w)$ is in an existential state; and

(ii) All children of an existential configuration are universal configurations, and all children of a universal configuration are existential configurations.

It is left to the reader to check that a linear space ATM can be easily modified into an equivalent linear space ATM having the above properties.

As in Cook's Theorem, we define Boolean variables to encode the configurations of the ATM $M$ on input $w$. We let $X = \{x_{i,a} : 1 \leq i \leq n, a \in \Gamma'\}$ and $Y = \{y_{i,a} : 1 \leq i \leq n, a \in \Gamma'\}$, where $\Gamma' = \Gamma \cup (Q \times \Gamma)$. The intended interpretation of the above variables is, as in Cook's Theorem, that $x_{i,a} = 1$ if and only if the $i$th character of the configuration $\beta$ is $a$, and $y_{i,a} = 1$ if and only if the $i$th character of the configuration $\alpha$ is $a$, where $\alpha$ and $\beta$ are two configurations of the computation tree of $M(w)$. Based on this interpretation, we define a formula $F_w$ to denote the following properties on the configurations $\alpha$ and $\beta$ (we say $\beta$ is a *rejecting configuration* if it is a halting configuration but is not in an accepting state):

(i) If $\tau(t) = 1$ then $\alpha \vdash_M \beta$ and $\beta$ is not a rejecting configuration; and

(ii) If $\tau(t) = 0$ then $\beta \vdash_M \alpha$ and $\alpha$ is not in an accepting state.

To do so, we let $\phi_1$ be the Boolean formula that asserts that $\alpha$ and $\beta$ are indeed configurations of $M$. To be more precise, we let $\phi_1 = \phi_{1,1}\phi_{1,2}\phi_{1,3}\phi_{1,4}$, where $\phi_{1,1}$ asserts that for each $i$, $1 \leq i \leq n$, there is a unique $a \in \Gamma'$, such that $x_{i,a} = 1$, $\phi_{1,2}$ asserts that for each $i$, $1 \leq i \leq n$, there is a unique $b \in \Gamma'$, such that $y_{i,b} = 1$, $\phi_{1,3}$ asserts that there is a unique $i$, $1 \leq i \leq n$, such that $x_{i,a} = 1$ for some $a \in Q \times \Gamma$, and $\phi_{1,4}$ asserts that there is a unique $i$, $1 \leq i \leq n$, such that $y_{i,b} = 1$ for some $b \in Q \times \Gamma$. The formulas $\phi_{1,1}$ and $\phi_{1,2}$ are just like formula $\phi_1$ in Cook's Theorem. Formulas $\phi_{1,3}$ and $\phi_{1,4}$ can be similarly defined. For instance, we can define $\phi_{1,3}$ by

$$\phi_{1,3} = \left( \sum_{i=1}^{n} \sum_{a \in Q \times \Gamma} x_{i,a} \right) \prod_{1 \leq i < j \leq n} \prod_{a,b \in Q \times \Gamma} (\bar{x}_{i,a} + \bar{x}_{j,b}).$$

Next, let $\phi_2$ be the formula for the property $\alpha \vdash \beta$, and $\phi_3$ be the formula for the property $\beta \vdash \alpha$. Again, they can be defined as in Cook's Theorem. Finally, let $\phi_4$ denote the property that $\beta$ is not a rejecting configuration, and $\phi_5$ denote the property that $\alpha$ is not in an accepting state. These are also easy to define:

$$\phi_4 = \prod_{i=1}^{n} \prod_{a \in \Gamma_0} \bar{x}_{i,a}, \qquad \phi_5 = \prod_{i=1}^{n} \prod_{b \in \Gamma_1} \bar{y}_{i,b},$$

where $\Gamma_0 = \{\langle q, s \rangle : q \in Q - F, s \in \Gamma, \Delta(q,s) = \emptyset\}$ and $\Gamma_1 = \{\langle q, s \rangle : q \in F, s \in \Gamma\}$.

Using these components, we define

$$F_w = \phi_1(\bar{t} + \phi_2\phi_4)(t + \phi_3\phi_5).$$

We also define an initial assignment $\tau_0$ as follows: For each $i$, $1 \le i \le n$, and each $a \in \Gamma'$, $\tau_0(y_{i,a}) = \tau_0(x_{i,a}) = 1$ if and only if the $i$th character of the initial configuration of $M(w)$ is $a$, and $\tau_0(t) = 0$.

Assume that $\langle F_w, \tau \rangle$ is a game configuration for BFG. If the assignment $\tau$ on variables in $Y$ defines an existential configuration $\alpha$ of $M(w)$ (based on the above interpretation of the variables $y_{i,a} \in Y$), and the player 0 is to make the next move, then the only moves he/she can make are to define an assignment $\tau'$ on variables in $X$ such that, with respect to $\tau'$, $\beta$ is a successor of $\alpha$. After that, the only moves player 1 can make are to make $\alpha$ a successor of $\beta$. Thus, starting with $\tau_0$ that defines $\alpha$ to be the initial configuration of $M(w)$, the sequence of moves made by the two players is a path of the computation tree of $M(w)$. At the end of the path, the player 0 cannot move (and loses) if all next moves in the computation tree of $M(w)$ reject, and the player 1 loses if all next moves accept. This shows that the game tree of BFG on $\langle F, \tau_0 \rangle$ coincides with the computation tree of $M(w)$, and we conclude that the player 0 has a winning strategy on $\langle F_w, \tau_0 \rangle$ if and only if $M$ accepts $w$. ∎

In the above, we have demonstrated some complete problems for the class *EXP*. In addition to these problems, there are problems which are known to have even higher complexity, for instance, complete for the classes *NEXP*, *EXPSPACE*, and so on. In the following, we list a few of these complete problems, and leave their proofs as exercises.

First, using the technique of succinct encoding for the problem EXP-CVP, we can encode a Boolean formula of exponential size in a DTM and the corresponding satisfiability problem becomes *NEXP*-complete. We say a 3-CNF formula is of *size* $(n, m)$ if it has $n$ Boolean variables and $m$ clauses. We say that a DTM $M$ *generates a 3-CNF formula $F$* of size $(n, m)$ in time $t$ if for any $1 \le j \le m$, $M(j)$ outputs in $t$ moves a 6-tuple $\langle i_1, b_1, i_2, b_2, i_3, b_3 \rangle$, with $i_1, i_2, i_3 \in \{1, \dots, n\}$ and $b_1, b_2, b_3 \in \{0, 1\}$, such that (i) the three variables occurring in clause $C_j$ are $x_{i_1}, x_{i_2}, x_{i_3}$, and (ii) for each $k = 1, 2, 3$, $x_{i_k}$ occurs in $C_j$ positively if and only if $b_k = 1$.

> EXP-SAT: Given a DTM $M$, integers $n, m$ written in the binary form, and an integer $t$ in unary form, such that $M$ generates a 3-CNF formula $F$ of size $(n, m)$ in time $t$, determine whether $F$ is satisfiable.

**Theorem 3.33** EXP-SAT *is NEXP-complete.*

In Theorem 3.27, we showed that the problem of determining the totality of a given regular expression (TRE) is *PSPACE*-complete. The idea of the proof is that regular expressions can be used to encode computation paths of a DTM in such a way that a computation path of length exponential in $n$ can be encoded by a regular expression of length polynomial in $n$. We say a string is an *extended regular expression* if it is a regular expression with an additional intersection operation $\cap$. For any two languages $A$ and $B$, $A \cap B$ simply denotes the set intersection of $A$ and $B$. The following result indicates that even more succinct representations of a computation path of a DTM can be achieved by extended regular expressions.

> TOTALITY OF EXTENDED REGULAR EXPRESSION (TERE): Given an extended regular expression $R$ over set $\Sigma$, determine whether $L(R) = \Sigma^*$.

**Theorem 3.34** TERE *is EXPSPACE-complete.*

## Exercises

**3.1** We say set $A$ is *strong-NP reducible* to set $B$ $(A \leq_T^{SNP} B)$ if there is a polynomial-time oracle NTM $M$ such that for each $x$, at least one computation path of $M^B(x)$ halts in an accepting state, and all computation paths of $M^B(x)$ halting in an accepting state output value 1 if $x \in A$, and all such paths output value 0 if $x \notin A$.

(a) Prove that $\leq_T^{SNP}$ is indeed a reducibility (i.e., it satisfies the transitivity property).

(b) Prove that $A \leq_T^{SNP} B$ if and only if $[A \in NP^B$ and $\bar{A} \in NP^B]$ if and only if $NP^A \subseteq NP^B$.

(c) Prove that $A \leq_T^{SNP} B$ and $B \in NP \cap coNP$ imply $A \in NP \cap coNP$.

(d) Let $\Sigma_1^{P,A}$ denote the class $NP^A$ and, for $k \geq 1$, let $\Sigma_{k+1}^{P,A}$ denote the class $NP(\Sigma_k^{P,A})$. Prove that, for $k \geq 1$, if $A$ is $\leq_T^{SNP}$-complete for the class $\Sigma_k^P$, then $\Sigma_k^{P,A} = \Sigma_{k+1}^P$.

**3.2** Prove that there exist sets $A, B$, such that $A \leq_T^{SNP} B$ but not $A \leq_T^P B$.

**3.3** In this exercise, we study the complexity class $DP$. A language $A$ is in the class $DP$ ($D$ stands for *difference*) if there exist sets $B \in NP$ and $C \in coNP$ such that $A = B \cap C$.

(a) The problem SAT-UNSAT asks for two given Boolean formulas $\phi$ and $\psi$, whether it is true that $\phi$ is satisfiable and $\psi$ is unsatisfiable. Prove that SAT-UNSAT is $DP$-complete under $\leq_m^P$-reducibility.

(b) Recall the problem EXACT-CLIQUE defined in Example 2.20(b). Show that EXACT-CLIQUE is $DP$-complete.

(c) The problem CRITICAL HC asks for a given graph $G = (V, E)$ whether it is true that $G$ does not have a Hamiltonian circuit but every graph $G'$ obtained from $G$ by adding a new edge has a Hamiltonian circuit. Prove that CRITICAL HC is $DP$-complete.

**3.4** In this exercise, we extend the class $DP$ to the *Boolean hierarchy* of sets between $P$ and $\Delta_2^P$. For each $n \geq 0$, define $DP_n$ as follows: $DP_0 = P$, $DP_{n+1} = \{A \cup B : A \in DP_n, B \in NP\}$ if $n$ is even, and $DP_{n+1} = \{A \cap B : A \in DP_n, B \in coNP\}$ if $n$ is odd. Then, define $BH = \bigcup_{n \geq 0} DP_n$.

(a) In Exercise 2.14, we defined the polynomial-time truth-table reducibility $\leq_{tt}^P$. We say set $A$ is *polynomial-time bounded truth-table reducible* to set $B$, written $A \leq_{btt}^P B$, if $A \leq_{tt}^P B$ via a reduction function $f$ and a set $C \in P$ such that, for any $x$, $f(x) = \langle y_1, \ldots, y_m \rangle$ for some constant $m \geq 1$. Show that a set $A$ is in $BH$ if and only if $A$ is the union of a finite number of sets in $DP$ if and only if $A \leq_{btt}^P B$ for some $B \in NP$.

(b) The classes $DP_n$ have complete sets. Define GCOLOR$_k$ be the following problem: Given a graph $G$, determine whether the chromatic number $\chi(G)$ of graph $G$ (the minimum number of colors necessary to color graph $G$) is odd and is between $3k$ and $4k$. Prove that for each odd $k$, GCOLOR$_k$ is $DP_k$-complete.

(c) Show that the Boolean hierarchy $BH$ is properly infinite (i.e., $DP_n \subsetneq DP_{n+1}$ for each $n \geq 0$) unless the polynomial-time hierarchy collapses.

**3.5** An *integer expression $E$* is defined like a regular expression. It contains integer constants and the $+$ and $\cup$ operations. Each integer expression $E$ represents a set $L(E)$ of integers: $L(k) = \{k\}$, $L(E+F) = \{m+n : m \in L(E), n \in L(F)\}$, and $L(E \cup F) = L(E) \cup L(F)$. The problem INTEGER EXPRESSION EQUIVALENCE (IEE) asks, for two given integer expressions $\langle E_1, E_2 \rangle$, whether $L(E) = L(F)$. Prove that IEE is $\Pi_2^P$-complete.

**3.6** The problem GRAPH CONSISTENCY (GC) asks, for two given sets $A$ and $B$ of graphs, whether there exists a graph $G$ such that each graph $g \in A$ is isomorphic to a subgraph of $G$ and each graph $h \in B$ is not isomorphic to any subgraph of $G$. Prove that the problem GC is $\Sigma_2^P$-complete.

**3.7** In the following, each problem is formulated to be a minimax function $f$. Prove that the corresponding decision problem of determining whether $f(G) \geq K$ for some given graph $G$ and constant $K$ is $\Pi_2^P$-complete. In the following, we write $F_{n,m}$ to mean the set of all functions $t : \{1, \ldots, n\} \to \{1, \ldots, m\}$.

(a) MINIMAX-CLIQUE: For a given graph $G = (V, E)$ with its vertices $V$ partitioned into subsets $V_{i,j}$, for $1 \leq i \leq I$, $1 \leq j \leq J$, find $f_{\text{CLIQUE}}(G) = \min_{t \in F_{I,J}} \max_Q \{|Q| : Q \subseteq V \text{ is a clique in } G_t\}$, where $G_t$ is the induced subgraph of $G$ on the vertex set $V_t = \bigcup_{i=1}^I V_{i,t(i)}$.

(b) MINIMAX-CIRCUIT: Given a graph $G$ with its vertex set $V$ partitioned into subsets $V_{i,j}$, for $1 \leq i \leq I$, $1 \leq j \leq J$, find $f_{\text{CIRCUIT}}(G) = \min_{t \in F_{I,J}} \max_{V'} \{|V'| : V' \subseteq V_t, G_t \text{ has a circuit on } V'\}$, where $G_t$ is defined as in (a) above.

(c) MINIMAX-3DM: Given mutually disjoint finite sets $W_{i,j}$, $1 \leq i \leq I$, $1 \leq j \leq J$, and a set $S$ of 3-element subsets of $W = \bigcup_{i=1}^I \bigcup_{j=1}^J W_{i,j}$, find $f_{\text{3DM}}(W, S) = \min_{t \in F_{I,J}} \max_{S'} \{|S'| : S' \subseteq S, S' \text{ is a matching in } W(t)\}$, where $W(t)$ means the set $\bigcup_{i=1}^I W_{i,t(i)}$.

(d) LONGEST DIRECT CIRCUIT: Given a digraph $G = (V, E)$ and a subset $E'$ of $E$, find $f_{\text{LDC}}(G, E') = \min_{D \subseteq E'} \max_{V'} \{|V'| : V' \subseteq V, G_D \text{ has a circuit}$

on $V'$}, where $G_D$ is the subgraph of $G$ with vertex set $V$ and edge set $(E - D) \cup \{(s,t) \mid (t,s) \in D\}$.

**3.8** Let *CFL* denote the class of context-free languages, and let *LOGCFL* be the class of sets that are reducible to the class *CFL* under the logarithmic-space reductions.

(a) Prove that $LOGCFL \subseteq ATIME((\log n)^2)$.

(b) Prove that $NSPACE(\log n) \subseteq LOGCFL$.

**3.9** The *transitive closure* of a relation $R$ is the set $R^* = \{\langle x,y \rangle :$ there exists a sequence of strings $x = z_0, z_1, \ldots, z_m = y$, such that $\langle z_i, z_{i+1} \rangle \in R$ for all $i$, $0 \leq i \leq m - 1$.

(a) Prove that for any relation $R \in P$ with the property $\langle x,y \rangle \in R \Rightarrow |x| = |y|$, $R^* \in PSPACE$.

(b) Prove that there exists a relation $R \in P$ with the property $\langle x,y \rangle \in R \Rightarrow |x| = |y|$, such that $R^*$ is *PSPACE*-complete.

**3.10** The problem DETERMINISTIC FINITE AUTOMATA INTERSECTION (DFA-INT) asks, for a given sequence of deterministic finite automata $M_1, M_2, \ldots, M_k$ over a common alphabet $\Sigma$, whether there exists a string $w \in \Sigma^*$ that is accepted by each $M_i$, $1 \leq i \leq k$. Prove that the problem DFA-INT is *PSPACE*-complete.

**3.11** The problem MINIMAL NONDETERMINISTIC FINITE AUTOMATON (MIN-NFA) asks, for a given deterministic finite automaton $M$ and an integer $n$, whether there exists an equivalent nondeterministic finite automaton that has at most $n$ states. Prove that the problem MIN-NFA is *PSPACE*-complete.

**3.12** Define formally the set $G_{QBF}$ of game configurations and the relations $R_0$ and $R_1$ for the game QBF. Prove that it is a polynomially bounded game.

**3.13** Prove that the game GEOGRAPHY played on undirected graphs is also *PSPACE*-complete.

**3.14** The game HEX played between two players 0 and 1 can be defined as follows: The game is played on a graph $G = (V, E)$ with four explicitly marked nodes $s_0$, $t_0$, $s_1$, $t_1$. The goal of the game for player 0 is to find a path from $s_0$ to $t_0$, and for player 1 to find a path from $s_1$ to $t_1$ by player 1. At the beginning, all but the four special nodes are uncolored, and the four special nodes are colored by $c(s_0) = c(t_0) = 0$ and $c(s_1) = c(t_1) = 1$. A legal move of player 0 (or, player 1) is to mark any uncolored node with the color 0 (or, color 1, respectively). The game ends when there is a path from $s_0$ to $t_0$ all colored with 0 (and player 0 wins) or when there is a path from $s_1$ to $t_1$ all colored with 1 (and player 1 wins).

(a) Define formally the set $G_{HEX}$ of game configurations and the relations $R_0$ and $R_1$. Prove that HEX is a polynomially bounded game.

(b) Prove that game HEX is *PSPACE*-complete.

**3.15** Give a proof for Corollary 3.20 without using the notion of alternating Turing machines (instead, using the idea of the proof of Theorem 3.18 directly).

**3.16** Verify that the size of the machine $M_C$ of Theorem 3.31 is bounded by $p(|\langle M, w \rangle|)$ for some polynomial $p$.

**3.17** Consider the following variation of the game BFG:

a) The variables of the formula $F$ are partitioned into two subsets, $X$ and $Y$.

b) In one move, player 0 (player 1) can change the value of at most one variable in $X$ (respectively, in $Y$).

c) The initial assignment $\tau_0$ makes $F$ false.

d) A player wins if the formula becomes true after his/her move.

Prove that this version of the game BFG is also *EXP*-complete.

**3.18** Prove Theorem 3.33.

**3.19** Prove Theorem 3.34.

**3.20** The problem ORACLE-SAT is defined as follows: Let $G$ be a Boolean formula over five sets of variables $X_1, X_2, X_3, Y, Z$, where $|X_j| = |Y| = n$ for $j = 1, 2, 3$, and $Z = \{z_1, z_2, z_3\}$. Determine whether it is true that there exists a Boolean function $g : \{0,1\}^n \to \{0,1\}$ such that for all assignments $\pi : X_1 \cup X_2 \cup X_3 \cup Y \to \{0,1\}$, $G$ is satisfied by $\pi$ and $\tau$, where $\tau : Z \to \{0,1\}$ is defined by $\tau(z_j) = g(\pi(X_j))$. (If $X = \{x_1, x_2, \ldots, x_n\}$, then we write $\pi(X)$ to denote the string of $n$ bits: $\pi(x_1)\pi(x_2)\cdots\pi(x_n)$.) Prove that ORACLE-SAT is $\leq_m^P$-complete for *NEXP*.

## Historical Notes

The polynomial-time hierarchy was first introduced by Stockmeyer and Meyer [1973] as a polynomial-time analogue of the arithmetic hierarchy [Rogers, 1967]. Stockmeyer [1977] and Wrathall [1977] contain complete sets $\text{SAT}_k$ and the alternating quantifier characterizations. Stockmeyer [1977] also contains the $\Pi_2^P$-complete problem IEE (Exercise 3.5). The $\Pi_2^P$-complete problem GRN is from Ko and Lin [1995a]. Alternating Turing machines are first defined in Chandra et al. [1981]. The simulations between ATMs and DTMs and NTMs are proven there. The alternating quantifier characterization of *PSPACE* and the *PSPACE*-complete problems QBF are from Stockmeyer and Meyer [1973], and the *PSPACE*-completeness of TRE is from Meyer and Stockmeyer [1972]. The first natural *PSPACE*-complete two-person game is HEX (Exercise 3.14), proved in Even and Tarjan [1976]. Schaefer [1978b] contains many other *PSPACE*-complete games, including the game GEOGRAPHY. The idea of using Turing machines to encode objects of exponential size has been used in many applications; see, for instance, Ko [1992] for the *EXP*-complete problem EXP-CVP. The game BFG (Theorem 3.32 and Exercise 3.17) is from Stockmeyer and Chandra [1979]. The *EXPSPACE*-complete problem TERE is from Hunt [1973].

The notion of strong-$NP$ reducibility was introduced by Long [1982]. A related strong-$NP$ many-one reducibility was first used by Adleman and Manders [1977]. The class *DP* and the corresponding *DP*-complete problems were studied in Papadimitriou and Yannakakis [1984] and Papadimitriou and Wolfe [1987]. The Boolean hierarchy has been studied by a number of people. A comprehensive study was presented in Cai et al. [1988, 1989]. The $\Sigma_2^P$-complete problem GC (Exercise 3.6) is from Ko and Tzeng [1991]. The $\Pi_2^P$-complete problems in Exercise 3.7 are from Ko and Lin [1995a]. The class *LOGCFL* has been studied in connection with parallel complexity. Exercise 3.8(a) is from Ruzzo [1980] and Exercise 3.8(b) is from Sudborough [1978]. The complexity of transitive closure was studied in Book and Wrathall [1982] and Ko [1985a]. The DFA Intersection problem is from Kozen [1977] and the minimal NFA problem is from Jiang and Ravikumar [1993].

# 4

# *Structure of NP*

In order to understand the difficulty of solving the $P$ versus $NP$ problem, we study in this chapter the internal structure of the complexity class $NP$. We demonstrate some natural problems as candidates of incomplete problems in $NP - P$, and study the notion of one-way functions. We also introduce the notion of relativization to help us understand the possible relations between subclasses of $NP$. One of the main proof techniques used in this study is stage-construction diagonalization, which has been used extensively in recursion theory.

## 4.1 Incomplete Problems in $NP$

We have seen many $NP$-complete problems in Chapter 2. Many natural problems in $NP$ turn out to be $NP$-complete. There are, however, a few interesting problems in $NP$ that are not likely to be solvable in deterministic polynomial time but also are not known to be $NP$-complete. The study of these problems is thus particularly interesting, because it not only can classify the inherent complexity of the problems themselves, but also provides a glimpse of the internal structure of the class $NP$. We start with some examples in number theory.

**Example 4.1** PRIMALITY TESTING (PRIME): Given a positive integer $N$ in the binary form, determine whether $N$ is a prime number.

The complement of PRIME is easily seen in $NP$ (and, hence, PRIME is in *coNP*): for any given integer $n$, we can guess a number $d$ between 2 and $n - 1$

and verify that $d$ divides $n$ thus proving that $n$ is not a prime number. Since $d < n$, the length of the witness $d$ is shorter than the length of the input, and so this simple algorithm works in nondeterministic polynomial time.

In the following, we show that the problem PRIME is also in *NP*. The proof requires two important results in number theory: Fermat's Theorem and Euler's Theorem. We omit their proofs and refer the reader to any standard textbook on number theory (e.g., Niven and Zuckerman [1960]). For any number $n > 0$, let $Z_n^* = \{x \in \mathbf{N} : 0 < x < n, \gcd(x, n) = 1\}$, where $\gcd(x, n)$ denotes the *greatest common divisor* of $x$ and $n$. In particular, if $p$ is a prime, then $Z_p^* = \{1, 2, \ldots, p-1\}$. Note that $Z_n^*$ is a multiplicative group. The order of the group $Z_n^*$ is denoted by $\varphi(n)$, and it is called the *Euler function*.

**Lemma 4.2** *(a)* (Fermat's Theorem) *For any positive integer $m$ and any $a \in Z_m^*$, $a^{\varphi(m)} \equiv 1 \bmod m$.*

*(b)* (Euler's Theorem) *For any positive integer $m$, the sum of $\varphi(d)$ over all divisors $d$ of $m$ is equal to $m$.*

**Lemma 4.3** *An odd integer $n > 2$ is a prime if and only if there exists a number $a \in Z_n^*$ such that*

*(i) $a^{n-1} \equiv 1 \bmod n$, and*

*(ii) $a^{(n-1)/q} \not\equiv 1 \bmod n$, for all prime divisors $q$ of $n - 1$.*

*Proof.* First assume that $n$ is a prime. Let the *order* of $a \in Z_n^*$ be the least number $k$ such that $a^k \equiv 1 \bmod n$. Note that each $a \in Z_n^*$ having the order $n - 1$ satisfies conditions (i) and (ii). So we only need to verify that there must be an $a \in Z_n^*$ having order $n - 1$. Let $S_d$ denote the set of all $a \in Z_n^*$ having order $d$. Then $S_d = \emptyset$ if $d$ does not divide $n - 1$. If $d$ is a divisor of $n - 1$, what is the size of $S_d$? First, the equation $x^d \equiv 1 \bmod n$ has at most $d$ solutions, because a polynomial of degree $d$ has at most $d$ zeros. Therefore, $S_d$ has at most $d$ elements. In addition, there is an integer $a$ such that $S_d \subseteq \{a, a^2, \ldots, a^d\}$ (if $b$ is a solution then $b^k$ is also a solution for any $k > 0$). Second, for each $a \in S_d$, we claim that $a^c \in S_d$ if and only if $\gcd(c, d) = 1$. To see this, we observe that $(a^c)^{d/\gcd(c,d)} \equiv 1 \bmod n$, and $a^c \in S_d$ implies $d/\gcd(c, d) \geq d$, or $\gcd(c, d) = 1$. Conversely, if $a^{cm} \equiv 1 \bmod n$ then $d$ divides $cm$, and so $\gcd(c, d) = 1$ implies that $d$ divides $m$ and $d$ is the least number $m$ satisfying $a^{cm} \equiv 1 \bmod n$. Combining the above two observations together, we conclude that $|S_d| = \varphi(d)$. In particular, $|S_{n-1}| = \varphi(n - 1) > 0$.

For the backward direction, suppose $a \in Z_n^*$ satisfies (i) and (ii). Condition (ii) implies that $a$ has order $n - 1$. Together with Fermat's Theorem, we know that $\varphi(n) = n - 1$ and so $n$ is a prime. ∎

**Lemma 4.4** *The exponentiation function modulo an integer $n$, $x^y \bmod n$, is computable in deterministic polynomial time.*

*Proof.* (Sketch) To find $x^y \bmod n$, we write $y$ in the binary form as $y = \sum_{i=0}^k y_i 2^i$, where $y_i \in \{0, 1\}$. Then, $x^y \bmod n = \prod_{0 \leq i \leq k, y_i = 1}(x^{2^i} \bmod n)$.

That means we only need to find $x^{2^i} \bmod n$ for $0 \le i \le \lceil \log n \rceil$, and then perform at most $\lceil \log n \rceil$ modulo multiplications. Each $x^{2^i} \bmod n$ can itself be computed using $i$ modulo multiplications, and so altogether we only need $O(\log n)$ modulo multiplications. For a detailed analysis of the complexity of modulo arithmetic operations, see, for instance, Knuth [1981].                   ∎

**Theorem 4.5** *The problem* PRIME *is in NP.*

*Proof.* We present a recursive, nondeterministic algorithm for PRIME based on Lemma 4.3.

*Nondeterministic Algorithm for* PRIME.

For any given odd integer $n > 2$, we guess an integer $a$, $1 \le a \le n-1$, and a sequence of numbers $q_1, q_2, \ldots, q_k$ between 2 and $n-2$. Then, verify the following conditions:

   (1) Each $q_i$, $1 \le i \le k$, is a prime number;
   (2) $q_i$'s are the only prime factors of $n - 1$;
   (3) $a^{n-1} \equiv 1 \bmod n$;
   (4) $a^{(n-1)/q_i} \not\equiv 1 \bmod n$, for all $1 \le i \le k$.

Note that the verification of condition (1) is to be done recursively using the above algorithm, since all $q_i$'s are less than $n-1$. The correctness of the above algorithm follows from Lemma 4.3. As an example, consider the prime number $311 = 2 \cdot 5 \cdot 31 + 1$. One of the accepting computations of the above algorithm on 311 works as follows: The first round guesses $(17; 2, 5, 31)$, and verifies that (2) $311 - 1 = 2 \cdot 5 \cdot 31$, (3) $17^{310} \equiv 1 \bmod 311$, and (4) $17^{155} \equiv 310 \bmod 311$, $17^{62} \equiv 36 \bmod 311$, and $17^{10} \equiv 260 \bmod 311$. Next, for number 31, the guess is $(3; 2, 3, 5)$, and the verifications are (2) $31 - 1 = 2 \cdot 3 \cdot 5$, (3) $3^{30} \equiv 1 \bmod 31$, and (4) $3^{15} \equiv 30 \bmod 31$, $3^{10} \equiv 25 \bmod 31$, and $3^6 \equiv 16 \bmod 31$.

In the following, we verify that the above algorithm accepts a prime number $n$ in time polynomial in $\log n$ and, hence, is a polynomial-time algorithm, since the size of the binary representation of $n$ is of length $\le \log n + 1$. First, we observe that for any prime $n$, if the algorithm guesses the correct prime factors $q_1, \ldots, q_k$ of $n - 1$ then, by Lemma 4.4, steps (2), (3), and (4) can be done by $O((\log n)^2)$ modulo multiplications, because $n - 1$ has at most $\log n$ prime factors. Therefore, the total number $m(n)$ of modulo multiplications done by the algorithm satisfies

$$ m(n) \le c(\log n)^2 + \sum_{i=1}^{k} m(q_i), $$

for some constant $c > 0$. Note that each $q_i$, $1 \le i \le k$, is between 2 and $(n-1)/2$, and that $\sum_{i=1}^{k} \log q_i \le \log n$. It follows that

$$ \sum_{i=1}^{k} (\log q_i)^3 \le (\log n)^3 - (\log n)^2 $$

and, hence, $c(\log n)^3$ is an upper bound of $m(n)$.                                                ∎

The above example shows that PRIME is in $NP \cap coNP$. An immediate consequence is that it is not complete for $NP$ or $coNP$ unless $NP = coNP$ (since the $\leq^P_m$-reduction preserves the membership in $NP$). So, what is the exact complexity of the problem PRIME? Is it actually in $P$ or is it complete for the class $NP \cap coNP$?

Unfortunately, we are not able to answer this question. In fact, we don't even know a single complete problem for $NP \cap coNP$. In our construction of complete sets for $NP$ and $PSPACE$, we always find a *generic* complete problem BHP or SBHP. A similar generic complete problem also exists for $\Sigma^P_k$, for each $k > 1$. These generic complete problems exist because of two properties of the complexity classes $NP$, $\Sigma^P_k$, and $PSPACE$: (1) We can recursively generate the underlying complexity-bounded machines that accept all sets in these classes; and (2) There is a universal machine to simulate these machines in polynomial time and/or polynomial space. It is difficult to construct even the first complete problem for the class $NP \cap coNP$ because it does not seem to have these two properties. The straightforward way to present a machine model for the class $NP \cap coNP$ is to use a pair of NTMs $(M_0, M_1)$ so that $L(M_0) = \overline{L(M_1)}$. However, the condition of $L(M_0) = \overline{L(M_1)}$ is a *semantic* property (meaning a property that cannot be determined by simply inspecting the codes of machines $M_0$ and $M_1$) and is, in general, unsolvable. Hence this approach does not give a generic complete problem.

Besides the above difficulty of finding a complete problem for $NP \cap coNP$, other evidence also suggests that the problem PRIME is not likely to be the hardest problem in $NP \cap coNP$ (see Chapter 8 for more discussion). The exact complexity of PRIME remains an interesting open question.

In addition to the problem PRIME, there are many number-theoretic problems in $NP$ that are not known to be $NP$-complete nor known to be in $P$. We list three of them that have major applications in cryptography. An integer $x \in Z^*_n$ is called a *quadratic residue* modulo $n$ if $x \equiv y^2$ mod $n$ for some $y \in Z^*_n$. We write $x \in QR_n$ to denote this fact.

**Example 4.6** (a) INTEGER FACTORING (FACTOR): Given a positive integer $n$, find its prime factors.

(b) QUADRATIC RESIDUOSITY (QR): Given two integers $n$ and $x \in Z^*_n$, determine whether $x \in QR_n$.

(c) SQUARE ROOT MODULO AN INTEGER (SQRT): Given two integers $n$ and $x \in QR_n$, find a $y \in Z^*_n$ such that $y^2 \equiv x$ mod $n$.

Similar to the complement of PRIME, all these three problems are easily seen to be in $NP$. For the problems FACTOR and SQRT, we need to reformulate them as decision problems:

> FACTOR: Given integers $(n, a, b)$, determine whether there is an integer $d$, $a \leq d \leq b$, that divides $n$.

SQRT: Given integers $n$, $x \in QR_n$, and $z \leq n - 1$, determine whether there is a $y \in Z_n^*$ such that $y \leq z$ and $y^2 \equiv x \bmod n$.

However, unlike the problem $\overline{\text{PRIME}}$, we do not know whether they are in *coNP*. We are going to see in Chapter 8 that if $n$ is a prime, then the problem SQRT has a polynomial-time *probabilistic* algorithm and, hence, not likely to be *NP*-complete. For the case of composite $n$, there is some evidence showing that the problem SQRT is solvable in deterministic polynomial time if and only if FACTOR is solvable in deterministic polynomial time (see Rabin [1979]).  □

Our last example is a natural graph-theoretic problem.

**Example 4.7** GRAPH ISOMORPHISM (GIso): Given two graphs $G_1 = (V_1, E_1)$ and $G_2 = (V_2, E_2)$, determine whether they are isomorphic, that is, whether there is a bijection $f : V_1 \rightarrow V_2$ such that $\{u, v\} \in E_1$ if and only if $\{f(u), f(v)\} \in E_2$.

The problem SUBGRAPH ISOMORPGISM, which asks whether a given graph $G_1$ is isomorphic to a subgraph of another given graph $G_2$, can be proved to be *NP*-complete easily. However, the problem GIso is not known to be *NP*-complete nor known to be in *P*, despite extensive studies in recent years. We will prove in Chapter 10, through the notion of interactive proof systems, that GIso is not *NP*-complete unless the polynomial-time hierarchy collapses to the level $\Sigma_2^P$. This result suggests that GIso is, like the problem PRIME, not *NP*-complete, but the supporting evidence is a little weaker than that for PRIME.  □

The above examples suggest that there are problems in $NP - P$ that are not complete for *NP*. However, we don't even know that such sets exist at all. In the following we show, in an abstract form, that such sets indeed exist, if $P \neq NP$. To prepare for the proof of the theorem, we recall from Chapter 1 that there exists an effective enumeration $\{M_i\}$ of all polynomial-time DTMs, as well as an effective enumeration $\{N_i\}$ of all polynomial-time oracle DTMs. For the following theorem, we need an even stronger property of these enumerations; that is, the enumerations can be done in polynomial time in the sense that the functions $g(i) = [\text{code of } M_i]$ and $h(i) = [\text{code of } N_i]$ are polynomial-time computable. It is easy to see that this stronger property follows from our enumeration method in Section 1.5. Also, in the following theorem, we let $A \triangle B$ denote the set difference between two sets $A$ and $B$; that is, $A \triangle B = (A - B) \cup (B - A)$.

**Theorem 4.8** *Assume that $P \neq NP$. Then, there exists a set $A \in NP - P$ such that $A$ is not $\leq_T^P$-complete for NP.*

*Proof.* Let $\{M_i\}$ and $\{N_i\}$ be the enumerations of machines as described above. Also let $M_{\text{SAT}}$ be a *deterministic* TM accepting SAT. We want to construct a set $A$ satisfying (1) $A \in NP$, (2) $A \notin P$, and (3) SAT $\not\leq_T^P A$. In terms of the enumerations $\{M_i\}$ and $\{N_i\}$, we can restate these requirements as follows:

$R_1$: $A \leq_m^P$ SAT,

$R_{2,i}$: $A \neq L(M_i)$, $i \geq 1$,

$R_{3,i}$: SAT $\neq L(N_i, A)$, $i \geq 1$,

and we need to satisfy $R = R_1 \wedge (\bigwedge_{i=1}^{\infty} R_{2,i}) \wedge (\bigwedge_{i=1}^{\infty} R_{3,i})$. To satisfy $R_1$, we will construct a set $S \in P$ and define $A = $ SAT $\cap S$. Then, it is easy to see that $A \leq_m^P$ SAT. (Why?) The set $S$ will be constructed in stages. At an even stage $2i$, we will construct $S$ to satisfy requirement $R_{2,i}$ and at stage $2i + 1$, we will construct $S$ to satisfy requirement $R_{3,i}$.

Assume that in stage $2i - 1$, we have defined a number $n_{2i-1}$ such that the membership of strings of length $\leq n_{2i-1}$ in $S$ have been determined. (Let $n_1$ be an arbitrary number, say, 0, and let $\lambda \notin S$.) Then, at stage $2i$, we will determine the integer $n_{2i} > n_{2i-1}$ and let $x \in S$ for all $x$ of length $n_{2i-1} < |x| \leq n_{2i}$. To determine $n_{2i}$, we note that we need $n_{2i}$ be large enough so that the requirement $R_{2,i}$ is satisfied by set SAT $\cap \{x : n_{2i-1} < |x| \leq n_{2i}\}$ in the sense that there exists a string $x$ of length $n_{2i-1} < |x| \leq n_{2i}$ such that $x \in $ SAT $\Delta L(M_i)$. By the assumption that $P \neq NP$, such an $x$ must exist. We consider a TM $U$, which operates as follows: On input $i$, it first simulates the stages $1, 2, \ldots, 2i - 1$ to get integers $n_j$, for $j = 1, \ldots, 2i - 1$. Then, it simulates $M_i$ and $M_{\text{SAT}}$ on $x$ of length $|x| > n_{2i-1}$, in the increasing order, until an $x$ is found in SAT $\Delta L(M_i)$, and it outputs this $x$. We define $n_{2i}$ to be the number of moves of $U$ taken on input $i$. Note that if $x$ is the output of $U(i)$, then $n_{2i-1} < |x| \leq n_{2i}$. Therefore, the requirement $R_{2,i}$ is satisfied.

At stage $2i + 1$, we assume that $n_{2i}$ is already determined by stage $2i$. We will determine $n_{2i+1}$ and let $x \notin S$ for all $x$ of length $n_{2i} < |x| \leq n_{2i+1}$. Our requirement here is SAT $\neq L(N_i, A)$ and we need a witness $x \in $ SAT $\Delta L(N_i, A)$. However, set $A$ is not completely defined yet. What we can do is to pretend that $A_{2i} = \{x \in $ SAT $: n_{2j-1} < |x| \leq n_{2j}, 1 \leq j \leq i\}$ is equal to $A$. That is, we will satisfy the requirement SAT $\neq L(N_i, A_{2i})$ instead of $R_{3,i}$, and in later stages, make sure that the computation of $N_i^{A_{2i}}$ on $x$ is the same as the computation of $N_i^A$ on $x$, where $x$ is the witness found in stage $2i + 1$.

To do this, we consider the following TM $V$: On input $i$, it first simulates the stages $1, 2, \ldots, 2i$ to get integers $n_j$, for $j = 1, \ldots, 2i$. Then, it simulates machines $M_{\text{SAT}}$ and $N_i^{A_{2i}}$ on input $x$ of length $> n_{2i}$, in the increasing order. For each $x$, if machine $N_i$ queries whether a string $y$ is in $A_{2i}$, it answers YES if and only if $n_{2j-1} < |y| \leq n_{2j}$ for some $j \leq i$ and $y \in $ SAT. Finally, the machine $V$ on input $i$ outputs the smallest $x$ of length $> n_{2i}$ which is in SAT$\Delta L(N_i, A_{2i})$. (Since $A_{2i}$ is a finite set, such an $x$ exists by the assumption that SAT $\notin P$.) We let $n_{2i+1}$ be the runtime of the machine $V$ on input $i$. Note that $n_{2i+1} \geq |x|$, where $x$ is the output of $V(i)$, and all queries made by $N_i^{A_{2i}}$ on $x$ are of length $\leq n_{2i+1}$. By setting this $n_{2i+1}$ and noticing that, in later stages, we will not change the membership of strings of length $\leq n_{2i+1}$, we conclude that $x \in L(N_i, A) \iff x \in L(N_i, A_{2i})$, and the requirement $R_{3,i}$ is satisfied.

Finally, we let $S$ be the set $\{x : n_{2i-1} < |x| \leq n_{2i}, i \geq 1\}$. We need to prove that set $S$ is computable in polynomial time. To see this, we first observe that the simulation of stage $j$ can be done in time $O(n_j)$, since $n_j$ is the runtime of stage $j$ to find $n_j$. Now, for any $x$ we can determine whether it is in $S$ as follows: We simulate stages $j$, $j \geq 1$, in the increasing order, and keep a counter to count the number of moves simulated up to that point. Before the simulation of stage $j$, we know that $n_{j-1} < |x|$. During the simulation of stage $j$, if the value of the counter becomes bigger than $|x|$, then we know that $|x| \leq n_j$, and we halt and output that $x \in S$ if and only if $j$ is even. This completes the proof of the theorem. ∎

The above proof method is called the *delayed diagonalization*, since the diagonalization requirements $R_{2,i}$ and $R_{3,i}$ are satisfied by a search process that always halts but the amount of time for the search is not known in advance.

The following theorem, an analog of Post's problem in the classical recursion theory, reveals further structural properties of languages in *NP*. It can be proved using the technique of delayed diagonalization. We leave its proof as an exercise.

**Theorem 4.9** *Assume that $P \neq NP$. Then, there exist sets $A, B \in NP$ such that $A \not\leq_T^P B$ and $B \not\leq_T^P A$.*

## 4.2 One-Way Functions and Cryptography

One-way functions are a fundamental concept in cryptography, having a number of important applications, including public-key cryptosystems, pseudorandom generators, and digital signatures. Intuitively, a one-way function is a function that is easy to compute but its inverse is hard to compute. Thus it can be applied to develop cryptosystems that need easy encoding but difficult decoding. If we identify the intuitive notion of "easiness" with the mathematical notion of "polynomial-time computability," then one-way functions are subproblems of *NP*, since the inverse function of a polynomial-time computable function is computable in polynomial-time relative to an oracle in *NP*, assuming that the functions are polynomially honest. Indeed, all problems in *NP* may be viewed as one-way functions.

**Example 4.10** Define a function $f_{\text{SAT}}$ as follows: For each Boolean function $F$ over variables $x_1, \ldots, x_n$, and each Boolean assignment $\tau$ on $x_1, \ldots, x_n$,

$$f_{\text{SAT}}(F, \tau) = \begin{cases} \langle F, 1 \rangle & \text{if } \tau \text{ satisfies } F, \\ \langle F, 0 \rangle & \text{otherwise.} \end{cases}$$

It is easily seen that $f_{\text{SAT}}$ is computable in polynomial time. Its inverse mapping $\langle F, 1 \rangle$ to $\langle F, \tau \rangle$ is exactly the search problem of finding a truth assignment for a given Boolean formula. Using the notion of polynomial-time Turing reducibility and the techniques developed in Chapter 2, we can see that the inverse function of $f_{SAT}$ is polynomial-time equivalent to the decision problem SAT. Thus, the inverse of $f_{SAT}$ is not polynomial-time computable if $P \neq NP$.                                                                                □

Strictly speaking, function $f_{SAT}$ is, however, not really a one-way function because it is not a one-to-one function and its inverse is really a *multi-valued* function. In the following, we define one-way functions for one-to-one functions. We say that a function $f : \Sigma^* \to \Sigma^*$ is *polynomially honest* if there is a polynomial function $q$ such that for each $x \in \Sigma^*$, $|f(x)| \leq q(|x|)$ and $|x| \leq q(|f(x)|)$.

**Definition 4.11** *A function $f : \Sigma^* \to \Sigma^*$ is called a* (weak) one-way function *if it is one-to-one, polynomially honest, polynomial-time computable, and there is no polynomial-time computable function $g : \Sigma^* \to \Sigma^*$ satisfying $g(f(x)) = x$ for all $x \in \Sigma^*$.*

The above definition did not define the notion of inverse function of $f$, since $f$ is not necessarily a surjection. We may arbitrarily define the inverse function $f^{-1}$ of a one-to-one function $f$ to be

$$f^{-1}(y) = \left\{ \begin{array}{ll} x & \text{if } y \in Range(f) \text{ and } f(x) = y, \\ \lambda & \text{if } y \notin Range(f), \end{array} \right.$$

and have the following equivalent definition.

**Proposition 4.12** *A function $f$ is one-way if and only if it is one-to-one, polynomially honest, polynomial-time computable and its inverse $f^{-1}$ is not polynomial-time computable.*

*Proof.* If $g$ satisfies $g(f(x)) = x$ for all $x$, then $y \in Range(f)$ if and only if $f(g(y)) = y$.                                                                                                      ∎

Following the example $f_{SAT}$, it is easy to see that if $P = NP$ then one-way functions do not exist. Indeed, the complexity of weak one-way functions can be characterized precisely by a subclass *UP* of *NP*.

**Definition 4.13** *An NTM $M$ is said to be* unambiguous *if for each input $x \in L(M)$, there is exactly one accepting computation path for $M(x)$. The class UP is the class of sets that are accepted by polynomial-time unambiguous NTMs.*

It is clear that $P \subseteq UP \subseteq NP$. The following characterization of *UP* is a simple extension of Theorem 2.1.

**Proposition 4.14** *A set $A \subseteq \Sigma^*$ is in UP if and only if there exist a set $B \in P$ and a polynomial $p$ such that for each $x \in \Sigma^*$,*

$$x \in A \Longleftrightarrow (\exists y, |y| \leq p(|x|)) \langle x, y \rangle \in B$$
$$\Longleftrightarrow (\exists \text{ unique } y, |y| \leq p(|x|)) \langle x, y \rangle \in B.$$

**Proposition 4.15** *(a) There exists a one-way function if and only if $P \neq UP$.*

*(b) There exists a one-way function whose range is in $P$ if and only if $P \neq UP \cap coUP$.*

*Proof.* (a): First assume that $f$ is a one-way function. Define $A_f = \{\langle y, u \rangle : (\exists v) \ f(uv) = y\}$. Then, from the polynomial honesty of $f$, $A_f$ is in *NP*, since we only need to guess a string $v$ of length at most $q(|y|)$, for some fixed polynomial $q$, and verify that $f(uv) = y$. In addition, $A_f \in UP$ since there is at most one string $x$ such that $f(x) = y$, and so the witness $v$, if exists, is unique. Finally, we observe that the inverse function $f^{-1}$ is polynomial-time computable using $A_f$ as an oracle (cf. Exercise 2.2). Thus, $A_f$ is not in $P$.

For the converse, we assume that $A \in UP - P$; or, equivalently, there exist a set $B \in P$ and a polynomial $p$ such that for each $x$,

$$x \in A \Longleftrightarrow (\exists y, |y| \leq p(|x|)) \ \langle x, y \rangle \in B$$
$$\Longleftrightarrow (\exists \text{ unique } y, |y| \leq p(|x|)) \ \langle x, y \rangle \in B.$$

Define a function

$$f_A(\langle x, y \rangle) = \begin{cases} \langle x, 1 \rangle & \text{if } \langle x, y \rangle \in B, \\ \langle x, 0y \rangle & \text{otherwise.} \end{cases}$$

Then, $f_A$ is a one-to-one function, since for each $x$ there is at most one string $y$ satisfying $\langle x, y \rangle \in B$. It is also easy to see that $f_A$ is polynomially honest. Finally, suppose, for the sake of contradiction, that there is a polynomial-time computable function $g$ such that $g(f_A(\langle x, y \rangle)) = \langle x, y \rangle$ for all $\langle x, y \rangle$. Then, the following algorithm solves the problem $A$ in polynomial time, and gives us a contradiction:

For any input $x$, compute $z = g(\langle x, 1 \rangle)$;
If $f(z) = \langle x, 1 \rangle$, then accept $x$, else reject it.

Part (b) can be proved in a similar way. We leave it as an exercise (Exercise 4.6). ∎

A one-way function $f$ satisfying the conditions of Definition 4.11 is called a *weak* one-way function because its inverse is not computable in polynomial time in the very weak worst-case complexity measure. That is, for every polynomial-time computable function $g$ attempting to compute the inverse of $f$, there are infinitely many instances $y_1, y_2, \ldots \in Range(f)$, such that $f(g(y_i)) \neq y_i$ for all $i \geq 1$. However, this function $g$ could be a very good approximation to $f$ in the sense that such errors occur only on some very sparsely distributed instances $y_1, y_2, \ldots$ . In the applications of cryptography, a cryptosystem based on such a one-way function is not secure since the approximation function $g$ allows a fast decoding for a large portion of the input instances. Instead, stronger one-way functions whose inverses are hard to compute for the majority of input instances are needed for a secure cryptosystem. We formulate this notion in the following. We only consider functions defined on alphabet $\{0, 1\}$.

**Definition 4.16** *A function* $f : \{0,1\}^* \rightarrow \{0,1\}^*$ *is a* strong one-way func-tion *if* $f$ *is one-to-one, polynomially honest, polynomial-time computable, and for any polynomial-time computable function* $g$, *and for any polynomial function* $p$, *the set* $S_g = \{x \in \Sigma^* : g(f(x)) = x\}$ *has the density* $|S_g \cap \{0,1\}^n| \le 2^n/p(n)$ *for almost all* $n > 0$.

In other words, a function $f$ is strongly one-way if any polynomial-time function approximating its inverse succeeds only on a small (or, negligible) amount of input instances.[1] Since this notion of strong one-way functions is much stronger than that of Definition 4.11, it is questionable whether strong one-way functions really exist. Most researchers in cryptography, however, be-lieve that many strong one-way functions do exist. Several natural candidates for strong one-way functions have been found in number-theoretic problems. In particular, the problems FACTOR and SQRT are believed to be so hard to solve that any attack to them can only solve a negligible portion of the input instances. That is, the functions *multiplication* and *square modulo an integer* are candidates of strong one-way functions.[2,3]

Corresponding to the notion of strong one-way functions, there are some natural decision problems that are also considered difficult to solve even for a small portion of the input instances. The following definition captures this notion in the spirit of Definition 4.16.

**Definition 4.17** *A set* $S \subseteq \{0,1\}^*$ *is said to be* strongly unpredictable *if for any polynomial-time computable function* $g : \{0,1\}^* \rightarrow \{0,1\}$ *and any polynomial function* $p$,

$$\left| \Pr_{|x|=n} [g(x) = \chi_S(x)] - \frac{1}{2} \right| \le \frac{1}{p(n)} ,$$

*for almost all* $n > 0$.

For instance, let $n$ be an integer which is a product of two prime numbers, $p$ and $q$. Then, in general, it is considered as an extremely difficult problem to determine whether a given integer $x$, $1 \le x \le n - 1$, is in $QR_n$. In theoretical cryptography, it is often assumed that sets $QR_n$ are strongly unpredictable in the sense that there exists an infinite set $S$ of positive integers such that

---

[1]In the literature, there are many different formulations for the notion of *negligibility*. The above formulation of Definition 4.16 is widely used.

[2]Strictly speaking, these functions are not one-to-one. They are, nevertheless, $k$-to-one for some small $k$. We say they are one-way if the problem of finding any of the $k$ inverse images is difficult.

[3]In a new computational model, the quantum Turing machine model, some of these number-theoretic problems are known to be polynomial-time invertible. These results raise the question of whether these number-theoretic functions are really one-way functions. See, for instance, Deutsch [1985] and Bernstein and Vazirani [1997] for the introduction to quantum complexity theory, and Shor [1997] for the polynomial-time quantum algorithms for some number-theoretic problems.

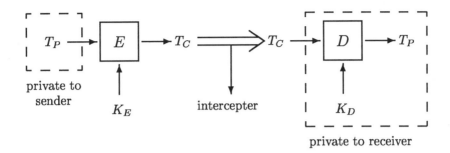

Figure 4.1: A public-key cryptosystem

for any polynomial-time computable function $g : \{0,1\}^* \to \{0,1\}$ and any polynomial function $p$,

$$\left| \Pr_{|x|=n} [g(\langle n,x \rangle) = \chi_{QR_n}(x)] - \frac{1}{2} \right| \leq \frac{1}{p(n)} ,$$

for almost all $n \in S$. (See an application in Exercise 4.9.)

In the rest of this section, we present a public-key cryptosystem based on the assumption that the function *multiplication* is a strong one-way function. We first introduce some terminology for public-key cryptosystems. In a public-key cryptosystem, a *sender $S$* needs to send a *message $T_P$* to a *receiver $R$* via a communication channel that is open to the public (hence not secure), and yet the system needs to protect the message from the unauthorized users of the system. To ensure the security, the sender $S$ first applies an *encryption* algorithm $E$ to map the message $T_P$ (the *plaintext*) to an encrypted message $T_C$ (the *ciphertext*), and then send the message $T_C$ to the receiver $R$. The receiver $R$ then applies a *decryption* algorithm $D$ to map $T_C$ back to $T_P$. Usually, all users in the system use the same encryption and decryption algorithms but each user uses different parameters when applying these algorithms. That is, the encryption and decryption algorithms compute functions of two parameters: $E(K_E, T_P) = T_C$ and $D(K_D, T_C) = T_P$. The parameters $K_E$ and $K_D$ are called the *encryption key* and the *decryption key*, respectively. Each pair of users $S$ and $R$ may be required to set up their own keys for communication. The transmission of the message is illustrated in Figure 4.1.

In a *private-key* cryptosystem, both keys $K_E$ and $K_D$ are kept secret from the public. Each pair of users $S$ and $R$ first, through some other secret channel, develop the encryption and decryption keys and keep them to themselves. Such a system is *secure* if it is hard to compute a plaintext from the ciphertext alone.

In a *public-key* cryptosystem, each user $R$ establishes his/her own pair of keys $K_E$ and $K_D$, and makes the encryption key $K_E$ open to the public, but keeps the decryption key $K_D$ secret to him/herself. Since the encryption key

$K_E$ is known not only to the sender $S$ but also known to all other users, the system is secure only if it is hard to compute the plaintext from the ciphertext *and* the encryption key. In particular, the decryption key must be hard to compute from the encryption key. In other words, the function $f(T_P) = E(K_E, T_P)$ needs to satisfy the following conditions:

(1) $f$ is easy to compute;

(2) $f^{-1}$ is hard to compute even if $K_E$ is known; and

(3) $f^{-1}$ becomes easy to compute if $K_D$ is available.

Such a function is called a *trapdoor one-way function*. So, a one-way function $f$ can be applied to public-key cryptosystems only if it is also a trapdoor function (i.e., to satisfy condition (3) above).

The *Rivest-Shamir-Adleman (RSA)* cryptosystem is based on the one-way function *exponentiation modulo an integer*. In this system, each pair of keys $K_E$ and $K_D$ are selected as follows: First, find two distinct prime numbers $p, q$, and let $n = p \cdot q$. Next, find a pair of integers $e, d < \varphi(n)$ such that

$$gcd(e, \varphi(n)) = 1, \qquad e \cdot d \equiv 1 \bmod \varphi(n), \tag{4.1}$$

where $\varphi(n)$ is the Euler function. Note that for $n = p \cdot q$, $\varphi(n) = (p-1)(q-1)$. Let $K_E = \langle e, n \rangle$ and $K_D = \langle d, n \rangle$. The encryption function $E$ and the decryption function $D$ are defined as follows:

$$E(\langle e, n \rangle, x) = x^e \bmod n, \qquad D(\langle d, n \rangle, y) = y^d \bmod n.$$

The following lemma shows that the above functions $E$ and $D$ are inverses of each other.

**Lemma 4.18** *Assume that $p, q$ are two primes, and $n = p \cdot q$. Also assume that $e, d < (p-1)(q-1)$ and satisfy (4.1). Then, for any $x$, $1 \le x < n$, $D(\langle d, n \rangle, E(\langle e, n \rangle, x)) = x$.*

*Proof.* Note that $D(\langle d, n \rangle, E(\langle e, n \rangle, x)) = x^{ed} \bmod n$. Since $ed \equiv 1 \bmod \varphi(n)$, we have $ed = k\varphi(n) + 1$ for some integer $k$, and

$$x^{ed} \bmod n = x^{k\varphi(n)+1} \bmod n = (x \bmod n) \cdot (x^{k\varphi(n)} \bmod n) \bmod n.$$

*Case 1.* $x \in Z_n^*$ (i.e., $gcd(x, n) = 1$). By Fermat's Theorem, $x^{\varphi(n)} \bmod n = 1$ and $x^{ed} \bmod n = (x \bmod n) \cdot (x^{k\varphi(n)} \bmod n) \bmod n = x \bmod n$.

*Case 2.* $x \notin Z_n^*$ and $x = ph$ for some $h < q$. Then, by Fermat's Theorem, $x^{q-1} \equiv 1 \bmod q$. It follows that $x^{k\varphi(n)} = x^{k(p-1)(q-1)} \equiv 1 \bmod q$, or $x^{k\varphi(n)} = qm + 1$ for some integer $m$. Thus,

$$\begin{aligned} x^{ed} \bmod n &= (x \bmod n) \cdot (x^{k\varphi(n)} \bmod n) \bmod n \\ &= (ph \bmod n) \cdot ((qm+1) \bmod n) \bmod n \\ &= (phqm + ph) \bmod n = ph = x. \end{aligned}$$

*Case 3.* $x \notin Z_n^*$ and $x = qj$ for some $j < p$. Similar to Case 2.  ∎

To see whether the RSA system is a good public-key cryptosystem, we need to verify whether the encryption function is indeed a trapdoor one-way function. First, we observe that if $p, q$ are known, then the integers $e$ and $d$ satisfying (4.1) are easy to find and, hence, the system is easy to set up. To find the pair $(e, d)$, we first select a prime number $e$ between $\max(p, q)$ and $(p-1)(q-1)$. Since $n = p \cdot q$, we have $\varphi(n) = (p-1)(q-1)$, and so $e$ is relatively prime to $\varphi(n)$. Then, using the Euclidean algorithm, we can find the integer $d$ such that $e \cdot d \equiv 1 \bmod \varphi(n)$ (see, e.g., Knuth [1981]). Next, we can see from Lemma 4.4 that both $E$ and $D$ are easy to compute. So it is left to determine how hard it is to compute the decryption key $\langle d, n \rangle$ from the encryption key (the public key) $\langle e, n \rangle$. If the factors $p$ and $q$ of $n$ are known, then $\varphi(n)$ is known and the integer $d$ can be found easily from $e$. Thus, the security of the RSA system depends on the hardness of factoring an integer into two primes. Conversely, suppose that integer factoring is known to be an intractable problem. Does it mean that the RSA system is secure? In other words, if we have a fast algorithm breaking the RSA system, do we then have a fast factoring algorithm? The answer is yes if we make some modification on the above scheme (see Rabin [1979]). Thus, the complexity of breaking the (modified) RSA system is equivalent to the complexity of factoring the products of two primes, and the RSA system is considered secure, assuming that the function *multiplication* is a strong one-way function.[4]

## 4.3 Relativization

The concept of relativization originates from recursive function theory. Consider, for example, the halting problem. We may formulate it in the following form: $K = \{x \mid M_x(x) \text{ halts}\}$, where $M_x$ is the $x$th Turing machine in a standard enumeration of all TMs. Now, if we consider all oracle Turing machines, we may ask whether the set $K_A = \{x \mid M_x^A(x) \text{ halts}\}$ is recursive relative to $A$. This is the halting problem relative to set $A$. It is easily seen from the original proof for the nonrecursiveness of $K$ that $K_A$ is nonrecursive relative to $A$ (i.e., no oracle TM can decide $K_A$ using $A$ as an oracle). Indeed most results in recursive function theory can be extended to hold relative to any oracle set. We say that such results *relativize*. In this section, we investigate the problem of whether $P = NP$ in the relativized form. First we need to define what is meant by relativizing the question of whether $P = NP$. For any set $A$, recall that $P^A$ (or $P(A)$) is the class of sets computable in polynomial time by oracle DTMs using $A$ as the oracle and, similarly, $NP^A$ (or $NP(A)$)

---

[4]In the real implementation, there are many other practical issues involved in the design of a good public-key cryptosystem. For instance, what is the best size for primes $p$ and $q$ so that the system is secure and the time and space complexity for encryption and decryption is tolerable? Here, we are concerned only with the theoretical relation between strong one-way functions and public-key cryptosystems. The reader is referred to textbooks on cryptography for more studies on implementation issues.

is the class of sets accepted in polynomial time by oracle NTMs using oracle set $A$. Using these natural relativized forms of the complexity classes $P$ and $NP$, we show that the relativized $P =?NP$ question has both the positive and negative answers, depending on the oracle set $A$.

**Theorem 4.19** (a) There exists a recursive set $A$ such that $P^A = NP^A$.
   (b) There exists a recursive set $B$ such that $P^B \neq NP^B$.

*Proof.* (a) Let $A$ be any set which is $\leq_m^P$-complete for *PSPACE*. Then, by Savitch's Theorem, we have

$$NP^A \subseteq NPSPACE = PSPACE \subseteq P^A.$$

(The first inclusion follows from the fact that all the query to set $A$ can be simulated in polynomial space.)
   (b) First define, for any set $B$, a set

$$L_B = \{0^n : (\exists x)\,[|x| = n, x \in B]\}.$$

Then we can see easily that $L_B \in NP^B$ because for any $0^n$ we can guess a string $x$ of length $n$ and check that $x$ is in $B$ by making a single query to $B$.
   We are going to construct a set $B \subseteq \{0,1\}^*$ such that $L_B \notin P^B$; thus $L_B$ is a set in $NP^B - P^B$. Intuitively, in polynomial time we can only query the set $B$ a polynomial number of times and leave a lot of strings $x$ of length $n$ unqueried and so cannot decide whether $0^n$ is in $L_B$ or not. Formally, we use a stage construction like that in Theorem 4.8 to define set $B$. Recall that the class of polynomial-time oracle TMs are effectively enumerable. We let them be $\{M_i\}_{i=1}^{\infty}$. In each stage $i \geq 1$, we need to construct a portion of set $B$ that satisfies the requirement

   $R_i$:  $(\exists x_i)\,[x_i \in L_B \iff x_i \notin L(M_i, B)]$.

If all requirements $R_i$, $i \geq 1$, are satisfied, then $L_B \neq L(M_i, B)$ for any $i \geq 1$ and, hence, $L_B \notin P^B$. We now describe each stage $i \geq 1$ as follows:
   Prior to Stage 1, we set $n_0 = 0$ and $B_0 = \emptyset$. We assume that by the end of stage $i - 1$, we have already defined integer $n_{i-1}$ and a finite set $B_{i-1} \subseteq \{x \in \{0,1\}^* : |x| \leq n_{i-1}\}$.
   *Stage i.* We choose the least integer $n$ such that (i) $2^n > p_i(n)$, where $p_i$ is the polynomial that bounds the runtime of machine $M_i$, and (ii) $n > 2^{n_{i-1}}$. Then, let $n_i = n$, and $x_i = 0^{n_i}$. Simulate $M_i$ on $x_i$ with oracle $B_{i-1}$. That is, when $M_i$ asks whether $y$ is in the oracle, we simulate with the answer YES if $y \in B_{i-1}$, and with answer NO otherwise. When $M_i$ halts, if it accepts $x_i$ then we let $B_i = B_{i-1}$ (and so $B_i \cap \{y : |y| = n_i\} = \emptyset$). If it rejects $x_i$, then we find a string $y$ of length $n_i$ that was not queried in the computation of $M_i(x_i)$, and let $B_i = B_{i-1} \cup \{y\}$. (Such a string $y$ always exists, since there are $2^{n_i}$ strings of length $n_i$, and $M_i$ on $x_i$ can query at most $p_i(n_i) < 2^{n_i}$ times.) Note that $B_i \subseteq \{x \in \{0,1\}^* : |x| \leq n_i\}$.

The above completes the description of each stage $i$. We let $B = \bigcup_{i=1}^{\infty} B_i$, and claim that each requirement $R_i$ is satisfied by $B$.

For each $i \geq 1$, we note that $M_i^B(x_i)$ accepts if and only if the simulation of $M_i^{B_{i-1}}$ on $x_i$ in stage $i$ accepts. This is true because we set $n_{i+1} > 2^{n_i} > p_i(n_i)$ and so we never add any string to $B$ of length shorter than or equal to $p_i(n_i)$ in later stages. In addition, if we add a string $y$ to $B_i$ in stage $i$, we have made sure that the computation of $M_i^{B_{i-1}}(x_i)$ never queries $y$. Thus, the queries made by $M_i$ to $B$ have the same membership in $B_{i-1}$ as in $B$. This shows that the simulation in stage $i$ is correct and, hence, $x_i \in L_B$ if and only if $M_i^B(x_i)$ rejects. ∎

It is interesting to know that the question of whether $P = NP$ can be answered either way relative to different oracles. What does this mean to the original unrelativized version of the question of whether $P$ is equal to $NP$? Much research has been done on this subject, and yet we don't have a consensus. We summarize in the following sections some of the interesting results in this study.

## 4.4 Unrelativizable Proof Techniques

First, a common view is that the question of whether $P$ is equal to $NP$ is a difficult question in view of Theorem 4.19. Since most proof techniques developed in recursion theory, including the basic diagonalization and simulation techniques, relativize, any attack to the $P =? NP$ question must use a new, unrelativizable proof technique. Many more contradictory relativized results like Theorem 4.19 (including some in Section 4.8) on the relations between complexity classes tend to support this viewpoint. On the other hand, some unrelativizable proof techniques do exist in complexity theory. For instance, we will apply an algebraic technique to collapse the complexity class $PSPACE$ to a subclass $IP$ (see Chapter 10). Since there exists an oracle $X$ that separates $PSPACE^X$ from $IP^X$, this proof is indeed unrelativizable. Though this is a breakthrough in the theory of relativization, it seems still too early to tell whether such techniques are applicable to a wider class of questions.

## 4.5 Independence Results

One of the most interesting topics in set theory is the study of independence results. A statement $A$ is said to be *independent* of a theory $T$ if there exist two models $M_1$ and $M_2$ of $T$ such that $A$ is true in $M_1$ and false in $M_2$. If a statement $A$ is known to be independent of the theory $T$, then neither $A$ nor its negation $\neg A$ is provable in theory $T$. The phenomenon of contradictory relativized results looks like a *mini*-independent result: neither the statement $P = NP$ nor its negation $P \neq NP$ is provable by relativizable techniques. This observation raises the question of whether they are provable within a formal proof system. In the following, we present a simple argument showing that this is indeed possible.

To prepare for this result, we first briefly review the concept of a formal proof system. An *axiomatizable theory* is a triple $F = (\Sigma, W, T)$, where $\Sigma$ is a finite alphabet, $W \subseteq \Sigma^*$ is a recursive set of *well-formed formulas*, and $T \subseteq W$ is an r.e. set. The elements in $T$ are called *theorems*. If $T$ is recursive, we say the theory $F$ is decidable. We are only interested in a *sound* theory in which we can prove the basic properties of Turing machines. In other words, we assume that Turing machines form a submodel for $F$, all basic properties of Turing machines are provable in $F$, and all theorems in $F$ are true in the Turing machine model. In the following we let $\{M_i\}$ be a fixed enumeration of multi-tape DTMs.

**Theorem 4.20** *For any formal axiomatizable theory $F$ for which Turing machines form a submodel, we can effectively find a DTM $M_i$ such that $L(M_i) = \emptyset$ and neither $"P^{L(M_i)} = NP^{L(M_i)}"$ nor $"P^{L(M_i)} \neq NP^{L(M_i)}"$ is provable in $F$.*

*Proof.* Let $A$ and $B$ be two recursive sets such that $P^A = NP^A$ and $P^B \neq NP^B$. Define a TM $M$ such that $M$ accepts $(j, x)$ if and only if among the first $x$ proofs in $F$ there is a proof for the statement $"P^{L(M_j)} = NP^{L(M_j)}"$ and $x \in B$, or there is a proof for the statement $"P^{L(M_j)} \neq NP^{L(M_j)}"$ and $x \in A$. By the recursion theorem, there exists an index $j_0$ such that $M_{j_0}$ accepts $x$ if and only if $M$ accepts $(j_0, x)$. (See, e.g., Rogers [1967] for the recursion theorem.)

Now, if there is a proof for the statement $"P^{L(M_{j_0})} = NP^{L(M_{j_0})}"$ in $F$ then for almost all $x$, $M$ accepts $(j_0, x)$ if and only if $x \in B$. That is, the set $L(M_{j_0})$ differs from the set $B$ by only a finite set and, hence, $P^B \neq NP^B$ implies $P^{L(M_{j_0})} \neq NP^{L(M_{j_0})}$. Similarly, if there exists a proof for the statement $"P^{L(M_{j_0})} \neq NP^{L(M_{j_0})}"$, then $L(M_{j_0})$ differs from $A$ by only a finite set and, hence, $P^A = NP^A$ implies $P^{L(M_{j_0})} = NP^{L(M_{j_0})}$. By the soundness of the theory $F$, we conclude that neither $"P^{L(M_{j_0})} = NP^{L(M_{j_0})}"$ nor $"P^{L(M_{j_0})} \neq NP^{L(M_{j_0})}"$ is provable in $F$.

Furthermore, since neither $"P^{L(M_{j_0})} = NP^{L(M_{j_0})}"$ nor $"P^{L(M_{j_0})} \neq NP^{L(M_{j_0})}"$ is provable in $F$, the machine $M_{j_0}$ does not accept any input $x$; that is, $L(M_{j_0}) = \emptyset$. ∎

We remark that although the above machine $M_{j_0}$ accepts an empty set, this does not imply that the statement $P = NP$ is independent of $F$ because the equivalence of $P^{L(M_{j_0})}$ and $P$ and the equivalence of $NP^{L(M_{j_0})}$ and $NP$ are not necessarily provable in the system $F$. So, the above result proves that for any *reasonable* formal proof system $F$, there are some statements of interests to complexity theory which are independent of $F$. Whether the statement $P = NP$ is independent of any specific formal proof system is yet still unknown.

## 4.6   Positive Relativization

Still another viewpoint is that the formulation of the relativized complexity class $NP^A$ used in Theorem 4.19 does not reflect correctly the concept of nondeterministic computation. Consider the set $L_B$ in the proof of Theorem 4.19. Note that although each computation path of the oracle NTM $M$ which accepts $L_B$ asks only one question to determine whether $0^n$ is in $B$, the whole computation tree of $M^B(x)$ makes an exponential number of queries. While it is recognized that this is the distinctive property of an NTM to make, in the whole computation tree, an exponential number of moves, the fact that $M$ can access an exponential amount of information about the oracle $B$ immediately makes the oracle NTMs much stronger than oracle DTMs. To make the relation between oracle NTMs and oracle DTMs close to that between regular NTMs and regular DTMs, we must not allow the oracle NTMs to make arbitrary queries. Instead, we would like to know whether an oracle NTM which is allowed to make, in the whole computation tree, only a polynomial number of queries is stronger than an oracle DTM. When we add these constraints to the oracle NTMs, it turns out that the relativized $P =? NP$ question is equivalent to the unrelativized version. This result supports the viewpoint that the relativized separation of Theorem 4.19 is due to the extra information that an oracle NTM can access, rather than the nondeterminism of the NTM and, hence, this kind of separation results bear no relation to the original unrelativized questions. This type of relativization is called *positive relativization*. We present a simple result of this type in the following.

**Definition 4.21**  *For each set $A$, let $NP_b^A$ denote the class of sets $B$ for which there exist a polynomial-time oracle NTM $M$ and a polynomial function $p$ such that for every $x$, $x \in B$ if and only if $M^A$ accepts $x$ and the total number of queries made by $M$ in the computation tree of $M^A(x)$ is bounded by $p(|x|)$.*

**Theorem 4.22**  $P = NP$ *if and only if* $(\forall A)$ $P^A = NP_b^A$.

*Proof.* The backward direction is trivial since $NP_b^\emptyset = NP$. Conversely, assume that $P = NP$ and let $L \in NP_b^A$. Let $M$ be a polynomial-time oracle NTM and $p$ a polynomial such that $M^A$ accepts $L$ in time $p$ and that the whole computation tree of $M^A(x)$ contains at most $p(|x|)$ queries. We are going to describe a polynomial-time algorithm for $L$.

Let $\alpha$ be a configuration in the computation of $M^A(x)$, $\vec{y} = \langle y_1, \ldots, y_m \rangle$ be a sequence of strings of length $\leq p(|x|)$, and $\vec{b} = \langle b_1, \ldots, b_m \rangle$ be a sequence of bits in $\{0, 1\}$. We say that a configuration $\beta$ is a *$q$-successor* of $\alpha$ with respect to $(\vec{y}, \vec{b})$ if the following conditions hold:

(a)  $\beta$ is an accepting configuration or a configuration in a query state whose query string $z$ is different from all $y_i$, $1 \leq i \leq m$; and

(b)  There exists a sequence of computation of $M$ of size $\leq p(|\alpha|)$, relative to an oracle $X$, from the configuration $\alpha$ to $\beta$ without making any queries other than $y_1, \ldots, y_m$ and, in the case that $y_i$ is queried, the answer to it is $b_i$, $1 \le i \le m$ (i.e., $X \cap \{y_1, \ldots, y_m\} = \{y_j : 1 \le j \le m, b_j = 1\}$).

We define a function $f$ that maps a triple $(\alpha, \vec{y}, \vec{b})$ to the *least* q-successor $\beta$ of $\alpha$ with respect to $(\vec{y}, \vec{b})$ (under the lexicographic ordering over all configurations), or, if such a q-successor does not exist, to the empty string $\lambda$. We claim that $f$ is polynomial-time computable.

To see this, let $B = \{\langle \alpha, \gamma, \vec{y}, \vec{b} \rangle$ : there exists a configuration $\beta$ which is a q-successor of $\alpha$ with respect to $(\vec{y}, \vec{b})$ such that $\gamma$ is a prefix of $\beta\}$. Apparently $B \in NP$. By our assumption of $P = NP$, $B \in P$. Thus we can compute $f(\alpha, \vec{y}, \vec{b})$ by the prefix search technique (see Exercise 2.2).

Now we describe a deterministic simulation of the computation of $M^A(x)$, based on the depth-first search of the computation tree of $M^A(x)$. The algorithm is written as a recursive procedure.

*Deterministic Algorithm for $L(M, A)$:*

On input $x$, let $\alpha_0$ be the initial configuration of $M^A(x)$, and let $\vec{y}$ and $\vec{b}$ be the empty list. Call the procedure *Search*$(\alpha_0)$. If the procedure *Search* returns FAIL, then reject $x$ and halt.

Procedure *Search*$(\alpha)$;
{In the following, $\vec{y}$ and $\vec{b}$ are global variables.}
Loop forever
    Let $\beta = f(\alpha, \vec{y}, \vec{b})$;
    *Case* 1. $\beta$ is an accepting configuration: accept $x$ and halt;
    *Case* 2. $\beta = \lambda$: return (FAIL);
    *Case* 3. $\beta$ is a query configuration with the query string $z$:
        Then, query $A$ to get $c := \chi_A(z)$;
        let $\alpha_1$ be the next configuration following $\beta$ using answer $c\}$;
        let $\vec{y}$ be the concatenation of $\vec{y}$ and $z$;
        let $\vec{b}$ be the concatenation of $\vec{b}$ and $c$;
        Call the procedure *Search*$(\alpha_1)$;
        If *Search*$(\alpha_1)$ returns FAIL then continue
End Loop

We note that *Search*$(\alpha)$ halts only when (a) an accepting configuration has been found (and $x$ is accepted), or (b) when the lists $(\vec{y}, \vec{b})$ becomes so big that $f(\alpha_0, \vec{y}, \vec{b}) = \lambda$ (and $x$ is rejected). We notice that whenever a query $z$ is added to the list $\vec{y}$, the corresponding answer $\chi_A(z)$ is added to the list $\vec{b}$. Therefore, in case (a), the accepting configuration found must be correct, and so $x$ is correctly accepted. In case (b), $f(\alpha_0, \vec{y}, \vec{b}) = \lambda$ implies that $\vec{y}$ contains all queries in the whole computation tree of $M^A(x)$ and that there is no accepting computation in this tree. So the algorithm rejects $x$ correctly.

It remains to show that the algorithm always halts within polynomial time. To analyze the above algorithm, let $\vec{y}_0$ and $\vec{b}_0$ be the empty list. Notice that the recursive calls generated by the procedure $Search(\alpha_0)$, with the initial lists $\vec{y}_0$ and $\vec{b}_0$, form an *ordered* tree $T_x$ with the following properties:

(1) Each node of the tree $T_x$ has a label $\alpha$ that is a configuration of $M^A(x)$. The label of the root of the tree $T_x$ is $\alpha_0$.

(2) Each edge of the tree is labeled with a pair of lists $(\vec{y}, \vec{b})$. The leftmost edge from the root $\alpha_0$ has label $(\vec{y}_0, \vec{b}_0)$.

(3) The child $\alpha_1$ of node $\alpha$ under the edge $(\vec{y}, \vec{b})$ is either (a) $\beta = f(\alpha, \vec{y}, \vec{b})$ if $\beta$ is an accepting configuration or if $\beta = \lambda$, and in this case $\alpha_1$ is a leaf, or (b) the next configuration of $\beta = f(\alpha, \vec{y}, \vec{b})$ if $\beta$ is a querying configuration. In the case (b), the leftmost edge of $\alpha_1$ has label $(\vec{y}_1, \vec{b}_1)$, where $\vec{y}_1$ is the concatenation of $\vec{y}$ and the query $z$ of $\beta$, and $\vec{b}_1$ is the concatenation of $\vec{b}$ and $\chi_A(z)$.

(4) Assume that an internal node $\alpha$ has $k > 1$ children, and assume that the $j$th leftmost edge, $1 \leq j < k$, has label $(\vec{y}_j, \vec{b}_j)$. Then, the label of the $(j+1)$st leftmost edge from $\alpha$ is $(\vec{y}_{j+1}, \vec{b}_{j+1})$, where $\vec{y}_{j+1}$ is the concatenation of list $\vec{y}_j$ and all query strings in the $j$th subtree of $\alpha$, and $\vec{b}_{j+1}$ is the concatenation of $\vec{b}_j$ and all the answers to those queries from oracle $A$.

Now we observe that since the computation tree $M^A(x)$ contains at most $p(|x|)$ queries, and since each internal node of $T_x$ corresponds to a unique query, there are at most $p(|x|)$ nodes in $T_x$. Our deterministic algorithm actually performs a depth-first search of this tree $T_x$. For each edge, it calls the function $f$ once to create the corresponding child node. Thus, the algorithm must halt in polynomial time. ∎

The above theorem shows that we are not able to separate $NP_b^A$ from $P^A$ for any oracle $A$ unless we can show $P \neq NP$. On the other hand, we note that $P^A = NP^A = PSPACE$ for all $PSPACE$-complete sets $A$. Thus, the relativized collapsing of $NP_b^A$ to $P^A$ still works but it only demonstrates the power of a $PSPACE$-complete set as an oracle, and does not provide much information about the unrelativized $P =? NP$ question.

## 4.7 Random Oracles

Consider the class $\mathcal{C}$ of all subsets of $\{0,1\}^*$ and define subclasses $\mathcal{A} = \{A \in \mathcal{C} : P^A = NP^A\}$ and $\mathcal{B} = \{B \in \mathcal{C} : P^B \neq NP^B\}$. One of the approaches to study the relativized $P =? NP$ question is to compare the two subclasses $\mathcal{A}$ and $\mathcal{B}$ to see which one is *larger*. In this subsection, we give a brief introduction to this study based on the probability theory on the space $\mathcal{C}$.

To define the notion of probability on the space $\mathcal{C}$, it is most convenient to identify each element $X \in \mathcal{C}$ with its characteristic sequence $\alpha_X = \chi_X(\lambda)\chi_X(0)\chi_X(1)\chi_X(00)\ldots$ (i.e., the $k$th bit of $\alpha_X$ is 1 if and only if the $k$th string of $\{0,1\}^*$, under the lexicographic ordering, is in $X$), and treat $\mathcal{C}$ as the set of all infinite binary sequences, or, equivalently, the Cartesian product $\{0,1\}^\infty$. We can define a topology on $\mathcal{C}$ by letting the set $\{0,1\}$ have the discrete topology, and forming the product topology on $\mathcal{C}$. This is the well-known *Cantor space*. We now define the uniform probability measure $\mu$ on $\mathcal{C}$ as the product measure of the simple equiprobable measure on $\{0,1\}$, that is, $\mu\{0\} = \mu\{1\} = 1/2$. In other words, for any integer $n \geq 1$, the $n$th bit of a random sequence $\alpha \in \mathcal{C}$ has the equal probability to be 0 or 1. If we identify each real number in $[0,1]$ with its binary expansion, then this measure $\mu$ is the Lebesgue measure on $[0,1]$.[5]

For any $u \in \{0,1\}^*$, let $\Gamma_u$ be the set of all infinite binary sequences which begin with $u$, that is, $\Gamma_u = \{u\beta : \beta \in \{0,1\}^\infty\}$. Each set $\Gamma_u$ is called a *cylinder*. All cylinders $\Gamma_u$, $u \in \{0,1\}^*$, together form a basis of open neighborhoods of the space $\mathcal{C}$ (under the product topology). It is clear that $\mu(\Gamma_u) = 2^{-|u|}$ for all $u \in \{0,1\}^*$. The smallest $\sigma$-field containing all $\Gamma_u$, for all $u \in \{0,1\}^*$, is called the *Borel field*.[6] Each set in the Borel field, called a *Borel set*, is measurable.

The question of which of the two subclasses $\mathcal{A}$ and $\mathcal{B}$ is larger can be formulated, in this setting, as to whether $\mu(\mathcal{A})$ is greater than $\mu(\mathcal{B})$. In the following, we show that $\mu(\mathcal{A}) = 0$.

An important idea behind the proof is Kolmogorov's zero-one law of tail events, which implies that if an oracle class is insensitive to a finite number of bit changes then its probability is either zero or one. This property holds for the classes $\mathcal{A}$ and $\mathcal{B}$ as well as most other interesting oracle classes.

**Theorem 4.23** (Zero-One Law) *Assume that a Borel set $\Gamma \subseteq \mathcal{C}$ is invariant under the finite changes; that is, if $\alpha \in \Gamma$ and $\alpha$ and $\beta$ differ on only a finite number of bits, then $\beta \in \Gamma$. Then, $\mu(\Gamma)$ is either 0 or 1.*

*Proof.* (Sketch) The main idea here is that if $\Gamma$ is invariant under the finite changes, then the distribution of $\Gamma$ in the whole space $\mathcal{C}$ is identical to that in the subspace $\Gamma_u$ for any $u \in \{0,1\}^*$. To see this, we observe that for any two strings $u, v$ in $\{0,1\}^*$ of the same length, the mapping $f(u\alpha) = v\alpha$ maps the subspace $\Gamma_u$ to subspace $\Gamma_v$ preserving the membership in $\Gamma$ and preserving the induced conditional probability. That is, $\mu(\Gamma \cap \Gamma_u) = \mu(\Gamma \cap \Gamma_v)$. It follows that

$$\mu(\Gamma \cap \Gamma_u) = \mu(\Gamma)\mu(\Gamma_u), \qquad (4.2)$$

for all $u \in \{0,1\}^*$. Since equation (4.2) holds for all $\Gamma_u$, $u \in \{0,1\}^*$, and since $\{\Gamma_u : u \in \{0,1\}^*\}$ generates all Borel sets, it also holds for all Borel sets, including $\Gamma$ itself. Thus, we get $\mu(\Gamma) = \mu(\Gamma)^2$ and, hence, $\mu(\Gamma)$ is equal to either 0 or 1. ∎

Now we prove the main result of this section.

---

[5] Note that $\mu\{w1^\infty : w \in \{0,1\}^*\} = 0$ and so we may ignore the problem that some real numbers may have two equivalent binary expansions.

[6] A $\sigma$-field is a class of sets that is closed under the complementation and countably infinite union operations.

**Theorem 4.24** *With probability one, $P^B \neq NP^B$; that is, $\mu(\mathcal{B}) = 1$.*

*Proof.* Let $L_B = \{0^n : (\exists x, |x| = n) \; x1, x1^2, \ldots x1^n \in B\}$. It is obvious that $L_B \in NP^B$ for all oracles $B$. We need to show that $L_B \in P^B$ only for a small set of oracles $B$ of probability zero. In the following, for any event $E$, we write $\Pr[E]$ to denote $\mu(E)$.

In order to show that $\Pr[L_B \in P^B] = 0$, it suffices to show that for any fixed polynomial-time deterministic oracle TM $M$, the probability of $M^B$ correctly computing $L_B$ is zero, since there are only countably many such TMs. Assume that the time complexity of $M$ is bounded by a polynomial function $p$, and assume that $p(n) > 2n$ for all $n \geq 0$. For any integer $n \geq 0$, we let $\mathcal{B}_n$ denote the set of oracles $B$ such that $M^B$ computes correctly the membership of $0^n$ in $L_B$; that is, $\mathcal{B}_n = \{B : 0^n \in L_B \iff M^B \text{ accepts } 0^n\}$. We are going to show that for any $\epsilon > 0$ there exist integers $n_1, n_2, \ldots, n_k$ such that $\Pr[\mathcal{B}_{n_1} \cap \mathcal{B}_{n_2} \cap \cdots \cap \mathcal{B}_{n_k}] < \epsilon$. This implies $\Pr[L_B = L(M, B)] = 0$ and, hence, the theorem follows.

We identify each set $X \subseteq \{0,1\}^*$ with its characteristic sequence $\alpha_X$. Let $u$ be a string of length $2^n - 1$. Then, the cylinder $\Gamma_u$ is the class of sets $X \subseteq \{0,1\}^*$ such that $\{x \in X : |x| < n\}$ is fixed (by $u$). Note that the machine $M$, on input $0^n$, can only query the oracle about the membership of strings of length $\leq p(n)$. Therefore, $\mathcal{B}_n$ is the union of a finite number of cylinders $\Gamma_u$ with $|u| = 2^{p(n)+1} - 1$.

Now we claim the following:

*Claim.* For any integer $n$ such that $2^n > 8p(n)$, the conditional probability $\Pr[\mathcal{B}_n \mid \Gamma_w]$ is at most $3/4$ for all $w$ of length $2^n - 1$.

In other words, if we fix the membership in the oracle $B$ for all strings of length less than $n$, the probability is at least $1/4$ that $M^B$ will make a mistake on input $0^n$.

We first show why the theorem follows from the above claim. Let $n$ and $m$ satisfy the conditions $2^n > 8p(n)$, $2^m > 8p(m)$ and $m > p(n)$, and consider the probability of the event $\mathcal{B}_n \cap \mathcal{B}_m$ with respect to the condition $\Gamma_u$, where $u$ is a string of length $2^m - 1$. We have observed that event $\mathcal{B}_n$ is the union of a finite number of cylinders $\Gamma_w$ with $|w| = 2^{p(n)+1} - 1$. Thus, for each $u$ of length $2^m - 1 \geq 2^{p(n)+1} - 1$, either $\Gamma_u \subseteq \mathcal{B}_n$ or $\Gamma_u \cap \mathcal{B}_n = \emptyset$. Let $A = \{u : |u| = 2^m - 1, \Gamma_u \subseteq \mathcal{B}_n\}$. Then, the claim implies that $\Pr[\mathcal{B}_n] \leq 3/4$, or, $|A| \leq (3/4)2^{2^m-1}$. Now, we have

$$
\begin{aligned}
\Pr[\mathcal{B}_n \cap \mathcal{B}_m] &= \sum_{|u|=2^m-1} \Pr[\mathcal{B}_n \cap \mathcal{B}_m \mid \Gamma_u] \cdot \Pr[\Gamma_u] \\
&= \sum_{u \in A} \Pr[\mathcal{B}_m \mid \Gamma_u] \cdot \Pr[\Gamma_u] \\
&\leq \left(\frac{3}{4}\right) \cdot 2^{2^m-1} \cdot \left(\frac{3}{4}\right) \cdot 2^{-(2^m-1)} = \left(\frac{3}{4}\right)^2.
\end{aligned}
$$

Let $k$ be large enough so that $(3/4)^k < \epsilon$. We can choose $k$ integers $n_1, n_2, \ldots, n_k$ with $2^{n_i} > 8p(n_i)$ for each $i = 1, \ldots, k$, and $n_{i+1} > p(n_i)$ for each $i = 1, \ldots, k - 1$, and repeat the above argument inductively to get

$$\Pr[\mathcal{B}_{n_1} \cap \mathcal{B}_{n_2} \cap \cdots \cap \mathcal{B}_{n_k}] \leq \left(\frac{3}{4}\right)^k < \epsilon.$$

Therefore, the theorem follows from the claim.

It is left to prove the claim. Let $n$ be an integer satisfying $2^n > 8p(n)$ and $w$ be a string of length $2^n - 1$. Define two sets of oracles $\mathcal{D}_0$ and $\mathcal{D}_1$ as follows: $\mathcal{D}_0 = \{B \in \Gamma_w : (\forall x, |x| = n)\ (\exists j, 1 \leq j \leq n)\ x1^j \notin B\}$, and $\mathcal{D}_1 = \{B \in \Gamma_w : (\exists \text{ unique } x, |x| = n)\ (\forall j, 1 \leq j \leq n)\ x1^j \in B\}$. Note that $0^n \in L_B$ for all $B \in \mathcal{D}_1$ and $0^n \notin L_B$ for all $B \in \mathcal{D}_0$. For $i = 0, 1$, let $r_i = \Pr[M^B \text{ accepts } 0^n \mid \mathcal{D}_i]$. Then, the error probability of $M$ on input $0^n$, relative to the condition $\Gamma_w$ is at least $(1 - r_1)\Pr[\mathcal{D}_1 \mid \Gamma_w] + r_0 \Pr[\mathcal{D}_0 \mid \Gamma_w]$; that is,

$$\Pr[0^n \in L_B \Leftrightarrow M^B \text{ rejects } 0^n \mid \Gamma_w] \geq (1 - r_1)\Pr[\mathcal{D}_1 \mid \Gamma_w] + r_0 \Pr[\mathcal{D}_0 \mid \Gamma_w].$$

It is easy to check that

$$\Pr[\mathcal{D}_0 \mid \Gamma_w] = \prod_{|x|=n} \Pr[(\exists j, 1 \leq j \leq n)\ x1^j \notin B]$$

$$= \prod_{|x|=n} (1 - 2^{-n}) = (1 - 2^{-n})^{2^n},$$

and

$$\Pr[\mathcal{D}_1 \mid \Gamma_w] = \sum_{|x|=n} \Big( \Pr[(\forall j, 1 \leq j \leq n)\ x1^j \in B]$$

$$\cdot \prod_{|y|=n, y \neq x} \Pr[(\exists j, 1 \leq j \leq n)\ y1^j \notin B] \Big)$$

$$= 2^n \cdot 2^{-n} \cdot (1 - 2^{-n})^{2^n - 1} = (1 - 2^{-n})^{2^n - 1}.$$

Both values approach $e^{-1}$ as $n$ approaches infinity. In particular, since $2^n > 8p(n)$ and, hence, $n > 3$, we have $\Pr[\mathcal{D}_1 \mid \Gamma_w] > \Pr[\mathcal{D}_0 \mid \Gamma_w] > 1/3$.

What are the conditional accepting probabilities $r_0$ and $r_1$ of $M$? They depend on the machine $M$, and we cannot find an absolute upper or lower bound for them. However, we can show that for large $n$, $r_0$ and $r_1$ are very close to each other. Intuitively, the machine $M$ does not have enough resources to distinguish between an oracle $B$ in $\mathcal{D}_0$ from an oracle $B'$ in $\mathcal{D}_1$ if they differ only at strings $x1, x1^2, \ldots, x1^n$ for some string $x$ of length $n$, because $M$ can query at most $p(n)$ times and $p(n)$ is much smaller than $2^n$. Formally, we define a random mapping $f$ from $\mathcal{D}_0$ to $\mathcal{D}_1$ as follows: For each $B \in \mathcal{D}_0$, choose a random $y$ of length $n$ and let $B' = B \cup \{y1, y1^2, \ldots, y1^n\}$. Now, with probability $1 - p(n)/2^n$, the computation of $M^B(0^n)$ does not query any string of the form $y1^j$, $1 \leq j \leq n$, and so $\Pr[M^{B'}(0^n) \neq M^B(0^n)] \leq p(n)/2^n < 1/8$. This implies that $|r_1 - r_0| \leq 1/8$. More precisely, we treat $f$ as a deterministic mapping from $S_0 = \mathcal{D}_0 \times \{0,1\}^n$ to $S_1 = \mathcal{D}_1 \times (\{0,1\}^n - \{1\}^n)$: $f(\langle B, z \rangle) = \langle B', z' \rangle$, where $B' = B \cup \{z1, z1^2, \ldots, z1^n\}$ and $z' = \chi_B(z1)\chi_B(z1^2)\ldots\chi_B(z1^n)$, and consider the uniform probability distributions $\mathrm{Pr}_{S_0}$ and $\mathrm{Pr}_{S_1}$ on the spaces $S_0$ and $S_1$, respectively. This is a one-to-one mapping between the two spaces and, hence, preserving the probability measure. Let $Q(B, z)$ denote the event that the computation of $M^B(0^n)$ does not query any string of the form $z1^j$ for any $1 \le j \le n$. Note that

$$\mathrm{Pr}_{S_0}[Q(B, z)] \ge 1 - \frac{p(n)}{2^n} \ge \frac{7}{8}.$$

In addition, the event $Q(B, z)$ is obviously independent of the event of $M^B$ accepting (or, rejecting) the input $0^n$. Thus, we have

$$\begin{aligned}
r_1 &= \mathrm{Pr}_C[M^{B'}(0^n) = 1 \mid B' \in \mathcal{D}_1\} \\
&= \mathrm{Pr}_{S_1}[M^{B'}(0^n) = 1] \\
&\ge \mathrm{Pr}_{S_0}[M^B(0^n) = 1 \text{ and } Q(B, z)] \\
&\ge \frac{7}{8} \cdot \mathrm{Pr}_{S_0}[M^B(0^n) = 1] = \frac{7}{8} \cdot r_0.
\end{aligned}$$

Similarly, we have $1 - r_1 \ge (7/8)(1 - r_0)$.

Now, substitute this to the error probability, we get

$$(1 - r_1)\mathrm{Pr}[\mathcal{D}_1 \mid \Gamma_w] + r_0\mathrm{Pr}[\mathcal{D}_0 \mid \Gamma_w] \ge (1 - r_1 + r_0)\frac{1}{3} \ge \frac{7}{8} \cdot \frac{1}{3} > \frac{1}{4},$$

and the claim is proven. ∎

The above type of random oracle separation results holds for many complexity classes. See Exercise 4.16 and Section 9.5.

## 4.8 Structure of Relativized *NP*

Although the meaning of relativized collapsing and relativized separation results is still not clear, many relativized results have been proven. These results show a variety of possible relations between the well-known complexity classes. In this section, we demonstrate some of these relativized results on complexity classes within *NP* to show what the possible relations are between *P*, *NP*, *NP* ∩ *coNP*, and *UP*. The relativized results on the classes beyond the class *NP*, that is, those on *NP*, *PH*, and *PSPACE*, will be discussed in later chapters.

The proofs of the following results are all done by the stage construction diagonalization. The proofs sometimes need to satisfy simultaneously two or more potentially conflicting requirements that make the proofs more involved.

**Theorem 4.25** *(a)* $(\exists A)$ $P^A \ne NP^A \cap coNP^A \ne NP^A$.
  *(b)* $(\exists B)$ $P^B \ne NP^B = coNP^B$.
  *(c)* $(\exists C)$ $P^C = NP^C \cap coNP^C \ne NP^C$.

*Proof.* (a) This can be done by a standard diagonalization proof. We leave it as an exercise (Exercise 4.17(a)).

(b) Let $\{M_i\}$ be an effective enumeration of all polynomial-time oracle DTMs and $\{N_i\}$ an effective enumeration of all polynomial-time oracle NTMs. For any set $B$, let $K_B = \{\langle i, x, 0^j \rangle : N_i^B$ accepts $x$ in $j$ moves$\}$. Then, by the relativized proof of Theorem 2.11, $K_B$ is $\leq_m^P$-complete for the class $NP^B$. Let $L_B = \{0^n : (\exists x)\, |x| = n, x \in B\}$. Then, it is clear that $L_B \in NP^B$. We are going to construct a set $B$ to satisfy the following requirements:

$R_{0,t}$: for each $x$ of length $t$, $x \notin K_B \iff (\exists y, |y| = t)\, xy \in B$,

$R_{1,i}$: $(\exists n)\, 0^n \in L_B \iff M_i^A$ does not accept $0^n$.

Note that if requirements $R_{0,t}$ are satisfied for all $t \geq 1$, then $\overline{K_B} \in NP^B$ and, hence, $NP^B = coNP^B$, and if requirements $R_{1,i}$ are satisfied for all $i \geq 1$, then $L_B \notin P^B$ and, hence, $P^B \neq NP^B$.

We construct set $B$ by a stage construction. At each stage $n$, we will construct finite sets $B_n$ and $B_n'$ such that $B_{n-1} \subseteq B_n$, $B_{n-1}' \subseteq B_n'$, and $B_n \cap B_n' = \emptyset$ for all $n \geq 1$. Set $B$ is define as the union of all $B_n$, $n \geq 0$.

The requirement $R_{0,t}$, $t \geq 1$, is to be satisfied by direct diagonalization at stage $2t$. The requirements $R_{1,i}$, $i \geq 1$, are to be satisfied by a delayed diagonalization in the odd stages. We always try to satisfy the requirement $R_{1,i}$ with the least $i$ such that $R_{1,i}$ is not yet satisfied. We cancel an integer $i$ when $R_{1,i}$ is satisfied. Before stage 1, we assume that $B_0 = B_0' = \emptyset$ and that none of integers $i$ is cancelled.

At stage $n = 2t$, we try to satisfy $R_{0,t}$. For every $x$, $|x| = t$, we determine whether $x \in K_{B_{n-1}}$. If yes, we let $B_n = B_{n-1}$ and $B_n' = B_{n-1}'$; if not, we search for a string $xy$, $|y| = t$, which is not in $B_{n-1}'$, and let $B_n = B_{n-1} \cup \{xy\}$ and $B_n' = B_{n-1}'$.

At stage $n = 2t + 1$, we try to satisfy $R_{1,i}$ for the least $i$ which is not yet cancelled. If $p_i(n) \geq 2^t$ (where $p_i$ is the runtime of $M_i$) or if there exists a string of length $\geq n$ which is in $B_{n-1}'$ then we delay the diagonalization and go to the next stage (with $B_n = B_{n-1}$ and $B_n' = B_{n-1}'$). Otherwise, we simulate $M_i^{B_{n-1}}$ on input $0^n$. If $M_i^{B_{n-1}}$ accepts $0^n$, then we let $B_n = B_{n-1}$; if it rejects $0^n$, then we search for a sting $z$ of length $n$ which is not queried by the computation of $M_i^{B_{n-1}}(0^n)$ and let $B_n = B_{n-1} \cup \{z\}$. In either case, we add all strings of length $\geq n$ which are queried in the computation of $M_i^{B_{n-1}}(0^n)$ (and answered NO) to the set $B_{n-1}'$ to form $B_n'$. (Note that we add at most $p_i(n) < 2^t$ strings to $B_n'$.)

From the above construction, it is clear that $B_{n-1} \subseteq B_n$, $B_{n-1}' \subseteq B_n'$ and $B_n \cap B_n' = \emptyset$ for all $n \geq 1$. We define $B = \bigcup_{n=1}^{\infty} B_n$.

We first claim that the above construction is well defined; that is, the searches for strings of the form $xy$ in the even stages and searches for strings $z$ in the odd stages always succeed. To see this, we observe that by stage $n = 2t$, $B_{n-1}'$ is of size $< 2^t$, because in any earlier odd stage $k \leq n - 1$, we added less than $2^{(k-1)/2}$ strings to $B_k'$. Therefore, there must be at least a string of the form $xy$, with $|y| = t$, that is not in $B'_{n-1}$. Similarly, at the beginning of stage $n = 2t + 1$, $B'_{n-1} \cap \{z : |z| = n\}$ is empty, and the computation in stage $n$ can query at most $2^t$ strings of length $n$, and so there must be at least a string $z$ of length $n$ not queried and not in $B'_{n-1}$.

Next we claim that each requirement $R_{0,t}$ is satisfied in stage $n = 2t$. From our standard encoding of pairing functions, we know that if $x = \langle i, y, 0^j \rangle$, then $j < |x|$. Thus, whether $x \in K_B$ depends only on the set $B \cap \{w : |w| < |x|\}$. It follows that the simulation of $x \in K_{B_{n-1}}$ in stage $2t$ is the same as the computation of $x \in K_B$ because we never added new strings of length $\leq t$ to $B$ after stage $2t$. We conclude that requirement $R_{0,t}$ is satisfied in stage $2t$.

Finally, we check that each requirement $R_{1,i}$ is eventually satisfied. By induction, it is easy to see that each requirement $R_{1,i}$ eventually becomes *vulnerable* by some stage $n = 2t+1$ (i.e., $p_i(n) \leq 2^t$, $B'_{n-1} \cap \{w : |w| \geq n\} = \emptyset$ and $i$ is the least uncanceled integer by stage $n$). At this stage $n$, we made sure that $0^n \in L_{B_n} \iff 0^n \notin L(M_i, B_n)$, and also made sure that all strings of length $\geq n$ that were queried by $M_i^{B_n}(0^n)$ were reserved in $B'_n$. Therefore, $0^n \in L(M_i, B_n) \iff 0^n \in L(M_i, B)$. This shows that the requirement $R_{1,i}$ is satisfied by stage $n$.

(c) We let $\{M_i\}$ and $\{N_i\}$ be the enumerations of deterministic and nondeterministic oracle machines as defined in part (b) above. We need to construct a set $C$ to satisfy two requirements: (i) $P^C \neq NP^C$ and (ii) $NP^C \cap coNP^C = P^C$. The first requirement can, similarly to part (b) above, be divided into an infinite number of subrequirements $R_{\langle i,i \rangle}$, $i \geq 1$:

$$R_{\langle i,i \rangle}: (\exists n) \; 0^n \in L_C \iff M_i^C \text{ rejects } 0^n,$$

where $L_C = \{0^n : (\exists z) |z| = n, z \in C\}$. ($\langle r, s \rangle$ is a standard pairing function encoding two integers into one.) The second requirement, however, cannot be divided into simple subrequirements involving oracle NTMs, because whether two oracle NTMs accept complementary sets is not a *syntactic* property that can be determined simply from their machine descriptions. Therefore, we only try to satisfy as many subrequirements as possible and then, at the end, use an indirect argument to show that the requirement (ii) is also satisfied. We define subrequirements $R_{\langle j,k \rangle}$, $j \neq k$, as follows:

$$R_{\langle j,k \rangle}: (\exists x) \text{ neither } N_j^C \text{ nor } N_k^C \text{ accepts } x.$$

Note that if $R_{\langle j,k \rangle}$ is satisfied for some pair $\langle j, k \rangle$, with $j \neq k$, then $L(N_j, C) \neq \overline{L(N_k, C)}$, and so $N_j$ and $N_k$ do not accept complementary sets.

Let $e(0) = 1$ and $e(n + 1) = 2^{2^{e(n)}}$. As in part (b), we are going to define set $C_n$ in stage $n$ to have the property that $C_{n-1} \subseteq C_n$ for all $n \geq 1$. In addition, $C_n$ is either equal to $C_{n-1}$ or is $C_{n-1}$ plus a single string of length $e(n)$.

First, let $Q$ be a *PSPACE*-complete set that has no string of length $e(n)$ for any $n \geq 0$. Note that from Theorem 4.19(a), we know that $P^Q = NP^Q$. In addition, if $N_j^Q$ accepts some string $x$, then we can use oracle $Q$ to find an accepting computation of $N_j^Q(x)$ in deterministic polynomial time (see

Exercise 4.13). We let $C_0 = Q$.

For each integer $i > 0$, let $p_i(n)$ be a polynomial function bounding the runtimes of both $M_i$ and $N_i$. A requirement $R_{\langle i,i \rangle}$ is said to be *vulnerable* at stage $n$, if $p_i(e(n)) < 2^{e(n)}$, and a requirement $R_{\langle j,k \rangle}$, with $j \neq k$, is *vulnerable* at stage $n$ if there exists a string $x$ such that $e(n-1) < \log_2 |x| \leq e(n) \leq \max\{p_j(|x|), p_k(|x|)\} < e(n+1)$ and neither $N_j^{C_{n-1}}$ nor $N_k^{C_{n-1}}$ accepts $x$. At stage $n$, we try to satisfy the requirement $R_t$ with the least $t$ such that $R_t$ is uncanceled and vulnerable.

*Case 1.* No such $t$ exists. We just let $C_n = C_{n-1}$ and go to the next stage.

*Case 2.* Such a $t$ exists and $t = \langle i,i \rangle$ for some $i \geq 1$. Then, we simulate $M_i^{C_{n-1}}(0^{e(n)})$. If it accepts $0^{e(n)}$, then we let $C_n = C_{n-1}$; else we search for a string $z$ of length $e(n)$ that was not queried in the computation of $M_i^{C_{n-1}}(0^{e(n)})$, and let $C_n = C_{n-1} \cup \{z\}$. We cancel $R_t$.

*Case 3.* Such a $t$ exists and $t = \langle j,k \rangle$ with $j \neq k$. We simply let $C_n = C_{n-1}$ and cancel $R_t$.

The above completes the construction of set $C = \bigcup_{n=0}^{\infty} C_n$. As in part (b), it is easy to verify that the search in Case 2 of each stage always succeeds since $M_i$ only queries $p_i(e(n)) < 2^{e(n)}$ strings on input $0^{e(n)}$.

We first show that each requirement $R_{\langle i,i \rangle}$ is eventually satisfied. Suppose otherwise and let $i_0$ be the least integer such that $R_{\langle i_0,i_0 \rangle}$ is not satisfied. Then, by some stage $m$, all requirements $R_t$ with $t < \langle i_0,i_0 \rangle$ that are eventually satisfied are already cancelled and $\langle i_0,i_0 \rangle$ is vulnerable. So it will be attacked at stage $m$. By the construction in stage $m$, it is clear that $M_{i_0}^C(0^{e(m)})$ behaves exactly the same as $M_{i_0}^{C_{m-1}}(0^{e(m)})$, and so $R_{\langle i_0,i_0 \rangle}$ would be satisfied. This is a contradiction.

Next, we show that if $S = L(N_j, C) = \overline{L(N_k, C)}$, then $S \in P^C$. We note that if $R_{\langle j,k \rangle}$ had been cancelled, then there would be an $x$ such that $x \notin L(N_j, C)$ and $x \notin L(N_k, C)$ (since all later stages do not affect the computations of $N_j^{C_{n-1}}(x)$ and $N_k^{C_{n-1}}(x)$). Therefore, $L(N_j, C) \neq \overline{L(N_k, C)}$. Thus, we know that $R_{\langle j,k \rangle}$ has never been cancelled. We choose a large integer $m$ satisfying the following properties:

(i) For each $\ell \geq e(m)$, $\max\{p_j(\ell), p_k(\ell)\} < 2^\ell$ (therefore, there is at most one integer $n$ satisfying $e(n-1) < \log_2 \ell \leq e(n) \leq \max\{p_j(\ell), p_k(\ell)\} < e(n+1)$); and

(ii) All requirements $R_{\langle r,s \rangle}$, with $\langle r,s \rangle$ less than $\langle j,k \rangle$, that are eventually cancelled are already cancelled by stage $m$.

We now describe a deterministic algorithm for $S$, using oracle $C$.

*Algorithm for $S$.* For any input $x$ of length $\leq e(m)$, we use a finite table to decide whether $x \in S$. For input $x$ of length $|x| = \ell > e(m)$, we first find the least $n$ such that $e(n) \geq \log_2 \ell$. Note that $n \geq m$. Then, we find set $D$ of all strings $w$ that were added to $C$ in stages 1 to $n-1$. This can be done by querying oracle $C$ over all strings of length $e(0), e(1), \ldots, e(n-1)$. Note that we only make $O(\ell)$ queries since $2^{e(n-1)} \leq \ell$.

*Case* 1. $e(n) > \max\{p_j(\ell), p_k(\ell)\}$. Then, the computation of $N_j^C(x)$ is the same as $N_j^{C_{n-1}}(x)$, since the computation cannot query any string of length $\geq e(n)$. Thus, we can use oracle $Q \cup D = C_{n-1}$ to decide whether $N_j^C$ accepts $x$:

First, we construct a new oracle NTM $N_{j'}$ such that $N_{j'}^Q(x) = N_j^{Q \cup D}(x)$. Next notice that $K_Q = \{\langle s, z, 0^t \rangle : N_s^Q$ accepts $z$ in $t$ moves$\}$ is complete for $NP^Q$ and, hence, in $P^Q$. So, we can query oracle $Q$ to determine whether $\langle j', x, 0^{p_{j'}(\ell)} \rangle$ is in $K_Q$ (which is equivalent to $N_{j'}^Q$ accepting $x$, or $N_j^C$ accepting $x$).

*Case* 2. $e(n) \leq \max\{p_j(\ell), p_k(\ell)\}$. Then, we observe that either $N_j^{C_{n-1}}$ or $N_k^{C_{n-1}}$ accepts $x$, because otherwise $\langle j, k \rangle$ would be the least uncancelled integer for which $R_{\langle j,k \rangle}$ is vulnerable at stage $n$ and $R_{\langle j,k \rangle}$ would be cancelled. Note that $C_{n-1} = Q \cup D$, and we can, as in Case 1, determine whether $N_j^{C_{n-1}}$ accepts $x$ or $N_k^{C_{n-1}}$ accepts $x$ in time polynomial in $\ell$. Assume that $N_j^{C_{n-1}}$ accepts $x$. Then, as proved in Exercise 4.13, we can find an accepting computation of $N_j^{C_{n-1}}(x)$ in polynomial time. This computation is of length $p_j(\ell)$ and makes at most $p_j(\ell)$ queries. We query the oracle $C$ to find out whether it queries a string in $C_n - C_{n-1}$ or not.

*Subcase* 2.1. The accepting computation of $N_j^{C_{n-1}}(x)$ does not query a string in $C_n - C_{n-1}$. Then, $N_j^C(x)$ also accepts $x$ since the accepting computation path of $N_j^{C_{n-1}}(x)$ remains unchanged in the computation of $N_j^C(x)$. We conclude that $x \in S$.

*Subcase* 2.2. Not Subcase 2.1. Then, the accepting computation of $N_j^{C_{n-1}}(x)$ that we found is not necessarily correct relative to the oracle $C_n$. But we have found a string $z \in C_n - C_{n-1}$ and, by the construction, we know that this is the only string in $C_n - C_{n-1}$. Thus, we can add it to set $D$, and then, as in Case 1, use oracle $Q$ together with set $D$ to determine whether $N_j^C$ accepts $x$.

The above completes the deterministic algorithm for $S$, using oracle $C$, as well as the proof of the theorem. ∎

Similar relations between relativized $UP$ and the relativized $P$ and $NP$ can be proved using the same proof techniques. We leave them as exercises.

**Theorem 4.26** *(a)* $(\exists A) \; P^A \neq UP^A \neq NP^A$.
*(b)* $(\exists B) \; P^B \neq UP^B = NP^B$.
*(c)* $(\exists C) \; P^C = UP^C \neq NP^C$.

## Exercises

**4.1** Apply the algorithm of Theorem 4.5 to find a short *proof* that the integer 971 is a prime number.

**4.2** Let $k \geq 1$. An NTM $M$ is $k$-*ambiguous* if for any input $x$, there are at most $k$ distinct accepting computations in $M(x)$. Let $k$-*UP* denote the class of sets acceptable by polynomial-time $k$-ambiguous NTMs. Prove that $P = UP$ if and only if $P = k$-$UP$.

**4.3** Let DGIso be the problem of determining whether two directed graphs are isomorphic. Show that DGIso and GIso are equivalent under the $\leq_m^P$-reducibility.

**4.4** Prove Theorem 4.9.

**4.5** Recall the notion of strong NP reducibility $\leq_T^{SNP}$ defined in Exercise 3.1. Prove that if $NP \neq coNP$ then there is a set $A \in NP - coNP$ that is not $\leq_T^{SNP}$-complete for $NP$.

**4.6** (a) Prove Proposition 4.15(b).

(b) For any complexity class $\mathcal{C}$, let $\mathcal{C}_1$ denote the class $\{A \in \mathcal{C} : A \subseteq \{0\}^*\}$. Prove that there exists a one-way function $f : \{0,1\}^* \to \{0,1\}^*$ such that $f^{-1}$ restricted to $\{0\}^* \cap Range(f)$ is not polynomial-time computable if and only if $P_1 \neq UP_1$.

(c) Prove that there exists a one-way function $f : \{0,1\}^* \to \{0,1\}^*$ such that $\{0\}^* \subseteq Range(f)$ and that $f^{-1}$ restricted to $\{0\}^*$ is not polynomial-time computable if and only if $P_1 \neq UP_1 \cap coUP_1$.

**4.7** In electronic data communication, a *digital signature* sent by a party X to a party Y is a message $M$ that has the following properties: (i) Y must be able to verify that $M$ is indeed sent by X; (ii) It is not possible for anyone except X, including Y, to forge the message $M$; (iii) X cannot deny sending the message $M$. Describe how to implement digital signatures in a public-key cryptosystem (allowing a negligible error).

**4.8** Intuitively, a pseudorandom generator is a function $f : \{0,1\}^* \to \{0,1\}^*$ that takes a random seed $s$ of length $n$ and outputs a string $x$ of length $l(n)$, where $l(n) > n$ such that no feasible algorithm can predict the last bit of $x$ from the first $l(n) - 1$ bits of $x$. Formally, we say that a function $f : \{0,1\}^* \to \{0,1\}^*$ is a *pseudorandom generator* if

(i) For all inputs $x$ of length $n$, $|f(x)| = l(n) > n$; and

(ii) For any polynomial-time computable function $g : \{0,1\}^* \to \{0,1\}$, and any polynomial function $p$,

$$\left| \Pr_{s \in \{0,1\}^*} [g(f_1(s)) = f_2(s)] - 1/2 \right| \leq 1/p(n)$$

for almost all $n > 0$, where $f(s) = f_1(s)f_2(s)$ and $|f_2(s)| = 1$.

A commonly used function to generate pseudorandom numbers is the *linear congruence generator*: Let $m$ be a fixed integer and $a, b$ be two integers less than $m$. Let $LC_{m,a,b,k}(x) = \langle x_0, x_1, \ldots, x_k \rangle$, where $x_0 = x$ and $x_{i+1} = (ax_i + b) \bmod m$ for $i = 1, \ldots, k$. Prove that the function $LC_{m,a,b,k}$ is not a pseudorandom generator in the above sense if $k \geq 2 \log m$. In other words, find a polynomial-time algorithm $g$ that, on given $\langle x_0, x_1, \ldots, x_{k-1} \rangle$, can find $x_k$ if $LC_{m,a,b,k}(x) = \langle x_0, x_1, \ldots, x_k \rangle$ and if $k \geq 2 \log m$.

**4.9** We continue the last exercise. Let $n$ be a positive integer and $x \in QR_n$. Define $x_{n,i} = x^{2^i} \bmod n$ if $i \geq 0$, and $x_{n,i}$ be the square root of $x_{n,i+1}$ if $i < 0$. Let $b_{n,i}(x)$ be the parity of $x_{n,i}$; that is, it is 1 if $x_{n,i}$ is odd and is 0 if $x_{n,i}$ is even. The *quadratic residue generator* is the function $QRG(\langle n, x, j, k \rangle) = b_{n,j}(x)b_{n,j-1}(x) \ldots b_{n,j-k}(x)$. Prove that $QRG$ is a pseudorandom generator if $k \leq p(n)$ for some fixed polynomial $p$, assuming that the sets $QR_n$ are strongly unpredictable in the sense of Section 4.2. That is, prove that if $QR_n$ are strongly unpredictable, then there exists an infinite set $N$ of integers such that for any polynomial-time computable function $g : \{0,1\}^* \to \{0,1\}$ and any polynomial functions $p$ and $q$,

$$\left| \Pr_{x \in QR_n} [g(b_{n,q(n)}(x)b_{n,q(n)-1}(x) \cdots b_{n,0}(x)) = b_{n,-1}(x)] - \frac{1}{2} \right| \leq 1/p(n)$$

for almost all $n \in N$.

**4.10** In this exercise, we study the application of one-way functions to the problem of finding a random bit to be shared by two parties. Suppose two parties X and Y need to have an unbiased random bit to be shared by both parties. One way to do it is to let each of X and Y pick a random bit $b_X$ and $b_Y$, respectively, write them down on a piece of paper, and then let them exchange the bits $b_X$ and $b_Y$ and use $b_X \oplus b_Y$ as the random bit, where $\oplus$ is the exclusive-or operation. Since X and Y must commit their own bit before seeing the other party's bit, neither of the parties can create a biased bit $b_X \oplus b_Y$ as long as the other bit is unpredictable. Assume, however, that the two parties are in two remote locations and cannot exchange their random bits $b_X$ and $b_Y$ both efficiently and reliably. Then, we need to use one-way functions to implement this idea. In each of the following communication environments, design a protocol to implement the above idea to find the shared random bit efficiently and reliably (assuming the existence of strong one-way functions).

(a) X and Y communicate in a computer network with the software capable of efficiently generating large primes and performing multiplications of large integers.

(b) X and Y communicate through a telephone line. They have two identical copies of a 1000-page telephone directory. (Note that this scheme should work even if $P = NP$. Can this scheme be implemented in a computer network?)

**4.11** For each set $A$, let $PSPACE_b^A$ denote the class of sets $B$ for which there exist a polynomial-space oracle DTM $M$ and a polynomial function $p$ such that for every $x$, $x \in B$ if and only if $M^A$ accepts $x$ and the total number

of queries made by $M$ in the computation tree of $M^A(x)$ is bounded by $p(|x|)$.
For any two sets $X, Y \subseteq \{0,1\}^*$, let $X \; join \; Y = \{0x : x \in X\} \cup \{1y : y \in Y\}$.
   (a) Prove that for each set $A$, $PSPACE_b^A = P^{\text{QBF} \; join \; A}$.
   (b) Prove that $P = PSPACE$ if and only if $P^A = PSPACE_b^A$.

**4.12** Recall the definition of tally and sparse sets in Exercise 1.30. Also recall the relativized polynomial-time hierarchy defined in Exercise 3.1. In this exercise, we study the power of tally and sparse oracles in relativization.
   (a) Prove that $P = NP$ if and only if for all tally sets $T$, $P^T = NP^T$.
   (b) Prove that for $k \geq 2$, $PH = \Sigma_k^P$ if and only if for all sparse sets $S$, $PH^S = \Sigma_k^{P,S}$.
   (c) Prove that $PH = PSPACE$ if and only if for all sparse sets $S$, $PH^S = PSPACE^S$.

**4.13** Assume that $P^A = NP^A$. Prove that for any polynomial-time oracle NTM $M$, there exists a polynomial-time oracle DTM $M_1$ such that for any input $x \in L(M, A)$, $M_1^A(x)$ outputs an accepting computation of $M^A(x)$, and for any input $x \notin L(M, A)$, $M_1^A$ rejects $x$.

**4.14** Use the relativized separation results of Section 4.8 to show that the relation $\leq_T^{NP}$ is not transitive.

**4.15** (a) In the following, we let $w$ be a fixed finite binary string in $\{0,1\}^*$ and $\beta$ be a fixed infinite binary sequence in $\{0,1\}^\infty$. For any infinite binary sequence $\alpha \in \{0,1\}^\infty$, we write $\alpha(k)$ to denote the $k$th bit of $\alpha$ and write $\alpha^k$ to denote the string of the first $k$ bits of $\alpha$. We say an infinite binary sequence $\alpha \in \{0,1\}^\infty$ is *recursive* if the set $X$ with $\alpha_X = \alpha$ is recursive. For each set $\Lambda_i$, $i = 1, 2, 3, 4$, of infinite binary sequences defined below, verify that it satisfies the assumption of the zero-one law and, hence, $\Pr[\Lambda_i]$ is either 0 or 1:

$\Lambda_1 = \{\alpha \in \{0,1\}^\infty : w \text{ occurs in } \alpha \text{ infinitely often}\}.$

$\Lambda_2 = \{\alpha \in \{0,1\}^\infty : \beta^k \text{ occurs in } \alpha \text{ for all } k \geq 1\}.$

$\Lambda_3 = \{\alpha \in \{0,1\}^\infty : \alpha \text{ is recursive}\}.$

$\Lambda_4 = \{\alpha \in \{0,1\}^\infty : \text{the series } \sum_{n=1}^\infty \alpha(n)/n \text{ converges}\}.$

   (b) For each $i = 1, 2, 3, 4$, decide whether $\Pr[\Lambda_i]$ is equal to 0 or 1.

**4.16** Prove that with probability one, (a) $NP^A \neq PSPACE^A$; and (b) $PSPACE^A \subsetneq EXP^A$.

**4.17** (a) Prove Theorem 4.25(a).
   (b) Prove that there exists an oracle $A$ relative to which $\Sigma_2^{P,A} \neq \Pi_2^{P,A}$.

**4.18** Prove Theorem 4.26

# Historical Notes

The existence of problems in $NP - P$ that are not $NP$-complete was first suggested by Karp [1972], who listed the problems PRIME, GIso, and Linear Programming as candidates. (The problem of linear programming later was shown to be in $P$; see Khachiyan [1979] and Karmarkar [1984].) Theorem 4.8 is from Ladner [1975a]. Schöning [1982] discussed the technique of delayed diagonalization. The complexity of problems PRIME and QR has been widely studied; see, for instance, Kranakis [1986], Loxton [1990], and Bach and Shallit [1996] for more references. The fact that PRIME is in $NP$ was proved by Pratt [1975]. The complexity of the problem GIso will be further studied in Chapter 10. The class $UP$ was defined by Valiant [1976]. The relation between the class $UP$ and one-way functions (Proposition 4.15) was first proved by Ko [1985a] and Grollman and Selman [1988]. The notion of public-key cryptography was introduced by Diffie and Hellman [1976]. The RSA cryptosystem was designed by Rivest et al. [1978]. Rabin [1979] relates the security of the RSA cryptosystem and the complexity of integer factoring. The relativization results on $P$, $NP$, and $coNP$ (Theorems 4.19 and 4.25) were first presented by Baker et al. [1975]. The independence result of Theorem 4.20 is from Hartmanis and Hopcroft [1976], which includes some other independence results in computer science. The positive relativization result of Theorem 4.22 is from Book et al. [1984]. Book et al. [1985] contains results on different types of positive relativization. Random oracles were first studied by Bennett and Gill [1981]. Theorem 4.24 is from there. For more results on random oracles, see Chapters 8 and 9, and the survey paper by Vollmer and Wagner [1997]. Relativization results on the relation between $UP$ and $NP$ (Theorem 4.26) were proved in Rackoff [1982].

Digital signatures and pseudorandom generators are important topics in cryptography; see any cryptography textbook for more references. For the linear congruence generator, see Plumstead [1982] and Frieze et al. [1984]. For the quadratic residue generator, see Blum et al. [1986]. In general, pseudorandom generators exist if strong one-way functions exist. See Hastad et al. [1999] for the detailed discussion. Exercise 4.12 is from Long and Selman [1986]. The relativized separation of the second level of $PH$ (Exercise 4.17(b)) was first proved by Heller [1984]. The relativized separation of the higher levels of $PH$ will be studied in Chapter 9.

# Part II

# Nonuniform Complexity

*The Way that can be described [with a finite number of words] is not the eternal Way.*

— Lao Tsu

# 5

# *Decision Trees*

The depth of a decision tree is a fundamental complexity measure for Boolean functions. We study the algebraic and topological proof techniques to establish lower bounds for the depth of decision trees, particularly for monotone graph properties. Randomized decision trees and branching programs are also studied.

## 5.1  Graphs and Decision Trees

We first review the notion of graphs and the Boolean function representations of graphs.[1] A graph is an ordered pair of disjoint sets $(V, E)$ such that $E$ is a set of pairs of elements in $V$ and $V \neq \emptyset$. The elements in the set $V$ are called *vertices* and the elements in the set $E$ are called *edges*. Two vertices

---

[1]In this chapter, we will mostly deal with Boolean functions. The reader should review Section 1.1 for basic notions and definitions about Boolean functions.

are *adjacent* if there is an edge between them. Two graphs are *isomorphic* if there exists a one-to-one correspondence between their vertex sets which preserves adjacency.

A *path* is an alternating sequence of distinct vertices and edges starting and ending with vertices such that every vertex is an endpoint of its neighboring edges. The length of a path is the number of edges appearing in the path. A graph is *connected* if every pair of vertices are joined by a path.

Let $V = \{1, \ldots, n\}$ be the vertex set of a graph $G = (V, E)$. Then its *adjacency matrix* $[x_{ij}]$ is defined by

$$x_{ij} = \begin{cases} 1 & \text{if } \{i, j\} \in E, \\ 0 & \text{otherwise.} \end{cases}$$

Note that $x_{ij} = x_{ji}$ and $x_{ii} = 0$. So, the graph $G$ can be determined by $n(n-1)/2$ independent variables $x_{ij}$, $1 \le i < j \le n$. For any property $\mathcal{P}$ of graphs with $n$ vertices, we define a Boolean function $f_{\mathcal{P}}$ over $n(n-1)/2$ variables $x_{ij}$, $1 \le i < j \le n$, as follows:

$$f_{\mathcal{P}}(x_{12}, \ldots, x_{n-1,n}) = \begin{cases} 1 & \text{if the graph } G \text{ represented by } [x_{i,j}] \text{ has} \\ & \text{the property } \mathcal{P}, \\ 0 & \text{otherwise.} \end{cases}$$

Then $\mathcal{P}$ can be determined by $f_{\mathcal{P}}$. For example, the property of connectivity corresponds to the Boolean functions $f_{con}$ of $n(n-1)/2$ variables such that $f_{con}(x_{12}, \ldots, x_{n-1,n}) = 1$ if and only if the graph $G$ represented by $[x_{ij}]$ is connected.

Not every Boolean function of $n(n-1)/2$ variables represents a graph property because a graph property should be invariant under graph isomorphism. A Boolean function $f$ of $n(n-1)/2$ variables is called a *graph property* if for every permutation $\sigma$ on the vertex set $\{1, \ldots, n\}$,

$$f(x_{12}, \ldots, x_{n-1,n}) = f(x_{\sigma(1)\sigma(2)}, \ldots, x_{\sigma(n-1)\sigma(n)}).$$

Consider two assignments for a Boolean function $f$ of $n$ variables, $t_1, t_2 : \{x_1, \ldots, x_n\} \to \{0, 1\}$. If $t_1(x_i) \le t_2(x_i)$ for all $1 \le i \le n$, then we write $t_1 \le t_2$. A Boolean function $f$ is *monotone increasing* if

$$f|_{t_1} = 1, t_1 \le t_2 \implies f|_{t_2} = 1.$$

A Boolean function $f$ is *monotone decreasing* if

$$f|_{t_1} = 0, t_1 \le t_2 \implies f|_{t_2} = 0.$$

For example, $f_{con}$ is monotone increasing because a graph having a connected subgraph on all vertices must be connected. A Boolean function is *monotone* if it is either monotone increasing or monotone decreasing.

For a Boolean function $f$, define its *dual* (or *contravariant*) $f^*$ by

$$f^*(x_1, x_2, \ldots, x_n) = \neg f(\bar{x}_1, \bar{x}_2, \ldots, \bar{x}_n).$$

For example, the dual of $f(x,y) = x \vee y$ is $f^*(x,y) = \neg(\bar{x} \vee \bar{y}) = x \wedge y$. Clearly, $f^{**} = f$. So, the dual of $x \wedge y$ is $x \vee y$. It is worth mentioning that a Boolean function is monotone increasing (or, monotone decreasing) if and only if its dual is monotone increasing (or, respectively, monotone decreasing).

It is interesting to see what the minterms are of a monotone increasing graph property. Recall that a minterm $p$ of a Boolean function $f$ is a minimal elementary product $p$ such that $p = 1$ implies $f = 1$. For instance, a minterm of $f_{con}$ is a spanning tree on $n$ vertices, since a spanning tree is a minimal connected graph. In general, a minterm of a monotone increasing graph property is a minimal graph having the property.

A Boolean function is *trivial* if it is a constant function, that is, if it equals to 0 for all assignments or to 1 for all assignments. Hence, a graph property is *trivial* if it holds for either all graphs or no graph.

A graph is a *bipartite* graph if its vertices can be partitioned into two sets such that every edge is between these two sets. A bipartite graph is denoted by $(A, B, E)$, where $A$ and $B$ are the two vertex sets and $E$ is the edge set. A property of bipartite graphs $(A, B, E)$ with $A = \{1, \ldots, m\}$ and $B = \{1, \ldots, n\}$ can be represented by a Boolean function $f$ of $mn$ variables $x_{ij}$, where $i \in A$ and $j \in B$. A bipartite graph property $f$ satisfies that for every permutation $\sigma$ on $A$ and every permutation $\tau$ on $B$,

$$f(x_{11}, \ldots, x_{m,n}) = f(x_{\sigma(1)\tau(1)}, \ldots, x_{\sigma(m),\tau(n)}).$$

A *digraph* (*directed graph*) is an ordered pair of disjoint sets $(V, E)$ such that $E$ is a set of ordered pairs of elements in $V$ and $V \neq \emptyset$. In general, a digraph of $n$ variables may have up to $n^2$ edges and its properties may be represented as Boolean functions of $n^2$ variables which are invariant under permutation of vertices, that is, a Boolean function $f(x_{11}, \ldots, x_{1n}, x_{21}, \ldots, x_{nn})$ is a digraph property if for every permutation $\sigma$ on $\{1, 2, \ldots, n\}$,

$$f(x_{11}, \ldots, x_{1n}, x_{21}, \ldots, x_{nn})$$
$$= f(x_{\sigma(1)\sigma(1)}, \ldots, x_{\sigma(1)\sigma(n)}, x_{\sigma(2)\sigma(1)}, \ldots, x_{\sigma(n)\sigma(n)}).$$

It is easy to see that if a Boolean function $f$ over variables $x_{ij}$, with $1 \le i \le n$ and $1 \le j \le n$, represents a bipartite graph property for bipartite graphs $G = (\{1, \ldots, n\}, \{1, \ldots, n\}, E)$, then it can also be considered as a digraph property for digraphs $G = (\{1, \ldots, n\}, E)$. However, the inverse is not necessarily true. For example, consider the Boolean function $f(x_{11}, x_{12}, x_{21}, x_{22}) = x_{12}x_{21}$. It can be checked that it is a digraph property, but not a bipartite graph property.

A *loop* in a digraph is an edge from a vertex to itself. Every isomorphism of digraphs must map a loop to another loop. In other words, loops must also be invariant under digraph isomorphisms. This leads to a decomposition of digraph properties. That is, each digraph property can be decomposed into a property of digraphs that have no loop plus a loop structure. More precisely, let $s_k = \sum_{1 \le i_1 < i_2 < \cdots < i_k \le n} x_{i_1 i_1} x_{i_2 i_2} \cdots x_{i_k i_k}$. Then, every digraph property

$f$ can be decomposed as

$$f = \sum_{k=0}^{n} s_k f|_{x_{11}=\cdots=x_{kk}=1, x_{k+1,k+1}=\cdots=x_{nn}=0}.$$

Note that each function $f|_{x_{11}=\cdots=x_{kk}=1, x_{k+1,k+1}=\cdots=x_{nn}=0}$ has $n(n-1)$ variables, and it can be considered as a property of digraphs without any loop.

A *cycle* in a graph is a path with an identical starting and ending vertex. A *tree* is a connected graph without a cycle. When a tree has a special vertex named *root*, it is called a *rooted* tree. In a rooted tree, all vertices of degree one (except for the root itself if the tree has at least two vertices) are called *leaves*, while other vertices are called *internal vertices*. All vertices in a rooted tree can be divided into different generations according to their distance from the root. If a vertex $u$ is adjacent to another vertex $v$ which is in a higher generation, then $u$ is called a *child* of $v$ and $v$ is called the *parent* of $u$. A *binary* tree is a rooted tree in which every internal vertex has exactly two children.

A *decision tree* of a Boolean function $f$ is a binary tree whose internal vertices are labeled by variables, leaves are labeled by 0 and 1, and edges are labeled also by 0 and 1 such that (a) every pair of edges from an internal vertex to its two children are labeled by 0 and 1, respectively; and (b) any variable appears at most once in any path from the root to a leaf. Given an assignment $t$ to the variables of a Boolean function $f$, we can compute the function value by its decision tree as follows:

> current-vertex := root;
> while current-vertex is not a leaf do
> begin
>     $x$ := label of the current-vertex;
>     if $t(x) = 0$ then let current-vertex be the child of the
>                 current-vertex along the edge with label 0
>           else  let current-vertex be the child of the
>                 current-vertex along the edge with label 1
> end;
> value-of-$f$ := the label of the current-vertex.

For instance, Figure 5.1 presents a decision tree which computes the function $f(x_1, x_2, x_3) = x_1(x_3 + \bar{x}_2 \bar{x}_3) + \bar{x}_1(\bar{x}_2 + x_2 x_3)$. The path marked with the dotted line on the decision tree shows the computation of $f(0, 1, 0)$. Since the leaf is labeled by 0, we have $f(0, 1, 0) = 0$.

A Boolean function may have a number of different decision trees, each defining a different procedure to compute the function. For each decision tree, an assignment $t$ to variables defines a path from the root to a leaf as follows: At each node with label $x_i$, follow the edge with label $t(x_i)$ to the next node. The computation time of a decision tree with respect to an assignment is the length of the path defined by the assignment, that is, the number of

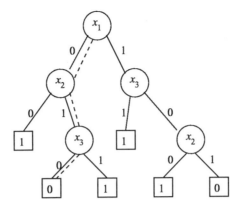

Figure 5.1: A decision tree.

variables in the path. The *depth* of a decision tree is the maximum length of paths from the root to leaves. For a Boolean function $f$, we denote by $D(f)$ the minimum depth of decision trees computing $f$. $D(f)$ is called the *decision tree complexity* of $f$. Clearly, $D(f) \le n$ when $f$ has $n$ variables. Every constant function $f$ ($f \equiv 0$ or $f \equiv 1$) has a decision tree consisting of only a leaf labeled by the constant. Therefore, we have $D(f) = 0$. For a nonconstant function $f$, at least one variable should be assigned to determine the function value. Therefore, $D(f) \ge 1$.

As an example, let us consider the parity function $p_n(x_1, \ldots, x_n) = x_1 \oplus \cdots \oplus x_n$; that is, $p_n(x_1, \ldots, x_n) = 1$ if and only if there are an odd number of 1's in $x_1, \ldots, x_n$. Note that, for any variable $x_i$, the function value $p_n$ varies as $x_i$ varies, no matter what values the other variables take. This means that in any decision tree computing $p_n$, every path from the root to a leaf must have length $n$. Thus, $D(p_n) = n$.

In a decision tree of a Boolean function $f$, a path from the root to a leaf with label 0 corresponds to a clause of $f$ and a path from the root to a leaf with label 1 corresponds to an implicant of $f$. For the parity function $p_n$, every clause is of size $n$ and every implicant is of size $n$. This gives an alternative way to prove that $D(p_n) = n$.

Recall that for any Boolean function $f$, $D_0(f)$ denotes the maximum size of a prime clause of $f$ and $D_1(f)$ denotes the maximum size of a minterm of $f$. Is it true that for any Boolean function $f$ there exists a decision tree of which each path from the root to a leaf corresponds to either a prime clause or a minterm? If the answer is yes, then we would have $D(f) \le \max\{D_1(f), D_0(f)\}$ for any Boolean function $f$. Unfortunately, this is not true and the inequality $D(f) \le \max\{D_1(f), D_0(f)\}$ does not hold for some function $f$. For instance, consider the function

$$f(x_1, x_2, x_3, x_4) = x_1 x_2 + x_2 x_3 + x_3 x_4 + x_4 x_1 = (x_1 + x_3)(x_2 + x_4).$$

Clearly, $D_0(f) = D_1(f) = 2$. However, we can prove that $D(f) = 4$. (The

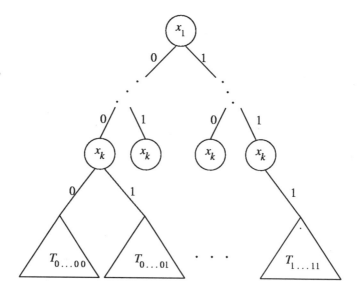

Figure 5.2: The decision tree computing $f$.

detail of the proof is left to the reader as an exercise.)

On the other hand, the inequality $D(f) \leq D_1(f)D_0(f)$ does hold.

**Theorem 5.1** $D(f) \leq D_1(f)D_0(f)$.

*Proof.* We prove it by induction on the number $n$ of variables in $f$. For $n = 1$, $D(f) = 0$ or 1. If $D(f) = 0$, then the inequality holds trivially. If $D(f) = 1$, then $f$ is not a constant function. Thus, both $D_0(f)$ and $D_1(f)$ are equal to 1, and so the inequality holds.

Now we consider the general case of $n > 1$. If $f$ is a constant function, then again the inequality holds trivially. So, assume that $f$ is a nonconstant function and $D(f) > 0$. Choose a minterm $p$ of $f$ that has size $D_1(f) = k \geq 1$. Without loss of generality, assume that the variables that appear in $p$ are $x_1, \ldots, x_k$. Note that every clause contains at least one variable in $\{x_1, \ldots, x_k\}$ (Proposition 1.3). Therefore, we claim that for any (partial) assignment $t$ on variables $x_1, \ldots, x_k$, the Boolean function $f|_t$ has prime clauses of size at most $D_0(f) - 1$. To see this, consider a prime clause $c$ of size $D_0(f|_t)$ for $f|_t$. Denote $x_{i,t} = x_i$ if $t(x_i) = 0$ and $x_{i,t} = \bar{x}_i$ if $t(x_i) = 1$. Then, $c' = c + x_{1,t} + \cdots + x_{k,t}$ is a clause of $f$. Thus $c'$ must have a subsum $c''$ which is a prime clause of $f$. First, we note that $c$ must belong to $c''$. In fact, if $c$ does not belong to $c''$, then $c''|_t$ is a proper subsum of $c$ which is a clause of $f|_t$, contradicting the assumption that $c$ is a prime clause of $f|_t$. Thus, $c$ must belong to $c''$. Moreover, $c''$ must contain at least one variable in $\{x_1, \ldots, x_k\}$. Therefore, $c''$ has size at least $D_0(f|_t) + 1$. Hence $D_0(f|_t) + 1 \leq D_0(f)$.

On the other hand, it is clear that $D_1(f|_t) \leq D_1(f)$. So, by the inductive hypothesis,

$$D(f|_t) \leq D_1(f)(D_0(f) - 1).$$

Now, we construct a decision tree as follows: First, construct a balanced complete binary tree of $k+1$ levels such that all vertices in the $i$th $(1 \leq i \leq k)$ level are labeled by $x_i$ and all $2^k$ paths correspond to $2^k$ assignments for $x_1, \ldots, x_k$. Each leaf is labeled by an assignment corresponding to the path from the root to the leaf. Then for each assignment $t$ with $t(x_i) = e_i$ for $1 \leq i \leq k$, we replace the leaf with label $e_1 \cdots e_k$ by a decision tree $T_{e_1 \cdots e_k}$ of depth at most $D_1(f)(D_0(f) - 1)$ which computes $f|_t$ (see Figure 5.2). The resulting decision tree computes $f$ and its depth $D(f)$ is at most

$$k + D_1(f)(D_0(f) - 1) = D_1(f)D_0(f). \qquad \blacksquare$$

## 5.2 Examples

A Boolean function $f$ of $n$ variables is called *elusive* if $D(f) = n$. For example, the parity function $p_n$ is elusive. Elusiveness is an interesting subject on decision trees with a number of deep results. Before exploring these results, we first look at some examples.

**Example 5.2** In a tournament, there are $n$ players $1, \ldots, n$. Let $x_{ij}$ be the result of the match between players $i$ and $j$, that is,

$$x_{ij} = \begin{cases} 1 & \text{if } i \text{ beats } j, \\ 0 & \text{if } j \text{ beats } i. \end{cases}$$

(Note that this is not necessarily a transitive relation.) Consider the following function:

$$t(x_{12}, \ldots, x_{n-1,n}) = \begin{cases} 1 & \text{if there is a player who beats all other players,} \\ 0 & \text{otherwise.} \end{cases}$$

We show that $D(t) \leq 2(n-1) - \lfloor \log_2 n \rfloor$. In other words, we need to design a tournament such that within $2(n-1) - \lfloor \log_2 n \rfloor$ matches, the value of function $t$ can be determined. To this end, we first use a balanced binary tree to design a knockout tournament. Let $i$ be the winner of the knockout tournament. Then let player $i$ play against everyone with whom he has not played yet. If $i$ wins all the matches, then $t$ equals 1; otherwise, $t$ equals 0. Note that each match in a knockout tournament knocks out exactly one player. A knockout tournament for $n$ players contain $n - 1$ matches, in which the winner $i$ played at least $\lfloor \log_2 n \rfloor$ times. So, the total number of matches is at most $2(n-1) - \lfloor \log_2 n \rfloor$. Thus, $t$ is not elusive. $\qquad \square$

**Example 5.3** Consider the following function:

$$m(x_{11}, \ldots, x_{1n}, x_{21}, \ldots, x_{nn}) = \bigwedge_{i=1}^{n} \bigvee_{j=1}^{n} x_{ij}.$$

We show that $m$ is elusive. To do so, let us consider a decision tree computing $m$ and look for a path in the following way. Starting from the root, we look at the variable $x_{ij}$ labeling the vertex. If all other variables $x_{ik}$, $k \neq j$, have been assigned value 0, then we assign 1 to $x_{ij}$; otherwise, assign 0 to $x_{ij}$. Then, we go to the next vertex along the edge with the value that was assigned to $x_{ij}$. In this way, it is easy to see that before all variables are assigned we cannot know the value of $m$. This means that the path we traveled through must contain all variables. Thus, $m$ is elusive.                                                    □

The above argument can be generalized to other Boolean functions. A Boolean function is called a *tree function* if it is monotone increasing and has an expression with no negation symbol in which each variable appears exactly once. The above function $m$ is an example of tree functions.

**Proposition 5.4** *Every tree function $f$ is elusive.*

*Proof.* Assume that $f$ has an expression in which each variable occurs exactly once, in the positive form. For any decision tree of $f$, consider the following strategy to find the longest path:

> If the current variable $x_i$ appears in a sum in the expression, then set $x_i = 0$; otherwise, $x_i$ must appear in a product in $E$, set $x_i = 1$. After assigning $x_i$ a value, simplify the expression $E$.

This strategy will eliminate one variable from the expression at each step. Therefore, all variables have to be assigned by this strategy in order to get a function value, and hence $f$ is elusive.                                              ∎

Note that $D(f)$ can be expressed as

$$D(f) = \min_{T} \max_{q \in p(T)} length(q),$$

where $T$ ranges over all decision trees computing $f$ and $p(T)$ is the set of all root-to-leaf paths in $T$. This expression tells us that $D(f)$ is a *minimax* value of path lengths. It suggests us to use a game-theoretical argument to prove the lower bounds of $D(f)$. That is, consider a game consisting of two players, a verifier and a prover. At each round, the verifier chooses a variable and the prover assigns a value to the variable. Then the verifier evaluates the function $f$. If he/she can determine the value of $f$, then the game ends; otherwise, the game goes to a new round. The verifier's goal is to choose the variables such that the value of $f$ can be determined as early as possible. On the contrary, the prover wants to assign the values such that the game lasts as long as possible. If the prover has a strategy to make the game to last for $k$ rounds, then it follows that $D(f) \geq k$. This type of game-theoretic argument is called the *adversary argument*.

**Example 5.5** The graph property connectivity $f_{con}$ is elusive. To show this, consider the following strategy of the prover:

At each step, let $G$ be the graph which contains the edge $\{p, q\}$ as long as the variable $x_{pq}$ has not been assigned the value 0. Then, for the variable $x_{ij}$ asked by the verifier, if deleting the edge $\{i, j\}$ would disconnect the graph $G$, then set $x_{ij} = 1$; otherwise, set $x_{ij} = 0$.

With this strategy, if the verifier finds the value of $f_{con}$, then this value must be 1. Suppose on the contrary that this strategy does not work, that is, there is a variable $x_{st}$ which has not been asked by the verifier but the verifier finds that all edges $\{i, j\}$ with $x_{ij} = 1$ form a connected graph $G$ already. Adding the edge $\{s, t\}$ to the graph, we will obtain a cycle containing $\{s, t\}$. From this cycle, choose an edge $\{h, k\}$ other than $\{s, t\}$. Then the graph $G'$ obtained by deleting $\{h, k\}$ and adding $\{s, t\}$ is still a connected graph, contradicting the strategy for setting $x_{hk} = 1$.                                          □

The strategy of the prover in the above example is called a *simple* strategy. In general, a simple strategy works as follows: A variable is assigned with value 0 whenever it would not make the underlying Boolean function equal to 0; and it is assigned with value 1 otherwise.

Consider an assignment $\tau$ for a Boolean function. Let $true(\tau) = \{x_i : \tau(x_i) = 1\}$ and $false(\tau) = \{x_i : \tau(x_i) = 0\}$. Using the simple strategy, we can prove the following.

**Theorem 5.6** *Assume that a Boolean function $f$ satisfies that for any truth assignment $\tau$ and any variable $x_i$ in $false(\tau)$, there exist a variable $x_j$ in $true(\tau)$ and a truth assignment $\pi$ such that $(false(\tau) - \{x_i\}) \cup \{x_j\} = false(\pi)$. Then $f$ is elusive.*

*Proof.* With the simple strategy, if the verifier finds the value of $f$ then this value must be 1. Suppose on the contrary that this strategy does not work, that is, there is a variable $x_i$ which has not been asked by the verifier but the verifier finds the value of function $f$ to be 1. Then there exists a truth assignment $\tau$ which coincides with values assigned by the prover and $\tau(x_k) = 0$ for all variables $x_k$ which have not been assigned a value by the prover (in particular, $\tau(x_i) = 0$). By the given condition on $f$, there exists a variable $x_j$ in $true(\tau)$ and a truth assignment $\pi$ such that $(false(\tau) - \{x_i\}) \cup \{x_j\} = false(\pi)$. Note that $f|_\pi = 1$ implies that all variables $x_k \in true(\tau)$ asked by the verifier before $x_j$ must be in $true(\pi)$ since, by the simple strategy, $\pi(x_k) = 0$ would make $f|_\pi = 0$. This means that the path corresponding to $\pi$ agrees with the path corresponding to $\tau$ up till $x_j$. Now, when the verifier asks for the value of $x_j$, the prover could assign value 0 to $x_j$ without making the value of $f$ to be 0, as $\pi(x_j) = 0$ and $f|_\pi = 1$. This contradicts the simple strategy of the prover when he assigns $x_j$ with value 1.                                          ■

Note that $D(f) = D(f^*)$. Applying Theorem 5.6 to $f^*$, we obtain the following.

**Corollary 5.7** *Assume that a Boolean function $f$ satisfies that for any truth assignment $\tau$ and any variable $x_i$ in $true(\tau)$, there exist a variable $x_j$ in $false(\tau)$ and a truth assignment $\pi$ such that $(true(\tau) - \{x_i\}) \cup \{x_j\} = true(\pi)$. Then $f$ is elusive.*

**Example 5.8** Planarity of graphs is elusive for $n \geq 5$, where $n$ is the number of vertices. By Corollary 5.7, it suffices to prove that for any planar graph $G$ and an edge $e$ of $G$, we can find an edge $e'$ not in $G$ such that replacing $e$ with $e'$ preserves planarity. If $G$ is not a maximal planar graph, then this condition holds trivially. In fact, suppose $e'$ is an edge not in $G$ and $G \cup \{e'\}$ is still planar. Then for any edge $e$ of $G$, replacing $e$ with $e'$ preserves planarity.

Now, we assume that $G$ is a maximal planar graph. That is, when $G$ is embedded into a plane, every face is a triangle. (If not, we can add an edge.) Let us fix an embedding of $G$ in the plane. Consider an edge $ab$ which is the boundary of two faces $abc$ and $abd$. If $cd$ is not an edge of $G$, we can replace $ab$ with $cd$ to keep the planarity of $G$. Thus, we may assume that $cd$ is an edge of $G$.

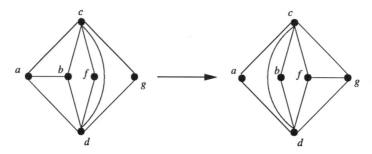

Figure 5.3: Replacing $ab$ with $fg$.

Let $cdf$ and $cdg$ be two faces containing $cd$. If $\{f, g\} = \{a, b\}$, then $abc$, $abd$, $cdf$, and $cdg$ are all the faces. Thus, $n = 4$, contradicting $n \geq 5$. Hence, we may assume $a \notin \{f, g\}$. This implies that $fg$ is not an edge of $G$ since, otherwise, $fg$ would intersect either the edge $cd$ or the path $cad$. Now, we add $fg$ to cross with $cd$, move $cd$ to cross with $ab$, and finally remove $ab$ (see Figure 5.3). The resulting graph is still planar. This means that replacing $ab$ in $G$ with $fg$ preserves planarity.                                                                                    □

The condition in Theorem 5.6 is actually necessary and sufficient for the simple strategy to be a "winning" strategy of the prover to show the elusiveness. We state below this result on monotone increasing functions.

**Corollary 5.9** *Let $f$ be a nontrivial monotone increasing Boolean function. Then, $f$ can be shown to be elusive by using the simple strategy if and only if for any truth assignment $\tau$ and any $x_i$ in $false(\tau)$, we can find a variable $x_j$ in $true(\tau)$ such that exchanging values of $x_i$ and $x_j$ would not change the function value.*

## 5.3 Algebraic Criterion

In this section we provide some tools to prove the elusiveness of a Boolean function. We begin with a simple one.

**Theorem 5.10** *A Boolean function with an odd number of truth assignments is elusive.*

*Proof.* The constant functions $f \equiv 0$ and $f \equiv 1$ have 0 and $2^n$ truth assignments, respectively. Hence, a Boolean function with an odd number of truth assignments must be a nonconstant function. If $f$ has at least two variables and $x_i$ is one of them, then the number of truth assignments of $f$ is the sum of those of $f|_{x_i=0}$ and $f|_{x_i=1}$. Therefore, either $f|_{x_i=0}$ or $f|_{x_i=1}$ has an odd number of truth assignments and is not a constant function. Thus, for any decision tree computing $f$, tracing the subtrees with an odd number of truth assignments, we will meet all variables in a path from the root to a leaf. ∎

Define
$$p_f(k) = \sum_{t \in \{0,1\}^n} f(t) k^{\|t\|},$$

where $\|t\|$ is the number of 1's in string $t$ (recall that a Boolean assignment may be viewed as a binary string). It is easy to see that $p_f(1)$ is the number of truth assignments for $f$. The following theorem is an extension of Theorem 5.10.

**Theorem 5.11** *For a Boolean function $f$ of $n$ variables, $(k+1)^{n-D(f)} \mid p_f(k)$.*

*Proof.* We prove this theorem by induction on $D(f)$. First, we note that if $f \equiv 0$ then $p_f(k) = 0$ and if $f \equiv 1$ then $p_f(k) = (k+1)^n$ (by the binomial theorem). This means that the theorem holds for $D(f) = 0$. Now, consider $f$ with $D(f) > 0$ and a decision tree $T$ of depth $D(f)$ computing $f$. Without loss of generality, assume that the root of $T$ is labeled by $x_1$. Denote $f_0 = f|_{x_1=0}$ and $f_1 = f|_{x_1=1}$. Then

$$
\begin{aligned}
p_f(k) &= \sum_{t \in \{0,1\}^n} f(t) k^{\|t\|} \\
&= \sum_{s \in \{0,1\}^{n-1}} f(0s) k^{\|s\|} + \sum_{s \in \{0,1\}^{n-1}} f(1s) k^{1+\|s\|} \\
&= p_{f_0}(k) + k p_{f_1}(k).
\end{aligned}
$$

Note that $D(f_0) \leq D(f) - 1$ and $D(f_1) \leq D(f) - 1$. By the induction hypothesis, $(k+1)^{n-1-D(f_0)} \mid p_{f_0}(k)$ and $(k+1)^{n-1-D(f_1)} \mid p_{f_1}(k)$. Thus, $(k+1)^{n-D(f)} \mid p_f(k)$. ∎

An important corollary is as follows. Denote $\mu(f) = p_f(-1)$.

**Corollary 5.12** *If $\mu(f) \neq 0$, then $f$ is elusive.*

Next, we present an application of the above algebraic criterion.

Let $H$ be a subgroup of the permutation group $S_n$ on $\{1, \ldots, n\}$. We say $H$ is *transitive* if for any $i, j \in \{1, \ldots, n\}$, there exists a permutation $\sigma \in H$ such that $\sigma(i) = j$.

Let $f$ be a Boolean function of $n$ variables. We say $f$ is *weakly symmetric* if there exists a transitive subgroup $H$ of $S_n$ such that for any $\sigma \in H$, $f(x_1, \ldots, x_n) = f(x_{\sigma(1)}, \ldots, x_{\sigma(n)})$. $f$ is *symmetric* if $f(x_1, \ldots, x_n) = f(x_{\sigma(1)}, \ldots, x_{\sigma(n)})$ for all $\sigma \in S_n$. For example,

$$f(x_1, \ldots, x_n) = x_1 x_2 + x_2 x_3 + \cdots + x_{n-1} x_n + x_n x_1$$

is weakly symmetric (let $H$ be the subgroup generated by the permutation $\sigma$ with $\sigma(x_i) = x_{i+1}$ for $1 \leq i < n$ and $\sigma(x_n) = x_1$) but is not symmetric. Every graph property is weakly symmetric. To see this, consider the set $H$ of all permutations on edges that are induced by all permutations on vertices. However, in general, a graph property is not necessarily symmetric because when a vertex is transformed to another one, all edges at the vertex have to move together. Denote $\mathbf{0} = (0, \ldots, 0)$ and $\mathbf{1} = (1, \ldots, 1)$.

**Theorem 5.13** *Let $n$ be a prime power, that is, $n = p^m$ for some prime $p$ and some positive integer $m$. If $f$ is weakly symmetric on $n$ variables and $f(\mathbf{0}) \neq f(\mathbf{1})$, then $f$ is elusive.*

*Proof.* Consider all truth assignments with exactly $k$ 1's. We count the total number of 1's in those truth assignments in the following two ways.

1. Let $n_k$ be the number of truth assignments with exactly $k$ 1's. Then the total number of 1's in those $n_k$ truth assignments is $k n_k$.

2. Since $f$ is weakly symmetric, the number of 1's taken by each variable in these truth assignments is the same; let it be $a_k$. Then the total number of 1's in those $n_k$ truth assignments is $n a_k$.

The counting results from the above two ways must be equal. Therefore, $p^m a_k = n a_k = n_k k$. If $0 < k < n$, then $p \mid n_k$. Thus,

$$\mu(f) = \sum_{k=0}^{n} n_k (-1)^k \equiv f(\mathbf{0}) + f(\mathbf{1})(-1)^n \ (\bmod \ p).$$

Since $f(\mathbf{0}) \neq f(\mathbf{1})$, $\mu(f) \equiv 1$ or $(-1)^n \pmod{p}$. Thus, $\mu(f) \neq 0$. By Corollary 5.12, $f$ is elusive. ∎

Recall that a bipartite graph property is a Boolean function $f(x_{11}, \ldots, x_{1n}, x_{21}, \ldots, x_{mn})$ satisfying the condition that for any permutation $\sigma$ on $\{1, 2, \ldots, m\}$ and any permutation $\tau$ on $\{1, 2, \ldots, n\}$,

$$f(x_{11}, \ldots, x_{1n}, x_{21}, \ldots, x_{mn})$$
$$= f(x_{\sigma(1)\tau(1)}, \ldots, x_{\sigma(1)\tau(n)}, x_{\sigma(2)\tau(1)}, \ldots, x_{\sigma(m)\tau(n)}).$$

Similar to graph properties, a bipartite graph property is weakly symmetric.

If a bipartite graph property is nontrivial, then there is a bipartite graph which has the property and there is also a bipartite graph which does not have the property. Note that the empty graph is a subgraph of every graph and every graph is a subgraph of the complete graph. Thus, if the bipartite graph property is nontrivial and monotone, then the complete bipartite graph and the empty bipartite graph cannot have the property simultaneously, that is, we must have $f(1) = 1$ and $f(0) = 0$. This observation leads to the following corollary.

**Corollary 5.14** *Let $\mathcal{P}$ be a nontrivial monotone property of bipartite graphs between vertex sets $A$ and $B$ with $|A| \cdot |B|$ a prime power. Then $\mathcal{P}$ is elusive.*

*Proof.* Note that the number of edges for a complete bipartite graph between $A$ and $B$ is a prime power $|A| \cdot |B|$. From the above observation, we see that all conditions of Theorem 5.13 are satisfied. Therefore, $\mathcal{P}$ is elusive. ∎

Actually, any nontrivial monotone bipartite graph property is elusive. We will prove this result in Section 5.6.

Could the condition that the number $n$ of variables is a prime power be dropped from Theorem 5.13? The answer is negative. The following is a counterexample.

**Example 5.15** Let $H$ be the subgroup generated by the cyclic permutation $\sigma = (1\ 2\ 3\ 4\ \cdots\ 12)$, that is, $H = \{\sigma^i : i = 0, 1, \ldots, 11\}$. Clearly, $H$ is transitive. Consider the following Boolean function of twelve variables:

$$f(x_1, \ldots, x_{12})$$

$$= \prod_{i=1}^{12} \bar{x}_i + \sum_{\tau \in H} \left( x_{\tau(1)} \prod_{i \neq 1} \bar{x}_{\tau(i)} + x_{\tau(1)} x_{\tau(4)} \prod_{i \neq 1,4} \bar{x}_{\tau(i)} + x_{\tau(1)} x_{\tau(5)} \prod_{i \neq 1,5} \bar{x}_{\tau(i)} \right.$$

$$+ x_{\tau(1)} x_{\tau(4)} x_{\tau(7)} \prod_{i \neq 1,4,7} \bar{x}_{\tau(i)} + x_{\tau(1)} x_{\tau(5)} x_{\tau(9)} \prod_{i \neq 1,5,9} \bar{x}_{\tau(i)}$$

$$\left. + x_{\tau(1)} x_{\tau(4)} x_{\tau(7)} x_{\tau(10)} \prod_{i \neq 1,4,7,10} \bar{x}_{\tau(i)} \right).$$

(In the above, we omit, in the products, the range of the index $i$ which is $\{1, \ldots, 12\}$; for instance, $\prod_{i \neq 1,4} \bar{x}_{\tau(i)}$ really means $\prod_{1 \leq i \leq 12, i \neq 1,4} \bar{x}_{\tau(i)}$.)

Clearly, $f$ is invariant under $H$. Thus, function $f$ is weakly symmetric. Next, we show that $f$ can be computed by the decision tree shown in Figure 5.4, which has depth 11. In Figure 5.4, $T_{i_1, \ldots, i_k}$ denotes a depth-$k$ subtree computing function $\bar{x}_{i_1} \cdots \bar{x}_{i_k}$.

First, we note that

$$f|_{x_1 = x_4 = 1} = \prod_{i \neq 1,4} \bar{x}_i + x_7 \prod_{i \neq 1,4,7} \bar{x}_i + x_{10} \prod_{i \neq 1,4,10} \bar{x}_i + x_7 x_{10} \prod_{i \neq 1,4,7,10} \bar{x}_i$$

$$= \prod_{i \neq 1,4,7,10} \bar{x}_i. \tag{5.1}$$

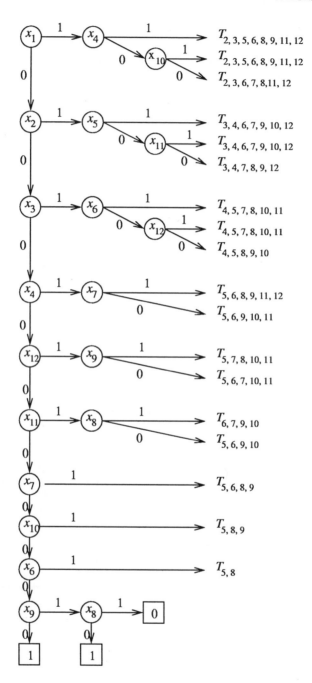

Figure 5.4: A decision tree for $f$.

(The third term $x_{10} \prod_{i \neq 1,4,10} \bar{x}_i$ came from $x_{\tau(1)} x_{\tau(4)} x_{\tau(7)} \prod_{i \neq 1,4,7,10} \bar{x}_{\tau(i)}$ with $\tau(1) = 10$.) Similarly,

$$f|_{x_1 = x_{10} = 1} = \prod_{i \neq 1,4,7,10} \bar{x}_i.$$

Hence,

$$f|_{x_1 = x_{10} = 1, x_4 = 0} = \prod_{i \neq 1,4,7,10} \bar{x}_i. \qquad (5.2)$$

Moreover,

$$f|_{x_1 = 1, x_4 = x_{10} = 0}$$

$$= \prod_{i \neq 1,4,10} \bar{x}_i + x_5 \prod_{i \neq 1,4,5,10} \bar{x}_i + x_9 \prod_{i \neq 1,4,9,10} \bar{x}_i + x_5 x_9 \prod_{i \neq 1,4,5,9,10} \bar{x}_i$$

$$= \prod_{i \neq 1,4,5,9,10} \bar{x}_i. \qquad (5.3)$$

The above showed that the top subtree is correct.

Similar to (5.1), (5.2), and (5.3), we have

$$f|_{x_2 = x_5 = 1} = f|_{x_2 = x_{11} = 1, x_5 = 0} = \prod_{i \neq 2,5,8,11} \bar{x}_i, \text{ and}$$

$$f|_{x_2 = 1, x_5 = x_{11} = 0} = \prod_{i \neq 2,5,6,10,11} \bar{x}_i.$$

Therefore,

$$f|_{x_2 = x_5 = 1, x_1 = 0} = f|_{x_2 = x_{11} = 1, x_1 = x_5 = 0} = \prod_{i \neq 1,2,5,8,11} \bar{x}_i,$$

$$f|_{x_2 = 1, x_1 = x_5 = x_{11} = 0} = \prod_{i \neq 1,2,5,6,10,11} \bar{x}_i,$$

and the second subtree is correct. The rest of the tree in Figure 5.4 can be verified analogously. Thus, $D(f) \leq 11$. □

## 5.4 Monotone Graph Properties

In this section, we prove a general lower bound for the decision tree complexity of nontrivial monotone graph properties.

First, let us analyze how to use Theorem 5.13 to study nontrivial monotone graph properties. Note that every graph property is weakly symmetric and for any nontrivial monotone graph property, the complete graph must have the property and the empty graph must not have the property. Therefore, if we want to use Theorem 5.13 to prove the elusiveness of a nontrivial monotone graph property, we need to verify only one condition that the number of variables is a prime power. For a graph property, however, the number of variables is the number of possible edges, which equals $n(n-1)/2$ for $n$ vertices and it is not a prime power for $n > 3$. Thus, Theorem 5.13 cannot be used directly to show elusiveness of graph properties. However, it can be used to establish a little weaker results by finding a partial assignment such that the number of remaining variables becomes a prime power. The following lemmas are derived from this idea.

**Lemma 5.16** *If $\mathcal{P}$ is a nontrivial monotone property of graphs of order $n = 2^m$ then $D(\mathcal{P}) \geq n^2/4$.*

*Proof.* Let $H_i$ be the disjoint union of $2^{m-i}$ copies of the complete graph of order $2^i$. Then $H_0 \subset H_1 \subset \cdots \subset H_m = K_n$. Since $\mathcal{P}$ is nontrivial and is monotone, $H_m$ has the property $\mathcal{P}$ and $H_0$ does not have the property $\mathcal{P}$. Thus, there exists an index $j$ such that $H_{j+1}$ has the property $\mathcal{P}$ and $H_j$ does not have the property $\mathcal{P}$. Partition $H_j$ into two parts with vertex sets $A$ and $B$, respectively, each containing exactly $2^{m-j-1}$ disjoint copies of the complete graph of order $2^j$. Let $K_{A,B}$ be the complete bipartite graph between $A$ and $B$. Then $H_{j+1}$ is a subgraph of $H_j \cup K_{A,B}$. So, $H_j \cup K_{A,B}$ has the property $\mathcal{P}$. Now, let $f$ be the function on bipartite graphs between $A$ and $B$ such that $f$ has the value 1 at a bipartite graph $G$ between $A$ and $B$ if and only if $H_j \cup G$ has the property $\mathcal{P}$. Then $f$ is a nontrivial monotone weakly symmetric function with $|A| \cdot |B| \ (= 2^{2m-2})$ variables. By Theorem 5.13, $D(\mathcal{P}) \geq D(f) = 2^{2m-2} = n^2/4$. ∎

**Lemma 5.17** *Let $\mathcal{P}$ be a nontrivial monotone property of graphs of order $n$. If $2^m < n < 2^{m+1}$, then*

$$D(\mathcal{P}) \geq \min\{D(\mathcal{P}'), 2^{2m-2}\}$$

*for some nontrivial monotone property $\mathcal{P}'$ of graphs of order $n - 1$.*

*Proof.* We will reduce the computation of $\mathcal{P}$ on $n$ vertices to the computation of a nontrivial monotone graph property on $n - 1$ vertices (in the following Cases 1 and 2) or to the computation of a nontrivial monotone bipartite graph property of graphs of order $2^m$ (in the following Case 3). Let $K_{n-1}$ be the complete graph on the vertex set $\{2, \ldots, n\}$ and $K_{1,n-1}$ the complete bipartite graph between vertex set $\{1\}$ and $\{2, \ldots, n\}$.

*Case* 1. $\{1\} \cup K_{n-1}$ has property $\mathcal{P}$. In this case, let $\mathcal{P}'$ be the property that a graph $G$ on the vertex set $\{2, \ldots, n\}$ has the property $\mathcal{P}'$ if and only if $\{1\} \cup G$ has the property $\mathcal{P}$. Then the empty graph does not have the property $\mathcal{P}'$ and $K_{n-1}$ has. So, $\mathcal{P}'$ is nontrivial. Clearly, $\mathcal{P}'$ is monotone since $\mathcal{P}$ is. Now, in a decision tree computing $\mathcal{P}$, we assign value 0 to all edges in $K_{1,n-1}$. Then we get a decision tree computing $\mathcal{P}'$. Thus, $D(\mathcal{P}) \geq D(\mathcal{P}')$.

*Case* 2. $K_{1,n-1}$ does not have property $\mathcal{P}$. In this case, let $\mathcal{P}'$ be the property that a graph $G$ on vertex set $\{2, \ldots, n\}$ has the property $\mathcal{P}'$ if and only if $K_{1,n-1} \cup G$ has the property $\mathcal{P}$. Then the empty graph on $\{2,\ldots,n\}$ does not have the property $\mathcal{P}'$ and $K_{n-1}$ has the property $\mathcal{P}'$ since $K_{1,n-1} \cup K_{n-1} = K_n$ has the property $\mathcal{P}$. Thus, $\mathcal{P}'$ is a nontrivial monotone property of graphs of order $n-1$ and, similar to Case 1, $D(\mathcal{P}) \geq D(\mathcal{P}')$.

*Case 3.* $K_{1,n-1}$ has the property $\mathcal{P}$ and $\{1\} \cup K_{n-1}$ does not have the property $\mathcal{P}$. Let $A = \{1,\ldots,2^{m-1}\}$, $B = \{n - 2^{m-1} + 1,\ldots,n\}$ and $C = \{2^{m-1} + 1,\ldots,n - 2^{m-1}\}$. Let $K_{B\cup C}$ denote the complete graph on vertex set $B \cup C$. Then $A \cup K_{B\cup C}$ is a subgraph of $\{1\} \cup K_{n-1}$. Since $\{1\} \cup K_{n-1}$ does not have the property $\mathcal{P}$ and $\mathcal{P}$ is monotone, $A \cup K_{B\cup C}$ does not have the property $\mathcal{P}$. Let $K_{A,B}$ be the complete bipartite graph between $A$ and $B$. Then $K_{A,B} \cup K_{B\cup C}$ contains all edges between $n$ and $i$ for $i = 1, 2,\ldots,n-1$; these edges form a subgraph isomorphic to $K_{1,n-1}$. Since $K_{1,n-1}$ has the property $\mathcal{P}$, $K_{A,B} \cup K_{B\cup C}$ also has the property $\mathcal{P}$. Now, let $\mathcal{P}'$ be the property that a bipartite graph $G$ between $A$ and $B$ has the property $\mathcal{P}'$ if and only if $G \cup K_{B\cup C}$ has the property $\mathcal{P}$. Then $\mathcal{P}'$ is a nontrivial monotone property of bipartite graphs between $A$ and $B$ with $|A| = |B| = 2^{m-1}$. By Corollary 5.14, $D(\mathcal{P}') = 2^{2m-2}$. In a decision tree computing $\mathcal{P}$, we assign value 1 to all edges in $K_{B\cup C}$ and assign value 0 to all edges in $K_A$ and all edges between $A$ and $C$. Then we get a decision tree for $\mathcal{P}'$ and, hence, $D(\mathcal{P}) \geq D(\mathcal{P}') = 2^{2m-2}$. ∎

**Theorem 5.18** *If $\mathcal{P}$ is a nontrivial monotone property of graphs of order $n$, then $D(\mathcal{P}) \geq n^2/16$.*

*Proof.* It follows immediately from Lemmas 5.16 and 5.17. ∎

The recursive formula in Lemma 5.17 can be improved to

$$D(\mathcal{P}) \geq \min\left\{D(\mathcal{P}'), \frac{n^2 - 1}{4}\right\},$$

using the topological approach in the next section.

The lower bound in Theorem 5.18 can also be improved. In fact, the *Karp conjecture* states that every nontrivial monotone graph property is elusive, that is, $D(\mathcal{P}) = n(n-1)/2$ for every nontrivial monotone graph property $\mathcal{P}$. The current best known lower bound for $D(\mathcal{P})$ is $n^2/4 + o(n^2)$. There are some positive results supporting this conjecture. For example, every nontrivial monotone property of graphs of prime power order is elusive. These results will be established in the later sections.

A generalization of the Karp conjecture is that every nontrivial monotone weakly symmetric function is elusive. This stronger conjecture remains open for functions of at least 12 variables.

## 5.5 Topological Criterion

In this section, we introduce a powerful tool to study the elusiveness of monotone Boolean functions. We start with some concepts in topology.

A *triangle* is a two-dimensional polygon with the minimum number of vertices. A *tetrahedron* is a three-dimensional polytope with the minimum number of vertices. They are the simplest polytopes with respect to the specific dimensions. They are both called *simplexes*. The concept of simplexes is a generalization of the notions of triangles and tetrahedrons. In general, a *simplex* is a polytope with the minimum number of vertices among all polytopes with certain dimension. For example, a point is a zero-dimensional simplex and a straight line segment is a one-dimensional simplex. The convex hull of linearly independent $n + 1$ points in a Euclidean space is an $n$-dimensional simplex.

A *face* of a simplex $S$ is a simplex whose vertex set is a subset of vertices of $S$. A *geometric simplicial complex* is a family $\Gamma$ of simplexes satisfying the following conditions:

(a) For $S \in \Gamma$, every face of $S$ is in $\Gamma$.

(b) For $S, S' \in \Gamma$, $S \cap S'$ is a face for both $S$ and $S'$.

(c) For $S, S' \in \Gamma$, $S \cap S'$ is also a simplex in $\Gamma$.

In Figure 5.5, there are three examples; first two are not geometric simplicial complexes; the last one is.

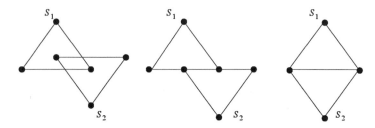

Figure 5.5: Three examples of families of simplexes.

Consider a set $X$ and a family $\Delta$ of subsets of $X$. $\Delta$ is called an (*abstract*) *simplicial complex* if for any $A$ in $\Delta$ every subset of $A$ also belongs to $\Delta$. Each member of $\Delta$ is called a *face* of $\Delta$. The *dimension* of a face $A$ is $|A| - 1$. Any face of dimension 0 is called a *vertex*. For example, consider a set $X = \{a, b, c, d\}$. The following family is a simplicial complex on $X$:

$$\Delta = \{\emptyset, \{a\}, \{b\}, \{c\}, \{d\}, \{a, b\}, \{b, c\},$$
$$\{c, d\}, \{d, a\}, \{a, c\}, \{a, b, c\}, \{a, c, d\}\}.$$

The set $\{a, b, c\}$ is a face of dimension 2 and the empty set $\emptyset$ is a face of dimension $-1$.

A *geometric representation* of a simplicial complex $\Delta$ on $X$ is a one-to-one, onto mapping from $X$ to the vertex set of a geometric simplicial complex $\Gamma$ such that this mapping induces a one-to-one, onto mapping from faces of $\Delta$ to simplexes in $\Gamma$. For example, consider a simplicial complex $\Delta$ of order $n$

and associate each vertex $x_i$ in $\Delta$ with the standard basis vector $e_i$ in the Euclidean space $\mathbf{R}^n$ (i.e., $e_i = (0,\ldots,0,1,0,\ldots,0)$ with the $i$th component equal to 1 and the others 0). Then the union $\widehat{\Delta}$ of all the convex hulls $\widehat{A} = conv(\{e_i : x_i \in A\})$ for $A \in \Delta$ is a geometric representation of $\Delta$. A geometric representation of the simplicial complex of the example above is shown in Figure 5.6.

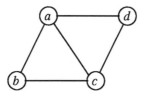

Figure 5.6: The geometric representation of a simplicial complex.

It is often helpful to think of a simplicial complex in terms of its geometric representations. For instance, the definition of the dimension of a face is very meaningful in the geometric representation. Extending this meaning, we define naturally the *Euler characteristic* of a simplicial complex $\Delta$ to be

$$\chi(\Delta) = \sum_{A \in \Delta, A \neq \emptyset} (-1)^{|A|-1} = \sum_{A \in \Delta} (-1)^{|A|-1} + 1.$$

A monotone Boolean function $f(x_1, x_2, \ldots, x_n)$ can be closely associated with a simplicial complex

$$\Delta_f = \{false(t) : t \text{ is a truth assignment of } f\}$$

if $f$ is monotone increasing, and

$$\Delta_f = \{true(t) : t \text{ is a truth assignment of } f\}$$

if $f$ is monotone decreasing. In fact, the monotonicity property of $f$ implies that every subset of $A$ in $\Delta_f$ is still in $\Delta_f$. The reader may question whether $\Delta_f$ is well defined, when $f$ is both monotone increasing and monotone decreasing. The answer is yes. In fact, a Boolean function which is both monotone increasing and monotone decreasing must be a trivial function. Now, it is easy to check that if $f \equiv 0$, then $\Delta_f = \emptyset$ and if $f \equiv 1$, then $\Delta_f = 2^X$ in either definition of $\Delta_f$, where $X = \{x_1, \ldots, x_n\}$ and $2^X$ is the power set of $X$ (the family of all subsets of $X$). For convenience, we say $\Delta_f$ *contains only one point* if $\Delta_f = 2^X$ with $|X| = 1$.

What is the relation between $\Delta_f$ and $\Delta_{f^*}$? Let $X$ be the vertex set of a simplicial complex $\Delta$. Define

$$\Delta^* = \{A \subseteq X : X - A \notin \Delta\}.$$

Then $\Delta^*$ is also a simplicial complex. $\Delta^*$ is the dual of $\Delta$. It is easy to see that $\Delta_{f^*} = \Delta_f^*$.

Now, let us come back to the notion of the Euler characteristic. Clearly, if $f$ is monotone increasing, then

$$
\begin{aligned}
\chi(\Delta_f) &= \sum_{t\in\{0,1\}^n} f(t)(-1)^{n-\|t\|-1} + 1 \\
&= (-1)^{n-1} \sum_{t\in\{0,1\}^n} f(t)(-1)^{\|t\|} + 1 \;=\; (-1)^{n-1}\mu(f) + 1;
\end{aligned}
$$

and if $f$ is monotone decreasing, then

$$
\begin{aligned}
\chi(\Delta_f) &= \sum_{t\in\{0,1\}^n} f(t)(-1)^{\|t\|-1} + 1 \\
&= -\sum_{t\in\{0,1\}^n} f(t)(-1)^{\|t\|} + 1 \;=\; -\mu(f) + 1.
\end{aligned}
$$

We know that if $\mu(f) \neq 0$, then $f$ is elusive. An alternative way to state this criterion is that if $\chi(\Delta_f) \neq 1$, then $f$ is elusive.

We have checked that if $f \equiv 0$, then $\Delta_f = \emptyset$ and if $f \equiv 1$, then $\Delta_f = 2^X$, where $X = \{x_1,\ldots,x_n\}$. In both cases, $\mu(f) = 0$. Thus, $\chi(\Delta_f) = 1$. Also note that $\emptyset$ is different from $\{\emptyset\}$. In fact, $\Delta_f = \{\emptyset\}$ if and only if $f$ has only one truth assignment in which either all variables have value 1 (when $f$ is monotone increasing) or all variables have value 0 (when $f$ is monotone decreasing). In the former case, we have $\mu(f) = (-1)^n$ and $\chi(\Delta_f) = 0$. In the latter case, we have $\mu(f) = 1$ and $\chi(\Delta_f) = 0$.

Another simple observation is that if $\Delta_f$ is a tree, then $\chi(\Delta_f) = 1$. In fact, a tree can be reduced to a point by a sequence of operations, each of which deletes a leaf together with the edge incident on the leaf. These operations preserve the Euler characteristic since each edge contributes $-1$ and each vertex contributes 1 to the Euler characteristic. This leaf-elimination operation on trees can be generalized to the following operation on simplicial complexes: A *maximal* face is a face that is not contained properly by another face. A *free* face is a nonmaximal face that is contained by only one maximal face. An *elementary collapse* is an operation that deletes a free face together with all faces containing it. The elementary collapse helps to simplify the computation of the Euler characteristic.

**Theorem 5.19** *The elementary collapse of a nonempty free face preserves the Euler characteristic.*

*Proof.* Suppose that $A$ is a free face of a simplicial complex $\Delta$. Let $B$ be the maximal face containing $A$. Let $a = |A|$ and $b = |B|$. Then, applying the elementary collapse on $A$ will delete, for each $i = 0, 1, \ldots, b-a$, $\binom{b-a}{i}$ faces of dimension $a + i - 1$ from $\Delta$. Therefore, the resulting simplicial complex has the Euler characteristic

$$
\chi(\Delta) - \sum_{i=0}^{b-a} \binom{b-a}{i}(-1)^{a+i-1} = \chi(\Delta).
$$

This proves the theorem. ∎

When can we apply the above theorem? Or, when does a simplicial complex not have a free face? The following theorem gives a necessary and sufficient condition.

**Theorem 5.20** *A simplicial complex $\Delta$ has no free face if and only if for each $A \in \Delta$ and each $x \in A$, there exists $y \notin A$ such that $(A-\{x\})\cup\{y\} \in \Delta$.*

*Proof.* First, suppose $\Delta$ has a free face $B$. Let $A$ be the maximal face containing $B$ and $x$ a vertex in $A - B$. Then we cannot find $y \notin A$ such that $(A - \{x\}) \cup \{y\}$ belongs to $\Delta$, since $(A - \{x\}) \cup \{y\} \in \Delta$ implies that there is another maximal face containing $B$.

Next, suppose $\Delta$ has no free face. For any $A \in \Delta$ and $x \in A$, we can find two maximal faces $B$ and $C$ containing $A - \{x\}$. Note that $B - (A - \{x\})$ and $C - (A - \{x\})$ are different and both are nonempty. So, there must exist an element $y$ in either $B - (A - \{x\})$ or $C - (A - \{x\})$ such that $y \neq x$. This implies that $y \notin A$ and $(A - \{x\}) \cup \{y\} \in \Delta$. ∎

In Theorem 5.6 and Corollary 5.7, we showed that a Boolean function $f$ whose associated simplicial complex $\Delta_f$ satisfies the condition of Theorem 5.20 is elusive. Thus, if we cannot elementarily collapse $\Delta_f$ then $f$ is elusive. We now further develop this idea.

A simplicial complex $\Delta$ is said to be *collapsible* if it can be elementarily collapsed to the empty simplicial complex. For example, the simplicial complex shown in Figure 5.6 is collapsible and Figure 5.7 shows a sequence of elementary collapses for it; on the other hand, the simplicial complex shown in Figure 5.8 is noncollapsible. It is worth mentioning that elementary collapse does not change the connectivity of the simplicial complex. Thus, any collapsible simplicial complex is connected, that is, its geometric representation is connected. In fact, the empty face is a free face if and only if the simplicial complex contains only one point. Therefore, if the simplicial complex is not connected, the empty face cannot be removed.

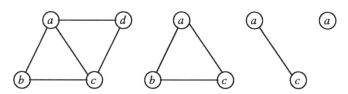

Figure 5.7: A sequence of elementary collapses.

The following is an important criterion.

**Theorem 5.21** (Topological Criterion). *If $f$ is a monotone Boolean function and $\Delta_f$ is noncollapsible, then $f$ is elusive.*

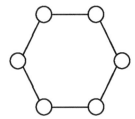

Figure 5.8: A noncollapsible simplicial complex.

In order to prove this theorem, we introduce two more concepts on the simplicial complex. Given a point $x$ in $X$ and a simplicial complex $\Delta$ on $X$, the *link* and *contrastar* of $x$ are defined, respectively, by

$$\begin{aligned} \text{LINK}(x, \Delta) &= \{A - \{x\} : x \in A \in \Delta\} \\ \text{CONT}(x, \Delta) &= \{A : x \notin A \in \Delta\}. \end{aligned}$$

**Lemma 5.22** *If* $\text{LINK}(x, \Delta)$ *and* $\text{CONT}(x, \Delta)$ *are collapsible, then* $\Delta$ *is collapsible.*

*Proof.* Note that $A$ is a free face of $\text{LINK}(x, \Delta)$ if and only if $A \cup \{x\}$ is a free face of $\Delta$. Thus, the sequence of elementary collapses which reduce $\text{LINK}(x, \Delta)$ to the empty simplicial complex would reduce $\Delta$ to $\text{CONT}(x, \Delta)$. Since $\text{CONT}(x, \Delta)$ can be further collapsed to the empty simplicial complex, $\Delta$ is collapsible.                                                                                  ■

Note that if $f$ is monotone increasing, then $\Delta_{f|_{x=0}} = \text{LINK}(x, \Delta_f)$ and $\Delta_{f|_{x=1}} = \text{CONT}(x, \Delta)$, and if $f$ is monotone decreasing, then $\Delta_{f|_{x=0}} = \text{CONT}(x, \Delta_f)$ and $\Delta_{f|_{x=1}} = \text{LINK}(x, \Delta)$. These relations will be used in the following proof.

*Proof of Theorem 5.21.* We prove Theorem 5.21 by induction on the number $n$ of variables in $f$. For $n = 1$, $\Delta_f$ equals $\emptyset$, or $\{\emptyset\}$, or $\{\emptyset, \{x_1\}\}$. Since $\Delta_f$ is noncollapsible, $\Delta_f$ equals $\{\emptyset\}$. Thus, $f(x_1) = x_1$ or $f(x_1) = \bar{x}_1$; so, it is elusive. For $n > 1$, consider any decision tree $T$ computing $f$. Suppose the root of $T$ is labeled by variable $x$. Then the two subtrees adjacent to $x$ compute $f|_{x=0}$ and $f|_{x=1}$, respectively. By Lemma 5.22, either $\Delta_{f|_{x=0}}$ or $\Delta_{f|_{x=1}}$ is noncollapsible. By the inductive hypothesis, either $D(f|_{x=0}) = n-1$ or $D(f|_{x=1}) = n-1$. Therefore, the depth of $T$ is $n$.                          ■

For monotone Boolean functions, the topological criterion is the strongest tool so far. First, we note that a simplicial complex without a free face is definitely noncollapsible. So, the topological criterion is able to establish the elusiveness of functions for which the simple strategy works.

Next, we note that the topological criterion is also stronger than the algebraic criterion. To see this, we note that the Euler characteristic of every collapsible function equals 1. Thus, a Boolean function $f$ with $\chi(\Delta_f) \neq 1$ must be noncollapsible. Moreover, if $\Delta_f$ is the disjoint union of a tree and several cycles, then we still have $\chi(\Delta_f) = 1$. However, $\Delta_f$ is noncollapsible since $\Delta_f$ is not connected.

An interesting question is whether the inverse of Theorem 5.21 also holds. That is, is the noncollapsibility of $\Delta_f$ also a necessary condition for $f$ to be elusive? Or, is it true that every collapsible function is not elusive? Before answering this question, let us look at a simple case when $\Delta_f$ is a tree.

**Example 5.23** If $f$ is a monotone Boolean function and $\Delta_f$ is a tree $T$, then $f$ is not elusive.

We assume that $f$ is monotone increasing. (The case of monotone decreasing $f$ is similar.) We note that $f = 1$ if and only if all variables assigned with value 0 are covered by a maximal face of $\Delta_f$. When $\Delta_f$ is a tree $T$, every edge of $T$ is a maximal face. Now, a decision tree of depth less than $n$ computing $f$ can be constructed as follows:

*Step* 1. Check all variables which do not appear in $T$. If one of them is assigned with 0, then the value of $f$ equals 0. If all of them are assigned with 1, then go to the next step.

*Step* 2. If $T$ contains only one vertex, then $f = 1$; otherwise, check a leaf $x_i$ on the tree $T$. If $x_i = 1$, then set $T := T - \{x_i\}$ and return to the beginning of this step. If $x_i = 0$, then go to the next step.

*Step* 3. Check all vertices in $T$ which are not adjacent to $x_i$. If one of them is assigned with 0, then $f = 0$; otherwise, $f = 1$.

The correctness of this decision tree is easy to verify. For instance, if we enter Step 3 with variable $x_i = 0$, and $x_j$ is the neighbor of $x_i$, then the value of $f$ is 1 if and only if all $x \in X - \{x_i, x_j\}$ are assigned with value 1, no matter whether $x_j$ is 1 or 0. We leave the other cases to the reader to verify.

Note that a tree contains at least one vertex. If the value of $f$ is obtained at Step 1, then at least one variable is not checked since all vertices of $T$ are not checked. If the value of $f$ is obtained in Step 2 or Step 3, then one of the vertices of $T$ is not checked. □

Since a collapsible simplicial complex with faces of dimension at most one must be a tree, for Boolean functions $f$, whose $\Delta_f$ have faces of dimension at most one, noncollapsibility is a necessary and sufficient condition for $f$ being elusive.

In general, however, the inverse of Theorem 5.21 does not hold. To see this, we note that it is known that there exists a collapsible complex $\Delta$ which may be collapsed to a noncollapsible complex $\Sigma$. (See, e.g., Glaser [1970].) Now, we define two functions $f$ and $g$ such that $\Delta_f = \Delta$ and $\Delta_g = \Sigma$. Note that

$$\Delta_{f^*} = \{A \subseteq X : X - A \notin \Delta_f\},$$

where $X = \{x_1, \ldots, x_n\}$. From the definition of collapsibility, it can be proved that $\Delta_{g^*}$ collapses to $\Delta_{f^*}$. Thus, either $\Delta_{g^*}$ is collapsible or $\Delta_{f^*}$ is noncollapsible. In either case, we found a function such that it is associated with a collapsible simplicial complex and its contravariant is associated with a noncollapsible simplicial complex. However, a Boolean function is elusive if and only if its contravariant is elusive. This implies that the inverse of Theorem 5.21 cannot be true.

Apparently, Theorem 5.21 implies that if either $\Delta_f$ or $\Delta_{f^*}$ is noncollapsible, then $f$ is elusive. Is the inverse of this stronger statement true? The above counterexample fails on this question, and the question remains open. More generally, finding a necessary and sufficient condition for the elusiveness of monotone Boolean functions is one of the major open questions in this area.

## 5.6    Applications of the Fixed Point Theorems

How can we tell whether a simplicial complex is noncollapsible? One of the tools to attack this problem is the fixed point theorems. To explain this, let us first introduce a concept. A simplicial complex $\Delta$ is *contractible* if its geometric representation $\widehat{\Delta}$ can be continuously shrunken into a point. It is important to note the following fact:

**Lemma 5.24** *Every collapsible simplicial complex is contractible.*

There exist several fixed point theorems on contractible simplicial complexes. Let us first consider the following one.

**Theorem 5.25** (Lefshetz' Fixed Point Theorem) *If a simplicial complex $\Delta$ is contractible, then any continuous mapping $\phi$ from $\widehat{\Delta}$ to itself has a fixed point $x$, that is, $\phi(x) = x$.*

According to Lefshetz' Theorem, if we can find a continuous mapping from the geometric representation $\widehat{\Delta}$ of a simplicial complex $\Delta$ to itself without a fixed point, then $\Delta$ is not contractible and hence noncollapsible. In the following, we will use this idea to get some new results.

Suppose that $\Delta$ and $\Delta'$ are simplicial complexes. A mapping $\phi$ from the vertex set of $\Delta$ to the vertex set of $\Delta'$ is called a *simplicial mapping* if for any $A \in \Delta$, $\phi(A) =_{\text{defn}} \{\phi(a) : a \in A\} \in \Delta'$. Every simplicial mapping $\phi$ yields a continuous linear mapping $\widehat{\phi}$ from $\widehat{\Delta}$ to $\widehat{\Delta'}$ by

$$\widehat{\phi}\Big( \sum_{v \in \Delta} \alpha_v \widehat{v} \Big) = \sum_{v \in \Delta} \alpha_v \widehat{\phi(v)}.$$

A simplicial one-to-one mapping from $\Delta$ to itself is called an *automorphism*.

Let $f(x_1, x_2, \ldots, x_n)$ be a monotone Boolean function and $\sigma$ a permutation of $\{1, 2, \ldots, n\}$. Suppose $f$ is invariant under $\sigma$, that is,

$$f(x_{\sigma(1)}, x_{\sigma(2)}, \ldots, x_{\sigma(n)}) = f(x_1, x_2, \ldots, x_n).$$

(Such a permutation is called a *symmetric permutation* of function $f$.) Then $\sigma$ is an automorphism of $\Delta_f$. To see this, we note that for each $A \in \Delta_f$, $\sigma(A) = \{x_{\sigma(i)} : x_i \in A\}$. If $f$ is monotone decreasing, then $A \in \Delta_f$ means that $A = true(t)$ for some truth assignment $t = t_1 t_2 \cdots t_n \in \{0, 1\}^n$. Note that $\sigma(A) = \{x_i : t_{\sigma^{-1}(i)} = 1\}$. To prove $\sigma(A) \in \Delta_f$, it suffices to show that

$$f(t_{\sigma^{-1}(1)}, t_{\sigma^{-1}(2)}, \ldots, t_{\sigma^{-1}(n)}) = f(t_1, t_2, \ldots, t_n).$$

This is clearly true since $\sigma^{-1} = \sigma^{n-1}$. Similarly, we can prove this property for monotone increasing functions $f$.

Let $f$ be a monotone Boolean function and $\sigma$ a symmetric permutation of $f$. An assignment $t$ is called an *invariant assignment* of $\sigma$ if $t_i = t_{\sigma(i)}$ for all $i$, that is, $\sigma$ does not change the value of each variable.

**Theorem 5.26** *Let $f$ be a nontrivial monotone function. Suppose that $f$ is not elusive. Then every symmetric permutation $\sigma$ on variables has an invariant truth assignment other than $\mathbf{0}$ and $\mathbf{1}$.*

*Proof.* We linearly extend $\sigma$ from $\Delta_f$ to $\widehat{\Delta_f}$ by $\sigma(\sum_i \alpha_i \widehat{x}_i) = \sum_i \alpha_i \widehat{x}_{\sigma(i)}$. Clearly, $\sigma$ is a continuous mapping from $\widehat{\Delta_f}$ to itself. Since $f$ is not elusive, $\Delta_f$ is contractible. Thus, $\sigma$ has a fixed point in $\widehat{\Delta_f}$. This fixed point must be of the form $\sum_{x_i \in A} \alpha_i \widehat{x}_i$ for some $A \in \Delta_f$ and $\alpha_i > 0$ for all $x_i \in A$. Note that in the geometric representation of $\Delta_f$, all points $\widehat{x}_i$ are linearly independent. Thus, from

$$\sigma\left(\sum_{i \in A} \alpha_i \widehat{x}_i\right) = \sum_{i \in A} \alpha_i \widehat{x}_{\sigma(i)}$$

we can obtain $A = \sigma(A)$. This means that the truth assignment $t_A$ associated with $A$ is invariant under $\sigma$. Since $A$ is clearly nonempty, $t_A$ cannot be $\mathbf{1}$ if $f$ is monotone increasing and cannot be $\mathbf{0}$ if $f$ is monotone decreasing. On the other hand, since $f$ is nontrivial, $\mathbf{0}$ cannot be a truth assignment if $f$ is monotone increasing, and $\mathbf{1}$ cannot be a truth assignment if $f$ is monotone decreasing. Thus, $t_A$ is neither $\mathbf{0}$ nor $\mathbf{1}$. ∎

Let us look at an example.

**Example 5.27** $f(x_1, x_2, \ldots, x_n) = x_1 x_2 \cdots x_k + x_2 x_3 \cdots x_{k+1} + \cdots + x_n x_1 \cdots x_{k-1}$ is elusive.

Consider a symmetric permutation $\phi$ of variables defined by $\phi(x_i) = x_{i+1}$ if $1 \le i < n$ and $\phi(x_n) = x_1$. Then $\phi$ clearly has no invariant truth assignment other than $\mathbf{1}$. Therefore, it is elusive. □

For a graph property $\mathcal{P}$, any permutation $\sigma$ on the vertex set of a graph induces a symmetric permutation on the possible edges (variables of the graph property as a Boolean function). Let us call a graph $G \in \mathcal{P}$ an *invariant graph* of $\sigma$ if all edges of $G$ are mapped to edges of $G$ under $\sigma$.

The following corollary follows immediately from Theorem 5.26.

**Corollary 5.28** *Let* $\mathcal{P}$ *be a nontrivial monotone graph property. If* $\mathcal{P}$ *is not elusive, then every permutation* $\sigma$ *of vertices has an invariant nonempty graph* $G$ *in* $\mathcal{P}$ *other than the complete graph.*

**Example 5.29** Let $\mathcal{P}$ be the graph property consisting of two-colorable graphs. Then $\mathcal{P}$ is elusive when the number $n$ of vertices is odd.

Let us label all vertices by elements in $\mathbf{Z}_n$. Consider the mapping $\phi : \mathbf{Z}_n \to \mathbf{Z}_n$ defined by $\phi(i) = (i+1) \bmod n$. For the sake of contradiction, suppose $\mathcal{P}$ is not elusive. By Corollary 5.28, $\phi$ has a nonempty invariant graph $G$ other than the complete graph. Consider an edge $\{i, j\}$ in $G$. $G$ must contain all edges $\{(i+k) \bmod n, (j+k) \bmod n\}$. These edges form a vertex-disjoint union of cycles in $G$. Since $n$ is odd, at least one of the cycles is of an odd order. But a cycle of an odd order is not two-colorable, and we have obtained a contradiction.                                                                                                 □

Next, let us introduce another result in fixed point theory.

An *orbit* of an automorphism $\phi$ on $\Delta$ is a minimal subset of vertices of $\Delta$ such that $\phi$ takes no vertex out of it. If an orbit $H$ is a face of $\Delta$, then the center of gravity of $\widehat{H}$ is a fixed point of $\widehat{\phi}$, since

$$\widehat{\phi}\left( \frac{1}{|H|} \sum_{v \in H} \widehat{v} \right) = \frac{1}{|H|} \sum_{v \in H} \widehat{\phi(v)} = \frac{1}{|H|} \sum_{v \in H} \widehat{v}.$$

If $H_1, \ldots, H_k$ are orbits and $H_1 \cup \cdots \cup H_k$ is a face of $\Delta$, then every point in the convex hull of the centers of gravity of $\widehat{H}_1, \ldots, \widehat{H}_k$ is a fixed point. Conversely, consider a fixed point $x$ of $\widehat{\phi}$ in $\widehat{\Delta}$. Write $x = \sum_v \alpha_v \widehat{v}$, where $v$ ranges over vertices of $\Delta$. Then

$$x = \widehat{\phi}(x) = \sum_v \alpha_v \widehat{\phi(v)}.$$

Since all vertices $\widehat{v}$ are linearly independent, we have $\alpha_v = \alpha_{\phi(v)}$ for every vertex $v$. This implies that all coefficients $\alpha_v$ for $v$ in the same orbit are equal. Suppose that $H_1, \ldots, H_k$ are all the orbits of $\phi$ such that for vertices $v$ in $H_i$, $\alpha_v = \beta_i > 0$. Then, $H_1 \cup \cdots \cup H_k \in \Delta$ and

$$x = \sum_{i=1}^{k} \beta_i |H_i| \, \widehat{c}_i,$$

where $\widehat{c}_i$ is the center of gravity of $\widehat{H}_i$. Therefore, the fixed point set of $\widehat{\phi}$ on $\widehat{\Delta}$ is exactly the union of convex hulls $conv(\widehat{c}_1, \ldots, \widehat{c}_k)$ for $H_1 \cup \cdots \cup H_k \in \Delta$, which can be seen as a geometric representation of the simplicial complex

$$\Delta^{\phi} = \{\{c_1, \ldots, c_k\} : H_1, \ldots, H_k \text{ are orbits and } H_1 \cup \cdots \cup H_k \in \Delta\}.$$

**Theorem 5.30** (Hopf Index Formula) *For any automorphism* $\phi$ *of a contractible simplicial complex* $\Delta$, *the Euler characteristic of the fixed point set of* $\widehat{\phi}$ *is 1, that is,* $\chi(\Delta^{\phi}) = 1$.

An interesting application of Hopf index formula is as follows:

**Theorem 5.31** *Every nontrivial monotone bipartite graph property is elusive.*

*Proof.* Let $f(x_{11}, \ldots, x_{1n}, x_{21}, \ldots, x_{mn})$ be a non-elusive monotone increasing bipartite graph property on bipartite graphs with vertex sets $A = \{1, \ldots, m\}$ and $B = \{1, \ldots, n\}$. (The case for monotone decreasing $f$ is symmetric.) Consider a permutation $\phi$ of edges corresponding to a cyclic permutation of the vertices in $B$ while leaving $A$ fixed.

Each orbit of $\phi$ corresponds to a vertex in $A$. In fact, it contains all edges incident on a single vertex in $A$. Let $H_i$, $1 \le i \le m$, denote the orbit consisting of all edges incident on $i \in A$. Since $f$ is a bipartite graph property, the existence of $k$ orbits whose union is a face of $\Delta_f$ implies that for any $k$ orbits, their union is a face of $\Delta_f$. Therefore, the Euler characteristic $\chi(\Delta_f^\phi)$ of the fixed point set of $\phi$ is equal to

$$\sum_{i=1}^{r} (-1)^{i+1} \binom{m}{i}$$

for some $1 \le r \le m$. Since $f$ is non-elusive, $\Delta_f$ must be contractible. Then, by the Hopf index formula, $\chi(\Delta_f^\phi) = 1$. Note that

$$
\begin{aligned}
\chi(\Delta_f^\phi) &= \binom{m}{1} - \binom{m}{2} + \cdots + (-1)^{r+1} \binom{m}{r} \\
&= \binom{m-1}{0} + \binom{m-1}{1} - \binom{m-1}{1} - \binom{m-1}{2} \\
&\quad + \cdots + (-1)^{r+1} \binom{m-1}{r-1} + (-1)^{r+1} \binom{m-1}{r} \\
&= 1 + (-1)^{r+1} \binom{m-1}{r}.
\end{aligned}
$$

So, $\chi(\Delta_f^\phi) = 1$ implies $r = m$. This, in turn, implies that $A \times B \in \Delta_f$. Since $f$ is monotone increasing, $f \equiv 1$. ∎

## 5.7 Applications of Permutation Groups

From the last section, we see that to show the elusiveness of a nontrivial monotone Boolean function $f$, it suffices to find a symmetric permutation $\sigma$ such that under its action, the complex $\Delta_f$ has only one orbit. (Why?) Even in the case that such a symmetric permutation cannot be found, it would still be helpful if we can find one with a small number of orbits. Motivated from this point, the group of symmetric permutations is considered to replace a single symmetric permutation.

An *orbit* of a group $\Gamma$ of automorphisms $\phi$ on $\Delta$ is a minimal subset of vertices of $\Delta$ such that any $\phi$ in $\Gamma$ takes no vertex out of it. If an orbit $H$ is a face of $\Delta$, then the center of gravity of $\hat{H}$ is a fixed point of $\hat{\phi}$ for every $\phi$ in $\Gamma$, which is called a *fixed point of* $\Gamma$. We can also see that all fixed points of $\Gamma$ form a geometric representation of the simplicial complex

$$\Delta^{\Gamma} = \{\{c_1, \ldots, c_k\} : H_1, \ldots, H_k \text{ are orbits and } H_1 \cup \cdots \cup H_k \in \Delta\}.$$

For group actions, the following fixed point theorem is useful to us.

**Theorem 5.32** *Let $\Gamma$ be a group of automorphisms on a contractible simplicial complex $\Delta$. Let $\Gamma_1$ be a normal subgroup of $\Gamma$ with $|\Gamma_1| = p^k$, a prime power such that $\Gamma/\Gamma_1$ is cyclic. Then there exists an $x \in \hat{\Delta}$ such that for any $\phi \in \Gamma$, $\phi(x) = x$. Furthermore, such fixed points form a geometric simplicial complex with the Euler characteristic 1, that is, $\chi(\Delta^{\Gamma}) = 1$.*

Similarly to Corollary 5.28, the following corollary follows immediately from Theorem 5.32.

**Corollary 5.33** *Consider a nontrivial monotone graph property $\mathcal{P}$. Let $\Gamma$ be a group of permutations on vertices which has a normal subgroup $\Gamma_1$ of prime power order such that $\Gamma/\Gamma_1$ is cyclic. If $\mathcal{P}$ is not elusive, then $\Gamma$ has a nonempty invariant graph in $\mathcal{P}$ other than the complete graph.*

An application of the above theorem is as follows:

**Theorem 5.34** *Suppose that $\mathcal{P}$ is a nontrivial monotone graph property on graphs with a prime power number of vertices. Then $\mathcal{P}$ is elusive.*

*Proof.* Suppose the number of vertices in the considered graph is $p^k$. Let us label the vertices of the graph by the elements of the Galois field $\mathrm{GF}(p^k)$. Consider the linear mapping $x \to ax + b$ from $\mathrm{GF}(p^k)$ to itself which induces an automorphism $\phi_{ab}$ of $\Delta_f$. All such automorphisms $\phi_{ab}$ form a group $\Gamma$. Let $\Gamma_1$ consist of all $\phi_{1b}$. Then $|\Gamma_1| = p^k$. Moreover, $\Gamma_1$ is a normal subgroup of $\Gamma$, since

$$(((ax + b) + b') - b)/a = x + b'/a,$$

and the quotient group $\Gamma/\Gamma_1$ is isomorphic to the multiplication group of $\mathrm{GF}(p^k)$ which is cyclic.

Suppose, to the contrary, that $\mathcal{P}$ is not elusive. Then, by Corollary 5.33, $\Gamma$ has a nonempty invariant graph $G$ in $\mathcal{P}$ other than the complete graph. Note that $\Gamma$ is transitive on the edges, because for any two edges $\{u_1, v_1\}$ and $\{u_2, v_2\}$, there exist $a, b$, such that $au_1 + b = u_2$ and $av_1 + b = v_2$. Thus, $G$ must contain all the edges. This is a contradiction. ∎

From the proof of the above result, we know that to prove the Karp conjecture, it suffices to find a group $\Gamma$ of automorphisms on vertices which satisfies the following conditions:

(a) $\Gamma$ has a normal subgroup $\Gamma_1$ such that $\Gamma/\Gamma_1$ is cyclic.

(b) $\Gamma$ is transitive on edges.

A group satisfying (a) is easy to find. For example, if $n = pm$ where $p$ is a prime, then we can just label all vertices by elements in $\mathbf{Z}_p \times \mathbf{Z}_m$ and then consider the group $\Gamma = G \times H$, where $G = \langle \tau \rangle$ and $H = \langle \sigma \rangle$ are generated, respectively, by the automorphism $\tau$ of $\mathbf{Z}_p$, defined by $\tau(x) = x + 1$, and the automorphism $\sigma$ of $\mathbf{Z}_m$, defined by $\sigma(y) = y + 1$. Clearly, $\Gamma$ satisfies condition (a). However, it does not satisfy condition (b).

Since it is hard to find an automorphism group satisfying both (a) and (b), let us try to find an automorphism group satisfying (a) with a small number of orbits. Such a candidate is $G \wr H$, the *wreath product* of the groups $G$ and $H$ above. The wreath product can be defined as follows:

$$G \wr H = \{(f; \pi) : f : \mathbf{Z}_m \to G, \pi \in H\},$$

where for $(i, j) \in V$ and $(f; \pi) \in G \wr H$,

$$(f; \pi)(i, j) = (f(\pi(j))(i), \pi(j)).$$

Recall that permutations $\tau$ and $\sigma$ are defined by $\tau(x) = x + 1$ and $\sigma(y) = y + 1$. The group $\Gamma = G \wr H$ contains the normal subgroup

$$\Gamma_1 = \{(f; 1_H) : f : \mathbf{Z}_m \to G\},$$

which is isomorphic to the $m$-fold direct product $G^m = G \times G \times \cdots \times G$. Hence, $\Gamma_1$ is a group of order $p^m$. In addition, we have $(G \wr H)/\Gamma_1 \simeq H$. Thus, $G \wr H$ satisfies condition (a).

The group $G \wr H$ has $\lceil(m-1)/2\rceil + \lceil(p-1)/2\rceil$ orbits. They can be described as follows:

For $k = 1, 2, \ldots, \lceil(m-1)/2\rceil$,

$$E_k = \{\{(x, i), (y, j)\} : i - j = k\}$$

is an orbit. To see this, consider two edges in $E_k$, $\{(x, i), (y, j)\}$ and $\{(x', i'), (y', j')\}$. Choose $f : \mathbf{Z}_m \to G$ such that $f(i')(x) = x'$ and $f(j')(y) = y'$. Then we have $(f, \sigma^{i'-i})(\{(x, i), (y, j)\}) = \{(x', i'), (y', j')\}$. Clearly, no automorphism in $G \wr H$ can change the difference of the second components. Therefore, $E_k$ is an orbit.

Similarly, we can show that for $h = 1, 2, \ldots, \lceil(p-1)/2\rceil$,

$$F_h = \{\{(x, i), (y, i)\} : x - y = h\}$$

is an orbit. When $p$ is an odd prime, each $F_h$ is a disjoint union of $m$ cycles

$$C_{p,i} = \{\{(x, i), (y, i)\} : x - y = h\},$$

for $i = 0, \ldots, m$. When $p = 2$, $F_1$ is a perfect matching.

The *girth* of a graph is the length of its shortest cycle. (The girth of an acyclic graph is $\infty$.) Using the group $G \wr H$, one can show the following result:

**Theorem 5.35** *Suppose $\mathcal{P}$ is a monotone decreasing nontrivial graph property such that every graph in $\mathcal{P}$ has girth greater than 4. Then $\mathcal{P}$ is elusive.*

*Proof.* If $n$ is a prime, then elusiveness follows from Theorem 5.34. Thus, we may assume $n = pm$ where $p$ is an odd prime. Label the graph and define groups $G$ and $H$ as above. Suppose, to the contrary, that $\mathcal{P}$ is not elusive. Then $\Delta_{\mathcal{P}}$ is contractible. By Theorem 5.33, $G \wr H$ has an nonempty invariant graph $T$ in $\mathcal{P}$. Moreover,

$$\chi(\Delta_{\mathcal{P}}^{G \wr H}) = 1.$$

Consider an edge in an invariant graph $T$. Denote $V_i = \mathbf{Z}_p \times \{i\}$. If the edge connects a vertex in $V_i$ and another vertex in $V_j$, with $i \neq j$, then the orbit of this edge contains a complete bipartite graph $K_{p,p}$ that has a cycle of size four. This contradicts the assumption. Thus, we may assume that every edge of the graph $T$ connects two vertices in the same $V_i$. The orbit of such an edge is isomorphic to $mC_{p,i}$, $m$ disjoint copies of the $p$-group $C_{p,i}$. If $p = 3$, then we obtain a contradiction again. So, we may assume $p \geq 5$, and the total number of such orbits is $(p-1)/2 > 1$.

Next, we show that the union of every two orbits must contain a 4-cycle. In fact, suppose the two orbits contain edges $\{(x,i),(x+j,i)\}$ and $\{(x,i),(x+k,i)\}$, respectively. Then they must also contain edges $\{(x+j,i),(x+k+j,i)\}$ and $\{(x+k,i),(x+j+k,i)\}$. These four edges form a 4-cycle.

This implies that the union of any two orbits is not in $\Delta_{\mathcal{P}}$. Therefore,

$$\Delta_{\mathcal{P}}^{G \wr H} = \{\{c_i\} : H_i \text{ is an orbit}\},$$

and the Euler characteristic $\chi(\Delta_{\mathcal{P}}^{G \wr H})$ is equal to the number of orbits $(p-1)/2 > 1$. This is a contradiction. ∎

## 5.8   Randomized Decision Trees[2]

A *randomized decision tree* is a decision tree with nodes labeled by random variables. That is, at each node, instead of querying a certain variable, it randomly chooses a variable to query according to a probability distribution which is dependent on the result of previous queries. Thus, for a given input, the number of queried variables is a random number. The complexity measure is the expectation of this number, called the *expected time*. The *expected depth* of a randomized decision tree is the maximum expected time over all inputs. The *randomized decision tree complexity* of a Boolean function $f$, denoted by $R(f)$, is the minimum expected depth over all randomized decision trees computing $f$.

---

[2]The general notion of randomized computation is studied in Chapter 8. The reader who is not familiar with this notion may want to study the examples of Section 8.1 first.

An alternative (and equivalent) model for the randomized decision tree is to make all random choices in advance and after that everything is deterministic. In this model, a randomized decision tree computing a Boolean function is specified by a probability distribution over all possible decision trees computing the function. Let $n(t,T)$ denote the number of queries made in a decision tree $T$ on an input $t$. Let $p_T$ denote the probability that the tree $T$ is chosen at the beginning. Then the expected depth with respect to the distribution $\{p_T\}$ is

$$\max_t \sum_T p_T n(t,T)$$

and the randomized decision tree complexity of function $f$ is

$$R(f) = \min_{p_T} \max_t \sum_T p_T n(t,T).$$

Does randomness increase the computational power? The answer is positive. To see this, let us study two examples.

**Example 5.36** Let us first define the majority function $m(x_1, x_2, x_3)$ as follows:

$$m(x_1, x_2, x_3) = \begin{cases} 1 & \text{if at least two of } x_1, x_2, x_3 \text{ equal } 1, \\ 0 & \text{otherwise.} \end{cases}$$

Clearly, $m$ is symmetric and nontrivial and, hence, elusive. So, $D(m) = 3$. We want to prove $R(m) \le 8/3$. To do so, we consider a simple randomized algorithm which randomly chooses a permutation of variables under the uniform distribution and then queries variables according to the order given by the permutation. With this randomized algorithm, on inputs 111 and 000 the expected depth is 2, and on all other inputs the expected depth equals $3 \cdot 2/3 + 2 \cdot 1/3 = 8/3$. Therefore, $R(m) \le 8/3 < D(m) = 3$.

Next, define

$$\begin{aligned}
f_1(x_1, x_2, x_3) &= m(x_1, x_2, x_3) \\
f_{k+1}(x_1, x_2, \ldots, x_{3^{k+1}}) &= m(f_k(x_1, \ldots, x_{3^k}), f_k(x_{3^k+1}, \ldots, x_{2 \cdot 3^k}), \\
&\qquad f_k(x_{2 \cdot 3^k+1}, \ldots, x_{3^{k+1}})).
\end{aligned}$$

Clearly, each $f_k$ is nontrivial, monotone, weakly symmetric and, hence, elusive. Thus, $D(f_k) = 3^k$.

If we query variables randomly under a uniform distribution, then by the above result, it is easy to see that $R(f_k) \le (8/3)^k$. Let $n = 3^k$. Then $(8/3)^k = n^{\log_3(8/3)} \approx n^{0.9}$. Thus, $R(f_k)$ is significantly smaller than $D(f_k)$. □

**Example 5.37** Consider the graph property $g$ of all graphs having at least $\lceil n(n-1)/4 \rceil$ edges where $n$ is the number of vertices. It is easy to show that $g$ is elusive. Next, we want to prove that

$$R(g) \le \frac{k+1}{k+2}(m+1) < m,$$

where $m = n(n-1)/2$ and $k = \lceil m/2 \rceil$. To do so, we consider a randomized algorithm which at each step chooses an unchecked edge randomly, under the uniform distribution, and checks whether it is an edge of the graph; the checking continues until either $k$ existing edges are found or $m-k$ nonexisting edges are found. Clearly, among the inputs which make $g = 1$, the greatest expected running time (expected number of checked edges at the end of the computation) is achieved by the inputs which contain exactly $k$ edges. In this case, the probability of making exactly $k+i$ queries to find all $k$ existing edges is $\binom{k+i-1}{k-1}/\binom{m}{k}$. Thus, the expected running time is

$$
\frac{\binom{k-1}{k-1}}{\binom{m}{k}}k + \frac{\binom{k}{k-1}}{\binom{m}{k}}(k+1) + \cdots + \frac{\binom{m-1}{k-1}}{\binom{m}{k}}m
$$

$$
= \frac{k}{\binom{m}{k}}\left(\binom{k}{k} + \binom{k+1}{k} + \cdots + \binom{m}{k}\right)
$$

$$
= \frac{k}{\binom{m}{k}}\left(\binom{k+1}{k+1} + \binom{k+2}{k+1} - \binom{k+1}{k+1} + \cdots + \binom{m+1}{k+1} - \binom{m}{k+1}\right)
$$

$$
= \frac{k}{\binom{m}{k}}\binom{m+1}{k+1} = \frac{k(m+1)}{k+1} \le \frac{k+1}{k+2}(m+1).
$$

Similarly, among the inputs which make $g = 0$, the greatest expected running time is

$$
\frac{(m-k+1)(m+1)}{m-k+2}.
$$

Note that $m - k + 1 \le k + 1$. We have $R(g) \le (m+1)(k+1)/(k+2)$.    □

To establish a general lower bound for the randomized decision tree complexity of weakly symmetric functions, let us first show an inequality about weakly symmetric functions.

**Lemma 5.38** *If $f$ is a weakly symmetric Boolean function, then*

$$
D_0(f)D_1(f) \ge n,
$$

*where $n$ is the number of variables of $f$.*

*Proof.* Let $\Gamma$ be the group of permutations on variables of $f$ which witnesses the weak symmetry of $f$. Consider the number

$$
|\{\sigma \in \Gamma : \sigma(x) = y\}|.
$$

This number is independent of the choices of $x$ and $y$, because for any $x$, $y$, and $z$, we can map $\{\sigma \in \Gamma : \sigma(x) = y\}$ one-to-one into $\{\sigma \in \Gamma : \sigma(x) = z\}$ in the following way: Choose $\tau \in \Gamma$ such that $\tau(y) = z$. Then, for each $\sigma$ with $\sigma(x) = y$, we have $\tau(\sigma(x)) = z$. That is, the mapping $\sigma \to \tau\sigma$ meets our requirement. Similarly, we can map $\{\sigma \in \Gamma : \sigma(x) = y\}$ one-to-one into $\{\sigma \in \Gamma : \sigma(z) = y\}$.

Let $q = |\{\sigma \in \Gamma : \sigma(x) = y\}|$. Then, for a fixed $x$,

$$qn = \sum_y |\{\sigma \in \Gamma : \sigma(x) = y\}| = |\Gamma|.$$

So, $q = |\Gamma|/n$.

Now, consider an implicant $E$ and a clause $F$. Let $E^\sigma$ denote the elementary product obtained from $E$ by replacing each variable $x$ with $\sigma(x)$. Since $f$ is invariant under $\Gamma$, $E^\sigma$ is still an implicant for any $\sigma \in \Gamma$. Thus, $E^\sigma$ and $F$ have at least one variable in common. This means that each $\sigma$ in $\Gamma$ maps a variable $x$ in $E$ to a variable $y$ in $F$. Note that for each fixed pair of $x$ and $y$, there are exactly $q$ $\sigma$'s in $\Gamma$ such that $\sigma(x) = y$. Therefore, at least $|\Gamma|/q = n$ pairs $(x, y)$ exist with $x \in E$ and $y \in F$. Thus, $|E| \cdot |F| \geq n$. It follows that $D_0(f)D_1(f) \geq n$.                     ∎

**Theorem 5.39** *If $f$ is a weakly symmetric Boolean function, then*

$$R(f) \geq \sqrt{n}.$$

*Proof.* For any minterm $x_{i_1} x_{i_2} \cdots x_{i_k}$ of $f$, define an input $t$ that assigns 1 to all variables in the minterm and 0 to all other variables. Then, the shortest decision tree on this input has depth $k$. So, $R(f) \geq D_1(f)$. Similarly, we can see that $R(f) \geq D_0(f)$. Now, $R(f) \geq \max(D_0(f), D_1(f)) \geq \sqrt{n}$.                     ∎

Next, we develop some tools for studying the lower bounds of random decision tree complexity.

**Lemma 5.40** *Assume that for $t \in \{0,1\}^n$, $q_t \geq 0$ and $\sum_{t \in \{0,1\}^n} q_t = 1$. Then,*

$$R(f) \geq \min_{p_T} \sum_T p_T \sum_t n(t, T)q_t,$$

*where $T$ ranges over all decision trees computing the Boolean function $f$ of $n$ variables.*

*Proof.* Note that the maximum of $\sum_t (\sum_T p_T n(t, T))q_t$ is achieved at some vertex of the simplex $\{\{q_t\}_{t \in \{0,1\}^n} : q_t \geq 0, \sum_{t \in \{0,1\}^n} q_t = 1\}$ and each vertex of this simplex has exactly one component equal to 1 and others equal to 0. Thus,

$$\max_{\{q_t\}} \sum_t \left( \sum_T p_T n(t, T) \right) q_t = \max_t \sum_T p_T n(t, T).$$

Since the number of inputs is finite, there exists an input $t_0$ such that

$$\min_{p_T} \sum_T p_T n(t_0, T) = \max_t \min_{p_T} \sum_T p_T n(t, T).$$

However,

$$\min_{p_T} \sum_T p_T n(t_0, T) \leq \min_{p_T} \max_t \sum_T p_T n(t, T).$$

Thus,

$$R(f) = \min_{p_T} \max_t \sum_T p_T n(t, T) \geq \min_{p_T} \sum_T p_T n(t_0, T)$$

$$= \max_t \min_{p_T} \sum_T p_T n(t, T) \geq \min_{p_T} \sum_T p_T \sum_t n(t, T) q_t. \qquad \blacksquare$$

Actually, one can obtain, from the von Neumann's minimax theorem, that

$$R(f) = \max_{\{q_t\}} \min_{p_T} \sum_T p_T \sum_t n(t, T) q_t.$$

However, it is worth mentioning that the following is not always true:

$$R(f) = \max_t \min_{p_T} \sum_T p_T n(t, T).$$

The above lemma suggests the following way to prove the lower bound for $R(f)$: Find a probability distribution $\{q_t\}$ of inputs $t$ such that every decision tree $T$ with this distribution has a large average-time complexity.

**Lemma 5.41** *If a monotone Boolean function $f$ has $k$ minterms of size $s$, then*

$$\binom{2R(f)}{s} \geq \frac{k}{2}.$$

*Proof.* For each minterm $F$, let $t_F$ denote the assignment that assigns all variables in $F$ with 1 and others with 0. Let $q_{t_F} = 1/k$ for every minterm $F$ of size $s$ and $q_t = 0$ for other inputs $t$.

Consider any decision tree $T$ computing the function $f$. Let $F$ and $F'$ be two minterms of size $s$. Then the computation paths of $T$ on inputs $t_F$ and $t_{F'}$ will lead to two different leaves. This is because for each input $t_F$ with a minterm $F$, $T$ has to find that all variables in $F$ equal 1 before it decides that the value of $f$ is 1. Since both minterms $F$ and $F'$ are of size $s$, they must contain different variables. At the variable which belongs to exactly one of $F$ or $F'$, two inputs lead to two different answers, that is, two different paths from the root to the leaves.

Note that for any minterm $F$ of size $s$ each path of $T$ on input $t_F$ contains exactly $s$ edges with label 1. Since $f$ has $k$ minterms of size $s$, $T$ has at least $k$ such paths. How many such paths have length at most $h$? Note that each path of exactly $s$ edges with label 1 corresponds to a binary string with exactly $s$ 1's and two such paths correspond to two binary strings with some different positions about 1. Thus, the number of paths of length at most $h$ having exactly $s$ edges with label 1 is at most $\binom{h}{s}$.

Now, suppose, to the contrary, that $\binom{2R(f)}{s} < k/2$. Then at least $k/2$ computation paths of $T$ on inputs $t_F$ for minterms $F$ of size $s$ have length at least $2R(f) + 1$. Thus,

$$\sum_t n(t, T) q_t \geq R(f) + 1/2.$$

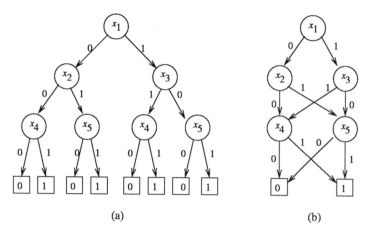

Figure 5.9: From a decision tree to a branching program

By Lemma 5.40,

$$R(f) \geq \min_{p_T} \sum_T p_T \sum_t n(t,T) q_t \geq R(f) + 1/2,$$

which is a contradiction. ∎

**Example 5.42** We consider the graph property $H$ of all graphs having a Hamiltonian circuit. Clearly, $H$ is monotone. Moreover, every minterm of $H$ is a Hamiltonian circuit. The total number of minterms of size $n$ is $(n-1)!/2$, where $n$ is the number of vertices. By the above lemma,

$$\binom{2R(H)}{n} \geq \frac{(n-1)!}{4}.$$

Note that $\binom{2R(H)}{n} \leq (2R(H))^n/n!$. Thus,

$$(2R(H))^n \geq \frac{(n!)^2}{4n} \geq \frac{\pi}{2}\left(\frac{n}{e}\right)^{2n}.$$

It follows that $R(H) \geq n^2/(2e^2)$. □

In general, the *Yao-Karp conjecture* states that for any nontrivial monotone graph property $\mathcal{P}$, $R(\mathcal{P}) = n^2/4 + o(n^2)$, where $n$ is the number of vertices.

## 5.9 Branching Programs

Consider a decision tree as shown in Figure 5.9(a). We see that some subtrees are identical. For such subtrees, we may use a single procedure to realize them. To exhibit this property explicitly, we may use an acyclic digraph instead of a tree to represent a decision process (see Figure 5.9(b)). Such a "decision graph" will be called a *branching program*. In general, a branching program of $n$ variables is an acyclic digraph which includes a single source node and some sink nodes. Each nonsink node is labeled by a variable and has exactly two out-edges labeled by 0 and 1, and each sink node is labeled by either 0 or 1. Note that in a decision tree, each variable appears at most once in any path from the root to a leaf. However, in a branching program, each variable may appear more than once in a path from the source to a sink. This relaxation provides a possibility to make a branching program scrawny.

The two most important complexity measures of a branching program are length (or, depth) and width. A *bounded-width* branching program of width $w$ and length $r$ is a branching program consisting of $r$ levels, each with at most $w$ nodes. Each directed edge goes from a node on one level to a node on the next level. Sink nodes exist only in the bottom level. The *size* of a branching program is the number of nodes in the program.

At the expense of increasing the size by a factor of $w$, a width-bounded branching program may sometimes be modified to have the property that all nodes in the same level have the same label. For such a branching program, one can give an algebraic interpretation: With a little modification on the first level, each level $k$ $(1 \leq k < r)$ can be seen as an instruction to read the input value $t_i$ of a variable $x_i$ where $x_i$ is the label of the nodes on this level and then emit a mapping $g_k(t_i) : \{1, \ldots, w\} \to \{1, \ldots, w\}$. The mappings emitted at all levels are composed and the Boolean function computed by the branching program has value 1 at the input assignment if and only if the composite mapping maps number 1 to a number in $\{1, \ldots, w\}$ which corresponds to a sink labeled by 1 at the bottom level. The set of mappings from $\{1, \ldots, w\}$ into itself forms a monoid $M$. Let $X$ be the set of mapping which map 1 to a number in $\{1, \ldots, w\}$ corresponding to a sink labeled by 1 at the bottom level. Then we obtain a *program over monoid $M$* consisting of a sequence of instructions

$$(g_1, i_1), \ldots, (g_r, i_r),$$

together with a subset $X$ of $M$, where each $g_k$ is a mapping from $\{0, 1\}$ to $M$ and each $i_k$ indicates the variable $x_{i_k}$ at the level $k$. On input assignment $t_1 \cdots t_n \in \{0, 1\}^*$, the program obtains the value 1 if and only if

$$g_1(t_{i_1}) \cdots g_r(t_{i_r}) \in X.$$

A *permutation branching program* is a program over a group, that is, the mapping emitted at each level is a permutation. A permutation branching program is given in Figure 5.10, which computes the parity function $p_n(x_1, \ldots, x_n) = x_1 \oplus \cdots \oplus x_n$.

When the width is too small, the power of the permutation branching program is limited. The following is a simple example.

**Example 5.43** We show that no permutation branching program of width two computes the function $x_1 \wedge x_2$.

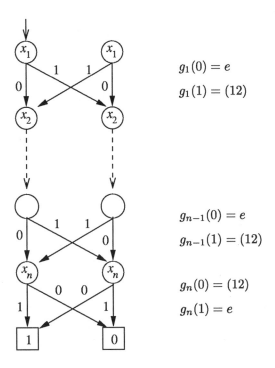

$g_1(0) = e$

$g_1(1) = (12)$

$g_{n-1}(0) = e$

$g_{n-1}(1) = (12)$

$g_n(0) = (12)$

$g_n(1) = e$

Figure 5.10: A permutation branching program $\{(g_1, 1), \ldots, (g_n, n)\}$.

There are only two permutations on two elements: the identity $e$ and $(12)$. By way of contradiction, suppose that $x_1 \wedge x_2$ can be computed by a permutation branching program of width two $\{(g_1, i_1), (g_2, i_2), \ldots, (g_k, i_k)\}$ such that $x_1 \wedge x_2 = 1$ if and only if $g_1(x_{i_1})g_2(x_{i_2})\cdots g_k(x_{i_k}) = e$. Note that if $g_j(0) = g_j(1)$, then removing this level does not affect the computation. Thus, we may assume that for any $j$, $g_j(0) \neq g_j(1)$. Since $x_1 \wedge x_2 = 1$ when $x_1 = x_2 = 1$, we have $g_1(1) \cdots g_k(1) = e$. Thus, there are an even number of $j$'s such that $g_j(1) = (12)$. Since $x_1 \wedge x_2 = 0$ when $x_1 = x_2 = 0$, we have $g_1(0) \cdots g_k(0) = (12)$. Thus, there are an odd number of $j$'s such that $g_j(0) = (12)$. This implies that there are an odd number of $j$'s such that $g_j(1) = e$. Hence, $k$ is an odd number.

Next, we consider two assignments $x_1 = 0$, $x_2 = 1$ and $x_1 = 1$, $x_2 = 0$. In both cases, we can show that there are an odd number of $(12)$'s. It follows that $k$ is an even number, a contradiction.                              □

Given a Boolean function, how do we design a branching program computing it? In the following, we study a method of constructing permutation branching programs. This method was first introduced by Barrington [1990].

Let $\sigma$ be a permutation which is not the identity. We say a permutation branching program $\{(g_1, i_1), \ldots, (g_r, i_r)\}$ $\sigma$-*computes* a Boolean function $f$ if

$$f(x) = 1 \iff g_1(x_{i_1}) \cdots g_r(x_{i_r}) = e$$
$$f(x) = 0 \iff g_1(x_{i_1}) \cdots g_r(x_{i_r}) = \sigma.$$

For example, the permutation branching program in Figure 5.10 (12)-computes parity function $p_n$.

The main idea of Barrington's method is to build a branching program recursively by following the *negation* and *disjunction* operations in the formula. In fact, constructing a branching program computing a literal is easy and any Boolean function can be built up based on literals through the negation and disjunction operations because $xy = \neg(\bar{x} + \bar{y})$, that is, conjunction can be represented through negation and disjunction. Let us first prove some properties of a permutation branching program which $\sigma$-computes a Boolean function for some cyclic $\sigma$. (A permutation is *cyclic* if it is composed of a single cycle on all of its elements.)

**Lemma 5.44** *Let $\sigma$ and $\tau$ be two cyclic permutations on the set $\{1, \ldots, w\}$. If there is a width-$w$ permutation branching program $p$ which $\sigma$-computes a Boolean function $f$, then there is a permutation branching program of the same width and length which $\tau$-computes $f$.*

*Proof.* Let $\{(g_1, i_1), \ldots, (g_r, i_r)\}$ $\sigma$-compute $f$. Since $\sigma$ and $\tau$ both are cyclic, there is a permutation $\gamma$ such that $\tau = \gamma^{-1}\sigma\gamma$. Suppose $\sigma = (a_1 a_2 \cdots a_n)$ and $\tau = (b_1 b_2 \cdots b_n)$; then, let $\gamma = \binom{a_1, a_2, \ldots, a_n}{b_1, b_2, \ldots, b_n}$. Define $h_1(b) = \gamma^{-1}g_1(b)$, $h_r(b) = g_r(b)\gamma$, for $b \in \{0,1\}$, and $h_k = g_k$ for $k = 2, \ldots, r-1$. Then $\{(h_1, i_1), \ldots, (h_r, i_r)\}$ $\tau$-computes $f$.                    ∎

**Lemma 5.45** *Let $\sigma$ be a cyclic permutation on the set $\{1, \ldots, w\}$. If there is a width-$w$ permutation branching program $p$ which $\sigma$-computes a Boolean function $f$, then there is a permutation branching program of the same width and length which $\sigma$-computes $\bar{f}$.*

*Proof.* Let $r$ be the length of the program $p$. Since $\sigma$ is cyclic, so is $\sigma^{-1}$. By the previous lemma, there exists a permutation branching program $\{(g_1, i_1), \ldots, (g_r, i_r)\}$ $\sigma^{-1}$-computing $f$. Let $h_1 = g_1, \ldots, h_{r-1} = g_{r-1}$ and $h_r(b) = g_r(b)\sigma$ for $b \in \{0,1\}$. Then $\{(h_1, i_1), \ldots, (h_r, i_r)\}$ $\sigma$-computes $\bar{f}$.                    ∎

**Lemma 5.46** *Let $\sigma$ and $\tau$ be two cyclic permutations on the set $\{1, \ldots, w\}$. If there are two width-$w$ permutation branching programs $p$ and $q$ $\sigma$-computing $f_1$ and $\tau$-computing $f_2$, respectively, then there is a width-$w$ permutation branching program of length $2(length(p) + length(q))$ which $(\sigma\tau\sigma^{-1}\tau^{-1})$-computes $f_1 + f_2$.*

*Proof.* Let $\{(g_1, i_1), \ldots, (g_r, i_r)\}$ $\sigma$-compute $f_1$ and $\{(h_1, j_1), \ldots, (h_t, j_t)\}$ $\tau$-compute $f_2$. By Lemma 5.44, there exists a program $\{(\hat{g}_1, i_1), \ldots, (\hat{g}_r, i_r)\}$

$(\sigma^{-1})$-computing $f_1$, and there exists a program $\{(\hat{h}_1, j_1), \ldots, (\hat{h}_t, j_t)\}$ $(\tau^{-1})$-computing $f_2$. Now, we compose four programs together in the order $\sigma, \tau, \sigma^{-1}, \tau^{-1}$. It is easy to verify that $\{(g_1, i_1), \ldots, (g_r, i_r), (h_1, j_1), \ldots, (h_t, j_t), (\hat{g}_1, i_1), \ldots, (\hat{g}_r, i_r), (\hat{h}_1, j_1), \ldots, (\hat{h}_t, j_t)\}$ $(\sigma\tau\sigma^{-1}\tau^{-1})$-computes $f_1 + f_2$. ∎

The above lemmas suggest a recursive procedure to build up a permutation branching program computing a Boolean function. In order to guarantee that this procedure always works, we need to make sure that $\sigma\tau\sigma^{-1}\tau^{-1}$ is cyclic.

**Lemma 5.47** *There are cyclic permutations $\sigma$ and $\tau$ in $S_5$ such that $\sigma\tau\sigma^{-1}\tau^{-1}$ is cyclic.*

*Proof.* $(12345)(13542)(12345)^{-1}(13542)^{-1} = (13254)$. ∎

Now, let us look at an example.

**Example 5.48** Consider function $f(x_1, \ldots, x_n) = \bigvee_{i=1}^{n} x_i$. We show that this function can be computed by a permutation branching program of width 5 and length at most $O(n^2)$. First, we use permutation branching programs of length one to (12345)-compute $x_1, x_3, \ldots$ and to (13542)-compute $x_2, x_4, \ldots$, respectively. Then, we construct, by Lemma 5.46, programs of length 4 to (13254)-compute $x_1 + x_2, x_3 + x_4, \ldots$. In order to continue the construction, we need to transfer, by Lemma 5.44, these programs to programs of the same length which (12345)-compute $x_1 + x_2, x_5 + x_6, \ldots$ and (13542)-compute $x_3 + x_4, x_7 + x_8, \ldots$, respectively. Next, we use Lemma 5.46 again to find programs of length 16 which (13254)-compute $x_1 + x_2 + x_3 + x_4, \ldots$, and so on. In this construction, each time an OR operation is performed, the total length of the involved programs is doubled. Since each variable is involved in at most $\lceil \log_2 n \rceil$ such length-doubling processes, the final program is of length at most $2^{\lceil \log_2 n \rceil} n = O(n^2)$. □

Barrington's method is simple and elegant. However, the branching program obtained by his method is a special kind of branching program. So, it is not necessarily of the minimum length. In fact, the permutation branching programs usually require length longer than the general branching programs. For example, consider function $\bigwedge_{i=1}^{n} x_i$. Clearly, a branching program of width 2 and length $n$ could compute it. However, it requires length $\Omega(n \log \log n)$ for any permutation branching program of a constant width $w$. To show this result, we first prove two combinatorial lemmas.

**Lemma 5.49** *In any sequence of at least $mn + 1$ distinct numbers, there is either an increasing subsequence of at least $m + 1$ numbers or a decreasing subsequence of at least $n + 1$ numbers.*

*Proof.* Suppose, by way of contradiction, that in some sequence $\{a_k\}$ of $mn+1$ distinct numbers, there is no increasing subsequence of $m+1$ numbers and no decreasing subsequence of $n + 1$ numbers. For each $a_k$, let $p_k$ be the length of the longest increasing subsequence whose last element is $a_k$ and $q_k$ the length of the longest decreasing subsequence whose last element is $a_k$. Then, $1 \leq p_k \leq m$ and $1 \leq q_k \leq n$. Since there are only $mn$ pairs $(i, j)$ with $1 \leq i \leq m$ and $1 \leq j \leq n$ and there are $mn + 1$ pairs $(p_k, q_k)$, there must exist two indices $k < k'$ such that $(p_k, q_k) = (p_{k'}, q_{k'})$. But this is not possible: If $a_k < a_{k'}$, then adding $a_{k'}$ to the increasing subsequence with $a_k$ as the end element, we see that $p_{k'}$ is at least $p_k + 1$, a contradiction. If $a_k > a_{k'}$, then adding $a_{k'}$ to the decreasing subsequence with the end element $a_k$, we see that $q_{k'}$ is at least $q_k + 1$, also a contradiction.                                              ∎

Consider a sequence $S$ of elements from $\{1, \ldots, n\}$. For $I \subseteq \{1, \ldots, n\}$, let $S_I$ denote the subsequence of $S$ consisting of all occurrences of elements in $I$.

**Lemma 5.50** *Let $S$ be a sequence of elements from $\{1, \ldots, n\}$. Assume that each $i \in \{1, \ldots, n\}$ occurs in $S$ exactly $k$ times, and that $n \geq 2^{3^k}$. Then there exists $I \subseteq \{1, \ldots, n\}$ with $|I| \geq n^{1/3^k}$ such that $S_I$ factors into $k'$ segments $\tau_1 \tau_2 \cdots \tau_{k'}$, where $k' \leq k$ and each $\tau_j$ is monotone with all $i \in I$ occurring an equal number of times in $\tau_j$.*

*Proof.* We prove Lemma 5.50 by induction on $k$. For $k = 1$, it follows from Lemma 5.49. For $k > 1$, let $i^{(j)}$ denote the $j$th occurrence of $i$ in the sequence. Applying the induction hypothesis to the subsequence $S'$ of all elements $i^{(j)}$'s for $j = 1, \ldots, k-1$ and $i = 1, \ldots, n$, we get a subset $I$ of $\{1, \ldots, n\}$ with $|I| \geq n^{1/3^{k-1}}$ such that $S_I'$ factors into at most $k-1$ monotone segments $\tau_1 \tau_2 \cdots \tau_{k'}$ where $\tau_j$ is monotone and all $i \in I$ occur an equal number of times in $\tau_j$. For the simplicity of notations, we may assume that $I = \{1, \ldots, n_1\}$. Clearly, all $i^{(k-1)}$, for $i = 1, \ldots, n_1$, must occur in $\tau_{k'}$. Since $i^{(k)}$ appears after $i^{(k-1)}$, $i^{(k)}$ does not appear in the segments $\tau_1 \cdots \tau_{k'-1}$. So, we need only to consider $\tau_{k'}$. Suppose that $\tau_{k'}$ is

$$1^{(\ell)}, 1^{(\ell+1)}, \ldots, 1^{(k-1)}, 2^{(\ell)}, 2^{(\ell+1)}, \ldots, 2^{(k-1)}, \ldots, n_1^{(\ell)}, n_1^{(\ell+1)}, \ldots, n_1^{(k-1)}.$$

Let $m$ be the maximum integer satisfying $m^3 \leq n_1$. We will show that there is a subset $J$ of at least $m$ elements from $\{1, \ldots, n_1\}$ such that $S_J$ factors into at most $k$ monotone segments.

Now, consider the subsequences $S_I''$ of $S_I$, consisting of all the $k$th occurrences, that is, elements from $\{1^{(k)}, \ldots, n_1^{(k)}\}$.

*Case 1.* There is a decreasing subsequence of $m$ elements, $i_m^{(k)} > i_{m-1}^{(k)} > \cdots > i_1^{(k)}$. Let $J = \{i_m, i_{m-1}, \ldots, i_1\}$. Since $i_m^{(k)}$ occurs after $i_m^{(k-1)}$, the sequence $S_J$ factors into $k' + 1$ monotone segments $(\tau_1)_J \cdots (\tau_{k'})_J S_J''$, meeting the requirement.

*Case 2.* Not Case 1. Then, by Lemma 5.49, there is an increasing subsequence of $m' =_{\text{defn}} m^2 - m + 1$ elements $i_1^{(k)} < i_2^{(k)} < \cdots < i_{m'}^{(k)}$. (Note that $m \geq 2$ implies $(m^2 - m + 1)m < m^3 \leq n$.) For each $j$, $1 \leq j \leq m$, let $q(j)$ be the largest $q$ such that $i_q^{(\ell)}$ occurs before $i_j^{(k)}$.

*Case* 2.1. If, for any $j$, $q(j) - j \geq m$, then choose $J = \{i_j, i_{j+1}, \ldots, i_{q(j)-1}\}$ so that $|J| \geq m$. Note that $i_{q(j)-1}^{(k-1)}$ occurs before $i_{q(j)}^{(\ell)}$ and, hence, before $i_j^{(k)}$. Therefore, like Case 1, $S_J = (\tau_1)_J \cdots (\tau_{k'})S_J''$ meets our requirement.

*Case* 2.2. Note Case 2.1. Choose $J = \{i_1, i_{1+q(1)}, i_{1+q(1+q(1))}, \ldots\}$. Since $q(j) - j < m$ for all $1 \leq j \leq m$, $|J| \geq m$. Furthermore, $S_J$ can be factored as $(\tau_1)_J \cdots (\tau_{k'-1})_J \sigma$, where $\sigma$ is

$$i_1^{(\ell)}, \ldots, i_1^{(k)}, i_{1+q(1)}^{(\ell)}, \ldots, i_{1+q(1)}^{(k)}, \ldots, i_{1+q(1+q(\cdots))}^{(\ell)}, \ldots, i_{1+q(1+q(\cdots))}^{(\ell)}.$$

Therefore, $S_J$ satisfies our requirement. ∎

**Theorem 5.51** *For any constant $w$, a width-$w$ permutation branching program for $\bigwedge_{i=1}^n x_n$ must have length $\Omega(n \log \log n)$.*

*Proof.* Let $L$ denote the length of the program. We want to show $L = \Omega(n \log \log n)$. Let $n_k$ be the number of variables that occur exactly $k$ times in the branching program. Then $\sum_{k \geq 1} n_k = n$ and $\sum_{k \geq 1} k n_k = L$.

Let $q = \log_3 \log_3 n$. If $\sum_{k \geq q/2} n_k \geq n/2$, then $L = \Omega(n \log \log n)$. So, we may assume that $\sum_{k \geq q/2} n_k < n/2$. In this case, there exists an integer $k < q/2$ such that $n_k \geq n/q$. Setting all other variables to true, we obtain a branching program on $n_k$ variables; each of them occurs exactly $k$ times. By Lemma 5.50, we can further choose $m \geq n_k^{1/3^k} > (n/q)^{1/\sqrt{\log_3 n}}$ variables and set all other variables to true to obtain a branching program on $m$ variables with the property that it factors into $k' \leq k$ monotone segments and, in each segment, variables occur an equal number of times. Now, for each segment, collapsing adjacent levels labeled by the same variable into one level, we obtain a new segment $S_j$ either in the form

$$(g_{1j}, i_1), (g_{2j}, i_2), \ldots, (g_{mj}, i_m)$$

or in the form

$$(g_{mj}, i_m), (g_{m-1,j}, i_{m-1}), \ldots, (g_{1j}, i_1),$$

where $i_1 < i_2 < \cdots < i_m$. The new program still computes the AND function of $m$ variables.

Define $F(\ell) = (s_1(\ell), s_2(\ell), \ldots, s_{k'}(\ell))$, where $s_j(\ell) = g_{1j}(0) \cdots g_{\ell j}(0)$. Then $F(\ell)$ is a vector whose components are permutations of $w$ numbers. Thus, there are at most $(w!)^{k'}$ different $F(\ell)$'s. Note that for sufficiently large $n$, $m > (w!)^{k'}$. So, there exist $\ell$ and $\ell'$, with $1 \leq \ell < \ell' \leq m$, such that $F(\ell) = F(\ell')$. This means that for $j = 1, \ldots, k'$, $g_{\ell+1,j}(0) \cdots g_{\ell' j}(0) = e$. Therefore, setting variables $x_{i_{\ell+1}}, \ldots, x_{i_{\ell'}}$ to 0 and others to 1, we will get value 1 from this program, contradicting the assumption that the program computes the AND function. ∎

The above is an explicit lower bound for permutation branching programs. What can we say about the lower bound for general branching programs? By a counting argument, it is known that most Boolean functions have exponential lower bounds. However, the best known lower bound for an explicit function is only $\Omega(n^2/\log^2 n)$ due to Neciporuk [1966].

## Exercises

**5.1** Nonplanarity is a monotone increasing graph property. Characterize its minterms by using Kuratowski's Theorem. (*Kuratowski's Theorem*: A graph is not planar if and only if it has a subgraph isomorphic to $K_{3,3}$ or $K_5$, that is, if the graph can be obtained from the complete bipartite graph $K_{3,3}$ or the complete graph $K_5$ by adding some vertices and some edges.)

**5.2** Prove that a graph property is monotone increasing if and only if there exists a family of graphs such that a graph has the property if and only if the graph contains a subgraph in the family.

**5.3** Given a digraph $G(V,E)$, construct a bipartite graph $G'(V,V',E')$ by setting $V' = \{v' : v \in V\}$ and $E' = \{(u,v') : (u,v) \in E\}$. For any bipartite graph property $\mathcal{P}$, define $\mathcal{P}' = \{G : G' \in \mathcal{P}\}$. Is $\mathcal{P}'$ a digraph property?

**5.4** Suppose $V = \{1,2,\ldots,n\}$ and $V' = \{1',2',\ldots,n'\}$. For each bipartite graph $G(V,V',E)$, construct a digraph $G'(V,E')$ by setting $E' = \{(u,v) : (u,v') \in E\}$. Define

$$\mathcal{P}_1 = \{G(V,V',E) : G'(V,E') \text{ has a loop}\}$$
$$\mathcal{P}_2 = \{G(V,V',E) : G'(V,E') \text{ is acyclic}\}$$
$$\mathcal{P}_3 = \{G(V,V',E) : G'(V,E') \text{ has a 1-factor}\}.$$

(A 1-*factor* is a union of vertex-disjoint cycles passing through all vertices.) For each $i = 1,2,3$, determine whether $\mathcal{P}_i$ is a bipartite graph property.

**5.5** Prove that in a binary tree, the sum of distances from internal vertices to the closest leaves is less than the total number of edges in the binary tree.

**5.6** Prove that $D(f) = D(\bar{f}) = D(f^*)$, $D_0(f) = D_1(f^*)$, and $D_1(f) = D_0(f^*)$.

**5.7** If $x$ is the variable labeling the root of a decision tree that is of depth $D(f)$ and computes function $f$, then $D(f|_{x=0}) \le D(f) - 1$ and $D(f|_{x=1}) \le D(f) - 1$. If $x$ is any variable of $f$, does the inequality still hold? Give a proof or show a counterexample.

**5.8** Prove that $D_0(y+zp_n) = D_1(y+zp_n) = n+1$ and $D(y+zp_n) = n+2$, where $y$ and $z$ are two variables other than the variables in the parity function $p_n$.

**5.9** Let $f(x_1,x_2,x_3,x_4) = x_1x_2 + x_2x_3 + x_3x_4 + x_4 + x_1$. Find two ways to prove that $D(f) = 4$.

**5.10** Let $f(x_1, x_2, \ldots, x_n)$ be a Boolean function with minterms of size at least $n - 1$. Suppose that for any variable $x_i$ there exists a minterm not containing $x_i$. Prove that $f$ is elusive.

**5.11** Let $H$ be a graph with $n$ vertices and $\mathcal{P}$ the graph property consisting of all graphs isomorphic to $H$. Determine $D_0(\mathcal{P})$, $D_1(\mathcal{P})$, and $D(\mathcal{P})$.

**5.12** Let $H_1$ and $H_2$ be two different graphs on the same set of $n$ vertices. Let $\mathcal{P}$ be the graph property consisting of all graphs isomorphic to either $H_1$ or $H_2$. Prove that $\mathcal{P}$ is elusive.

**5.13** Let $H_1$, $H_2$, and $H_3$ be three graphs as shown in Figure 5.11. Let $\mathcal{P}$ be the graph property consisting of all graphs isomorphic to $H_1$, $H_2$, or $H_3$. Prove that $\mathcal{P}$ is not elusive.

Figure 5.11: Three graphs $H_1$, $H_2$, and $H_3$.

**5.14** Draw a decision tree for Example 5.2 with five players and a decision tree for Example 5.5 with four vertices.

**5.15** A graph $G$ of $n$ vertices is a *scorpion graph* if it contains a vertex $t$ of degree one, a vertex $b$ of degree $n - 2$, and a vertex $a$ of degree two adjacent to $t$ and $b$. Suppose that $\mathcal{P}$ is the property of a graph being scorpion. Then $D(\mathcal{P}) \leq 6n$.

**5.16** Is the Boolean function $t$ in Example 5.2 a graph property? Is it a digraph property? Explain each of your answers by either a proof or a disproof.

**5.17** If all graphs satisfying nontrivial property $\mathcal{P}$ form a matroid, then $\mathcal{P}$ is elusive. (A collection $\mathcal{G}$ of graphs on the same vertex set $V$ is a *matroid* if the following two conditions hold: (1) $G \in \mathcal{G}, G' \subset G \Rightarrow G' \in \mathcal{G}$. (2) For any $G \in \mathcal{G}$ and any edge $e$ not in $G$, there exists an edge $e'$ in $G$ such that $(G - \{e'\}) \cup \{e\} \in \mathcal{G}$.)

**5.18** Prove that the following graph properties are elusive:
(a) The graph is acyclic.
(b) The graph is 2-connected.
(c) The graph is a spanning tree.
(d) The graph has at most $k$ edges.

**5.19** Let $\mathcal{P}$ be the following property of graphs of order $n$: there exists an isolated star of $n - 4$ leaves. Prove that $\mathcal{P}$ is not elusive.

**5.20** Prove that the assumption condition in Theorem 5.6 is necessary.

**5.21** Use the algebraic criterion to prove that Boolean function $x_1 x_2 + x_2 x_3 + x_3 x_4 + x_4 x_5 + x_5 x_6 + x_6 x_1$ is elusive.

**5.22** Prove that every nontrivial symmetric Boolean function is elusive.

**5.23** Define $f(x_1, \ldots, x_n) = 1$ if and only if $x_1 \cdots x_n$ contains three consecutive 1's. Prove the following:
(a) $f(x_1, x_2, x_3, x_4, x_5)$ is not elusive.
(b) For $n \equiv 0$ or $3 \ (\mathrm{mod}\ 4)$, $f(x_1, \ldots, x_n)$ is elusive.

**5.24** Define $p_f(t_1, \ldots, t_n) = \sum_{x \in \{0,1\}^n} f(x) t_1^{x_1} \cdots t_n^{x_n}$. Prove that $p_f(t_1, \ldots, t_n)$ is in the ideal generated by $(t_{i_1} + 1) \cdots (t_{i_{n-D(f)}} + 1)$, $1 \leq i_1 < \cdots < i_{n-D(f)} \leq n$ in the polynomial ring.

**5.25** Let $f$ be a Boolean function. Function $M_f(y) = \sum_{x \geq y} (-1)^{\|x - y\|} f(x)$ is called the *Mobius transform* of $f$. Prove that if $\|y\| < n - D(f)$, then $M_f(y) = 0$.

**5.26** Show that $D_0(f) = n$ implies $D(f) = n$, where $n$ is the number of variables. If we replace $n$ by $k < n$, is the above relation still true?

**5.27** Prove that if the number of truth assignments for Boolean function $f$ is odd, then $\mu(f) \neq 0$.

**5.28** Prove or disprove that if $\mathcal{P}$ is a non-elusive property of graphs of order $n$ then the number of graphs satisfying $\mathcal{P}$ can be divided by four.

**5.29** Let $2 \leq r \leq n$. Then the property of all graphs of order $n$ which contains a complete subgraph of order $r$ is elusive.

**5.30** Let $2 \leq r \leq n$. Then the property of $r$-colorable graphs of order $n$ is elusive.

**5.31** Let $H$ be a given graph of order $r$ and $\mathcal{P}$ the property of graphs of order $n$ containing a subgraph isomorphic to $H$. Prove that $D(\mathcal{P}) \geq (n - r + 1)(n - r + 3)/2$.

**5.32** Prove
$$\chi(\Delta^*) = (-1)^{n-1}(\chi(\Delta) - 1) + 1,$$
where $n$ is the number of vertices in $\Delta$.

**5.33** Let $f(x_1, x_2, x_3, x_4, x_5, x_6) = x_1 x_2 + x_2 x_3 + x_3 x_4 + x_4 x_5 + x_5 x_6 + x_6 x_1$. Draw a geometric representation of $\Delta_f$ and compute $\chi(\Delta_f)$.

**5.34** Prove that the simple strategy is a winning strategy for proving the elusiveness of a monotone increasing Boolean function $f$ if and only if $\Delta_f$ has no free face.

**5.35** Prove the following:
(a) If $f$ and $g$ are monotone increasing Boolean functions with the same variables, then $\Delta_{f \wedge g} = \Delta_f \cap \Delta_g$.
(b) If $\Delta_f$ or $\Delta_g$ is noncollapsible, then $\Delta_{f \wedge g}$ is noncollapsible.

**5.36** Compute the Euler characteristic of the simplicial complex whose maximal faces are the following:

$$\{a, b, e\}, \{b, c, e\}, \{c, d, e\}, \{d, a, e\}, \{a, b, d\}$$
$$\{a, b, f\}, \{b, d, f\}, \{d, c, f\}, \{c, a, f\}, \{a, b, c\}.$$

Prove that this simplicial complex is noncollapsible. (This is an example that a connected noncollapsible simplicial complex has the Euler characteristic 1.)

**5.37** Prove that if $\Delta$ is collapsed to $\Sigma$ by a sequence of elementary collapses, then $\Sigma^*$ can be collapsed to $\Delta^*$ by a sequence of elementary collapses.

**5.38** Let $f(x_1, \ldots, x_n)$ be a Boolean function. Let $g(x_1, \ldots, x_n, x_{n+1}) = f(x_1, \ldots, x_n)$. Prove the following:
(a) $\Delta_g = \Delta_f \cup \{\{x_{n+1}\} \cup A : A \in \Delta_f\}$.
(b) $\Delta_g$ is collapsible. (It is easy to prove it by the topological criterion. Can you also prove it by the definition of collapsibility?)

**5.39** Prove that the linear extension of an automorphism has a fixed point on the geometric representation of the simplicial complex if and only if the automorphism has a nonempty fixed face in the simplicial complex.

**5.40** Consider graphs $G$ of order $n$ for some even number $n$. Let $\mathcal{P}$ be the property of $G$ not being bipartite. Prove that every automorphism of $\Delta_{\mathcal{P}}^*$ has a nonempty fixed face.

**5.41** Let vertices of a simplicial complex be labeled by elements in $\mathbf{Z}_n$ and $\phi$ an automorphism defined by $\phi(i) \equiv i + 1 \bmod n$. Prove that if $n$ is odd, then all edges can be decomposed into $(n-1)/2$ orbits of cardinality $n$ and if $n$ is even, then all edges can be decomposed into $(n/2 - 1)$ orbits of cardinality $n$ and one orbit of cardinality $n/2$. Moreover, the set of edges $\{(0, j) : 1 \leq j \leq \lfloor n/2 \rfloor\}$ is a system of distinct representatives for the orbits.

**5.42** Prove that the property of graphs being bipartite is elusive.

**5.43** Let $\mathcal{P}_k$ be the set of graphs with $n$ vertices whose maximum degree is at most $k$. Then $\mathcal{P}_k$ is elusive for $0 \leq k \leq n - 2$.

**5.44** Let $\mathcal{P}$ be a nontrivial monotone property of graphs of order $n$. Apply the topological approach of Sections 5.5–5.7 to prove that $D(\mathcal{P}) \geq q(q+2)/4$, where $q$ is the largest prime power less than $n$.

**5.45** A one-dimensional complex is a simplicial complex such that all its nonempty faces are edges and vertices. Prove that a collapsible one-dimensional complex having a transitive group of automorphisms is either a single vertex or an edge.

**5.46** Let $\Delta$ be a nonempty simplicial complex over $X$, where $X = X_1 \times X_2$ with $|X_1| = p^k$, a prime power. Prove that if $X_1 \times \{x\} \not\subseteq \Delta$ for every $x \in X_2$ and $\Delta$ is invariant under $\Gamma$, a transitive group of permutations on $X_1$, then $\Delta$ is elusive. ($\Delta$ is elusive if the function $f$ with $\Delta_f = \Delta$ is elusive.)

**5.47** Prove that acyclic property of digraphs is elusive.

**5.48** Suppose that $H$ is a graph whose automorphism group is transitive on its vertices. Then each component of $H$ is either factor-critical or contains a 1-factor. (A graph is called *factor-critical* if, after deleting a vertex, it has a 1-factor.)

**5.49** Is every program of length $r$ over a monoid $M$ equivalent to a branching program of width $|M|$ and length $r$?

**5.50** Prove that $\bigwedge_{i=1}^{n} x_i$ $(n \geq 2)$ can be computed by a branching program of width two and cannot be computed by any permutation branching program of width two.

**5.51** Design a permutation branching program computing $x_1 \oplus \cdots \oplus x_n$. Is your program the shortest?

**5.52** Let $r, n, k$ be three positive numbers. Consider a $k$-coloring of the set $\{(i,j) : 1 \leq i < j \leq n\}$. If $n > r^k$ then there exists a sequence

$$1 \leq i_1 < i_2 < \cdots < i_r \leq n$$

of integers such that all $(i_j, i_{j+1})$ have the same color.

## Historical Notes

The study of decision tree complexity was started with research on graph and digraph properties by Holt and Reingold [1972], Best et al. [1974], and Kirkpatrick [1974]. Milner and Welsh [1976] discovered the simple strategy for the elusiveness of many graph properties. Aanderaa and Rosenberg (see Rosenberg [1973] and Best et al. [1974]) proposed the conjecture that for every nontrivial monotone graph property $\mathcal{P}$, $D(\mathcal{P}) \geq \varepsilon n^2$. This conjecture was confirmed by Rivest and Vuillemin [1975, 1976] (our Theorem 5.18). Meanwhile, they proposed a new one, that every nontrivial, weakly symmetric function is elusive. Illies [1978] found a counterexample to the Rivest-Vuillemin conjecture. This counterexample suggests a modified conjecture: every nontrivial, weakly symmetric, monotone function is elusive. Gao, Hu and Wu [1999] and Gao, Wu, Du and Hu [1999] showed that this modified conjecture is true for $n = 6$ and $n = 10$. The lower bound in Theorem 5.18 was improved subsequently to $n^2/9 + o(n^2)$ by Kleitman and Kwiatkowski [1980] and to $n^2/4 + o(n^2)$ by Kahn et al. [1984]. Karp conjectured that every nontrivial monotone graph property is elusive (see Kahn et al. [1984] and King [1989, 1990]). This conjecture was also proposed independently by other people (see Best et al. [1974] and Bollobas [1978]). The topological approach was first introduced by Kahn et al. [1984]. They used this approach to prove Theorem 5.34. Theorem 5.31 is due to Yao [1986], and Theorem 5.35 is from Triesch [1994]. The fixed-point theorem on group actions on complexes (Theorem 5.32) came from Oliver [1975].

For random decision tree complexity, Yao [1977] conjectured that for any nontrivial monotone graph property $\mathcal{P}$, $R(\mathcal{P}) = cn^2 + o(n^2)$ for some constant $c$. Karp conjectured the bound can be improved to $R(\mathcal{P}) = n^2/4 + o(n^2)$. Yao [1987] and King [1989, 1990] made progress on this conjecture. Branching programs were first introduced by Lee [1959] as a model for switching functions. Borodin, Dolev et al. [1983] studied branching programs of width two. The work on permutation branching programs is due to Barrington [1990] and Barrington and Straubing [1991]. Theorem 5.51 is from Cai and Lipton [1994].

# 6

# *Circuit Complexity*

We introduce in this chapter Boolean circuits as a nonuniform computational model. We demonstrate a close relation between the notion of circuit size complexity and Turing machine time complexity. Two important exponential lower bound results are proved. The exponential lower bound for monotone circuits is proved by the technique of combinatorial approximation, and the exponential lower bound for constant-depth circuits is proved by the probabilistic method. The $NC$ hierarchy is introduced as a model for computational complexity of parallel computation. The related notion of $P$-completeness is also presented.

## 6.1   Boolean Circuits

A *Boolean circuit* is an acyclic digraph of which each node with no in-edge is labeled by a variable or a Boolean constant 0 or 1, and each other node is labeled by a Boolean operator: negation, disjunction (OR), or conjunction (AND). The nodes labeled by variables are the *input* nodes, and the nodes labeled by Boolean operators are called *logical gates*. For each logical gate, its *fanin* is the number of its in-edges (or inputs) and its *fanout* is the number of its out-edges. We do not restrict the fanout of a gate, but we sometimes restrict the fanin of a gate. For a circuit computing a Boolean function, we assume that there exists a unique node of the circuit that has no out-edge; this node gives the output of the circuit and is called the *output gate*. Given a Boolean assignment to the variables, the circuit computes the function value in the following way: Each gate reads the values from its in-edges, applies the

195

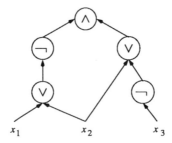

Figure 6.1: A circuit for $\neg(x_1 \vee x_2) \wedge (x_2 \vee \neg x_3)$.

Boolean operation on them, and then sends the resulting value through its out-edges to other gates. Figure 6.1 is an example.

The *size* of a circuit is the number of logical gates in the circuit. The *depth* of a circuit is the length of a longest path in the circuit from a variable to the output gate.

Using de Morgan's law, every Boolean circuit can be transformed to an equivalent one with the same depth and at most twice the size such that all negation gates occur at the bottom (i.e., they only take inputs from variables). This can be done as follows: Consider the highest conjunction or disjunction gate $G$ which has an output going to a negation gate. If this gate $G$ also has an output going to other non-negation gates (i.e., OR or AND gates), then we make a copy $G'$ of the gate $G$ and use this copy $G'$ to output to those non-negation gates. We keep, from the gate $G$, only those outputs going to negation gates. Then, by de Morgan's law, we may remove those negation gates and add new negation gates to all inputs of $G$ while changing the operation of $G$ from conjunction to disjunction or from disjunction to conjunction. We repeat this process until all negation gates only have in-edges from variables. An example is shown in Figure 6.2. Since negation gates of the new circuit only have in-edges from variables, we may, for simplicity, replace a negation by a literal, as shown in Figure 6.2. From now on, we will consider only this type of circuit which has no negation gate but might use negative literals. The *size* of a circuit of this type will be defined to be the number of AND and OR gates.

A given Boolean function $f$ can be computed by many different circuits. The *circuit complexity* $C_t(f)$ of $f$ is the minimum size of a circuit $C$ computing $f$ in which the fanins of logical gates are at most $t$. When $t = 2$, we simply write $C(f)$ instead of $C_2(f)$.

What is the maximum $C(f)$ for a Boolean function $f$ of $n$ variables? Note that since

$$f(x_1,\ldots,x_n) = (x_1 \wedge f|_{x_1=1}) \vee (\bar{x}_1 \wedge f|_{x_1=0}),$$

we have

$$C(f) \leq C(f|_{x_1=0}) + C(f|_{x_1=1}) + 3.$$

Based on this recurrence relation, it is easy to prove by induction on the

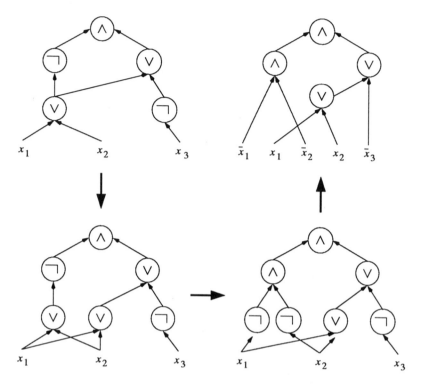

Figure 6.2: Transformation of a circuit.

number of variables that

$$C(f) \leq 3 \cdot 2^{n-1} - 3$$

for $n \geq 1$. Indeed, for $n = 1$, we have $C(f) = 0$ and so the inequality holds. Suppose the inequality holds for functions $f$ of $n - 1$ variables. Then, for functions $f$ of $n$ variables, we have

$$C(f) \leq 2 \cdot (3 \cdot 2^{n-2} - 3) + 3 = 3 \cdot 2^{n-1} - 3.$$

With a more complicated argument, the general upper bound for $C(f)$ can be improved to

$$C(f) \leq \left(1 + O\left(\frac{1}{\sqrt{n}}\right)\right) \frac{2^n}{n}.$$

This is almost the best we can get. In fact, one can prove that for sufficiently large $n$, there exists a Boolean function $f$ of $n$ variables such that $C(f) > 2^n/n$. In the following, we present a simple information-theoretic proof for a little weaker result.

**Theorem 6.1** *Almost all Boolean functions $f$ of $n$ variables have $C(f) = \Omega(2^n/n)$.*

*Proof.* We first count the number of circuits with $n$ variables that have size $s$. Note that each logical gate is assigned with an AND or OR operator that acts on two previous nodes. Each previous node could be a previous gate (at most $s$ choices), a literal ($2n$ choices), or a constant (2 choices). Thus, there are at most $s + 2n + 2$ choices for a previous node. Therefore, there are at most $(2(2 + s + 2n)^2)^s$ circuits of $n$ variables and size $s$. So, the number of circuits of $n$ variables and size at most $s$ is bounded by $s(2(2 + s + 2n)^2)^s$.

For $s = 2^n/(8n)$, $s(2(2 + s + 2n)^2)^s = O(2^{2^n/4+n}) \ll 2^{2^n}$. Since there are $2^{2^n}$ Boolean functions of $n$ variables, the probability of a Boolean function having a circuit of size at most $2^n/(8n)$ goes to zero as $n$ goes to $\infty$. ■

The above theorem shows that very few Boolean functions are computable by circuits of small size. However, no explicit function has been found so far that even has a provable super-linear lower bound for its circuit size complexity. The best known result is a lower bound $3n - o(n)$. In the following, we demonstrate a weaker result that the circuit size complexity of the following threshold function is at least $2n - 4$ when $m = 2$:

$$TH_{m,n}(x_1, x_2, \ldots, x_n) = \begin{cases} 1 & \text{at least } m \text{ of } n \text{ variables are 1,} \\ 0 & \text{otherwise.} \end{cases}$$

**Example 6.2** $C(TH_{2,n}) \geq 2n - 4$.

*Proof.* We will use the gate-elimination method. That is, given a circuit for $TH_{2,n}$, we first argue that the fanout of some variable must be greater than or equal to 2. Setting this variable to constant 0 will eliminate at least two gates and reduce the circuit to a circuit for $TH_{2,n-1}$.

More formally, we prove the inequality by induction on $n$. For the cases $n = 2$ and $n = 3$, the inequality holds trivially. Now, let $C$ be an optimal circuit for $TH_{2,n}$. Then every logical gate in $C$ must have two inputs coming from different nodes. Suppose that $G$ is a gate at the bottom; that is, its two inputs come from two literals, say $y_i$ and $y_j$ ($y_i$ is either $x_i$ or $\bar{x}_i$). Note that when we fix the input values of $x_i$ and $x_j$, $TH_{2,n}$ is reduced to one of the three possible functions: $TH_{2,n-2}$, $TH_{1,n-2}$, or $TH_{0,n-2}$, depending on the four possible settings of $x_i$ and $x_j$. Therefore, either $x_i$ or $x_j$ (or one of their their negations) has output to another gate $G'$; for, otherwise, circuit $C$ receives only two possible outputs from $G$ but could compute three different functions, $TH_{2,n-2}$, $TH_{1,n-2}$, and $TH_{0,n-2}$, which is not possible. Without loss of generality, assume that $x_i$ or $\bar{x}_i$ has outputs to another gate $G'$. We now set $x_i = 0$. This will eliminate two gates $G$ and $G'$ and the resulting circuit $C'$ will compute the function $TH_{2,n-1}$. By the inductive hypothesis, the size of $C'$ is at least $2(n - 1) - 4$ and, hence, the size of $C$ is at least $2 + 2(n - 1) - 4 = 2n - 4$. ■

Let $A$ be a language over $\{0, 1\}$ and $n$ a natural number. We define sets $A_{=n}$ and $A_{\leq n}$ as follows: $A_{=n} = \{x \in A : |x| = n\}$, $A_{\leq n} = \{x \in A : |x| \leq n\}$. The set $A$ can be represented by a sequence of Boolean functions $\{\chi_{n,A}\}$: $\chi_{n,A}$ has

$n$ variables such that $\chi_{n,A}(x_1, x_2, \ldots, x_n) = 1$ if and only if $x = x_1 x_2 \cdots x_n \in A$ (i.e., $\chi_{n,A}$ is the characteristic function of $A_{=n}$.) Now, we define the *circuit size complexity* of $A$ to be $CS_{t,A}(n) = C_t(\chi_{n,A})$. When $t = 2$, we write $CS_A(n)$ instead of $CS_{2,A}(n)$. A set $A$ is said to have *polynomial-size circuits* if there exists a polynomial $p$ such that $CS_A(n) \leq p(n)$ for all $n$.

**Proposition 6.3** *$A$ has polynomial-size circuits if and only if there exists a polynomial $p$ such that $CS_{\infty,A}(n) \leq p(n)$.*

*Proof.* Since $CS_{\infty,A}(n) \leq CS_A(n)$, the necessity direction is obvious. To see the sufficiency, consider a circuit of size at most $p(n)$ computing the Boolean function $\chi_{n,A}$. Although the fanin of each gate is not restricted, it is at most $p(n) + 2n + 1$ because there are at most $p(n) + 2n + 2$ nodes in the circuit. So, each gate can be simulated by at most $p(n) + 2n$ gates of fanin 2. The resulting circuit has size at most $(p(n) + 2n)p(n)$. Thus, $CS_A(n)$ is still bounded by a polynomial $q(n) = (p(n) + 2n)p(n)$. ∎

## 6.2 Polynomial-Size Circuits

We now investigate the relation between the nonuniform model of Boolean circuits for a language and the uniform model of Turing machines.

**Theorem 6.4** *Every set in $P$ has polynomial-size circuits.*

*Proof.* Consider a set $A \in P$. Then, there exists a one-tape DTM $M$ with time bound $p(n)$ accepting $A$ for some polynomial $p$. The computation of $M$ on an input $x$, with $|x| = n$, can be described by exactly $p(n) + 1$ configurations $\alpha_0, \alpha_1, \ldots, \alpha_{p(n)}$, each of size exactly $p(n) + 1$. We may also assume that the accepting configuration is unique, containing only blank symbols and the tape head pointing to cell one. (See the proof of Cook's Theorem in Section 2.2.) Between two consecutive configurations $\alpha_i$ and $\alpha_{i+1}$, we note that the $j$th symbol of $\alpha_{i+1}$ is determined by the transition function of $M$ from the $(j-1)$st, $j$th, and $(j+1)$st symbols of $\alpha_i$ (see Cook's Theorem). We use $f$ to represent the function that computes the $j$th symbol of $\alpha_{i+1}$ from the $(j-1)$st, $j$th, and $(j+1)$st symbols of $\alpha_i$. Suppose we use $k$ Boolean bits to represent one symbol of the configuration, then the function $f$ can be computed by a subcircuit of size at most $O(2^{3k})$.

Now, we describe a Boolean circuit $C$ that computes the function $\chi_{n,A}$ as follows: The circuit $C$ has $p(n) + 2$ levels, and each level has $(p(n) + 1)k$ gates. The top level takes the input values and computes the initial configuration $\alpha_0$. From level $i$ to level $(i+1)$, it computes the configuration $\alpha_{i+1}$ from configuration $\alpha_i$, with $p(n) + 1$ subcircuits for function $f$ between the two levels. Finally, the output gate computes the value from the first symbol of $\alpha_{p(n)}$ which contains the accepting state symbol if $M$ accepts $x$. This computation also contains at most $O(2^k)$ symbols. Thus, the total number of

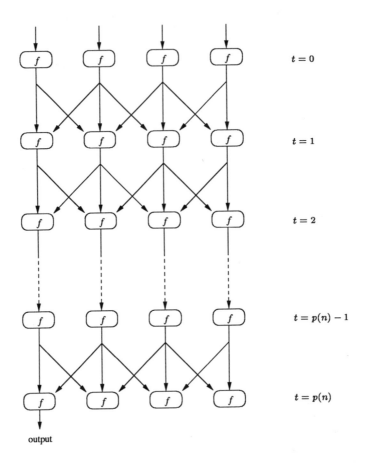

Figure 6.3: Simulating TM by a circuit.

gates of the circuit $C$ is bounded by $c \cdot (p(n)+1)^2 k 2^{3k}$ for some constant $c > 0$. Since $k$ is a constant depending only on $M$, this circuit is of polynomial size.

We show the layout of circuit $C$ in Figure 6.3, in which each box represents a group of at most $O(2^{3k})$ gates computing the function $f$.  ∎

The following extension of Theorem 6.4 will be used later.

**Lemma 6.5** *If $A \leq_T^P B$ and $B$ has polynomial-size circuits, then $A$ has polynomial-size circuits.*

*Sketch of Proof.* Consider a polynomial-time oracle DTM $M$ which, together with oracle $B$, accepts $A$. We need to construct polynomial-size circuits to simulate the computation of $M$. Now, we have to deal with an oracle DTM instead of a regular DTM. First, we note that an oracle DTM uses an additional tape as the query tape, and yet our simulation in Theorem 6.4 only simulates

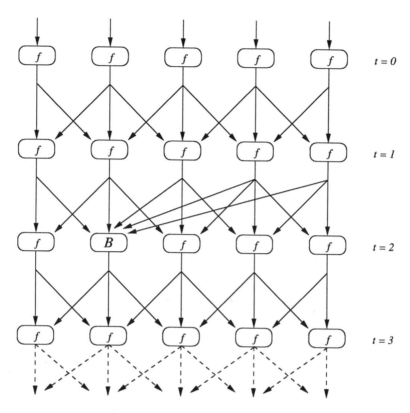

Figure 6.4: Simulating an oracle TM by a circuit.

DTMs with one tape. We can overcome this problem by encoding two tapes into one, as illustrated in Theorem 1.6. Second, when the oracle machine enters a query state, the next state is determined by the entire string on the query tape (or, more precisely, the second section of the encoded tape) instead of the single symbol scanned by the tape head. So, the subcircuit which outputs the next state has to be constructed by using the polynomial-size circuit computing $B$. This means that each constant-size circuit computing the function $f$ in Figure 6.3 needs to be augmented by a polynomial-size circuit computing $B$. (We show in Figure 6.4 this augmentation on one box. The complete circuit should replace every box labeled by $f$ by a box labeled by $B$.) However, the size of a circuit with polynomially many polynomial-size subcircuits is still bounded by a polynomial. Therefore, the resulting new circuit is still of polynomial size.                                                                 □

Next, we will present some characterization of the class of sets with polynomial-size circuits by uniform computational models. First, let us introduce some new notations.

**Definition 6.6** *Let $C$ be a class of languages and $\mathcal{F}$ a class of functions from nonnegative integers to strings. We define the class $C/\mathcal{F}$ to consist of all languages $A = \{x : \langle x, h(|x|)\rangle \in B\}$ for some $B \in C$ and $h \in \mathcal{F}$.*

In particular, let *poly* denote the set of functions $h$ from nonnegative integers to strings over $\{0,1\}$ such that for every $n$, $|h(n)| \leq p(n)$ for some fixed polynomial $p$. Then, the class $P/poly$ consists of sets $A$ for which there exist a set $B \in P$ and a function $h \in poly$ such that for every $x$ of length $n$,

$$x \in A \Longleftrightarrow \langle x, h(|x|)\rangle \in B.$$

We note the similarity between this class and the characterization of $NP$ in Theorem 2.1. That is, if $A \in P/poly$, then each string $x \in A$ has a *witness* $h(|x|)$ that can be verified in polynomial time. There is, however, an important difference between $P/poly$ and $NP$: the witness $h(|x|)$ of a string $x \in A$ must work for *all* strings of the same length; therefore, we cannot simply *guess* such a witness and instead have to store it in the program. In fact, it is known that $P/poly - NP \neq \emptyset$ (see Exercise 6.11).

We say a language $A$ is *sparse* if there exists a polynomial $p$ such that $|A_{=n}| \leq p(n)$ for all $n \geq 0$. For a language $S$, define $h_S(n) = \langle y_1, y_2, \ldots, y_k \rangle$, where $\{y_1, y_2, \ldots, y_k\} = S_{\leq n}$ and $y_1 < y_2 < \cdots < y_k$ in the lexicographic order. Then $h_S \in poly$ if and only if $S$ is sparse.

**Lemma 6.7** *Every sparse set has polynomial-size circuits.*

*Proof.* Let $S$ be a sparse set with $|S_{=n}| \leq p(n)$ for some polynomial $p$. For every string $w = w_1 \cdots w_n \in S$ with $|w| = n$ (each $w_i$ is a bit 0 or 1), define a Boolean function $f_w$ of $n$ variables by $f_w(x_1, \ldots, x_n) = x_1^{w_1} \wedge \cdots \wedge x_n^{w_n}$ where, for each $1 \leq i \leq n$,

$$x_i^{w_i} = \begin{cases} x_i & \text{if } w_i = 1, \\ \bar{x}_i & \text{if } w_i = 0. \end{cases}$$

The circuit $\chi_{n,S}(x_1, \ldots, x_n) = \bigvee_{w \in S_{=n}} f_w$ is of size at most $np(n) + 1$.    ∎

**Theorem 6.8** *The following statements are equivalent:*
  *(a) $A$ has polynomial-size circuits.*
  *(b) There exists a sparse set $S$ such that $A \leq_T^P S$.*
  *(c) $A \in P/poly$.*

*Proof.* We are going to prove the theorem in the directions (b) $\Rightarrow$ (a) $\Rightarrow$ (c) $\Rightarrow$ (b). The part (b) $\Rightarrow$ (a) follows immediately from Lemmas 6.5 and 6.7. Next, we prove the other two parts.

(a) $\Rightarrow$ (c): Encoding the polynomial-size circuit computing $\chi_{n,A}$ into a string $h(n)$ (see discussion in Section 6.5), we obtain a function $h$ in *poly*. Let $B$ be the set of all pairs $\langle x, y \rangle$ satisfying the condition that $y$ encodes a circuit that accepts $x$. Then, it is easy to see that $B$ is in $P$: we can construct a

TM working as a *circuit interpreter* which reads the code $y$ of a circuit and simulates it on the input $x$. Thus, $A = \{x : \langle x, h(|x|)\rangle \in B\}$ is in $P/poly$.

(c) $\Rightarrow$ (b): Let $A = \{x : \langle x, h(|x|)\rangle \in B\}$ for some $h \in poly$ and $B \in P$. Define a sparse set $S$ as follows:

$$S = \{0^n 1^i : \text{ the } i\text{th symbol of } h(n) \text{ is } 0\}$$
$$\cup \{1^n 0^i : \text{ the } i\text{th symbol of } h(n) \text{ is } 1\}.$$

Let $p$ be a polynomial such that $|h(n)| \leq p(n)$. Consider the following algorithm:

```
Input: x;
Oracle: S;
Begin
    w := λ;
    For i := 1 to p(|x|) do
        If 0^|x|1^i ∈ S then w := w0
        else if 1^|x|0^i ∈ S then w := w1;
    If ⟨x, w⟩ ∈ B then accept else reject
End.
```

Clearly, this algorithm with oracle $S$ accepts $A$ in polynomial time. Hence, $A \leq^P_T S$.                                                                            ∎

The above characterization of the class $P/poly$ can be extended so that the string $h(n)$ can be used to solve all inputs of length *at most* $n$. Furthermore, this characterization can be extended to the class $NP/poly$.

**Proposition 6.9** *A set $A$ is in $P/poly$ if and only if there exist a set $B \in P$ and a function $h \in poly$ such that for all $x$ of length at most $n$, $x \in A \Longleftrightarrow \langle x, h(n)\rangle \in B$.*

**Proposition 6.10** *The following statements are equivalent:*
*(a) $A \in NP/poly$.*
*(b) $A \in NP^S$ for some sparse set $S$.*
*(c) There exist a set $B \in NP$ and a function $h \in poly$ such that for all $x$ of length at most $n$, $x \in A \Longleftrightarrow \langle x, h(n)\rangle \in B$.*

*Proof.* See Exercise 6.12.                                                       ∎

We have seen that $P$ is included in $P/poly$. So, one possible way of separating $NP$ from $P$ is to prove that $NP$ is not included in $P/poly$, that is, to find a problem in $NP$ which does not have polynomial-size circuits. Is this a reasonable approach? The following theorem gives a somewhat positive evidence.

**Theorem 6.11** *If $NP \subseteq P/poly$, then the polynomial-time hierarchy collapses to the second level, that is, $PH = \Sigma^P_2$.*

This theorem follows immediately from the following two lemmas.

**Lemma 6.12** *For any $k > 2$, if $\Sigma_k^P \subseteq P/poly$, then $\Sigma_k^P = \Sigma_2^P$.*

*Proof.* Recall the $\leq_m^P$-complete problem $\mathrm{SAT}_k$ for $\Sigma_k^P$ that we studied in Section 3.3. We may restrict that an input instance $\phi$ to the problem $\mathrm{SAT}_k$ must be a quantified Boolean formula with at most $k$ quantifier changes such that the starting quantifier is $\exists$ if there are exactly $k$ quantifier changes. We say such a Boolean formula $\phi$ is of the $\Sigma_k$-form.

By the assumption, $\mathrm{SAT}_k \in P/poly$. By Proposition 6.9, there exist $B \in P$ and $h \in poly$ such that for all $\phi$ of the $\Sigma_k$-form and of length $\leq n$

$$\phi \in \mathrm{SAT}_k \iff \langle \phi, h(n) \rangle \in B.$$

Set $w = h(n)$. Then the string $w$ satisfies the following conditions with respect to all $\phi$ of the $\Sigma_k$-form that have $|\phi| \leq n$:

(a) $\langle 0, w \rangle \notin B$ and $\langle 1, w \rangle \in B$.

(b) If $\phi = (\exists x)\psi$, then $\langle \phi, w \rangle \in B$ if and only if $\langle \psi|_{x=0}, w \rangle \in B$ or $\langle \psi|_{x=1}, w \rangle \in B$.

(c) If $\phi = (\forall x)\psi$, then $\langle \psi, w \rangle \in B$ if and only if $\langle \psi|_{x=0}, w \rangle \in B$ and $\langle \psi|_{x=1}, w \rangle \in B$.

Conversely, we can show inductively that if $w$ satisfies the above conditions, then a Boolean formula $\phi$ of the $\Sigma_k$-form and of length $\leq n$ belongs to $\mathrm{SAT}_k$ if and only if $\langle \phi, w \rangle \in B$. In fact, if $\phi$ is a constant, then by (a), the above claim is true. Assume that it is true for all $\phi$ of the $\Sigma_k$-form that have at most $m - 1$ variables, with $m \leq n$. For every $\phi$ of the $\Sigma_k$-form with $m$ variables, $\phi$ is of the form either $\phi = (\exists x)\psi$ or $\phi = (\forall x)\psi$. So, we can apply (b) or (c) to finish the induction. In particular, note that $\psi$ has a single free variable $x$ and $\psi|_{x=0}$ and $\psi|_{x=1}$ are of the $\Sigma_k$-form and have $m - 1$ variables.

Thus, we have proved that for any $\theta$ of the $\Sigma_k$-form with length $\leq n$,

$$\theta \in \mathrm{SAT}_k \iff (\exists w, |w| \leq p(n))(\forall \phi, |\phi| \leq n)[(a) \wedge (b) \wedge (c) \wedge (\langle \theta, w \rangle \in B)].$$

This means that $\mathrm{SAT}_k \in \Sigma_2^P$. Since $\mathrm{SAT}_k$ is $\Sigma_k^P$-complete, we conclude that $\Sigma_k^P \subseteq \Sigma_2^P$. ∎

**Lemma 6.13** *If $NP \subseteq P/poly$, then $PH \subseteq P/poly$.*

*Proof.* We prove $\Sigma_k^P \subseteq P/poly$ by induction on $k$. For $k = 1$, it is trivial since $\Sigma_1^P = NP$. For $k > 1$, consider $A \in \Sigma_k^P$. Assume $A \in NP^B$ for some $B \in \Sigma_{k-1}^P$. By the inductive hypothesis, $\Sigma_{k-1}^P \subseteq P/poly$, and so $B \in P^S$ for some sparse set $S$.

Note that $A \in NP^B$ and $B \in P^S$ imply $A \in NP^S$. By Proposition 6.10, we know that there exist a set $C \in NP$ and a function $h \in poly$ such that for all $x$, $x \in A$ if and only if $\langle x, h(|x|) \rangle \in C$. By the assumption that $NP \subseteq P/poly$, we know that $C \in P/poly$. By Proposition 6.9, there exist a function $g$ in *poly* and a set $R$ in $P$ such that, for all $y$ of length $|y| \leq n$, $y \in C$ if and only if $\langle y, g(n) \rangle \in R$. Hence, for any $x$ of length $n$,

$$x \in A \iff \langle x, h(n) \rangle \in C$$
$$\iff \langle \langle x, h(n) \rangle, g(p(n)) \rangle \in R,$$

where $p(n)$ is a polynomial such that $|\langle x, h(n) \rangle| \leq p(n)$ for all $x$ of length $|x| \leq n$. Combining $h(n)$ and $g(p(n))$ together, we see that $A \in P/poly$. ∎

For relations between the class *P/poly* and other uniform classes such as *PSPACE* and *EXP*, see Exercise 6.13.

## 6.3 Monotone Circuits

From the results of the last section, we know that we could prove $P \neq NP$ by establishing a superpolynomial lower bound for the circuit-size complexity of an NP-complete problem. However, so far nobody has yet been able to prove even a superlinear lower bound. In this section, we turn our attention to a small subclass of circuits and prove an exponential lower bound for the circuit size complexity of an *NP*-complete problem.

A circuit is called *monotone* if it contains no negation gate nor does it use the negation of any variable. Not every Boolean function can be computed by monotone circuits. In fact, it is easy to characterize the class of Boolean functions computable by monotone circuits. Recall from Chapter 5 that a Boolean function $\phi$ is *monotone increasing* if it has the property that for any Boolean assignments $\tau$ and $\pi$ with $\tau \leq \pi$ (i.e., $\tau(x_i) = 1 \Rightarrow \pi(x_i) = 1$), $\phi|_\tau = 1$ implies $\phi|_\pi = 1$.

**Proposition 6.14** *A Boolean function is computable by a monotone circuit if and only if it is monotone increasing.*

*Proof.* We first show that a monotone increasing function $f$ can be computed by a monotone circuit. This fact can be proved by induction on the number of variables of $f$. Suppose $f$ has no variable. Then, $f$ is a constant function and is computable by a monotone circuit. Next, assume that $f$ has variables $x_1, x_2, \ldots, x_n$, $n \geq 1$. Since $f|_{x_1=0} \leq f|_{x_1=1}$, we know that

$$f = (x_1 \wedge f|_{x_1=1}) \vee f|_{x_1=0}.$$

By the inductive hypothesis, the functions $f|_{x_1=1}$ and $f|_{x_1=0}$ are computable by monotone circuits. It follows that $f$ is computable by a monotone circuit.

Conversely, it is clear that changing an input value of a variable of a monotone circuit from 0 to 1 does not decrease its value. Hence, the function computed by a monotone circuit is monotone increasing. ∎

Now, we consider a monotone increasing graph property $\text{CLIQUE}_{k,n}$ defined by

$$\text{CLIQUE}_{k,n}(G) = \begin{cases} 1 & \text{if graph } G \text{ has a complete subgraph of order } k, \\ 0 & \text{otherwise,} \end{cases}$$

where $G$ is a graph of order $n$ and $k \leq n$. Recall that the problem CLIQUE is an *NP*-complete problem. We are going to show the following exponential lower bound result on $\text{CLIQUE}_{k,n}$.

**Theorem 6.15** *For $k \leq n^{1/4}$, every monotone circuit computing $\text{CLIQUE}_{k,n}$ contains $n^{\Omega(\sqrt{k})}$ gates.*

The proof is mainly a counting argument involving two types of graphs called the positive test graphs and the negative test graphs. In the rest of the proof, we assume that $n$ and $k$ are fixed.

A *positive test graph* is a graph on $n$ vertices that consists of a clique on some set of $k$ vertices, and contains no other edge. All positive test graphs are isomorphic, but they are considered as different graphs. So, the total number of positive test graphs is $\binom{n}{k}$. $\text{CLIQUE}_{k,n}$ outputs 1 on each positive test graph. A *negative test graph* is a graph on $n$ vertices, together with a coloring of vertices by $k - 1$ possible colors such that there is no edge between two vertices if and only if they have the same color. For the sake of a simple counting, two graphs with the same topological structure and different colorings on vertices will be considered as different negative test graphs. Thus, the total number of negative test graphs is $(k - 1)^n$. $\text{CLIQUE}_{k,n}$ outputs 0 on every negative test graph. (See Figure 6.5.)

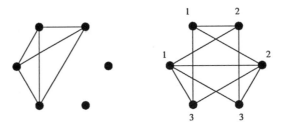

Figure 6.5: Positive and negative test graphs for $\text{CLIQUE}_{4,6}$. (The labels in the negative test graph denote the coloring.)

For any Boolean function $f$ on graphs of order $n$ (i.e., any Boolean function of $\binom{n}{2}$ variables $x_{i,j}$, $1 \leq i < j \leq n$, with each $x_{i,j}$ corresponding to the predicate that there is an edge between vertices $v_i$ and $v_j$), let $N_+(f)$ be the number of positive test graphs on which $f$ outputs 1 and $N_-(f)$ the number of negative test graphs on which $f$ outputs 1. Clearly, we have

$$N_+(\text{CLIQUE}_{k,n}) = \binom{n}{k}, \qquad N_-(\text{CLIQUE}_{k,n}) = 0.$$

We will use $N_+(f)$ and $N_-(f)$ as two coordinates for locating function $f$ on a plane (see Figure 6.6). Then, we will define a family of functions called *terminals* whose locations on this plane are far from the location of $\text{CLIQUE}_{k,n}$.

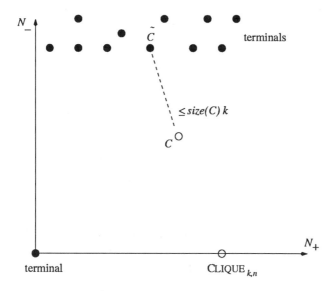

Figure 6.6: Basic idea of the proof.

For a subset $X$ of vertices, let the *clique indicator* $\lceil X \rceil$ of $X$ be the Boolean function on graphs of $n$ vertices which outputs 1 if and only if the induced subgraph of the input graph on vertex set $X$ is a complete graph. A *terminal* is a disjunction of at most $m$ clique indicators, each of whose underlying vertex sets has cardinality at most $\ell$, where the parameters $\ell$ and $m$ are defined as follows:[1]

$$\ell = \lfloor \sqrt{k} \rfloor, \quad p = \lceil 10\sqrt{k}\log n \rceil, \quad m = (p-1)^\ell \cdot (\ell!).$$

That is, a terminal $t$ is of the form $t = \bigvee_{i=1}^{r} \lceil X_i \rceil$, with $|X_i| \le \ell$ and $r \le m$.

**Lemma 6.16** *If $t$ is a terminal, then either $N_+(t) = 0$ or $N_-(t) \ge [1 - \binom{\ell}{2}/(k-1)] \cdot (k-1)^n$.*

*Proof.* Let $t = \bigvee_{i=1}^{r} \lceil X_i \rceil$ be a terminal. If $t$ is identical to 0, then $N_+(t) = 0$. If not, then $N_-(t) \ge N_-(\lceil X_1 \rceil)$. A negative test graph is rejected by the clique indicator $\lceil X_1 \rceil$ if and only if its associated vertex coloring assigns the same color to a pair of vertices in $X_1$. Thus, among $(k-1)^n$ negative test graphs, at most $\binom{|X_1|}{2}(k-1)^{n-1}$ graphs are rejected by $\lceil X_1 \rceil$. Hence,

$$
\begin{aligned}
N_-(\lceil X_1 \rceil) &\ge (k-1)^n - \binom{|X_1|}{2}(k-1)^{n-1} \\
&\ge (k-1)^n - \binom{\ell}{2}(k-1)^{n-1}.
\end{aligned}
$$

---

[1] Recall that we write $\log n$ to denote $\log_2 n$.

Thus, $N_-(t) \geq [1 - \binom{\ell}{2}/(k-1)] \cdot (k-1)^n$.                                              ∎

Now, we know that for any terminal $t$, either

$$N_+(\text{CLIQUE}_{k,n}) - N_+(t) = \binom{n}{k}$$

or

$$N_-(t) - N_-(\text{CLIQUE}_{k,n}) \geq \left(1 - \frac{\binom{\ell}{2}}{k-1}\right) \cdot (k-1)^n.$$

This means that every terminal $t$ is located far from $\text{CLIQUE}_{k,n}$ (Figure 6.6). Our next goal is to prove the following lemma.

**Lemma 6.17** (Distance Lemma) *For any monotone circuit $C$ computing a Boolean function on graphs of order $n$, there exists a terminal $\tilde{C}$ such that*

$$N_+(C) - N_+(\tilde{C}) \;\leq\; size(C) \cdot m^2 \cdot \binom{n-\ell-1}{k-\ell-1} \tag{6.1}$$

$$N_-(\tilde{C}) - N_-(C) \;\leq\; size(C) \cdot m^2 \cdot \binom{\ell}{2}^p \cdot (k-1)^{n-p}. \tag{6.2}$$

This lemma tells us that the distance from a monotone circuit to terminals is bounded above by a number proportional to the size of the circuit. Thus, any circuit computing a function far from terminals must have a large size.

Before proving the lemma, let us explain how to derive the main theorem from the lemma.

*Proof of Theorem 6.15.* Since $\ell = \lfloor \sqrt{k} \rfloor$, $p = \lceil 10\sqrt{k} \log n \rceil$, and $m = (p-1)^\ell \cdot (\ell!)$, we have $\binom{\ell}{2} < (k-1)/2$ and

$$m \leq (p-1)^\ell \ell^\ell \leq (10k \log n)^{\sqrt{k}} \leq n^{(0.3)\sqrt{k}}$$

for sufficiently large $n$ (note that $k \leq n^{1/4}$). Consider any circuit $C$ computing $\text{CLIQUE}_{k,n}$. By Lemma 6.16, either

$$N_+(C) - N_+(\tilde{C}) = \binom{n}{k} \tag{6.3}$$

or

$$N_-(\tilde{C}) - N_-(C) \geq \left(1 - \frac{\binom{\ell}{2}}{k-1}\right) \cdot (k-1)^n > \frac{(k-1)^n}{2}. \tag{6.4}$$

If (6.3) holds, then by Lemma 6.17 we have

$$size(C) \;\geq\; \frac{\binom{n}{k}}{m^2 \binom{n-\ell-1}{k-\ell-1}} \geq \frac{\left(\frac{n}{k}\right)^{\ell+1}}{m^2} \geq \frac{n^{(0.75)\sqrt{k}}}{m^2} = n^{\Omega(\sqrt{k})},$$

since $k \leq n^{1/4}$. If (6.4) holds, then by Lemma 6.17 we have

$$
\begin{aligned}
size(C) &\geq \frac{(k-1)^n \cdot 2^{-1}}{m^2 \binom{\ell}{2}^P (k-1)^{n-p}} \\
&\geq \frac{(k-1)^n \cdot 2^{-1}}{m^2 (k-1)^p 2^{-p} (k-1)^{n-p}} \geq \frac{2^{p-1}}{m^2} \geq \frac{n^{10\sqrt{k}}}{2m^2} = n^{\Omega(\sqrt{k})}. \qquad \blacksquare
\end{aligned}
$$

Let us call $\tilde{C}$ a *projection* of $C$. To show Lemma 6.17, we use the *bottom-up* approach to construct the projection $\tilde{C}$ of $C$. That is, we start from the bottom level and move up finding projections of subcircuits of $C$ one by one. At the beginning, any subcircuit at the bottom level contains only one variable $x_{i,j}$. The projection of $x_{i,j}$ is simply itself. Indeed, $x_{i,j} = \lceil \{v_i, v_j\} \rceil$ is a terminal.

Now, suppose that the projections of all proper subcircuits of a circuit $C$ have been found. Let us see how to find the projection for $C$.

First, we consider the case that the top gate of $C$ is an OR gate. Suppose that two inputs of the top gate come from two subcircuits $A$ and $B$, whose projections are $\tilde{A} = \bigvee_{i=1}^r \lceil X_i \rceil$ and $\tilde{B} = \bigvee_{j=1}^s \lceil Y_j \rceil$, respectively. The OR of $\tilde{A}$ and $\tilde{B}$ is an OR of $r + s$ clique indicators which may not be a terminal since it is possible that $r + s > m$.

So, we need to reduce the number of clique indicators in $\tilde{A} \vee \tilde{B}$. To do this, we introduce a new combinatorial object, called a sunflower. A *sunflower* is a collection of distinct sets $Z_1, Z_2, \ldots, Z_p$ such that the intersection of any two sets $Z_i \cap Z_j$ is the same, that is, $Z_i \cap Z_j = Z_1 \cap Z_2 \cap \cdots \cap Z_p$. Each set $Z_i$ in the sunflower is called a *petal* and the intersection $Z_1 \cap Z_2 \cap \cdots \cap Z_p$ is called the *center* of the sunflower (see Figure 6.7).

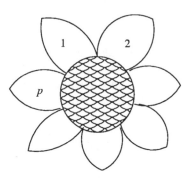

Figure 6.7: Sunflower.

To reduce the number of clique indicators, we choose a sunflower of size $p$ with its petals in $\{X_1, X_2, \ldots, X_r, Y_1, Y_2, \ldots, Y_s\}$ and replace all its petals by its center (and, hence, reduce the number of clique indicators by $p - 1$). This operation is called *plucking*. Repeat the plucking operation until no sunflower with $p$ petals exists. The following lemma shows that the resulting function, denoted by $\tilde{A} \sqcup \tilde{B}$, is a terminal.

**Lemma 6.18** (Erdös-Rado Sunflower Lemma) *Let $\mathcal{Z}$ be a family of distinct sets each with cardinality at most $\ell$. If $|\mathcal{Z}| > (p-1)^\ell \cdot \ell! = m$, then $\mathcal{Z}$ contains a sunflower with $p$ petals.*

*Proof.* The proof is by induction on $\ell$. For $\ell = 1$, any $p$ distinct sets with cardinality at most 1 form a sunflower with $p$ petals; the center is empty. Now, consider $\ell \geq 2$. Let $\mathcal{M}$ be a maximal subfamily of disjoint sets in $\mathcal{Z}$. If $|\mathcal{M}| \geq p$, then a sunflower with $p$ petals and the empty center can be found in $\mathcal{M}$. Thus, we may assume $|\mathcal{M}| \leq p - 1$. Let $S$ be the union of all sets in $\mathcal{M}$. Clearly, $|S| \leq (p-1)\ell$. For every $x \in S$, define $\mathcal{Z}_x = \{Z - \{x\} : Z \in \mathcal{Z} \text{ and } x \in Z\}$. Since $S \cap Z \neq \emptyset$ for every $Z \in \mathcal{Z}$, we have

$$\sum_{x \in S} |\mathcal{Z}_x| \geq |\mathcal{Z}| > (p-1)^\ell \cdot \ell!.$$

Therefore, by the pigeonhole principle, there exists an $x \in S$ such that

$$|\mathcal{Z}_x| > \frac{(p-1)^\ell \cdot \ell!}{(p-1)\ell} = (p-1)^{\ell-1} \cdot (\ell-1)!.$$

By the inductive hypothesis, $\mathcal{Z}_x$ contains a sunflower with $p$ petals. Adding $x$ back to all these $p$ petals, we get the required sunflower in $\mathcal{Z}$. ∎

In the case that the top gate of $C$ is an AND gate, we have

$$\tilde{A} \wedge \tilde{B} = \bigvee_{i=1}^r \bigvee_{j=1}^s (\lceil X_i \rceil \wedge \lceil Y_j \rceil).$$

This is not necessarily a terminal, for two reasons: (a) $\lceil X_i \rceil \wedge \lceil Y_j \rceil$ is not necessarily a clique indicator, and (b) there may be too many, possibly as many as $m^2$, terms.

To overcome problem (a), we replace each term $\lceil X_i \rceil \wedge \lceil Y_j \rceil$ by $\lceil X_i \cup Y_j \rceil$. This may cause another problem, that the cardinality of $X_i \cup Y_j$ may be larger than $\ell$. In such a case, we will simply delete this term. To overcome problem (b), we can use plucking again until there are at most $m$ terms. The final result is denoted by $\tilde{A} \sqcap \tilde{B}$.

Now, we need to show that by setting $\tilde{C}$ equal to $\tilde{A} \sqcup \tilde{B}$ in the first case or to $\tilde{A} \sqcap \tilde{B}$ in the second case, the final $\tilde{C}$ will satisfy the two inequalities in Lemma 6.17.

Before we start, we first note that the plucking operation can only enlarge the class of accepted graphs of the functions. That is, if $t_1$ is a disjunction of clique indicators and $t_2$ is the function obtained from $t_1$ after a plucking, then $t_1 \leq t_2$.

To prove inequality (6.1), let $P(S < T)$ be the set of positive test graphs for which the inequality $S < T$ holds, where $S$ and $T$ are two Boolean functions. Clearly,

$$N_+(C) - N_+(\tilde{C}) \leq |P(\tilde{C} < C)|.$$

If $C = A \vee B$, then

$$
\begin{aligned}
P(\tilde{C} < C) &\subseteq P(\tilde{A} \sqcup \tilde{B} < \tilde{A} \vee \tilde{B}) \cup P(\tilde{A} \vee \tilde{B} < A \vee B) \\
&\subseteq P(\tilde{A} < A) \cup P(\tilde{B} < B),
\end{aligned}
$$

since

$$
P(\tilde{A} \sqcup \tilde{B} < \tilde{A} \vee \tilde{B}) = \emptyset,
$$

from our observation above about plucking.

If $C = A \wedge B$, then

$$
\begin{aligned}
P(\tilde{C} < C) &\subseteq P(\tilde{A} \sqcap \tilde{B} < \tilde{A} \wedge \tilde{B}) \cup P(\tilde{A} \wedge \tilde{B} < A \wedge B) \\
&\subseteq P(\tilde{A} < A) \cup P(\tilde{B} < B) \cup P(\tilde{A} \sqcap \tilde{B} < \tilde{A} \wedge \tilde{B}).
\end{aligned}
$$

Note that for a positive test graph, $\lceil X_i \rceil \wedge \lceil Y_j \rceil$ outputs 1 if and only if $\lceil X_i \cup Y_j \rceil$ outputs 1. Moreover, the removal of a term with $|X_i \cup Y_j| \geq \ell + 1$ can only lose at most $\binom{n-\ell-1}{k-\ell-1}$ positive test graphs. Since at most $m^2$ terms may be removed, the total number of positive test graphs that are lost in the process of resolving problem (a) is no more than $m^2 \cdot \binom{n-\ell-1}{k-\ell-1}$. Finally, note again that the plucking procedures do not reduce any accepted graphs. Thus, we obtain the inequality

$$
|P(\tilde{A} \sqcap \tilde{B} < \tilde{A} \wedge \tilde{B})| \leq m^2 \cdot \binom{n - \ell - 1}{k - \ell - 1}.
$$

This means that in either case, we have

$$
P(\tilde{C} < C) \subseteq P(\tilde{A} < A) \cup P(\tilde{B} < B) \cup S_C,
$$

where $S_C$ is a set of at most $m^2 \cdot \binom{n-\ell-1}{k-\ell-1}$ positive test graphs. Now, we use a "top down" procedure to exploit the above relation. Note that at the bottom of the circuit $C$ each subcircuit is a variable $x_{i,j}$ and $P(\tilde{x}_{i,j} < x_{i,j}) = \emptyset$. So, expanding the above recurrence relation, we get

$$
P(\tilde{C} < C) \subseteq \bigcup_A S_A,
$$

where $A$ ranges over all subcircuits in $C$ which have at least one gate. Therefore,

$$
|P(\tilde{C} < C)| \leq \sum_A |S_A| \leq size(C) \cdot m^2 \binom{n - \ell - 1}{k - \ell - 1}.
$$

This proved the inequality (6.1).

To show the second inequality (6.2), let $Q(S < T)$ be the set of negative test graphs for which the inequality $S < T$ holds for Boolean functions $S$ and $T$. Clearly,

$$
N_-(\tilde{C}) - N_-(C) \leq |Q(C < \tilde{C})|.
$$

From an argument similar to the above, it is sufficient to prove the following two inequalities:

$$|Q(\tilde{A} \vee \tilde{B} < \tilde{A} \sqcup \tilde{B})| \;\leq\; m^2 \cdot \binom{\ell}{2}^p \cdot (k-1)^{n-p}, \tag{6.5}$$

$$|Q(\tilde{A} \wedge \tilde{B} < \tilde{A} \sqcap \tilde{B})| \;\leq\; m^2 \cdot \binom{\ell}{2}^p \cdot (k-1)^{n-p}. \tag{6.6}$$

Let us first study (6.5). From $\tilde{A} \vee \tilde{B}$ to $\tilde{A} \sqcup \tilde{B}$, only plucking procedures were employed. How many accepted negative test graphs can be increased in each plucking? To answer this question, we note that a negative test graph $G$ is added by a plucking procedure if and only if the underlying sunflower $\{Z_1, Z_2, \ldots, Z_p\}$ satisfies the following two conditions: $(\alpha)$ every petal $Z_i$ contains a pair of vertices with the same color in $G$, and $(\beta)$ the center $Z$ of the sunflower contains no pair of vertices with the same color in $G$. Thus, among $(k-1)^n$ negative test graphs, the probability of a negative test graph being added by each plucking is

$$\Pr[(\alpha) \wedge (\beta)] \leq \Pr[(\alpha) \mid (\beta)]$$

$$= \Pr[(\forall i)[Z_i - Z \text{ has a pair of the same color}$$
$$\qquad\qquad \text{or } Z_i - Z \text{ has a vertex of a color used in } Z]]$$

$$= \prod_{i=1}^{p} \Pr[Z_i - Z \text{ has a pair of the same color}$$
$$\qquad\qquad \text{or } Z_i - Z \text{ has a vertex of a color used in } Z]$$

$$\leq \prod_{i=1}^{p} \Pr[Z_i \text{ has a pair of the same color}]$$

$$\leq \prod_{i=1}^{p} \frac{\binom{|Z_i|}{2}}{k-1} \leq \frac{\binom{\ell}{2}^p}{(k-1)^p} \; .$$

Therefore, the number of accepted negative test graphs increased by each plucking is at most $\binom{\ell}{2}^p (k-1)^{n-p}$. Since at most $2m$ pluckings are performed, at most $2m\binom{\ell}{2}^p (k-1)^{n-p}$ new negative test graphs are accepted. Hence (6.5) holds.

To show (6.6), we note that $\lceil X_i \rceil \wedge \lceil Y_j \rceil \geq \lceil X_i \cup Y_j \rceil$. So, no new negative test graph is accepted from replacing the former by the latter. Similarly, no new negative test graph is accepted by removing a clique indicator of size larger than $\ell$. Finally, at most $m^2$ pluckings are performed. Therefore, (6.6) holds. This completes the proof of Lemma 6.17.

What is the gap between the monotone circuit complexity and the circuit complexity of a monotone increasing function? If the gap is small, then one can obtain a lower bound for general circuits through the study of monotone circuits. Unfortunately, the gap in general can be very big. It has been proved that there is a function computable by polynomial-size circuits which requires exponential-size monotone circuits. Whether the approach of monotone circuits is plausible remains an interesting question (cf. Exercise 6.14).

## 6.4 Circuits with Modulo Gates

An extension of the lower bound results on circuit size complexity is to study circuits with additional, potentially more powerful, logical operations. For example, we may add $\text{MOD}_k$ gates, for some $k \geq 2$, to the circuits. A $\text{MOD}_k$ function is the Boolean function which outputs 1 if and only if the total number of variables assigned with value 1 is not equal to 0 modulo $k$. Note that when the $\text{MOD}_k$ gate is involved, we may need to remove the convention on the location of NOT gates; that is, in the new circuits, NOT gates can occur everywhere and negations of variables are no longer allowed as inputs.

In this section, we show an exponential lower bound for the size of a constant-depth circuit that computes the parity function with unbounded fanin and with the AND, OR, NOT, and $\text{MOD}_3$ gates. We will use an approach similar to the one we used to establish the exponential lower bound for the monotone circuit size of the function $\text{CLIQUE}_{k,n}$.

In this approach, we first define the notion of terminals and then establish two lemmas about the distance between functions. The first lemma shows that a function computed by a constant-depth circuit with small size has a short distance from terminals. The second lemma shows that the parity function is far from terminals. Together, they imply that any constant-depth circuit with a small number of gates cannot compute the parity function.

In the following proof, we use the number of input settings on which the two functions disagree with each other to measure the distance between two Boolean functions. To set up the terminals, we consider polynomials over the finite field $GF(3)$. For any integer $b > 0$, we define a $b$-*terminal* to be a polynomial over $GF(3)$ which is of degree at most $b$ and which takes values from $\{0, 1\}$ on inputs from $\{0, 1\}$.

**Lemma 6.19** (Second Distance Lemma) *For every circuit $C$ of depth $d$ and every $\ell \geq 1$, there exists a $(2\ell)^d$-terminal $\tilde{C}$ which disagrees with $C$ on at most $size(C)2^{n-\ell}$ input settings.*

*Proof.* Similar to Lemma 6.17, we call $\tilde{C}$ a projection of $C$. First, let us explain how to obtain $\tilde{C}$ by a bottom-up construction.

Note that the inputs to $C$ are variables which are 1-terminals. Now, suppose $C'$ is a subcircuit of $C$ with depth $d' \geq 1$ such that projections of all of its proper subcircuits have been found. We describe $\tilde{C}'$ in four cases.

*Case* 1. The top gate of $C'$ is a NOT gate whose unique input comes from a subcircuit $D$. Then, we set $\tilde{C}' = 1 - \tilde{D}$. Let $I(E)$ denote the set of input settings satisfying condition $E$. Then, clearly, $I(C' \neq \tilde{C}') = I(D \neq \tilde{D})$.

*Case* 2. The top gate of $C'$ is a $\text{MOD}_3$ gate whose inputs come from subcircuits $D_1, D_2, \ldots, D_k$. Then, we set $\tilde{C}' = (\tilde{D}_1 + \tilde{D}_2 + \cdots + \tilde{D}_k)^2$. Since all $\tilde{D}_i$'s are $(2\ell)^{d'-1}$-terminals, $\tilde{C}'$ is a $(2\ell)^{d'}$-terminal. Clearly, $I(C' \neq \tilde{C}') \subseteq \bigcup_{i=1}^{k} I(D_i \neq \tilde{D}_i)$.

*Case* 3. The top gate of $C'$ is an OR gate whose inputs come from subcircuits $D_1, D_2, \ldots, D_k$. Then, we will choose $\tilde{C}'$ in the form

$$1 - \prod_{i=1}^{\ell}(1 - f_i),$$

where each $f_i = (\sum_{D \in F_i} \tilde{D})^2$ and $F_1, F_2, \ldots, F_\ell$ are $\ell$ subsets of $\{D_1, D_2, \ldots, D_k\}$. Clearly, such a choice is a $(2\ell)^{d'}$-terminal. In addition, we require that $\tilde{C}'$ satisfy

$$|I(\tilde{C}' \neq (\tilde{D}_1 \vee \tilde{D}_2 \vee \cdots \vee \tilde{D}_k))| \leq 2^{n-\ell}.$$

The existence of such a choice can be proved by a probabilistic argument as follows: First, we note that if $(\tilde{D}_1 \vee \cdots \vee \tilde{D}_k) = 0$ then $\tilde{C}' = 0$. Thus, $(\tilde{D}_1 \vee \cdots \vee \tilde{D}_k) \neq \tilde{C}'$ implies that $(\tilde{D}_1 \vee \cdots \vee \tilde{D}_k) = 1$ and $\tilde{C}' = 0$.

We choose each $F_i$ randomly such that each $D_j$ has $1/2$ chance to be in $F_i$, and choose each $F_i$ independently to form a random $\tilde{C}'$. We claim that for any fixed input setting $\tau$ that satisfies $(\tilde{D}_1|_\tau \vee \cdots \vee \tilde{D}_k|_\tau) = 1$, the probability of $\tilde{C}'|_\tau = 0$ is at most $2^{-\ell}$. Note that if $(\tilde{D}_1|_\tau \vee \cdots \vee \tilde{D}_k|_\tau) = 1$, then there must exist an index $j$, $1 \leq j \leq k$, such that $\tilde{D}_j|_\tau = 1$. It is not hard to see that, then, $\Pr[f_i|_\tau = 0] \leq 1/2$ for each $f_i$, because for any fixed $F_i$ with $f_i|_\tau = 0$, adding $D_j$ to $F_i$ or deleting $D_j$ from $F_i$ will change $f_i|_\tau$ to 1. Thus, $\Pr[\tilde{C}'|_\tau = 0] \leq 2^{-\ell}$, and the claim is proven.

From the above claim, we see that for any fixed input setting $\tau$, the probability of $\tilde{C}'$ being different from $\tilde{D}_1 \vee \cdots \vee \tilde{D}_k$ is at most $2^{-\ell}$. It follows that the expected value of $|I(\tilde{C}' \neq (\tilde{D}_1 \vee \cdots \vee \tilde{D}_k))|$ is at most $2^{n-\ell}$. Therefore, there exists a choice of $\tilde{C}'$ such that $|I(\tilde{C}' \neq (\tilde{D}_1 \vee \cdots \vee \tilde{D}_k))| \leq 2^{n-\ell}$.

*Case* 4. The top gate of $C'$ is an AND gate whose inputs come from subcircuits $D_1, D_2, \ldots, D_k$. Then, we set $\tilde{C}'$ in the form $\prod_{i=1}^{\ell}(1 - g_i)$, where each $g_i = (\sum_{D \in F_i}(1 - \tilde{D}))^2$ and $F_1, F_2, \ldots, F_\ell$ are $\ell$ subsets of $\{D_1, D_2, \ldots, D_k\}$. Applying de Morgan's law to the argument of Case 3, we can prove that there exists a choice of $\tilde{C}'$ that satisfies

$$|I(\tilde{C}' \neq \tilde{D}_1 \wedge \tilde{D}_2 \wedge \cdots \wedge \tilde{D}_k)| \leq 2^{n-\ell}.$$

In every case, when $C'$ has its inputs coming from subcircuits $D_1, D_2, \ldots, D_k$, we have

$$I(C' \neq \tilde{C}') \subseteq S_{C'} \cup \left(\bigcup_{i=1}^{k} I(D_i \neq \tilde{D}_i)\right),$$

where $S_{C'}$ is a set of at most $2^{n-\ell}$ input settings. Therefore, the projection $\tilde{C}$ satisfies

$$I(C \neq \tilde{C}) \subseteq \bigcup_{C'} S_{C'},$$

where $C'$ ranges over all subcircuits of $C$ of depth at least one. It follows that $|I(C \neq \tilde{C})| \leq size(C)2^{n-\ell}$. ∎

Next, we show that the parity function is far from $\sqrt{n}$-terminals.

**Lemma 6.20** *Any $\sqrt{n}$-terminal disagrees with the parity function on at least $\frac{1}{10}2^n$ input settings for sufficiently large $n$.*

*Proof.* Consider a $\sqrt{n}$-terminal $t$. Let $U = \{0,1\}^n$ be the set of $2^n$ possible input settings and $G$ the set of input settings on which $t$ and the parity function $p_n$ agree with each other. Note that $2 = -1$ in $GF(3)$, and that $1 + p_n$ agrees with $\prod_{i=1}^n (1 + x_i)$ on $U$ over the field $GF(3)$. For each variable $x_i$, let $y_i = 1 + x_i$. Consider $1 + t$ and $1 + p_n$ as functions over variables $y_1, \ldots, y_n$. Let $U' = \{-1, 1\}^n$ and $G'$ be the set of input settings, in $U'$, on which $1 + t$ agrees with $\prod_{i=1}^n y_i$. Then, it is clear that $|G| = |G'|$. Denote by $\mathcal{F}_{G'}$ the family of functions $f : G' \to \{-1, 0, 1\}$. Note that $|\mathcal{F}_{G'}| = 3^{|G'|}$. Thus, we may find an upper bound of $|G'|$ by counting $|\mathcal{F}_{G'}|$. To do so, we first show that for each $f \in \mathcal{F}_{G'}$, there exists a polynomial over $GF(3)$ of degree at most $\frac{1}{2}(n + \sqrt{n})$ which agrees with $f$ on $G'$.

Clearly, we can always find a polynomial $f'$ agree with $f$ on $G'$. Suppose $f'$ contains a term $c y_{i_1} \cdots y_{i_j}$ with $j > \frac{1}{2}(n + \sqrt{n})$. Note that, in $GF(3)$,

$$y_{i_1} \cdots y_{i_j} = \prod_{i=1}^n y_i \prod_{i \notin \{i_1, \ldots, i_j\}} y_i,$$

if $y_1, \ldots, y_n \in \{-1, 1\}$. Thus, we can replace this term by $(1+t) \prod_{i \notin \{i_1, \ldots, i_j\}} y_i$ and still preserve the agreement on $G'$. Since $t$ has degree at most $\sqrt{n}$, these two new terms have degree at most $n - \frac{1}{2}(n + \sqrt{n}) + \sqrt{n} = \frac{1}{2}(n + \sqrt{n})$. So, $f'$ can be transformed into a polynomial of degree at most $\frac{1}{2}(n + \sqrt{n})$.

Now, note that each term in a polynomial is multilinear. The number of multilinear monomials of degree at most $\frac{1}{2}(n + \sqrt{n})$ is $\sum_{0 \leq i \leq (n+\sqrt{n})/2} \binom{n}{i}$. For even $n$, by the Stirling approximation, we have

$$\sum_{0 \leq i \leq (n+\sqrt{n})/2} \binom{n}{i} \leq 2^{n-1} + \frac{\sqrt{n}}{2}\binom{n}{n/2}$$

$$\leq 2^{n-1} + \frac{1}{\sqrt{2\pi}}2^n e^{1/(12n)} < \frac{9}{10} \cdot 2^n$$

for sufficiently large $n$. Similarly, for odd $n$,

$$\sum_{0 \leq i \leq (n+\sqrt{n})/2} \binom{n}{i} \leq 2^{n-1} + \frac{\sqrt{n}}{2}\binom{n}{(n+1)/2} < \frac{9}{10} \cdot 2^n$$

for sufficiently large $n$. Thus, there exist at most $3^{(0.9)2^n}$ polynomials over $GF(3)$ of degree at most $\frac{1}{2}(n + \sqrt{n})$; that is, $|\mathcal{F}_{G'}| \leq 3^{(0.9)2^n}$ for sufficiently large $n$. It follows that $|G'| \leq \frac{9}{10}2^n$ for sufficiently large $n$ and, hence, $|U - G| \geq \frac{1}{10}2^n$ for sufficiently large $n$. ∎

**Theorem 6.21** *The parity function cannot be computed by a circuit of depth* $d$ *and with at most* $\frac{1}{20}2^{(0.5)n^{1/(2d)}}$ *gates of the types* AND, OR, NOT, *or* MOD$_3$.

*Proof.* Choose $\ell = \lfloor n^{1/(2d)}/2 \rfloor$. Then, $(2\ell)^d \le \sqrt{n}$. By the two lemmas above, if circuit $C$ computes the parity function, then

$$size(C) \ge \frac{\frac{1}{10}2^n}{2^{n-\ell}} = \frac{1}{10}2^\ell \ge \frac{1}{20} \cdot 2^{(0.5)n^{1/(2d)}}.$$                    ∎

## 6.5  NC

In this section, we turn our attention to the complexity classes of sets defined by polynomial-size, polylog-depth circuits. We define two types of complexity classes: uniform complexity classes and nonuniform complexity classes. By uniformity, we mean that the circuit family $\{C_n\}$ that computes functions $\{\chi_{n,A}\}$ can be generated by a deterministic Turing machine in logarithmic space.

For any fixed integer $i \ge 0$, we define the classes nonuniform-$NC^i$ and nonuniform-$AC^i$ as follows: A language $A$ is in the class *nonuniform-$NC^i$* if there is a polynomial $p(n)$ and a constant $k$ such that $\chi_{n,A}$ can be computed by a circuit $C_n$ with the following properties:

 (i) The gates in $C_n$ have fanin 2.

 (ii) $C_n$ has depth $k(\log n)^i$.

 (iii) The size of $C_n$ is at most $p(n)$.

A language $A$ is in the class *nonuniform-$AC^i$* if condition (i) above is relaxed to

 (i') The gates in $C_n$ have unbounded fanin.

To define the notion of uniformity, we need to fix a standard encoding of circuits by binary strings. For instance, we may encode a circuit of fanin at most $k$ by a sequence of $(k+2)$-tuples $\langle g, i, j_1, \ldots, j_k \rangle$, each representing one gate, where $g$ denotes which type of gate (OR or AND or input gate) it is, $i$ denotes the gate number, and $j_1, \ldots, j_k$ denote the gate numbers from which there is an edge to gate $i$. (For an input gate $i$ which reads variable $x_j$ (or $\bar{x}_j$), we may encode it by $\langle g, i, j, 1 \rangle$ (or, respectively, by $\langle g, i, j, 0 \rangle$).) We say a DTM *generates a circuit family* $\{C_n\}$ *in logarithmic space* if it computes on input $n$ the code of circuit $C_n$ as described above in space $c \log n$ for some constant $c$. For any $i \ge 0$, we say a language $A$ is in class $NC^i$ (or, in $AC^i$) if it is in nonuniform-$NC^i$ (or, respectively, in nonuniform-$AC^i$) and the circuits $C_n$ can be generated by a DTM in logarithmic space.[2]

---

[2]There are several different formulations of the notion of uniformity, depending on the resources required by the DTM to generate the codes of circuits $C_n$. Our definition here is termed *log-space uniformity*, and is the simplest and most commonly used notion of uniformity. For other definitions of uniformity and their relations with our definition, see, for instance, Cook [1985].

Note that a gate of a polynomial-size fanin can be simulated by a circuit of depth $O(\log n)$ with gates of fanin 2. It follows that $NC^i \subseteq AC^i \subseteq NC^{i+1}$ for all $i \geq 0$. We define the class *nonuniform-NC* as the union of *nonuniform-NC^i* over all $i \geq 0$, and class $NC$ as the the union of $NC^i$ over all $i \geq 0$.

Boolean circuits are often used as a theoretical model for parallel computation. Two most important complexity measures of parallel machines are parallel time and the number of processors. In the Boolean circuit model for parallel computation, the circuit depth corresponds to parallel time complexity and the maximum number of gates in the same level of the circuit corresponds to the number of processors. From this interpretation, we see that the class $NC^i$ consists of languages that can be recognized in parallel time $O((\log n)^i)$ with a polynomial number of processors.

The relation between the classes in the $NC$ hierarchy and the complexity classes defined by other parallel computational models, such as alternating Turing machines, SIMDAG machines, and parallel random access machines (PRAMs) have been studied extensively. Here we will not get into this area, but only point out that the machines in these models can be simulated by the Boolean circuit model within a constant factor of parallel time and a polynomial factor of the number of processors. Thus, $NC$ represents the class of languages that are solvable *efficiently* in parallel.

Before we study the $NC$ hierarchy and its relation with other uniform complexity classes, first let us look at some examples in classes $NC^1$ and $NC^2$.

**Example 6.22** (PARITY) The parity function of $n$ inputs is

$$p_n(x_1, x_2, \ldots, x_n) = x_1 \oplus x_2 \oplus \cdots \oplus x_n,$$

where $\oplus$ is the exclusive-or operation: $a \oplus b = 1$ if and only if exactly one of $a$ and $b$ is 1. Its corresponding language is

$$A = \{x \in \{0,1\}^* : x \text{ contains an odd number of 1's}\}.$$

We check that the parity function is in $NC^1$. It is easy to find a circuit $C_n$ of depth $\lceil \log n \rceil$ to compute function $p_n$ with $n-1$ $\oplus$-gates, each having fanin 2. Since each $\oplus$-gate can be replaced by three gates of AND and OR in two levels, $p_n$ can be computed by a circuit $D_n$ of depth $2\lceil \log n \rceil$ and of size at most $3n$. Finally, we note that both circuit $C_n$ and its conversion to $D_n$ can easily be described by a DTM in logarithmic space. So, set $A$ is in $NC^1$. $\square$

**Example 6.23** (EXTRACTING A BIT) Let $w$ be a word over $\{0,1\}$ of length $n$ and $k$ an integer, $0 \leq k \leq n-1$, written in the binary form. The problem here is to find the $(k+1)$st bit (starting from the left) of $w$. To simplify the argument, assume that $n = 2^m$ for some $m \geq 0$ and let $w = w_1 \cdots w_{2^m}$, where each $w_j$ is a bit 0 or 1. The circuit $C$ for this problem is the composition of $m$ subcircuits $C_1, \ldots, C_m$, where the input to $C_{i+1}$ consists of the $(i+1)$st bit of $k$ and the $2^{m-i}$ output bits of $C_i$ (when $i = 0$, the input to $C_1$ is the first bit of $k$ and $w$). Each subcircuit $C_i$ with input $(b, w_{j+1}, w_{j+2}, \ldots, w_{j+2^{m-i}})$ will output bits $(w_{j+1}, w_{j+2}, \ldots, w_{j+2^{m-i-1}})$ if $b = 0$ and output bits $(w_{j+2^{m-i-1}+1}, \ldots, w_{j+2^{m-i}})$ if $b = 1$. It is easy to see that each $C_i$ has only depth 2 and so $C$ has depth $2m = O(\log n)$. (For instance, the first output bit of $C_i$ is $(b \wedge w_{j+2^{m-i-1}+1}) \vee (\bar{b} \wedge w_{j+1})$.) Thus, the problem of extracting the $k$th bit of $w$ is in $NC^1$.  □

**Example 6.24** (BOOLEAN MATRIX MULTIPLICATION) Let $X$ and $Y$ be two Boolean matrices of size $n \times n$. Suppose $X = [x_{ij}]$ and $Y = [y_{ij}]$. Then, their Boolean product is $XY = [\sum_{k=1}^n x_{ik} y_{kj}]$, where the multiplication and addition are all Boolean operations. For each fixed pair $(i, j)$ with $1 \leq i, j \leq n$, the value $\sum_{k=1}^n x_{ik} y_{kj}$ can be computed by a circuit of depth 2, size $n+1$, with unbounded fanin. Thus, the $n^2$ values of $XY$ can be computed by a circuit of depth 2, size $O(n^3)$, with unbounded fanin. (Note that this circuit has $n^2$ output gates.) In other words, the Boolean matrix multiplication problem is in $AC^0$ and hence in $NC^1$. In particular, if we only use gates of fanin 2, then the circuit computing the multiplication has depth $\lceil \log n \rceil + 1$ and has size $O(n^3)$.  □

**Example 6.25** (GRAPH AACCESSIBILITY) The graph accessibility problem asks, for a given digraph $G$ and two vertices $s$ (*source*) and $t$ (*target*), whether there is a path in $G$ from $s$ to $t$. We let GAcc be the set $\{(G, s, t) : \text{there is a path in } G \text{ from } s \text{ to } t\}$.

Let the vertices of $G$ be named $1, 2, \ldots, n$, and let $M$ be the Boolean matrix obtained from the adjacency matrix of $G$ by adding 1 to all diagonal elements. Consider $M^k$ with Boolean multiplication and addition. Clearly, $M^k[i, j] = 1$ if and only if there is a path from vertex $i$ to vertex $j$ of length at most $k$ in $G$. Therefore, for $m \geq n$, $M^m[i, j] = 1$ if and only if there exists a path in $G$ from vertex $i$ to vertex $j$; that is, $M^m$ is the transitive closure of $M$.

Let $q = \lceil \log n \rceil$. Then $M^{2^q} = M^n$. Therefore, we can compute $M^n$ by computing $M^2, M^4, \ldots, M^{2^q}$ in turn. In Example 6.24, we have seen that $M^2$ is in $NC^1$, or, is computable by a circuit of depth $O(\log n)$ with $O(n^3)$ gates Thus, $M^{2^q}$ can be computed by a circuit of depth $O(q \log n) = O((\log n)^2)$ and size $O(qn^3) = O(n^3 \log n)$. This means that GAcc $\in NC^2$.  □

To compare the classes in the $NC$ hierarchy with the uniform complexity classes such as $P$ and $LOGSPACE$, we first introduce the notion of logarithmic-space (log-space) reducibility.

**Definition 6.26** *We say set $A$ is many-one reducible to set $B$ in logarithmic space, written as $A \leq_m^{\log} B$, if $A \leq_m B$ via a function $f$ that is computable in logarithmic space.*

**Proposition 6.27** (a) *Log-space many-one reducibility $\leq_m^{\log}$ is transitive.*
(b) *If $A \leq_m^{\log} B$ and $B \in LOGSPACE$, then $A \in LOGSPACE$.*

*Proof.* It suffices to prove that the composition of two log-space computable functions is still log-space computable. Assume that $M_1$ and $M_2$ are two log-space DTMs computing functions $f_1$ and $f_2$, respectively. Let $c_1 \log n$ and $c_2 \log n$ be the space bounds for $M_1$ and $M_2$, respectively. We want to show that the function $g(x) = f_2(f_1(x))$ is computable in logarithmic space.

First, we observe that on input $x$ of length $n$, the machine $M_1$ must halt in $p(n)$ steps for some polynomial $p$ (cf. Theorem 1.28(b)) and, hence, the output $y = f_1(x)$ is of length at most $p(n)$. It follows that $M_2(y)$ uses only space $c_2 \log(p(n)) \le c_3 \log n$ for some $c_3 > 0$. Now, it appears trivial to compute $g(x)$: first simulate $M_1(x)$ to get $y$ and then simulate $M_2(y)$ to get $g(x)$, and the total space used is bounded by $\max\{c_1 \log n, c_3 \log n\}$. However, if we examine this simulation idea more carefully, we see that it does not work— because we also need $p(n)$ cells to store the intermediate string $y$. (Note that the $p(n)$ cells of $y$ were not counted in the space complexity in the computation of $M_1(x)$ since $y$ is the output of $M_1$.)

A simple way to get around this trouble is as follows: We simulate $M_2(y)$ without writing down the whole string $y$ in the work tape. When we need to read the $k$th character of string $y$, we simulate $M_1(x)$ until it prints the $k$th character on the output tape. Note that during the simulation we do not write down all of the first $k$ characters of $y$ on the work tape; instead, we store the counter $k$ in the work tape (in the binary form) and count the number of output characters until it is the $k$th one. The total amount of space we need is $c_1 \log n$ for simulating $M_1(x)$, $c_3 \log n$ for simulating $M_2(y)$ plus $\log(p(n))$ for the storage of $k$, which is still bounded by $O(\log n)$. ∎

Using this stronger type of many-one reducibility, we can compare the complexity of problems within $P$. For instance, the following theorem establishes an *NLOGSPACE*-complete problem with respect to the $\le_m^{\log}$-reducibility.

**Theorem 6.28** GAcc *is* $\le_m^{\log}$*-complete for NLOGSPACE.*

*Proof.* First, we check that the problem GAcc is in *NLOGSPACE*: For a given digraph $G$ and two vertices $s$ and $t$, we can guess a sequence of vertices $s = v_0, v_1, \ldots, v_k = t$ and verify that, for each $i$, $0 \le i \le k-1$, there is an edge in $G$ from $v_i$ to $v_{i+1}$. Note that we do not need to guess the whole sequence of vertices at once; instead, at stage $i$, we can guess one vertex $v_{i+1}$, verify that there is an edge from $v_i$ to $v_{i+1}$, and then discard vertex $v_i$ and go to the next stage. This way, we only need to store two vertices' names at any stage. Since there are $n$ vertices, each vertex name is only of size $\lceil \log n \rceil$. Therefore, the above guess-and-verify algorithm can be implemented by an NTM using space $O(\log n)$.

Next, we show how to reduce a set $A$ in *NLOGSPACE* to GAcc. Assume that $A = L(M)$ for some NTM $M$ with space bound $c \log n$. Then, there is a polynomial function $p$ such that for any input $x$ of length $n$, the computation of $M$ on $x$ contains at most $p(n)$ different configurations. Define a digraph $G$ with $p(n)$ vertices as follows: Each vertex $v_i$ is labeled with one of $p(n)$

configurations $\alpha_i$. (More precisely, the name of $v_i$ is the concatenation of the contents of the work tape configuration and a string $h$ of $\lceil \log n \rceil$ bits indicating the position of the head of the input tape.) There is an edge from vertex $v_i$ to vertex $v_j$ if and only if $\alpha_i \vdash_M \alpha_j$. Also let $s$ be the vertex labeled with the initial configuration of $M(x)$, and $t$ be the vertex labeled with the (unique) halting configuration. Then, apparently, there is a path in $G$ from $s$ to $t$ if and only if $M(x)$ halts. Thus, this is a many-one reduction from $A$ to GACC.

Now we observe that the above reduction can be carried out by a log-space DTM: for any two vertices $v_i$, $v_j$, the question of whether there is an edge from $v_i$ to $v_j$ depends only on whether $\alpha_i \vdash_M \alpha_j$ and so can be determined in logarithmic space. Therefore, the adjacency matrix of $G$ can be generated in logarithmic space. It follows that this reduction is actually a $\leq_m^{\log}$-reduction.

In addition, we observe that this reduction function can be computed by an $NC^1$ circuit: First, from Example 6.23, we can extract the current bit of the input tape of $\alpha_i$ scanned by the tape head by a circuit of depth $O(\log n)$. Then, by the observation made in Cook's Theorem, we can compute all possible successor configurations of $\alpha_i$ by a circuit of depth $O(\log n)$. Finally, it is easy to construct a circuit of depth $O(\log \log n)$ to compare two configurations to determine whether they are equal.                                                      ∎

**Theorem 6.29** *(a) $NC \subseteq P$.*
  *(b) For each $i \geq 1$, $NC^i \subseteq DSPACE((\log n)^i)$.*
  *(c) $NC^1 \subseteq LOGSPACE \subseteq NLOGSPACE \subseteq NC^2$.*

*Proof.* Let $C$ be a Boolean circuit of size $s$, depth $d$, and with its fanin bounded by two. Assume that $C$ is represented by a sequence of quadruples $\langle g, i, j_1, j_2 \rangle$ as described at the beginning of this section. Also assume that the gates in $C$ are numbered in such a way that the inputs to gate $i$ are always from gates $j_1, j_2$ which are greater than $i$. In other words, we number the gates starting from the top to the bottom. We consider a simulation of $C$ by a DTM $M$ as follows: The input to the machine $M$ is the code of $C$ plus the input $x$ to $C$. The machine $M$ simulates circuit $C$ in the depth-first order: it starts from the top gate $i_1$ with the quadruple $\langle g, i_1, j_1, j_2 \rangle$, and then recursively evaluates the output value of the gates $j_1$ and $j_2$. A straightforward implementation of this depth-first simulation requires a stack of size $O(d \log s)$. A more clever implementation, however, can reduce the space requirement down to $O(d + \log s)$: At any moment, the machine $M$ only keeps in its work tape a copy of the current gate (which needs space $\log s$) and a binary string $y$ of length $d$ which indicates which branch of the circuit $C$ the machine $M$ is currently simulating. The machine $M$ uses string $y$ together with the code of $C$ (which does not count toward the space complexity measure) to perform the recursive simulation. The total amount of space needed by $M$ is, hence, $O(d + \log s)$ and the time complexity of $M$ is $O(s^2)$.

Now we check (a) and (b): For (a), assume that $A \in NC$ can be computed by a family of polynomial-size circuits $\{C_n\}$ which can be generated by a DTM

$M_1$ in logarithmic space. Then, the composition of $M_1$ and the machine $M$ above solves the problem $A$ in polynomial time.

(b) Suppose that $A$ is computable by a family of circuits $\{C_n\}$ of depth $c(\log n)^i$ and that $\{C_n\}$ can be generated by a DTM $M_1$ in logarithmic space. Then, we can use the above DTM $M$ to simulate the computation of $C_n$ in space $O((\log n)^i)$, provided that the quadruples of $C_n$ are given to $M$ as the input. Now, if we compose the machine $M_1$ with $M$, using the simulation technique of Proposition 6.27, we can simulate machines $M_1$ and $M$ within space $O((\log n)^i)$.

(c) The fact of $LOGSPACE \subseteq NLOGSPACE$ is trivial. The relation $NC^1 \subseteq LOGSPACE$ follows from part (b) above.

Next, we consider the relation $NLOGSPACE \subseteq NC^2$. We have proved in Theorem 6.28 that the problem GAcc is $\leq_m^{\log}$-complete for $NLOGSPACE$. Furthermore, we observed that the reduction function of Theorem 6.28 from $A \in NLOGSPACE$ to GAcc can actually be computed by $NC^1$ circuits. We have also seen in Example 6.25 that GAcc is computable by $NC^2$ circuits. Since the composition of two $NC^2$ circuits is still an $NC^2$ circuit (the depth only doubles), we can combine the construction of these two families of circuits into a single family of $NC^2$ circuits. So, every $A \in NLOGSPACE$ is actually in $NC^2$.

In fact, it can be proved that $NLOGSPACE \subseteq AC^1$. We leave it as an exercise (Exercise 6.23).                                                      ∎

**Corollary 6.30** (a) Let $i \geq 2$. If $A \leq_m^{\log} B$ and $B \in NC^i$, then $A \in NC^i$.
(b) If $A \leq_m^{\log} B$ and $B \in NC$, then $A \in NC$.

## 6.6  Parity Function

It is not known whether the hierarchy $NC^0 \subseteq AC^0 \subseteq NC^1 \subseteq \cdots$ is properly infinite. Theorem 6.29 actually showed that if $NC^i \neq NC^{i+1}$ for any $i \geq 2$, then $NLOGSPACE \neq P$. Thus, the separation of the $NC$ hierarchy is not easy. In this section, we prove that the first three classes $NC^0$, $AC^0$, and $NC^1$ of the hierarchy are all different. First, we give a simple characterization of $NC^0$.

**Theorem 6.31** *A language $A$ is in nonuniform-$NC^0$ if and only if there is a constant $K$ such that $CS_A(n) \leq K$ for all $n$.*

*Proof.* The backward direction is trivial. In the following, we prove the forward direction. Consider a language $A \in$ *nonuniform-$NC^0$*. Then there is constant $c$ such that each $\chi_{n,A}$ is computed by a circuit of depth at most $c$. Note that there exists only one output node. At most two nodes have out-edges going to the output node. At most four nodes have out-edges going to these two nodes. In general, at most $2^i$ nodes have out-edges going to the nodes of distance exactly $i-1$ from the output node. Therefore, there are at most $1 + 2 + \cdots + 2^{c-1}$ gates in the circuit, that is, the size of the circuit is bounded by the constant $2^c$.                                                                                     ∎

Note that $L = \{1, 11, 111, \ldots\}$ is in $AC^0$ and its characteristic function $\bigwedge_{i=1}^{n} x_i$ cannot be computed by a circuit of constant size and of fanin 2. We obtain the following corollary.

**Corollary 6.32** $NC^0 \neq AC^0$; in fact, $AC^0 \not\subseteq nonuniform\text{-}NC^0$.

Next, we consider the class $NC^1$.

**Theorem 6.33** *A language $A$ is in nonuniform-$NC^1$ if and only if $A$ has a branching program of constant width and polynomial length.*

*Proof.* For the backward direction we note that, using a fixed amount of circuits, one can compose two levels into one in a branching program. Doing this in parallel across the branching program of constant width and polynomial length and repeating it $O(\log n)$ times, we obtain the desired $NC^1$ circuit. For the forward direction we note that, by Barrington's method, a depth-$d$ circuit can be simulated by a branching program of width 5 and length at most $4^d$ (cf. Example 5.48). Hence, an $NC^1$ circuit can be simulated by a branching program of width 5 and polynomial length.                                        ∎

**Theorem 6.34** $AC^0 \neq NC^1$; in fact, $NC^1 \not\subseteq nonuniform\text{-}AC^0$.

We are going to prove this theorem by showing that the parity function is not in $AC^0$. Recall that we already proved that the parity function is in $NC^1$. To show that the parity function is not in $AC^0$, we will work with levelable circuits.[3] A circuit is *levelable* if its nodes can be divided into several levels such that an edge exists only from the $(i+1)$st level to the $i$th level for some $i \geq 1$ and gates at each level satisfy the following conditions:

(i) The first level of a circuit contains only the output node.

(ii) The gates at the same level must be of the same type.

(iii) Two gates at two adjacent levels are always of different types.

(iv) All input nodes (nodes labeled by literals) also lie at the same level.

As a convention, the level with input nodes do not count, and so by the bottom level we mean the lowest level that contains logical gates. It is easy to see that an $AC^0$ circuit can be simulated by an $AC^0$ levelable circuit. So, considering levelable circuits only does not lose the generality.

The proof will be done by induction on the number of levels of a circuit. First, let us look at the circuits of depth 2. A depth-2 circuit is called an AND-OR (or, OR-AND) circuit if the gate at the first level is the AND-gate (or, respectively, the OR-gate).

---

[3]In the rest of this section, all circuits have the unbounded fanin.

**Lemma 6.35** *Every depth-2 circuit computing $p_n$ or $\neg p_n$ is of size $2^{n-1}+1$.*

*Proof.* We first consider an OR-AND circuit $G$ computing $p_n$ (or $\neg p_n$). In an OR-AND circuit, the first level contains only one OR-gate and the second level contains a number of AND-gates. Each AND-gate corresponds to an implicant. Since $p_n$ and $\neg p_n$ take different values as a variable $x_i$ takes different values, every AND-gate of $G$ must have an input from either $x_i$ or $\bar{x}_i$. This means that every AND-gate must have $n$ inputs, representing a unique satisfying Boolean assignment. Moreover, there are exactly $2^{n-1}$ Boolean assignments satisfying $p_n$ (or $\neg p_n$). Therefore, $G$ must have at least $2^{n-1}+1$ gates. For an AND-OR circuit computing $p_n$ or $\neg p_n$, we can, by de Morgan's law, consider a corresponding OR-AND circuit computing $\neg p_n$ or, respectively, $\neg p_n$. ∎

Now we consider the general case of the induction argument. A simple idea for reducing the depth is to switch the bottom two levels of AND-OR (or, OR-AND) subcircuits to equivalent OR-AND (or, respectively, AND-OR) subcircuits and then merge the second and third bottom levels. This idea suggests that we study more carefully the switching between two kinds of depth-2 circuits.

The Switching Lemma below estimates the probability of the existence of such a switching when an AND-OR circuit is restricted by a random partial assignment. In order to state this lemma, we need some more terminologies.

The *bottom fanin* of a circuit is defined to be the maximum fanin of gates at the bottom level. An AND-OR circuit $G$ is said to be able to switch to an OR-AND circuit $G'$ (with bottom fanin $s$) if $G$ and $G'$ compute the same Boolean function. A *restriction* $\rho$ of a circuit $G$ is a partial assignment to variables of $G$ (i.e., $\rho(x) \in \{0, 1, *\}$ with $*$ to mean that $x$ is unchanged). We write $G|_\rho$ to denote the circuit $G$ with the restriction $\rho$. To measure the probability of the switching, each restriction is assigned with a probability. Here, we consider a probability space $R_p$, where $p$ is a real number in $[0, 1]$, in which a randomly chosen restriction $\rho$ from $R_p$ satisfies the following equations for any variable $x_i$:

$$\Pr[\rho(x_i) = 0] = \Pr[\rho(x_i) = 1] = (1-p)/2,$$
$$\Pr[\rho(x_i) = *] = p.$$

In other words, if, for each $j \in \{0, 1, *\}$, the number of variables $x$ such that $\rho(x) = j$ is $n_j$, then $\Pr(\rho) = ((1-p)/2)^{n_0+n_1} p^{n_*}$. The proof of the following lemma is quite involved. We will prove it at the end of this section.

**Lemma 6.36** (Switching Lemma) *Let $G$ be an AND-OR circuit with bottom fanin $t$ and $\rho$ a random restriction from $R_p$, where $t$ is a positive integer and $p$ is a real number in $[0, 1]$. Then, the probability that $G|_\rho$ cannot be switched to an OR-AND circuit with bottom fanin less than $s$ is bounded by $q^s$, where*

$$q = \frac{\alpha p t}{1+p} \quad and \quad \alpha = \frac{2}{\ln \frac{1+\sqrt{5}}{2}}.$$

Let us look at an example of this lemma. Consider $G = (x_1 + x_2)(\bar{x}_1 + \bar{x}_2)$. Let $s = t = 2$. First, we notice that $R_p$ consists of nine elements with probabilities as follows:

$$\Pr[x_1 = 0, x_2 = 0] = \Pr[x_1 = 0, x_2 = 1] = \Pr[x_1 = 1, x_2 = 0]$$

$$= \Pr[x_1 = 1, x_2 = 1] = \left(\frac{1-p}{2}\right)^2,$$

$$\Pr[x_1 = 0, x_2 = *] = \Pr[x_1 = 1, x_2 = *] = \Pr[x_1 = *, x_2 = 0]$$

$$= \Pr[x_1 = *, x_2 = 1] = \frac{p(1-p)}{2},$$

$$\Pr[x_1 = *, x_2 = *] = p^2.$$

Clearly, for any $\rho$ of the first eight restrictions, $G|_\rho$ contains at most one variable and, hence, can be switched to an OR-AND circuit of bottom fanin 1. For the ninth restriction $\rho$, $G|_\rho = G$ cannot be switched to an OR-AND circuit of bottom fanin 1, which happens with probability $p^2 < q$.

Directly using the Switching Lemma for our purpose would meet two troubles: (1) an $AC^0$ circuit has unbounded bottom fanin and so we cannot apply the Switching Lemma; and (2) after the switch, the size of the new circuit may increase too much because the Switching Lemma provides no control over the size of the new circuit.

In order to overcome these two troubles, we consider only the circuits of the following type: (a) the bottom fanin is bounded and fanins of gates in other levels are not bounded; and (b) the number of gates at all levels other than the bottom level are bounded by a polynomial. For an $AC^0$ circuit, we can add one level at the bottom such that gates at the new bottom level have fanin 1, thus satisfying condition (a) above. Now, we are ready to prove the following lower bound result.

**Theorem 6.37** *For sufficiently large $n$, the parity function $p_n$ cannot be computed by a depth-$d$ circuit containing at most $2^{(0.1)n^{1/(d-1)}}$ gates not at the bottom level and with bottom fanin at most $(0.1)n^{1/(d-1)}$. The same result holds for the function $\neg p_n$.*

*Proof.* We prove it by induction on $d$. For $d = 2$, it follows from Lemma 6.35. Next, we consider $d \geq 3$. By way of contradiction, suppose that there exists a depth-$d$ circuit $C$ with bottom fanin $t$ ($\leq (0.1)n^{1/(d-1)}$) and with $g$ ($\leq 2^{(0.1)n^{1/(d-1)}}$) gates not at the bottom level, which computes either $p_n$ or $\neg p_n$. Without loss of generality, assume that its bottom two levels are in the AND-OR form. (Otherwise, we can consider $\neg C$.) Let us first suppose that there is a restriction $\rho$ satisfying the following conditions:

($A$) All of depth-2 AND-OR circuit at the bottom of $C|_\rho$ can be switched to OR-AND circuits with bottom fanin less than $s = t$.

($B$) $C|_\rho$ contains $m$ ($\geq n^{(d-2)/(d-1)}$) remaining variables.

Let $G$ be the circuit obtained from $C|_\rho$ by switching all AND-OR circuits at the bottom two levels into OR-AND circuits with bottom fanin less than $s$.

Since the second bottom level and the third bottom level both consist of OR-gates, we can merge them into one level. Thus, $G$ is actually a depth-$(d-1)$ circuit satisfying the following conditions:

(i) It has $m$ $(\geq n^{(d-2)/(d-1)})$ variables.

(ii) It contains $\leq g$ $(\leq 2^{(0.1)n^{1/(d-1)}} \leq 2^{(0.1)m^{1/(d-2)}})$ gates not at the bottom level.

(iii) Its bottom fanin is less than $s = (0.1)n^{1/(d-1)} \leq (0.1)m^{1/(d-2)}$.

Moreover, $C|_\rho$ computes either $p_m$ or $\neg p_m$, so does $G$, contradicting the inductive hypothesis. Thus, to complete our induction argument, it suffices to prove the existence of a restriction $\rho$ satisfying $(A)$ and $(B)$. In the following, we prove it by a probabilistic argument.

Choose $p = n^{-1/(d-1)}$. Draw $\rho$ randomly from $R_p$. Since $C$ contains $g$ gates not at the bottom level, there are at most $g$ AND-OR circuits at the bottom two levels of $C|_\rho$. By the Switching Lemma, the probability is at most $q^s$ such that an AND-OR circuit cannot be switched to an OR-AND circuit of bottom fanin less than $s$. Thus, the event $(A)$ occurs with probability $\Pr[A] \geq (1 - q^s)^g$. Note that

$$\lim_{n\to\infty} q = \lim_{n\to\infty} \frac{\alpha pt}{1+p} = \frac{\alpha}{10} < \frac{1}{2}.$$

Thus, for sufficiently large $n$, $q < 1/2$. Hence,

$$\begin{aligned} \Pr[A] &\geq (1-q^s)^g \geq 1 - g \cdot q^s \\ &\geq 1 - (2q)^{(0.1)n^{1/(d-1)}} \to 1, \quad \text{as } n \to \infty. \end{aligned}$$

Thus, $\lim_{n\to\infty} \Pr[A] = 1$. Now, we consider the probability $\Pr[B]$. Note that $pn = n^{(d-2)/(d-1)}$. Since $pn$ is the expected number of remaining variables, we have $\Pr[B] > 1/3$ for sufficiently large $n$.[4] Hence, for sufficiently large $n$,

$$\Pr[A \wedge B] = \Pr[A] + \Pr[B] - \Pr[A \vee B] \geq \Pr[A] + \Pr[B] - 1 \geq 1/3.$$

This means that for sufficiently large $n$, there is a restriction $\rho$ such that events $(A)$ and $(B)$ both occur. ∎

Regarding an $AC^0$ circuit of depth $d$ as a depth-$(d+1)$ circuit of the type considered in Theorem 6.37, we obtain the following:

**Corollary 6.38** *For sufficiently large $n$, the parity function $p_n$ cannot be computed by a depth-$d$ circuit of size at most $2^{(0.1)n^{1/d}}$.*

This corollary completes the proof of Theorem 6.34. Actually, we can obtain a better lower bound than that in Corollary 6.38 by using the argument in the proof of Theorem 6.37 once again. We leave it as an exercise.

---

[4] According to DeMoive-Laplace Limit Theorem, if $0 < p < 1$ and $p(1-p)n \to \infty$ as $n \to \infty$, then $\sum_{k \geq pn} \binom{n}{k} p^k (1-p)^{n-k} \to 1/2$ as $n \to \infty$.

The rest of this section is devoted to the proof of the Switching Lemma.

First, we notice that each AND-OR circuit corresponds to a CNF and each OR-AND circuit corresponds to a DNF. Recall, from Chapter 5, that for any Boolean function $f$, $D(f)$ denotes the minimum depth of a decision tree computing $f$. Also recall that in a decision tree computing $f$, each path from root to a leaf with label 1 corresponds to an implicant. Therefore, if $D(f) \leq s$, then $f$ can be computed by an OR-AND circuit of bottom fanin at most $s$. The above observation shows that to prove the Switching Lemma, it suffices to prove that for any AND-OR circuit $G$ with bottom fanin at most $t$,

$$\Pr[D(G|_\rho) \geq s] \leq q^s.$$

The following is a stronger form.

**Lemma 6.39** *Let $G$ be an* AND-OR *circuit with bottom fanin $\leq t$. Let $\rho$ be a random restriction from $R_p$. Then, for any Boolean formula $F$ and any $s > 0$, we have*

$$\Pr[D(G|_\rho) \geq s \mid F|_\rho \equiv 1] \leq q^s,$$

*where*

$$q = \frac{\alpha p t}{1+p} \quad and \quad \alpha = \frac{2}{\ln \frac{1+\sqrt{5}}{2}}.$$

*Proof.* Write $G = \bigwedge_{i=1}^{w} G_i$. We prove the lemma by induction on $w$. For $w = 0$, $G \equiv 1$, hence the inequality holds trivially.

Next, we consider $w \geq 1$. First, note that

$$\Pr[D(G|_\rho) \geq s \mid F|_\rho \equiv 1]$$
$$= \Pr[D(G|_\rho) \geq s \mid F|_\rho \equiv 1 \wedge G_1|_\rho \equiv 1] \cdot \Pr[G_1|_\rho \equiv 1 \mid F|_\rho \equiv 1]$$
$$+ \Pr[D(G|_\rho) \geq s \mid F|_\rho \equiv 1 \wedge G_1|_\rho \not\equiv 1] \cdot \Pr[G_1|_\rho \not\equiv 1 \mid F|_\rho \equiv 1].$$

If $G_1|_\rho \equiv 1$, then $G|_\rho = \bigwedge_{i=2}^{w} G_i|_\rho$. Thus, by the inductive hypothesis, we have

$$\Pr[D(G|_\rho) \geq s \mid F|_\rho \equiv 1 \wedge G_1|_\rho \equiv 1]$$
$$= \Pr[D(G|_\rho) \geq s \mid (F \wedge G_1)|_\rho \equiv 1] \leq q^s.$$

Therefore, it suffices to prove

$$\Pr[D(G|_\rho) \geq s \mid F|_\rho \equiv 1 \text{ and } G_1|_\rho \not\equiv 1] \leq q^s.$$

Let $G_1 = \bigvee_{i \in T} y_i$, where $y_i = x_i$ or $\bar{x}_i$. Let $\rho(J) = *$ denote the event that for every $i \in J$, $\rho(y_i) = *$ and let $\rho(J) = 0$ denote the event that for every $i \in J$, $\rho(y_i) = 0$. Then

$$\Pr[D(G|_\rho) \geq s \mid F|_\rho \equiv 1 \wedge G_1|_\rho \not\equiv 1] = \sum_{J \subseteq T} a_J b_J,$$

where

$$a_J = \Pr[D(G|_\rho) \geq s \mid F|_\rho \equiv 1 \land \rho(J) = * \land \rho(T - J) = 0]$$
$$b_J = \Pr[\rho(J) = * \land \rho(T - J) = 0 \mid F|_\rho \equiv 1 \land G_1|_\rho \not\equiv 1].$$

We claim that for $J \subseteq T$,

$$a_J \leq (2^{|J|} - 1)q^{s-|J|}. \tag{6.7}$$

To prove this, we assume, without loss of generality, that $G_1 = \sum_{i \in T} x_i$. Note that for $\rho$ with $\rho(T - J) = 0$, $D(G|_\rho) \geq s$ implies that there exists a Boolean assignment $\sigma$ on variables in $J$ with $\sigma \not\equiv 0$ such that $D(G|_{\rho\sigma}) \geq s - |J|$. (Indeed, if this is not true, then we can build a decision tree for $G|_\rho$ of depth less than $s$ by concatenating each small tree for $G|_{\rho\sigma}$ to a balanced binary tree of depth $|J|$, each path corresponding to a $\sigma$. This would be a contradiction to the assumption of $D(G|_\rho) \geq s$.) Let $\rho_1$ and $\rho_2$ be assignments on $\{x_i : i \in T\}$ and $\{x_i : i \notin T\}$ induced from $\rho$, respectively. Let $F_J$ be the function $F$ with the restriction that all variables in $T - J$ are assigned 0. Then, it is easy to see that the condition $[F|_\rho \equiv 1 \land \rho(J) = * \land \rho(T - J) = 0]$ is equivalent to the condition $[F_J|_{\rho_2} \equiv 1]$. Thus, by the inductive hypothesis, we have

$$a_J \leq \sum_{\substack{\sigma \in \{0,1\}^{|J|} \\ \sigma \neq 0^{|J|}}} \Pr[D(G|_{\rho\sigma}) \geq s - |J| \mid F|_\rho \equiv 1 \land \rho(J) = * \land \rho(T - J) = 0]$$

$$\leq \sum_{\substack{\sigma \in \{0,1\}^{|J|} \\ \sigma \neq 0^{|J|}}} \Pr[D((G|_{\rho_1\sigma})|_{\rho_2}) \geq s - |J| \mid F_J|_{\rho_2} \equiv 1]$$

$$\leq (2^{|J|} - 1)q^{s-|J|}.$$

(In the above, we represented each assignment $\sigma$ on variables in $J$ by a string in $\{0,1\}^{|J|}$.) The above proves that (6.7) holds.

Next, we want to show

$$\sum_{J \subseteq I \subseteq T} b_I \leq \left(\frac{2p}{1+p}\right)^{|J|}. \tag{6.8}$$

Note that

$$\Pr[F|_\rho \equiv 1 \mid \rho(J) = * \land G_1|_\rho \not\equiv 1] \leq \Pr[F|_\rho \equiv 1 \mid G_1|_\rho \not\equiv 1]$$

since requiring some variables not to be assigned with value 0 or 1 cannot increase the probability that a function is determined. Thus, we have

$$\sum_{J \subseteq I \subseteq T} b_I = \Pr[\rho(J) = * \mid F|_\rho \equiv 1 \land G_1|_\rho \not\equiv 1]$$

$$= \frac{\Pr[\rho(J) = * \land F|_\rho \equiv 1 \land G_1|_\rho \not\equiv 1]}{\Pr[F|_\rho \equiv 1 \land G_1|_\rho \not\equiv 1]}$$

$$
\begin{aligned}
&= \frac{\Pr[F|_\rho \equiv 1 \mid \rho(J) = * \wedge G_1|_\rho \not\equiv 1]}{\Pr[F|_\rho \equiv 1 \mid G_1|_\rho \not\equiv 1]} \cdot \Pr[\rho(J) = * \mid G_1|_\rho \not\equiv 1] \\
&\le \Pr[\rho(J) = * \mid G_1|_\rho \not\equiv 1].
\end{aligned}
$$

Now, observe that the condition $G_1|_\rho \not\equiv 1$ means that $\rho(y_i)$ is either 0 or $*$ for all literals $y_i$ with $i \in J$. Thus, we get

$$
\Pr[\rho(J) = * \mid G_1|_\rho \not\equiv 1] = \left( \frac{p}{p + (1-p)/2} \right)^{|J|} = \left( \frac{2p}{1+p} \right)^{|J|}.
$$

This proves (6.8).

Now, how do we use (6.7) and (6.8) to get an upper bound for $\sum_{J \subseteq T} a_J b_J$? A simple way is just to use $(2p/(1+p))^{|J|}$ as an upper bound for $b_J$. In this way, we obtain

$$
\begin{aligned}
\sum_{J \subseteq T} a_J b_J &\le \sum_{J \subseteq T} (2^{|J|} - 1) q^{s - |J|} \left( \frac{2p}{1+p} \right)^{|J|} \\
&= \sum_{k=0}^{|T|} \binom{|T|}{k} (2^k - 1) q^{s-k} \left( \frac{2p}{1+p} \right)^k \\
&\le \sum_{k=0}^{t} \binom{t}{k} (2^k - 1) q^{s-k} \left( \frac{2p}{1+p} \right)^k \\
&= q^{s-t} \sum_{k=0}^{t} \binom{t}{k} \left[ q^{t-k} \left( \frac{4p}{1+p} \right)^k - q^{t-k} \left( \frac{2p}{1+p} \right)^k \right] \\
&= q^{s-t} \left[ \left( q + \frac{4p}{1+p} \right)^t - \left( q + \frac{2p}{1+p} \right)^t \right].
\end{aligned}
$$

Now, recall that $q = \alpha p t / (1+p)$ and so

$$
q^{-t} \left[ \left( q + \frac{4p}{1+p} \right)^t - \left( q + \frac{2p}{1+p} \right)^t \right] = \left( 1 + \frac{4}{\alpha t} \right)^t - \left( 1 + \frac{2}{\alpha t} \right)^t,
$$

which is close to and less than $e^{4/\alpha} - e^{2/\alpha}$. Also recall that $\alpha$ was defined to be $2/\ln((1 + \sqrt{5})/2)$ which satisfies $e^{4/\alpha} - e^{2/\alpha} - 1 = 0$. Thus, we get

$$
\sum_{J \subseteq T} a_J b_J \le q^s.
$$

A more clever way to use (6.8) employs the inversion formula to get a better bound for $q$ which, in turn, improves the bound of Corollary 6.38. We leave the detail as an exercise to the reader (see Exercises 6.16 and 6.17).  ∎

## 6.7 *P*-Completeness

From the discussion in Section 6.5, we see that the question of whether $NC = P$ is a major open question in complexity theory. Intuitively, this question asks whether all polynomial-time computable problems (representing problems with efficient sequential algorithms) can be recognized by circuits of poly-log depth (representing efficient parallel algorithms). Thus, it is a question of interest not only to the theoreticians who are concerned with the time/space tradeoff, but also to the algorithm designers who are interested in finding efficient parallel algorithms.

Similar to the study of the $P$ versus $NP$ question, we can use the notion of *P*-completeness to demonstrate that some problems are the *hardest* problems that can be solved in polynomial time by sequential machines. If a problem $A$ is shown to be *P*-complete, then it is in $NC$ if and only if $NC = P$. Thus, this fact demonstrates that $A$ is probably not in $NC$ under the (commonly believed) assumption that $NC \neq P$.

**Definition 6.40** *A language $B$ is said to be *P*-complete if it belongs to $P$ and every language $A$ in $P$ is logarithmic-space reducible to $B$.*

Our first *P*-complete problem is the following:

CIRCUIT VALUE PROBLEM (CVP): Given a circuit $C$ and an assignment $\tau$ to its variables, determine whether the circuit value $C|_\tau$ equals one.

**Theorem 6.41** CVP *is P-complete.*

*Proof.* Clearly, the problem CVP is in $P$ since each Boolean operation only takes a constant time to evaluate. To show its *P*-completeness, we need to prove that every language in $P$ is logarithmic-space reducible to CVP. Actually, such a reduction is already done in the proof of Theorem 6.4. We constructed, in that proof, for each polynomial-time DTM $M$ and each input $x$, a circuit $C_x$ (with no variable) such that $M$ accepts $x$ if and only if the output value of circuit $C_x$ equals 1. The only thing we need to check here is that the construction of $C_x$ can be done in space $O(\log n)$. This can be seen from the fact that $C_x$ is constructed through connecting many copies of a constant-size circuit computing a fixed mapping (called $f$ in that proof) which depends only on $M$ and the fact that the connection is not only independent of $x$ but actually independent of $M$. ∎

Some special subproblems of CVP remain *P*-complete. The following are two examples.

MONOTONE CIRCUIT VALUE (MCV): Given a monotone circuit $C$ and an assignment $\tau$ to its variables, determine whether the circuit value $C|_\tau$ equals one.

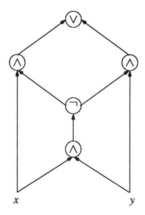

Figure 6.8: A planar circuit for $x \oplus y$.

**Corollary 6.42** MCV *is P-complete.*

*Proof.* We reduce CVP to MCV by pushing down negation gates as described in Section 6.1 and replacing each negation of a variable by a new variable. ∎

A circuit $C$ is a *planar* circuit if it can be laid out in the plane.

PLANAR CIRCUIT VALUE (PCV): Given a planar circuit $C$ and an assignment $\tau$ to its variables, determine whether the circuit value $C|_\tau$ equals one.

**Corollary 6.43** PCV *is P-complete.*

*Proof.* It suffices to explain how to construct a *crosser*, that is, a planar circuit that simulates the crossover of two edges in a circuit. First, the exclusive-or function can be computed by a planar circuit as shown in Figure 6.8. Next, using three exclusive-or gates, we can construct a crosser as shown in Figure 6.9. ∎

The next example is the odd maximum flow problem. A *network* $N = (V, E, s, t, c)$ is a digraph $G = (V, E)$ with two specific vertices, the *source* $s$ and the *sink* $t$, and a positive-integer capacity $c(i, j)$ for each edge $(i, j)$ (see Figure 6.10). The source $s$ has no incoming edge and the sink $t$ has no outgoing edge. A *flow* in the network is an assignment of a nonnegative integer value $f(i, j)$ to each edge $(i, j)$ such that $f(i, j) \le c(i, j)$ and that for each vertex other than the source and the sink the sum of the assigned values for incoming edges equals the sum of the assigned values for outgoing edges. The *value* of a flow $f$ equals the sum of the assigned values $f(s, j)$ for all outgoing edges at the source $s$ (or the sum of the assigned values $f(i, t)$ for all incoming edges at the sink $t$).

ODD MAXIMUM FLOW (OMF): Given a network $N = (V, E, s, t, c)$, determine whether the maximum flow of $N$ is an odd number.

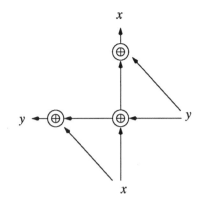

Figure 6.9: A crosser.

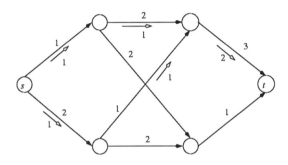

Figure 6.10: A network $N$ and a flow $f$.

**Theorem 6.44** OMF *is P-complete.*

*Proof.* First, we need to show that OMF belongs to $P$. To do so, it suffices to show that the maximum flow of any network $N = (V, E, s, t, c)$ can be computed in polynomial time. There are a number of efficient algorithms for computing the maximum network flow. The reader is referred to any standard algorithm textbook for a complete proof, for instance, Cormen et al. [1990].

Now, we show that the problem MCV is logarithmic-space reducible to the problem OMF. To do so, for each monotone circuit $C$ with variables already assigned values, we construct a network $N = (V, E, s, t, c)$ such that the output value of circuit $C$ is 1 if and only if the value of the maximum flow of the network $N$ is odd.

First, we assume that $C$ has the following two properties: (a) every OR gate has fanout at most two; and (b) the output gate is an OR gate. It is easy to see that a circuit $C$ which does not have these two properties can be transformed to one having the properties by adding some new OR-gates.

Now, for the circuit $C$ with the two properties, we construct a network $N_C$ based on the circuit $C$ as follows:

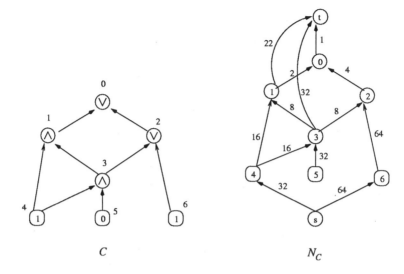

Figure 6.11: Reduction from MCV to OMF.

(1) We label the output gate with 0 and other vertices in the circuit $C$, consecutively, in the topological order starting from 1. By the topological order, we mean that no vertex gets an input from another one with a smaller label.

(2) For any edge coming out of the vertex with label $i$, we assign capacity $2^i$ to the edge.

(3) We add two special vertices, the source $s$ and the sink $t$, to the circuit. We connect the source $s$ to every variable assigned with value 1 and connect the output OR gate and each AND gate to the sink $t$.

(4) For any edge from $s$ to a variable with label $i$, we assign capacity $d \cdot 2^i$ to it, where $d$ is the number of outputs of the variable. For any edge from an AND gate to the sink $t$, its capacity is assigned so that the total capacity of incoming edges to this gate is equal to the total capacity of outgoing edges from this gate. The edge from the output OR gate to $t$ has capacity 1.

It is clear that the construction of the network $N_C$ can be done in logarithmic space. We show an example of the construction in Figure 6.11.

A gate or a variable in the network $N_C$ is said to be *false* if its value is 0 and is said to be *true* if its value is 1. Consider all edges from $s$ and true vertices in $N_C$ to $t$ and false vertices in $N_C$. They form a cut from $s$ to $t$. We claim that the value of the maximum flow of $N_C$ equals the total capacity of this cut. (In Figure 6.11, this value is equal to 33.)

Before proving our claim, we observe that the correctness of the reduction follows from the claim. Indeed, all the values of the capacity are even numbers except one, which is the capacity of the edge from the output gate of $C$ to the sink $t$. If the output value of $C$ is 1, then this edge belongs to the cut and,

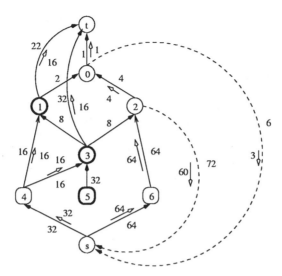

Figure 6.12: The maximum flow for $N_C$ with the imaginary edges. The dark vertices are the *false* ones.

hence, the maximum flow is odd by our claim. If the output value of $C$ is 0, then this edge is not in the cut and the maximum flow is even.

It is left to prove our claim. To do so, we first show that the cut is a minimal cut. We note that a false variable has no incoming edge and a false OR gate cannot have an incoming edge from any true vertex. Thus, only false AND gates can receive incoming edges from any true vertex. For each of these false AND gates, there is an edge directly connecting it to the sink $t$. Moreover, the circuit $C$ is monotone and, hence, every true gate is reachable from a true variable through a path consisting of only true gates. Therefore, removing any edge from the cut will result in a path from $s$ to $t$; that is, the cut is minimal.

Next, we modify the network $N_C$ by adding an imaginary edge from each OR gate to the source $s$, with the capacity equal to the total capacity of the incoming edges to the OR gate (see Figure 6.12). With this modification, it suffices to prove the existence of a flow $f$ on the modified $N_C$ such that all edges in our cut are *full* (i.e., having $f(i,j) = c(i,j)$) and that no edge from $t$ or a false vertex to $s$ or a true vertex is assigned with a positive value. Such a flow $f$ can be found as follows:

(1) For each edge $(s, i)$, set $f(s, i) = c(s, i)$.

(2) For each true variable vertex $i$, set $f(i, j) = c(i, j)$ for all $j$. For each false variable vertex $i$, set $f(i, j) = 0$ for all $j$.

(3) If $f$ has been set for all incoming edges of a gate, then $f$ is set for its outgoing edges as follows: For each true OR gate $i$, set $f(i, j) = c(i, j)$ for $j \neq s$ and $f(i, s) = \sum_k f(k, i) - \sum_{j \neq s} c(i, j)$. For each false OR gate $i$, set

$f(i,j) = 0$ for all $j$. For each true AND gate $i$, set $f(i,j) = c(i,j)$ for all $j$. For each false AND gate $i$, set $f(i,t) = \sum_k f(k,i)$ and $f(i,j) = 0$ for all $j \neq t$.

It is not hard to verify that the flow $f$ is well defined. In addition, $f$ satisfies the following properties:

(a) If $i$ is a false vertex, then for any $j \neq t$, $f(i,j) = 0$.

(b) If $i$ is a true vertex, then for any $j \neq s$, $f(i,j) = c(i,j)$.

From property (b), we know that every edge from $s$ or a true vertex to $t$ or a false vertex is full. From property (a), we see that the flow passing through an imaginary edge does not pass through any false vertex. Therefore, if we remove the imaginary edges and reduce the corresponding flow in the path, then the flow in each edge from $s$ or a true vertex to $t$ or a false vertex is still full. Moreover, property (a) also implies that all edges from false vertices to true vertices have zero flow. Thus, the total value over the cut is equal to the value of the flow. This completes the claim and, hence, our proof. ∎

We remark that there is a rich collection of $P$-complete problems in literature. As we mentioned before, these $P$-complete problems are not in $NC$ unless $P = NC$. In addition to these $P$-complete problems, there also exist a number of problems in $P$ for which one does not know whether it is $P$-complete or in $NC$. The following is an example.

A *comparator* gate has two output wires. In the first one, it outputs $x \wedge y$ and in the second one, it outputs $x \vee y$ where $x$ and $y$ are its two inputs. The problem COMPARATOR CIRCUIT VALUE is the circuit value problem on circuits consisting of comparator gates. Again, it is easy to see that this problem is in $P$. It is, however, not known whether it is $P$-complete. All problems that are logarithmic-space reducible to this problem form a complexity class, called $CC$. It is known that $NC \subseteq CC \subseteq P$, but not known whether the inclusions are proper or not.

## 6.8   Random Circuits and $RNC$

A random circuit is a circuit with random variables. Let $C(x,y)$ be a circuit of variables $x_1, \ldots, x_n$ and $y_1, \ldots, y_m$, where $y_1, \ldots, y_m$ are random variables with the distribution satisfying

$$\Pr[y_i = 0] = \Pr[y_i = 1] = \frac{1}{2}.$$

Let $a$ and $b$ be two real numbers satisfying $0 \leq a < b \leq 1$. We say a language $A \subseteq \{0,1\}^n$ is $(a,b)$-*accepted* by the random circuit $C(x,y)$ with $n$ input variables if for all $x \in A$, we have $\Pr[C(x,y) = 1] \geq b$, and for all $x \in \{0,1\}^n - A$, we have $\Pr[C(x,y) = 1] \leq a$.

In the following, we show that a random circuit can be simulated by a deterministic circuit with an increase of a constant number in depth and a polynomial factor in size. Let $C$ be a random circuit and $C_1, C_2, \ldots, C_k$ be copies of $C$ with different random variables. Define two new random circuits $C^{\wedge k} = \bigwedge_{i=1}^k C_i$ and $C^{\vee k} = \bigvee_{i=1}^k C_i$.

**Lemma 6.45** *Let $C$ be a random circuit. If $C$ $(a,b)$-accepts $A$, then, for each $k \geq 1$, $C^{\wedge k}$ $(a^k, b^k)$-accepts $A$, and $C^{\vee k}$ $(1 - (1-a)^k, 1 - (1-b)^k)$-accepts $A$.*

*Proof.* Obvious. ∎

**Theorem 6.46** *Assume that $A \subseteq \{0,1\}^n$, is $(1/2, (1+(\log n)^{-k})/2)$-accepted by a random depth-$d$ circuit of size $s$ for some $k \geq 1$. Then, for sufficiently large $n$, $A$ is accepted by a depth-$(d + 2k + 2)$ deterministic circuit of size polynomial in $s$ and $n$.*

*Proof.* Assume that the random circuit $C$ has size $s$, depth $d$, and $(1/2, (1 + (\log n)^{-k})/2)$-accepts $A$. Note that

$$\left(\frac{1}{2}\right)^{2 \log n} = n^{-2},$$

and

$$\left(\frac{1}{2}(1 + (\log n)^{-k})\right)^{2 \log n} > n^{-2}(1 + 2(\log n)^{-k+1}).$$

Thus, by Lemma 6.45, we know that circuit $C_1 = C^{\wedge 2 \log n}$ $(n^{-2}, n^{-2}(1 + 2(\log n)^{-k+1}))$-accepts $A$.[5] Next, we consider the circuit $C_2 = C_1^{\vee(n^2 - 1) \ln 2}$. We note that, for sufficiently large $n$,

$$1 - (1 - n^{-2})^{(n^2 - 1) \ln 2} < 1 - e^{-\ln 2} = \frac{1}{2}.$$

Also, note that $(3/2) \ln 2 > 1$. Thus, for sufficiently large $n$,

$$1 - (1 - n^{-2}(1 + 2(\log n)^{-k+1}))^{(n^2 - 1) \ln 2}$$
$$> 1 - \exp(-(\ln 2)(1 + 2(\log n)^{-k+1})(n^2 - 1)n^{-2})$$
$$> 1 - \exp(-(\ln 2)(1 + (3/2)(\log n)^{-k+1}))$$
$$= 1 - \tfrac{1}{2} \cdot \exp(-(3/2)(\ln 2)(\log n)^{-k+1})$$
$$> 1 - \tfrac{1}{2}(1 - (\log n)^{-k+1}) = \tfrac{1}{2}(1 + (\log n)^{-k+1}).$$

Thus, circuit $C_2$ $(1/2, (1 + (\log n)^{-k+1})/2)$-accepts $A$. Note that $C_2$ is a depth-$(d + 2)$ circuit of size $(2s \log n + 1)(n^2 - 1) \ln 2 + 1$, which is bounded by a polynomial in $s$ and $n$.

Now we apply the above construction for $k - 2$ more times, and we obtain a random circuit $Q$ of depth $(d + 2k - 2)$ and of polynomial size that $(1/2, (1 + (\log n)^{-1})/2)$-accepts $A$. Using a similar argument, we can show that the circuit

$$Q' = (((Q^{\wedge 2 \log n})^{\vee 2n^2 \ln n})^{\wedge n^3})^{\vee n}$$

---

[5]Strictly speaking, circuit $C_1$ is $C^{\wedge \lceil 2 \log n \rceil}$. Here, we use $C^{\wedge 2 \log n}$ to simplify the calculation of the accepting and rejecting probabilities.

$(2^{-n-1}, 1 - 2^{-n-1})$-accepts $A$. (See Exercise 6.31.)

Now, we have found a random depth-$(d+2k+2)$ circuit $Q'$ of size polynomial in $s$ and $n$ which $(2^{-n-1}, 1 - 2^{-n-1})$-accepts $A$. Thus, for any input $x$ of length $n$, $Q'$ makes a mistake with probability at most $2^{-n-1}$. Therefore, the probability that $Q'$ makes a mistake on at least one input is at most $2^{-n-1} \cdot 2^n = 1/2$. Thus, there exists an assignment for random variables such that, with this assignment, $Q'$ accepts $A$ without mistake. The theorem is proved by noticing that every random circuit with an assignment for its random variables is simply a deterministic circuit.                                      ∎

The following is an application of this result to the lower bound of the parity function. We say a (deterministic) circuit $C$ *c-approximates* a function $f$ if $C(x) = f(x)$ for at least $c \cdot 2^n$ inputs $x$.

**Theorem 6.47** *There does not exist a polynomial-size, constant-depth circuit* $(3/4)$-*approximating the parity function.*

*Proof.* Suppose otherwise that a polynomial-size, constant-depth circuit $C$ $(3/4)$-approximates the parity function $p_n$. We first convert it into a randomized circuit. Let $y_0, \ldots, y_n$ be $n + 1$ new random variables. Consider circuit

$$Q(x, y) = C(x_1 \oplus y_0 \oplus y_1, x_2 \oplus y_1 \oplus y_2, \ldots, x_n \oplus y_{n-1} \oplus y_n) \oplus y_0 \oplus y_n.$$

Clearly, $Q$ is also a polynomial-size constant-depth circuit. Suppose $y_0, \ldots, y_n$ are random variables with the uniform distribution; then so are $x_1 \oplus y_0 \oplus y_1$, $\ldots, x_n \oplus y_{n-1} \oplus y_n$. Thus, $Q$ makes a mistake within probability $1/4$, that is, $Q$ $(1/4, 3/4)$-computes the parity function. However, this implies, by Theorem 6.46, that there exists a polynomial-size constant-depth deterministic circuit computing the parity function, contradicting Theorem 6.37.                                      ∎

We now define a new complexity class based on random circuits. A language is in $RNC$ if there is a uniform family of $NC$ circuits, with polynomially many random variables, that $(0, 1/2)$-accepts the language. It is obvious that $NC \subseteq RNC$. It is not known whether the inclusion is proper. Note that Theorem 6.46 showed that a language in $RNC$ can be computed by a family of polynomial-size deterministic circuits. However, the proof was nonconstructive and so this family is a *nonuniform-NC* family of circuits. It is not known whether there is a way to find the deterministic circuits uniformly.

It is known that $NC \subseteq P$, but it is not clear whether $RNC \subseteq P$. (It is obvious that $RNC \subseteq R$, where $R$ is the class of languages accepted by polynomial-time probabilistic Turing machines with bounded errors on inputs in the language and no error on inputs not in the language. See Chapter 8.) On the other hand, it is unlikely that $P \subseteq RNC$, and so, a problem is probably not $P$-hard if it is in $RNC$. The following is an example.

PERFECT MATCHING (PM): Given a bipartite graph $G = (A, B, E)$ with vertex sets $A = \{u_1, \ldots, u_n\}$ and $B = \{v_1, \ldots, v_n\}$, determine whether $G$ has a perfect matching, that is, a permutation $\pi$ on $\{1,\ldots,n\}$ such that $\{u_i, v_{\pi(i)})\}$ is an edge of $G$.

It is well known that PM is polynomial-time solvable [Lawler, 1976]. Before we show that PM is in *RNC*, we first prove a probabilistic lemma which will also be used in Chapter 9.

**Lemma 6.48** (Isolation Lemma) *Let $S$ be a finite set $\{x_1,\ldots,x_n\}$ with a weight $w_i$, chosen independently and randomly from 1 to $2n$, assigned to each element $x_i$, for $1 \le i \le n$. Let $\mathcal{F}$ be a nonempty family of subsets of $S$. For each set $T$ in $\mathcal{F}$, its weight is the total weight of elements in $T$. Then, with probability at least $1/2$, there is a unique minimum-weight set $T$ in $\mathcal{F}$.*

*Proof.* We call an element $x_i$ in $S$ *bad* if it is in some minimum-weight subset $T$ in $\mathcal{F}$ but not in all of them. Clearly, a bad element exists if and only if the minimum-weight subset is not unique.

What is the probability that an element is bad? Consider an element $x_i$ in $S$. Suppose that all weights are chosen except the one for element $x_i$. Let $w^*[\bar{x}_i]$ be the smallest total weight among sets $T$ in $\mathcal{F}$ not containing $x_i$ and $w^*[x_i]$ the smallest total weight among all sets $T$ in $\mathcal{F}$ containing $x_i$ but not counting the weight of $x_i$. We claim that $x_i$ is bad if and only if the weight $w_i$ of $x_i$ is chosen to be $w^*[\bar{x}_i] - w^*[x_i]$. (If $w^*[\bar{x}_i] - w^*[x_i] < 1$, then $x_i$ cannot be bad.)

In fact, if $w_i > w^*[\bar{x}_i] - w^*[x_i]$, then the minimum-weight subsets in $\mathcal{F}$ do not contain $x_i$ and hence $x_i$ is not bad. If $w_i < w^*[\bar{x}_i] - w^*[x_i]$, then every minimum-weight subset $T$ in $\mathcal{F}$ contains $x_i$ and, hence, $x_i$ is not bad. If $w_i = w^*[\bar{x}_i] - w^*[x_i]$, then there exists a minimum-weight subset $T_1$ in $\mathcal{F}$ containing $x_i$, as well as a minimum-weight subset $T_0$ in $\mathcal{F}$ not containing $x_i$, and so $x_i$ is bad.

Since there are $2n$ choices for the weight $w_i$ of $x_i$ and at most one choice can make the element $x_i$ bad and since the weights $w^*[\bar{x}_i]$ and $w^*[x_i]$ are chosen independently of weight $w_i$, the probability that $x_i$ is bad is at most $1/(2n)$. Therefore, the probability that no element is bad is at least $1 - n \cdot (1/2n) = 1/2$. This completes the proof of the lemma. ∎

**Theorem 6.49** PM *is in RNC.*

*Proof.* To design a randomized algorithm for PM, we randomly choose weights from 1 to $2|E|$ for edges in the input bipartite graph $G$ where $E$ is the edge set of $G$. The Isolation Lemma states that if there is a perfect matching then, with probability at least $1/2$, the minimum-weight perfect matching is unique.

Now, for each set of randomly chosen weights $\{w_{ij}\}$ on edges $(u_j, v_j) \in E$, compute the determinant of the matrix $A[i,j] = 2^{w_{ij}}$ ($A[i,j] = 0$ if $(u_i, v_j) \notin E$):

$$\det(A) = \sum_{\pi} \sigma(\pi) \prod_{i=1}^{n} A[i, \pi(i)],$$

where $\pi$ ranges over all permutations on $\{1,\ldots,n\}$ and $\sigma(\pi) = 1$ if $\pi$ is an even permutation and it is $-1$ if $\pi$ is an odd permutation. Note that $\prod_{i=1}^{n} A[i,\pi(i)] \neq 0$ if and only if the permutation $\pi$ defines a perfect matching $\{(u_i, v_{\pi(i)}) : i = 1, 2, \ldots, n\}$. Thus, if the determinant of $A$ is nonzero, then $G$ must have a perfect matching. In addition, if with weights $\{w_{ij}\}$ the minimum-weight perfect matching is unique, say its weight is $w^*$, then

$$\det(A) = (1 + 2c)2^{w^*}$$

for some integer $c$; hence the determinant of $A$ must not equal zero. In other words, if $G$ has no perfect matching, then $\Pr[\det(A) \neq 0] = 0$, and if $G$ has a perfect matching, then $\Pr[\det(A) \neq 0] \geq 1/2$.

Finally, we note that it is well known that $\det(A)$ can be computed in $NC^2$ (see, e.g., Cook [1985]). This completes the proof of the theorem. ∎

## Exercises

**6.1** Design a monotone Boolean circuit with at most six gates and with fanin at most three that computes the following Boolean function:

$$f(x_1, x_2, x_3, x_4) = \begin{cases} 0 & \text{if there exists at most one } x_i \text{ equal to 1,} \\ 1 & \text{otherwise.} \end{cases}$$

**6.2** Construct a Boolean circuit computing the function $f_{con}$ on graphs of $n$ vertices (defined in Section 5.2). How many gates do you need?

**6.3** Design a multi-output circuit which combines two levels of a branching program into one.

**6.4** Prove that for a monotone increasing Boolean function $f$,

$$f(x_1, x_2, \ldots, x_n) = f(0, x_2, \ldots, x_n) \vee (x_1 \wedge f(1, x_2, \ldots, x_n)).$$

Using this formula, find an upper bound for $C(f)$ when $f$ is a monotone increasing Boolean function.

**6.5** (a) Prove that any planar circuit that computes the product of two $n$-bit binary integers must have at least $\lfloor n/2 \rfloor^2/162$ vertices.

(b) Prove that any planar circuit that computes the product of two $n \times n$ Boolean matrices must contain at least $cn^4$ vertices for some constant $c$.

**6.6** Show that if $A \in DTIME(t(n))$, then $CS_A(n) = O(t(n)\log t(n))$.

**6.7** Let $x$ and $y$ be Boolean variables, and let $u = \neg\max(x,y)$ and $v = \neg\min(x,y)$. Show that both $\neg x$ and $\neg y$ can be represented monotonely from $x$, $y$, $u$, and $v$ (with only AND and OR operations).

**6.8** The *inverter* is the collection $I_n$ of functions $f_1, \ldots, f_n$ of variables $x_1, \ldots, x_n$ such that $f_1(x_1, \ldots, x_n) = \neg x_1, \ldots, f_n(x_1, \ldots, x_n) = \neg x_n$. Design a circuit with $n$ inputs and $n$ outputs computing $I_n$ with at most $\lceil \log(n+1) \rceil$ negation gates.

**6.9** (a) Show that the problem of multiplying a given sequence of integers is in $NC^1$.

(b) Show that the problem of multiplying a given sequence of permutations in $S_n$ is in *LOGSPACE*.

**6.10** Suppose that $\mathcal{D} = P^{\mathcal{C}}$. Show that $A \in \mathcal{D}/poly$ if and only if there exist $B \in \mathcal{C}$ and $h \in poly$ such that

$$(\forall x, |x| \leq n)\, [x \in A \iff \langle x, h(n) \rangle \in B].$$

**6.11** Prove that $P/poly$ contains nonrecursive sets.

**6.12** Prove Propositions 6.9 and 6.10.

**6.13** Let *log* be the class of all functions $h$ from integers to strings over $\{0, 1\}$ such that $|h(n)| \leq c \cdot \log n$ for some constant $c > 0$. Prove the following statements:
(a) $PSPACE \subseteq P/poly \Rightarrow PSPACE = \Sigma_2^P$.
(b) $PSPACE \subseteq P/log \Rightarrow PSPACE = P$.
(c) $EXP \subseteq PSPACE/log \Rightarrow EXP = PSPACE$.
(d) $NP \subseteq P/log \Rightarrow P = NP$.

**6.14** A function $f$ is called a *slice* function if for some positive integer $k$,

$$f(x) = \begin{cases} 1 & \text{if } x \text{ contains more than } k \text{ 1's,} \\ 0 & \text{if } x \text{ contains less than } k \text{ 1's.} \end{cases}$$

(a) Show that a general circuit computing a slice function can be converted into a monotone circuit by adding only a polynomial number of extra gates.

(b) Show that it is NP-complete to determine whether a given graph $G$ with $2n$ vertices has a clique of size $n$.

(c) Show that it is also *NP*-complete to determine whether a given graph with $2n$ vertices has a clique of size $n$ or at least $\lceil n(2n-1)/2 \rceil + 1$ edges.

(d) Define a sequence of slice functions which represents the above *NP*-complete problem.

**6.15** Let $G$ be an AND-OR circuit with bottom fanin 1 and $\rho$ a random restriction from $R_p$. Prove that the probability that $G|_\rho$ cannot be switched to an OR-AND circuit with bottom fan-in less than $s$ is bounded by $q^s$ where $q = 2p/(1+p)$.

**6.16** We refer to the proof of Lemma 6.39. Prove that there exist values $u_I \geq 0$ for $I \subseteq T$ such that

$$(2^{|J|} - 1)q^{s-|J|} = \sum_{I \subseteq J} u_I.$$

**6.17** Prove that for sufficiently large $n$, the parity function $p_n$ cannot be computed by a depth-$d$ circuit of size at most $2^{0.1(0.33n)^{1/(d-1)}}$.

**6.18** Let $D_1(f)$ denote the maximum size of minterms of a Boolean function $f$. Let $G$ be an AND-OR circuit with bottom fanin $t$. Let $\rho$ be a random restriction from $R_p$. Then for any Boolean formula $F$ and $s > 0$, we have

$$\Pr[D_1(G|_\rho) \geq s \mid F|_\rho \equiv 1] \leq q^s,$$

where $q$ satisfies

$$[(1+p)q + 4p]^t - [(1+p)q + 2p]^t = [(1+p)q]^t.$$

**6.19** Let $\mathcal{B} = \{B_i\}$ be a partition of variables in a circuit. Let $R_{p,\mathcal{B}}^+$ be a probability space of restrictions defined as follows: For each $B_i$, first randomly pick one value, denoted by $s_i$, from $\{*, 0\}$ with probabilities

$$\Pr[s_i = *] = p \quad \text{and} \quad \Pr[s_i = 0] = 1 - p.$$

Then for every $x \in B_i$, define

$$\Pr[\rho(x) = s_i] = p \quad \text{and} \quad \Pr[\rho(x) = 1] = 1 - p.$$

For each $\rho \in R_{p,\mathcal{B}}^+$, define a (deterministic) restriction $g(\rho)$ as follows: For each $B_i$ with $s_i = *$, let $V_i$ be the set of all variables in $B_i$ which are given value $*$. The restriction $g(\rho)$ selects, for each $B_i$, a variable $y_i$ in $V_i$ and assigns value $*$ to $y_i$ and value 1 to all other variables in $V_j$.

Let $G$ be an AND-OR circuit with bottom fanin $\leq t$. Then from a random restriction $\rho$ from $R_{p,\mathcal{B}}^+$, the probability that $G|_{\rho g(\rho)}$ is not equivalent to an OR-AND circuit with bottom fanin $\leq s$ is bounded by $q^s$ where $q < 6pt$.

**6.20** Let $C_d^n$ be the depth-$d$ levelable circuit that has the following structure: (a) $C_d^n$ has a top OR gate, (b) the fanin of all gates is $n$, (c) $C_d^n$ has exactly $n^d$ variables, each variable occurring exactly once in a leaf in the positive form. Let the function computed by $C_d^n$ be $f_d^n$. Prove that, for any $k > 0$ there exists an integer $n_k$ such that if a depth-$d$ circuit $D$ computes the function $f_d^n$, with $n > n_k$, which has bottom fanin $\leq (\log n)^k$ and fanins $\leq 2^{(\log n)^k}$ for all gates at level 1 to $d-2$ (and with fanins of gates at level $d-1$ unrestricted), then $D$ must have size $2^{cn^{1/3}}$ for some constant $c > 0$. [*Hint*: Use the random restriction of the last exercise.]

**6.21** Let $\{C_i\}_{i=1}^n$ be a collection of AND-OR circuits of bottom fanin $\leq t$. Then for a random restriction $\rho$ from $R_{p,\mathcal{B}}^+$, the probability that every $C_i|_{\rho g(\rho)}$ can be switched to an OR-AND circuit with bottom fanin $< s$ is greater than $2/3$, if $n \leq 1/(3q^s)$, where $q < 6pt$.

**6.22** A language $A$ is said to be *NC-reducible* to another language $B$ if there exists a mapping $f$ from strings to strings computable by a uniform family of $NC^i$ circuits for some $i > 0$ such that $x \in A$ if and only if $f(x) \in B$. Prove the following:

(a) If $A$ is $NC$-reducible to $B$ and $B$ is $NC$-reducible to $C$, then $A$ is $NC$-reducible to $C$.

(b) If $A$ is $NC$-reducible to $B$ and $B$ is in $NC$, then $A$ is in $NC$.

**6.23** Prove that $NLOGSPACE \subseteq AC^1$.

**6.24** Prove $NC^1 \subseteq LOGSPACE$ by designing and combining three logarithmic-space algorithms which perform the following tasks:

(a) Generate a circuit in the given uniform family of circuits with logarithmic depth.

(b) Transform the generated circuit to an equivalent circuit with all outdegree one (i.e., a tree-like circuit).

(c) Evaluate the tree-like circuit.

**6.25** A Boolean circuit is called an AM2 *circuit* if, in the circuit, (a) on any path from an input to the output the gates are required to alternate between OR and AND gates, (b) inputs are connected to only OR gates, (c) the output also comes from an OR gate, and (d) every gate has fanout at most two. The problem AM2 CIRCUIT VALUE is defined as follows: Given an AM2 circuit and an assignment to its variables, determine whether the circuit value equals one. Prove that AM2 CIRCUIT VALUE is $P$-complete.

**6.26** An NAND gate can be obtained by putting a NOT gate at output of an AND gate. An NAND circuit is a Boolean circuit consisting of only NAND gates. The NAND CIRCUIT VALUE problem is to determine, for a given NAND circuit and a given assignment to its variables, whether the circuit value equals one. Prove that NAND CIRCUIT VALUE is $P$-complete.

**6.27** Show that the following problem is $P$-complete: Given a network $N = (V, E, s, t, c)$ and a nonnegative integer $k$, determine whether the maximum flow of $N$ has value $\geq k$.

**6.28** Prove than if a bipartite graph has a perfect matching, then a perfect matching can be found in $RNC$.

**6.29** Show that if $NC^i = NC^{i+1}$, then $NC^i = NC$.

**6.30** An $r$-circuit is a circuit with AND and OR gates of fanin two and at most $r$ negation gates. Let $C^r(f)$ denote the size of the smallest $r$-circuit computing $f$. Show that for any Boolean function $f$ of $n$ variables,

$$C^{\lceil \log(1+n) \rceil}(f) \leq 2C(f) + O(n \log n).$$

**6.31** Prove that if a random circuit $Q$ $(1/2, (1 + (\log n)^{-1})/2)$-accepts $A$ then $Q' = (((Q^{\wedge 2 \log n})^{\vee 2n^2 \ln n})^{\wedge n^3})^{\vee n}$ $(2^{-n-1}, 1 - 2^{-n-1})$-accepts $A$.

**6.32** Show that if $n \leq 2^{r+1} - 1$, then $C^r(p_n) = O(n \log n)$, where $p_n$ is the parity function of $n$ variables.

**6.33** Show that the problem of finding the determinant of an integer matrix is in $NC^2$.

## Historical Notes

Shannon [1949] first considered the circuit size of a Boolean function as a measure of its computational complexity. Lupanov [1958] showed the general upper bound $C(f) \leq (1 + O((\frac{1}{\sqrt{n}}))\frac{2^n}{n}$. Fischer [1974] proved the existence of Boolean functions with circuit complexity $\Omega(2^n/n)$. An explicit language with circuit complexity $3n + o(n)$ was found by N. Blum [1984]. See Dunne [1988] and Wegener [1987] for more complete treatment on Boolean circuits. Karp and Lipton [1982] studied the class of sets having polynomial-size circuits and obtained a number of results on *P/poly*, including Theorem 6.11 and Exercise 6.13. Sparse sets are first studied in Lynch [1975] and Berman and Hartmanis [1977]. They are further studied in Chapter 7. For the sets with succinct descriptions (such as sets with polynomial-size circuits), see the survey paper of Balcázar et al. [1997].

Razborov [1985a] obtained the first superpolynomial lower bound $n^{\Omega(\log n)}$ for the size of monotone circuits computing the function CLIQUE$_{k,n}$. The exponential lower bound of Theorem 6.15 was given by Alon and Boppana [1987]. Razborov [1985b] showed that the perfect matching problem for bipartite graphs also requires monotone circuits of superpolynomial size. Tardos [1988] showed an exponential lower bound for another problem in $P$. Berkowitz [1982] introduced slice functions and showed that circuits computing slice functions can be converted into monotone circuits with a polynomial number of additional gates (see Exercise 6.14). He also pointed out that there exist NP-complete slice functions. Yao [1994] adapted Razborov's technique to show that the connectivity requires depth $\Omega((\log n)^{3/2}/\log \log n)$ in a monotone circuit. Goldman and Hastad [1995] improved the bound to $\Omega((\log n)^2/\log \log n)$. There are also many references on circuits with limited number of negation gates; see, for instance, Beals et al. [1995].

The class $NC$ was first defined by Nick Pippenger [1979]; indeed, the term $NC$ was coined by Cook to mean "Nick's Class." There are, in literature, a number of models for parallel computation, including PRAMs [Fortune and Wyllie, 1978], SIMDAGs [Goldschlager, 1978], Alternating Turing machines [Chandra et al., 1981], and vector machines [Pratt et al., 1974]. See, for instance, van Emde Boas [1990] for a survey of these models and their relations to the Boolean circuit model. The fact that *LOGSPACE* and *NLOGSPACE* are in $NC^2$ was proved by Borodin [1977]. Exercise 6.9 is from Beame et al. [1984], Cook and McKenzie [1987], and Immerman and Landau [1989].

The fact that the parity function is not in $AC^0$ was first proved by Furst et al. [1984] and Ajtai [1983]. They proved that the parity function requires constant-depth circuits of super-polynomial size. Yao [1985] pushed up this lower bound to exponential size. Hastad [1986a, 1986b] simplified Yao's proof and proved the first Switching Lemma. Exercises 6.18 and 6.19 are from Hastad [1986a] and [1986b]. The idea of using the inversion formula in the proof of the Switching Lemma was initially proposed by Moran [1987]. However, in his proof, he used $D_1(G)$ instead of $D(G)$, so that the inequality corresponding to (6.7) was incorrect. The current idea of using $D(G)$ was suggested by Cai [1991]. For other applications of the random restriction and the Switching Lemma, see, for instance, Aiello et al. [1990] and Ko [1989b]. Exercise 6.21 is from Ko [1989a]. Smolensky [1987] extended Yao's result to modulo functions. Fortnow and Laplante [1995] (see also section 6.12 of Li and Vitanyi [1997]) offer a different proof of Hastad's Lemma using the notion of Kolmogorov complexity.

$P$-complete problems were first studied by Cook [1973b]. The $P$-completeness of the the problem CVP was established by Ladner [1975b]. The circuit value problem for monotone and planar circuits was studied by Goldschlager [1977]. Exercise 6.5 is from Lipton and Tarjan [1980]. The problem OMF is proved to be $P$-complete by Goldschlager et al. [1982]. Greenlaw et al. [1995] included a list of hundreds of $P$-complete problems. Mayr and Subramanian [1989] defined the class $CC$ and studied its properties.

Random circuits were studied by Ajtai and Ben-Or [1984]. Feather [1984] showed that finding the size of the maximum matching is in $RNC$. Karp et al. [1986] found the first $RNC$ algorithm for finding the maximum matching. Our proof using the Isolation Lemma is from Mulmuley et al. [1987]. The fact that the determinant of an integer matrix can be computed in $NC$ was first proved by Csansky [1976] and Borodin, Cook and Pippenger [1983]. See Cook [1985] for more discussions about the problem of computing integer determinants.

# 7

# *Polynomial-Time Isomorphism*

Two decision problems are said to be *polynomial-time isomorphic* if there is a membership-preserving one-to-one correspondence between the instances of the two problems which is both polynomial-time computable and polynomial-time invertible. Thus, this is a stronger notion than the notion of equivalence under the polynomial-time many-one reductions. It has, however, been observed that many reductions constructed between *natural NP*-complete problems are not only many-one reductions but can also be easily converted to polynomial-time isomorphisms. This observation leads to the question of whether two *NP*-complete problems must be polynomial-time isomorphic. Indeed, if this is the case then, within a polynomial factor, all *NP*-complete problems are actually identical, and any heuristics for a single *NP*-complete problem works for all of them. In this chapter we study this question and the related questions about the isomorphism of *EXP*-complete and *P*-complete problems.

## 7.1  Polynomial-Time Isomorphism

Let us start with a simple example.

**Example 7.1** Recall the decision problems IS and CLIQUE from Chapter 2:

> INDEPENDENT SET (IS): Given a graph $G$ and an integer $k > 0$, determine whether $G$ has an independent set of size at least $k$.

CLIQUE: Given a graph $G$ and an integer $k > 0$, determine whether $G$ has a clique of size at least $k$.

For any graph $G = (V, E)$, let the *complement graph* $G^c$ of $G$ be the graph that has the same vertex set $V$ but has the edge set $E' = \{\{u, v\} : u, v \in G, \{u, v\} \notin E\}$. It is clear that a subset $A$ of vertices is an independent set of a graph $G = (V, E)$ if and only if the complement graph $G^c$ of $G$ has a clique on subset $A$. Thus, a graph $G$ has an independent set of size $k$ if and only if its complement $G^c$ has a clique of size $k$. That is, there is a *bijection* (i.e., a one-to-one and onto mapping) $f$ between the inputs of these problems: $f(G, k) = (G^c, k)$ such that IS $\leq_m^P$ CLIQUE via $f$, and CLIQUE $\leq_m^P$ IS via $f^{-1}$. In other words, these two problems are not only reducible to each other by the polynomial-time many-one reductions, but are also equivalent to each other through polynomial-time computable bijections. We say that they are *polynomial-time isomorphic* to each other.                                    □

We now give the formal definition of the notion of polynomial-time isomorphism.

**Definition 7.2** *A language $A \subseteq \Sigma^*$ is said to be* polynomial-time isomorphic *(or, simply, p-isomorphic) to a language $B \subseteq \Gamma^*$, written as $A \equiv^P B$, if there exists a bijection $f : \Sigma^* \to \Gamma^*$ such that both $f$ and $f^{-1}$ are polynomial time computable and for all $x \in \Sigma^*$, $x \in A$ if and only if $f(x) \in B$.*

Clearly, if $A \equiv^P B$ and $B \equiv^P C$, then $A \equiv^P C$. For example, in addition to the problems IS and CLIQUE of Example 7.1, consider a third problem, VC:

VERTEX COVER (VC): Given a graph $G$ and an integer $k > 0$, determine whether $G$ has a vertex cover of size at most $k$.

It is clear that a subset $A$ of vertices is an independent set of a graph $G = (V, E)$ if and only if the complement set $V - A$ is a vertex cover of $G$. This suggests a $p$-isomorphism between IS and VC via the bijection $g(G, k) = (G, |V| - k)$. Combining the function $g$ with the bijection $f$ of Example 7.1, we get a $p$-isomorphism between CLIQUE and VC: $g \circ f^{-1}$.

Polynomial-time isomorphism builds a close relation between decision problems. However, it is not so easy to construct a polynomial-time isomorphism directly even if such a relation exists. This situation is similar to what happens in set theory, when one tries to prove that two infinite sets have the same cardinality. Instead of constructing a bijection between the two sets explicitly, one often employs the Bernstein-Schröder-Cantor theorem (see any textbook in set theory) to show the existence of such a bijection.

**Theorem 7.3** (Bernstein-Schröder-Cantor Theorem) *If there exist a one-to-one mapping from a set $A$ to another set $B$ and a one-to-one mapping from $B$ to $A$, then there is a bijection between $A$ and $B$.*

For any two *NP*-complete sets $A$ and $B$, we have $A \leq_m^P B$ and $B \leq_m^P A$. Does it follow, similar to the Bernstein-Schröder-Cantor theorem, that $A \equiv^P B$? The answer is a qualified yes, if stronger reductions between the two sets exist.

First, like the Bernstein-Schröder-Cantor theorem, we need to consider one-to-one reductions instead of many-one reductions: A language $A \subseteq \Sigma^*$ is said to be *polynomial-time one-one reducible* to $B \subseteq \Gamma^*$, denoted by $A \leq_1^P B$, if there exists a polynomial-time computable one-to-one mapping from $\Sigma^*$ to $\Gamma^*$ such that $x \in A$ if and only if $f(x) \in B$.[1]

Second, we want to have a bijection between $A$ and $B$ such that both the bijection and its inverse are polynomial-time computable. This requires that the one-to-one reduction be polynomial-time invertible. A one-to-one function $f : \Sigma^* \to \Gamma^*$ is said to be *polynomial-time invertible* if for any $y \in \Gamma^*$, there exists a polynomial-time algorithm which can either tell NO when $y \notin f(\Sigma^*)$ or find an $x \in \Sigma^*$ such that $f(x) = y$ when $y \in f(\Sigma^*)$. If $A \leq_1^P B$ via a polynomial-time invertible function $f$, then we say $A$ is *polynomial-time invertibly reducible* to $B$ and we write $A \leq_{inv}^P B$ to denote this fact.

The third requirement is that the reductions must be length-increasing. A reduction $f : \Sigma^* \to \Gamma^*$ is said to be *length-increasing* if for any $x \in \Sigma^*$, $|f(x)| > |x|$. We denote by $A \leq_{inv,li}^P B$ if the polynomial-time one-one reduction from $A$ to $B$ is polynomial-time invertible and length-increasing.

With these three additional requirements on reductions, we have a polynomial-time version of the Bernstein-Schröder-Cantor theorem:

**Theorem 7.4** *If $A \leq_{inv,li}^P B$ and $B \leq_{inv,li}^P A$, then $A \equiv^P B$.*

*Proof.* Suppose $A \subseteq \Sigma^*$ and $B \subseteq \Gamma^*$. Assume that $A \leq_{inv,li}^P B$ via $f : \Sigma^* \to \Gamma^*$ and $B \leq_{inv,li}^P A$ via $g : \Gamma^* \to \Sigma^*$. Define

$$
\begin{aligned}
R_1 &= \{(g \circ f)^k(x) : k \geq 0, x \in \Sigma^* - g(\Gamma^*)\} \\
R_2 &= \{(g \circ f)^h \circ g(y) : h \geq 0, y \in \Gamma^* - f(\Sigma^*)\}.
\end{aligned}
$$

We claim that $R_1 \cap R_2 = \emptyset$ and $R_1 \cup R_2 = \Sigma^*$.

To show $R_1 \cap R_2 = \emptyset$, we suppose, for the sake of contradiction, that

$$(g \circ f)^k(x) = (g \circ f)^h \circ g(y)$$

for some $x \in \Sigma^* - g(\Gamma^*)$, $y \in \Gamma^* - f(\Sigma^*)$, and some $k$ and $h \geq 0$. Note that $f$ and $g$ are one-to-one mappings. If $k \leq h$, then

$$x = (g \circ f)^{h-k} \circ g(y),$$

---

[1] Note that we are working, in this chapter, with a number of different types of reducibility, but when we use the term *completeness*, unless the type of reducibility is specified, we refer to completeness with respect to $\leq_m^P$-reductions.

contradicting the assumption of $x \in \Sigma^* - g(\Gamma^*)$. If $k > h$, then

$$f \circ (g \circ f)^{k-h-1}(x) = y,$$

contradicting the assumption of $y \in \Gamma^* - f(\Sigma^*)$.

To show $R_1 \cup R_2 = \Sigma^*$, we look at the following algorithm, which tells, for a given string $x \in \Sigma^*$, whether $x \in R_1$ or $x \in R_2$.

(1) Set $z := x$.

(2) Determine whether $z \in g(\Gamma^*)$.

(3) If the answer to step (2) is NO, then we conclude that $x \in R_1$.

(4) If the answer to step (2) is YES, then we set $z := g^{-1}(z)$ and determine whether $z \in f(\Sigma^*)$.

(5) If the answer to step (4) is NO, then we conclude that $x \in R_2$; otherwise, we set $z := f^{-1}(z)$ and go back to step (2).

First, we check the correctness of the algorithm. It is easy to see that if the algorithm stops then the conclusion of either $x \in R_1$ or $x \in R_2$ is correct. For instance, if the algorithm halts in step (5), concluding $x \in R_2$, then it must be the case that $z \notin f(\Sigma^*)$. From the algorithm, it is clear that $z$ is an *ancestor* of $x$ in the sense that $z = g^{-1} \circ (f^{-1} \circ g^{-1})^k(x)$ for some $k \geq 0$. It follows that $x = (g \circ f)^k \circ g(z)$ and so $x \in R_2$. In addition, we observe that each time we go through steps (2) to (5), we reduce the length of $z$ by at least two letters, because both $f$ and $g$ are length-increasing. Therefore, the algorithm can pass through the loop from step (2) back to step (2) at most $\lfloor |x|/2 \rfloor$ times before it halts. We conclude that for each $x \in \Sigma^*$, it must be in either $R_1$ or in $R_2$.

For the time complexity of the algorithm, we observe that since $f$ and $g$ are polynomial-time invertible, steps (2) and (4) can be done in time polynomial in $|x|$. Together with the earlier observation that the algorithm can pass through the loop at most $\lfloor |x|/2 \rfloor$ times, we see that the algorithm is a polynomial-time algorithm to determine whether a given $x \in \Sigma^*$ belongs to $R_1$ or $R_2$.

Next, define

$$
\begin{aligned}
S_1 &= \{f \circ (g \circ f)^k(x) : k \geq 0, x \in \Sigma^* - g(\Gamma^*)\}, \\
S_2 &= \{(f \circ g)^h(y) : h \geq 0, y \in \Gamma^* - f(\Sigma^*)\}.
\end{aligned}
$$

By a similar argument, we can show that $S_1 \cap S_2 = \emptyset$, $S_1 \cup S_2 = \Gamma^*$, and there exists a polynomial-time algorithm to determine whether a given $y \in \Gamma^*$ belongs to $S_1$ or $S_2$.

Clearly, $f(R_1) = S_1$ and $g^{-1}(R_2) = S_2$. Define

$$
\phi(x) = \begin{cases} f(x) & \text{if } x \in R_1, \\ g^{-1}(x) & \text{if } x \in R_2. \end{cases}
$$

Then, we can see that $\phi$ is one-to-one and onto and

$$
\phi^{-1}(y) = \begin{cases} f^{-1}(y) & \text{if } y \in S_1, \\ g(y) & \text{if } y \in S_2. \end{cases}
$$

Therefore, both $\phi$ and $\phi^{-1}$ are polynomial-time computable. Moreover, for $x \in R_1$, $x \in A$ if and only if $\phi(x) = f(x) \in B$, and for $x \in R_2$, $x \in A$ if and only if $\phi(x) = g^{-1}(x) \in B$. Thus, $\phi$ is a $p$-isomorphism between $A$ and $B$. ∎

We show a simple example of a $p$-isomorphism using the above theorem. More examples will be demonstrated using the padding technique of the next section.

**Example 7.5** Consider the problems SAT and 3-SAT. First, we note that 3-SAT $\leq_{inv,li}^{P}$ SAT via the simple function that maps each 3CNF formula $F$ to the 3CNF $F' = F \cdot F$. To show SAT $\leq_{inv,li}^{P}$ 3-SAT, recall the reduction $g$ from SAT to 3-SAT in Chapter 2. Note that in that reduction, $g$ maps each clause $C$ of the input instance $F$ for SAT to one or more 3-literal clauses with no repeated clauses. We can modify the reduction to add two copies of a dummy clause with three new variables at the end of each $g(C)$. Then, the reduction becomes polynomial-time invertible: We can separate each $g(C)$ from the image $g(F)$ using the duplicated clauses as the separators. Then, from the mapping $g$, it is clear that we can convert $g(C)$ back to $C$ easily. It is also clear that this modified reduction is length-increasing and so we get SAT $\leq_{inv,li}^{P}$ 3-SAT. By Theorem 7.4, we conclude that SAT and 3-SAT are $p$-isomorphic. □

## 7.2 Paddability

We have seen, in the last section, an example of improving the $\leq_m^P$-reductions to $\leq_{inv,li}^P$-reductions, which allows us to get the $p$-isomorphism between two $NP$-complete problems. This technique is, however, still not simple enough since, to prove that $n$ $NP$-complete problems are all $p$-isomorphic to each other, we need to construct $n(n-1)$ $\leq_{inv,li}^P$-reductions. In this section, we present a more general technique which further simplifies the task.

**Definition 7.6** (a) A language $A \subseteq \Sigma^*$ is said to be paddable *if there exists a polynomial-time computable and polynomial-time invertible function* $p : \Sigma^* \times \Sigma^* \to \Sigma^*$ *such that* $p(x,y) \in A$ *if and only if* $x \in A$. *The function* $p$ *is called a* padding function *of* $A$.

(b) A paddable language $A$ is said to be length-increasingly paddable *if it has a length-increasing padding function* $p$ *(i.e., padding function* $p$ *satisfies an additional condition of* $|p(x,y)| > |x| + |y|$*).*

There is a seemingly weaker but actually equivalent definition for paddability. We state it in the following proposition.

**Proposition 7.7** *A language* $A \subseteq \Sigma^*$ *is paddable if and only if there exist polynomial-time computable functions* $p : \Sigma^* \times \Sigma^* \to \Sigma^*$ *and* $q : \Sigma^* \to \Sigma^*$ *such that*

(a) $p(x,y) \in A$ *if and only if* $x \in A$, *and*

(b) $q \circ p(x,y) = y$.

*Proof.* If $A$ is paddable, then it is clear that such functions $p$ and $q$ exist. Now, suppose such functions $p$ and $q$ exist. Define $p_A : \Sigma^* \times \Sigma^* \to \Sigma^*$ by

$$p_A(x, y) = p(x, \langle x, y \rangle),$$

where $\langle \cdot, \cdot \rangle$ is a pairing function. Then $p_A$ is a padding function for $A$.    ∎

Now, we show an application of paddability to the construction of polynomial-time invertible reductions.

**Theorem 7.8** *If $A \leq_m^P B$ for a paddable language $B$, then $A \leq_{inv}^P B$. Moreover, if the padding function of $B$ is length-increasing, then $A \leq_{inv,li}^P B$.*

*Proof.* Let $A \subseteq \Sigma^*$ and $B \subseteq \Gamma^*$. We first define a length-increasing encoding function $e$ from $\Sigma^*$ to $\Gamma^*$: if $|\Gamma|^{c-1} < |\Sigma| \leq |\Gamma|^c$ for some integer $c \geq 1$, then $e$ maps the $k$th symbol of $\Sigma$ to the $k$th string in $\Gamma^c$. Suppose $p$ is a padding function of $B$ and $A \leq_m^P B$ via $f$. Then, we define $g : \Sigma^* \to \Gamma^*$ by

$$g(x) = p(f(x), e(x)).$$

Clearly, $x \in A$ if and only if $f(x) \in B$ if and only if $p(f(x), e(x)) \in B$. Moreover, since $p$ and $e$ are polynomial-time invertible, $g$ is one-to-one and polynomial-time invertible. Thus, $g$ is a $\leq_{inv}^P$-reduction from $A$ to $B$. Furthermore, if $p$ is length-increasing, then we have

$$|g(x)| > |f(x)| + |e(x)| \geq |x|.$$

That is, $g$ is also length-increasing.    ∎

**Corollary 7.9** *Suppose that $A$ and $B$ are length-increasingly paddable. If $A \leq_m^P B$ and $B \leq_m^P A$, then $A \equiv^P B$.*

*Proof.* It follows immediately from Theorems 7.4 and 7.8.    ∎

**Example 7.10** *The problem* SAT *is length-increasingly paddable.* First, let $b : \Sigma^* \to \{0, 1\}^*$ be a length-increasing binary encoding of strings over the input alphabet $\Sigma$; that is, if $\Sigma = \{s_1, s_2, \ldots, s_k\}$ and $m = \lceil \log_2 k \rceil$, then we use the $i$th string $t_i$ in $\{0, 1\}^m$ to encode the symbol $s_i$, for $1 \leq i \leq k$, and define $b(s_{i_1} s_{i_2} \cdots s_{i_r}) = t_{i_1} t_{i_2} \cdots t_{i_r}$. Now, we define a padding function $p : \Sigma^* \times \Sigma^* \to \Sigma^*$ as follows: Assume that $x$ is a CNF formula with $m$ clauses $C_1, C_2, \ldots, C_m$ over $n$ variables $v_1, \ldots, v_n$. Let $y$ be any string in $\Sigma^*$. The output $p(x, y)$ is a CNF formula of $m + |b(y)| + 2$ clauses $C_1, \ldots, C_m, C_{m+1}, \ldots, C_{m+|b(y)|+2}$ over $n + |b(y)| + 1$ variables $v_1, \ldots, v_n, v_{n+1}, \ldots, v_{n+|b(y)|+1}$, with $C_{m+1} = C_{m+2} = v_{n+1}$ and

$$C_{m+2+j} = \begin{cases} v_{n+j+1} & \text{if the } j\text{th bit of } b(y) = 1, \\ \bar{v}_{n+j+1} & \text{if the } j\text{th bit of } b(y) = 0, \end{cases}$$

for $j = 1, \ldots, |b(y)|$.

Clearly, $p$ is polynomial-time computable and $x \in$ SAT if and only if $p(x, y) \in$ SAT. Furthermore, it is easy to see that $p$ is polynomial-time invertible: for a CNF formula $C_1 C_2 \cdots C_r$, we look for the maximum $m < r - 1$ such that $C_{m+1} = C_{m+2}$; if it exists, then $C_1 C_2 \cdots C_m$ is $x$ and $C_{m+3} \cdots C_r$ encodes $y$. ☐

Corollary 7.9 and Example 7.10 imply that, in order to prove a problem being $p$-isomorphic to SAT, we only need to show that this problem is length-increasingly paddable. An interesting observation is that almost all known *natural NP*-complete problems are actually length-increasingly paddable. Thus, these problems are actually all $p$-isomorphic to each other. This observation leads to the following conjecture:

**Berman-Hartmanis Isomorphism Conjecture:** All *NP*-complete sets are $p$-isomorphic to each other.

Since there are thousands of natural *NP*-complete problems and we cannot demonstrate their paddability one by one here, we only study a few popular ones.

**Example 7.11** *The problem* VC *is length-increasingly paddable.* Each input instance of VC consists of a graph $G$ and a positive integer $k$. Assume that the input alphabet is $\Sigma = \{0, 1\}$. Now, we define a padding function $p$ as follows: For each input instance $x = \langle G, k \rangle$ and for each binary string $y$, we first add a path of $2|y| + 1$ new vertices $(v_0, u_1, v_1, \ldots, u_{|y|}, v_{|y|})$ to graph $G$ and connect $v_0$ to all vertices in $G$. Then, for each $i = 1, 2, \ldots, |y|$ such that the $i$th symbol of $y$ is 1, add a new vertex $w_i$ and a new edge $\{u_i, w_i\}$ into the graph. Let the resulting graph be $G_y$ (see Figure 7.1). We define $p(x, y) = \langle G_y, k + |y| + 1 \rangle$.

It is easy to see that a graph $G$ has a vertex cover of size $k$ if and only if $G_y$ has a vertex cover of size $k + |y| + 1$ for any $y$. It is also easy to see that $p$ is both polynomial-time computable and polynomial-time invertible. (The extra path encoding $y$ is easy to recognize as $v_0$ is the vertex of the highest degree if $G$ has at least three vertices.) Thus, $p$ is a padding function for VC. In addition, $p$ is length-increasing as long as the pairing function is length-increasing. Thus, $p$ is a length-increasing padding function for VC. ☐

**Example 7.12** *The problem* HC *is length-increasingly paddable.* An instance of HC is a graph $G$ of $n$ vertices. For any binary string $y$, we construct a graph $G_y$ as follows: Choose a vertex $v^*$ in $G$. Construct a path of $n + |y| + 1$ new vertices $(u_1, \ldots, u_n, v_1, \ldots, v_{|y|}, v_{|y|+1})$. For each $i = 1, 2, \ldots, |y|$, if the $i$th symbol of $y$ is 1, then add a new vertex $w_i$ and two new edges $(v_i, w_i)$ and $(w_i, v_{i+1})$. Finally, connect $u_1$ to $v^*$ and $v_{|y|+1}$ to all neighbors of $v^*$ in $G$. (See Figure 7.2.)

It is easy to check that a graph $G$ has a Hamiltonian circuit if and only if $G_y$ has a Hamiltonian circuit. It is also easy to recover the graph $G$ from $G_y$. So, HC is paddable. ☐

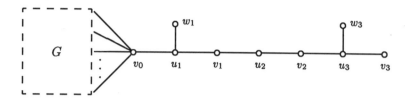

Figure 7.1: Padding for VC, with $y = 101$.

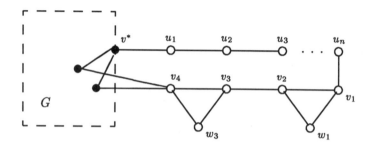

Figure 7.2: Padding for HC, with $y = 101$.

**Example 7.13** *The problem* PARTITION *is length-increasingly paddable.* Recall that the problem PARTITION asks, for a given list of $n$ positive integers $(a_1, a_2, \ldots, a_n)$, whether these numbers can be partitioned into two groups such that the sums of integers in two groups are equal. Let $s = a_1 + a_2 + \cdots + a_n$. For any binary string $y$, we first add $2n$ copies of $s$ to the input list and then for each $i$, $1 \leq i \leq |y|$, such that the $i$th symbol of $y$ is 1, we add two copies of $i \cdot s$ to the list. Clearly, the new list of integers can be partitioned into two equal subsums if and only if the original list can. The polynomial-time invertibility of this padding function is easy to check too. $\Box$

In the above examples, the padding functions are obviously length-increasing. In general, though, this property of the padding function is not necessary, as shown in the following improvement over Corollary 7.9.

**Theorem 7.14** *Suppose $A$ and $B$ are paddable sets. If $A \leq_m^P B$ and $B \leq_m^P A$, then $A \equiv^P B$.*

*Proof.* Let $A \subseteq \Sigma^*$ and $B \subseteq \Gamma^*$. Assume that both $\Sigma$ and $\Gamma$ contain the symbol 0. Suppose $p_A$ and $p_B$ are padding functions for $A$ and $B$, respectively. By Theorem 7.8, $A \leq_{inv}^P B$ and $B \leq_{inv}^P A$. Assume that $A \leq_{inv}^P B$ via function

$f_1$ and $B \leq_{inv}^P A$ via function $g_1$. Define $f : \Sigma^* \to \Gamma^*$ by

$$f(x) = \begin{cases} p_B(f_1(x), 0^{2k}) & \text{if } x = p_A(z, 0^k) \text{ for some } z \in \Sigma^* \\ & \text{and some } k \geq 1 \\ p_B(f_1(x), 0^3) & \text{otherwise,} \end{cases}$$

and $g : \Gamma^* \to \Sigma^*$ by

$$g(y) = \begin{cases} p_A(g_1(y), 0^{2k}) & \text{if } y = p_B(z, 0^k) \text{ for some } z \in \Gamma^* \\ & \text{and some } k \geq 1 \\ p_A(g_1(y), 0^3) & \text{otherwise.} \end{cases}$$

Clearly, $A \leq_{inv}^P B$ via $f$ and $B \leq_{inv}^P A$ via $g$.

Similarly to the proof of Theorem 7.4, we define

$$R_1 = \{(g \circ f)^k(x) : k \geq 0, x \in \Sigma^* - g(\Gamma^*)\},$$
$$R_2 = \{(g \circ f)^h \circ g(y) : h \geq 0, y \in \Gamma^* - f(\Sigma^*)\},$$

and we can prove that $R_1 \cap R_2 = \emptyset$. To prove $R_1 \cup R_2 = \Sigma^*$, we will also use the algorithm given in the proof of Theorem 7.4 and show that the algorithm runs in finitely many steps for any given input $x$. However, the argument will be different, since $f$ and $g$ may not be length-increasing.

By Proposition 7.7 there exist polynomial-time computable functions $q_A : \Sigma^* \to \Sigma^*$ and $q_B : \Gamma^* \to \Gamma^*$ such that

$$q_A(p_A(x, z)) = z, \quad q_B(p_B(y, u)) = u.$$

Suppose $4^{n-1} < |q_A(x)| \leq 4^n$. We will prove by induction on $n$ that the algorithm can pass through step (2) for at most $n + 1$ times.

For $n = 0$, note that $|q_A(g(y))| \geq 2$ for every $y \in \Gamma^*$. Thus, $|q_A(x)| = 4^0 = 1$ implies $x \notin g(\Gamma^*)$. This implies that the algorithm passes through step (2) only once and determines that $x \in R_1$.

Now, assume $n > 0$. If $x \notin g(\Gamma^*)$, then the algorithm passes through step (2) once and determines $x \in R_1$. Thus, we may assume $x \in g(\Gamma^*)$. By the definition of $g$, we see that either

(i) $x = p_A(g_1(y), 0^3)$, or

(ii) $x = p_A(g_1(y), 0^{2k})$ and $y = p_B(z, 0^k)$ for some $z \in \Gamma^*$ and some $k \geq 1$.

In Case (i), we know, from the definition of $g$, that $y = g^{-1}(x) \neq p_B(z, 0^k)$ for any $k > 0$ and $z \in \Gamma^*$. It follows that $y \notin f(\Sigma^*)$. Thus, the algorithm passes through step (2) only once and concludes that $x \in R_2$.

In Case (ii), we have $q_A(x) = 0^{2k}$ and $q_B(g^{-1}(x)) = 0^k$. Applying a similar argument to $y = g^{-1}(x)$, we can see that if $k$ is odd or $y \notin f(\Sigma^*)$ then $x$ can be determined to belong to $R_1$ or $R_2$ with the algorithm passing through step (2) at most twice. Now, we may assume $k$ is even and $y \in f(\Sigma^*)$. By the definition of $f$, we have $q_A(f^{-1}(g^{-1}(x))) = 0^{k/2}$. Note that $2k = |q_A(x)| \leq 4^n$. Thus, $|q_A(f^{-1}(g^{-1}(x)))| = k/2 \leq 4^{n-1}$. By the inductive hypothesis, with

$x' = f^{-1}(g^{-1}(x))$ as the input, the algorithm passes through step (2) at most $n$ times. Therefore, with input $x$, the algorithm passes through step (2) at most $n + 1$ times.

We complete the proof of this theorem by noting that the remaining part is the same as the corresponding part in the proof of Theorem 7.4. ∎

It would be interesting to know, without knowing that $A$ or $B$ is paddable, whether $A \leq_{inv}^{P} B$ and $B \leq_{inv}^{P} A$ are sufficient for $A \equiv^{P} B$. It is an open question.

## 7.3 Density of $NP$-Complete Sets

We established, in the last section, a proof technique with which we were able to show that many natural $NP$-complete languages are $p$-isomorphic to SAT. This proof technique is, however, not strong enough to settle the Berman-Hartmanis conjecture, as we do not know whether every $NP$-complete set is paddable. Indeed, we observe that if all $NP$-complete sets are $p$-isomorphic to each other, then $P \neq NP$. To see this, we observe that a finite set cannot be $p$-isomorphic to an infinite set. Thus, if the Berman-Hartmanis conjecture is true, then every nonempty finite set is in $P$ but is not $NP$-complete and hence is a witness for $P \neq NP$. From this simple observation, we see that it is unlikely that we can prove the Berman-Hartmanis conjecture without a major breakthrough. Nevertheless, one might still gain some insight into the structure of the $NP$-complete problems through the study of this conjecture. In this section, we investigate the density of $NP$-complete problems and provide more support for the conjecture.

The *density* (or, the *census function*) of a language $A$ is a function $C_A : N \to N$ defined by $C_A(n) = |\{x \in A : |x| \leq n\}|$; that is, $C_A(n) = |A_{\leq n}|$. Recall that a language $A$ is *sparse* if there exists a polynomial $q$ such that for every $n \in N$, $C_A(n) \leq q(n)$.

**Theorem 7.15** *If a set $A$ is paddable then it is not sparse.*

*Proof.* Let $A \subseteq \Sigma^*$ be a paddable set with the padding function $p : \Sigma^* \times \Sigma^* \to \Sigma^*$. For the sake of contradiction, suppose $A$ is sparse; that is, suppose that there exists a polynomial function $q$ such that for all $n \in N$, $C_A(n) \leq q(n)$. Since $p$ is polynomial-time computable, there exists a polynomial $r$ such that

$$|p(x, y)| \leq r(|x| + |y|).$$

For a fixed $x \in A$, since $p$ is one-to-one, we have

$$2^n \leq |\{p(x, y) : |y| \leq n\}| \leq C_A(r(|x| + n)) \leq q(r(|x| + n)).$$

Thus,

$$2^n / q(r(|x| + n)) \leq 1.$$

Letting $n$ approach the infinity, we obtain a contradiction. ∎

**Corollary 7.16** *If the Berman-Hartmanis conjecture is true, then all NP-complete sets and all coNP-complete sets are nonsparse.*

*Proof.* Suppose that the Berman-Hartmanis conjecture is true. Then, every NP-complete set $A$ is $p$-isomorphic to SAT. Let $A \equiv^P$ SAT via $f$ and $p_{\text{SAT}}$ be a padding function for SAT. Define

$$p_A(x, y) = f^{-1}(p_{\text{SAT}}(f(x), y)).$$

Then,

$$
\begin{aligned}
x \in A &\iff f(x) \in \text{SAT} \\
&\iff p_{\text{SAT}}(f(x), y) \in \text{SAT} \\
&\iff f^{-1}(p_{\text{SAT}}(f(x), y)) \in A.
\end{aligned}
$$

Moreover, since $p_{\text{SAT}}$ and $f$ are polynomial-time computable and polynomial-time invertible, $p_A$ is polynomial-time computable and polynomial-time invertible. Therefore, $p_A$ is a padding function for $A$; that is, $A$ is paddable. By Theorem 7.15, $A$ is nonsparse.

Note that each *coNP*-complete set is the complement of an *NP*-complete set and the complement of a paddable set is paddable. Hence, every *coNP*-complete set is nonsparse. ∎

In the following, we will show that the conclusions of Corollary 7.16 hold true even without the assumption of the Berman-Hartmanis conjecture. Thus, this provides a side evidence for the conjecture.

The proofs of these results use the self-reducibility property of *NP*-complete sets. We first investigate the notion of self-reducibility. This notion can be explained easily by the example of SAT.

**Example 7.17** Let $F$ be a quantifier-free Boolean formula with variables $x_1, x_2, \ldots, x_m$. Then, it is clear that $F \in$ SAT if and only if either $F|_{x_1=0} \in$ SAT or $F|_{x_1=1} \in$ SAT. In other words, the membership question about $F$ is *reduced* to the membership questions about $F|_{x_1=0}$ and $F|_{x_1=1}$. It is called a *self-reduction* because the new questions are also about memberships in SAT.

We note that the membership questions about formulas $F|_{x_1=0}$ and $F|_{x_1=1}$ can be further reduced to formulas $F|_{x_1=a, x_2=b}$ recursively, where $a$ and $b$ range over $\{0, 1\}$. These self-reductions provide a *self-reducing tree* for $F$: The root of the tree is $F$. Each node of the tree is a formula $F|_{x_1=a_1, \ldots, x_k=a_k}$ for some $a_1, \ldots, a_k \in \{0, 1\}$ and $1 \leq k \leq m$. If $k = m$, then this formula is a leaf. If $k < m$, then this node has two children: $F|_{x_1=a_1, \ldots, x_k=a_k, x_{k+1}=0}$ and $F|_{x_1=a_1, \ldots, x_k=a_k, x_{k+1}=1}$. This tree has the following properties:

(i) The depth of the tree is $m$.
(ii) $F \in$ SAT if and only if there is a path from the root to a leaf of which all nodes belong to SAT. □

In general, different types of self-reducibility can be defined from different types of reducibility. For instance, the above self-reducibility of the problem

SAT is a special case of $d$-self-reducibility. In order to define them, we first define a class of polynomial-time truth-table reducibilities (cf. Exercise 2.14). Recall that for a set $B$, $\chi_B$ is its characteristic function: $\chi_B(y) = 1$ if $y \in B$ and $\chi_B(y) = 0$ if $y \notin B$.

**Definition 7.18** *(a) A set $A \subseteq \Sigma^*$ is* polynomial-time truth-table reducible *to a set $B \subseteq \Gamma^*$, written as $A \leq_{tt}^P B$, if there exist two polynomial-time TMs $M_1$ and $M_2$ such that for any $x \in \Sigma^*$, $M_1$ outputs a list $\langle y_1, y_2, \dots, y_k \rangle$ of strings in $\Gamma^*$, and $M_2$ accepts $\langle x, \langle \chi_B(y_1), \chi_B(y_2), \dots, \chi_B(y_k) \rangle \rangle$ if and only if $x \in A$. The list $\langle y_1, y_2, \dots, y_k \rangle$ produced by $M_1(x)$ is called a* tt-condition.

*(b) Let $k$ be a fixed positive integer. A set $A$ is* polynomial-time $k$-truth-table reducible *to a set $B$, written as $A \leq_{ktt}^P B$, if $A \leq_{tt}^P B$ and the tt-conditions in the reduction have at most $k$ strings.*

*(c) A set $A$ is* polynomial-time conjunctive truth-table reducible *to a set $B$, written as $A \leq_c^P B$, if $A \leq_{tt}^P B$ via TMs $M_1$ and $M_2$ such that $M_2$ accepts $\langle x, \langle \tau_1, \dots, \tau_k \rangle \rangle$ if and only if $\tau_1 \tau_2 \cdots \tau_k = 1$ (i.e., $x \in A$ if and only if all $y_i$ in the tt-condition are in $B$).*

*(d) A set $A$ is* polynomial-time disjunctive truth-table reducible *to a set $B$, written as $A \leq_d^P B$, if $A \leq_{tt}^P B$ via TMs $M_1$ and $M_2$ such that $M_2$ accepts $\langle x, \langle \tau_1, \dots, \tau_k \rangle \rangle$ if and only if $\tau_1 + \tau_2 + \cdots + \tau_k = 1$ (i.e., $x \in A$ if and only if at least one $y_i$ in the tt-condition is in $B$).*

For instance, for $A \subseteq \{0, 1\}^*$, let

$$C = A \text{ join } \overline{A} = \{x1 : x \in A\} \cup \{y0 : y \notin A\}.$$

Then, $C \leq_{1tt}^P A$ via the following TMs $M_1$ and $M_2$: $M_1(za) = z$ for any $z \in \{0, 1\}^*$ and any $a \in \{0, 1\}$, and $M_2$ accepts $\langle x, \tau \rangle$, where $\tau \in \{0, 1\}$, if and only if the rightmost symbol of $x$ is equal to $\tau$.

It is easy to see that $A \leq_m^P B$ implies $A \leq_{1tt}^P B$ and $A \leq_{ktt}^P B$ implies $A \leq_{(k+1)tt}^P B$ for all $k \geq 1$.

Now, for each type of the reducibilities $\leq_{tt}^P$, $\leq_c^P$, and $\leq_d^P$, we define a corresponding notion of self-reducibility. We say a partial ordering $\prec$ on $\Sigma^*$ is *polynomially related* if there exists a polynomial $p$ such that (i) the question of whether $x \prec y$ can be decided in $p(|x| + |y|)$ moves; (ii) if $x \prec y$ then $|x| \leq p(|y|)$; and (iii) the length of any $\prec$-decreasing chain is shorter than $p(n)$, where $n$ is the length of its maximal element.

**Definition 7.19** *(a) A set $A$ is* tt-self-reducible *if there exist a polynomially related ordering $\prec$ on $\Sigma^*$ and two polynomial-time TMs $M_1$ and $M_2$, such that the following hold:*

*(i) For all inputs $x$, $M_1(x)$ generates a list $\langle y_1, y_2, \dots, y_k \rangle$ with the property that $y_j \prec x$ for all $1 \leq j \leq k$ (if $x$ is a minimal element of $\prec$, then $M_1(x)$ generates an empty list); and*

*(ii) $x \in A$ if and only if $M_2$ accepts $\langle x, \langle \chi_A(y_1), \chi_A(y_2), \dots, \chi_A(y_k) \rangle \rangle$, where $\langle y_1, \dots, y_k \rangle$ is the list generated by $M_1$.*

*(b) A set A is c-self-reducible if it is tt-self-reducible and the machine $M_2$ has the property that it accepts $\langle x, \langle \tau_1, \ldots, \tau_k \rangle \rangle$ if and only if $\tau_1 \tau_2 \cdots \tau_k = 1$.*

*(c) A set A is d-self-reducible if it is tt-self-reducible and the machine $M_2$ has the property that it accepts $\langle x, \langle \tau_1, \ldots, \tau_k \rangle \rangle$ if and only if $\tau_1 + \tau_2 + \cdots + \tau_k = 1$.*

Example 7.17 showed that SAT is $d$-self-reducible, with the natural ordering $\prec$: if $G$ is a formula obtained from $F$ by substituting constants 0 or 1 for some variables in $F$ then $G \prec F$. Let us look at another example.

**Example 7.20** Consider the problem QUANTIFIED BOOLEAN FORMULA (QBF). An instance of QBF is a quantified Boolean formula $F$. Assume that $F$ is in the normal form:

$$F = (Q_1 x_1)(Q_2 x_2) \cdots (Q_m x_m)\, G(x_1, x_2, \ldots, x_m),$$

where $Q_1, Q_2, \ldots, Q_m \in \{\exists, \forall\}$ and $G$ is a quantifier-free formula. Then, $F$ is reduced to

$$F_0 = (Q_2 x_2) \cdots (Q_m x_m)\, G(0, x_2, \ldots, x_m)$$

and

$$F_1 = (Q_2 x_2) \cdots (Q_m x_m)\, G(1, x_2, \ldots, x_m):$$

if $Q_1 = \exists$ then $F \in$ QBF if and only if $F_0 \in$ QBF or $F_1 \in$ QBF, and if $Q_1 = \forall$ then $F \in$ QBF if and only if $F_0 \in$ QBF and $F_1 \in$ QBF. Using a similar ordering $\prec$ as the one for SAT, we see that QBF is $tt$-self-reducible.

The self-reducing tree of an instance $F$ of QBF is similar to the one of SAT, except that property (ii) of Example 7.17 does not hold. Instead, for each $F \in$ QBF, there is a subtree that witnesses the fact that $F \in$ QBF. The root of the subtree is the root $F$. For each internal node $F'$ in the subtree, if the first quantifier of $F'$ is $\exists$ then one of its two children is in the subtree, and if the first quantifier of $F'$ is $\forall$ then both children are in the subtree. The leaves of the subtree are all true. $\qquad\square$

As shown in the above examples, each instance of a self-reducible set has a self-reducing tree. The structure of these self-reducing trees provides an upper bound for the complexity of the set.

**Proposition 7.21** *(a) A d-self-reducible set A must be in NP.*
*(b) A c-self-reducible set A must be in coNP.*
*(c) A tt-self-reducible set A must be in PSPACE.*

*Proof.* (a) The self-reducing tree of an instance $x$ of a $d$-self-reducible set $A$ has the structure just like that of SAT (with the depth of the tree of $x$ bounded by $p(|x|)$ for some polynomial $p$). Note that from each node $y$ of the tree, we can generate its children by machine $M_1$. Property (ii) of the self-reducing tree shows us that in order to decide whether $x \in A$, we only need to guess, nondeterministically, a path of the tree, starting from the root to the leaf, and verify that the leaf node $z$ is in $A$ (by running $M_2(z)$, as the $tt$-condition for $z$ is empty). Property (i) of the self-reducing tree guarantees that this path is of length at most $p(|x|)$. This shows that $A \in NP$.

(b) The argument is symmetric to part (a).

(c) The self-reducing tree of an instance $x$ of a $tt$-self-reducible set $A$ also has the depth bounded by $p(|x|)$ for some polynomial $p$. To decide whether the root $x$ is in $A$, we can evaluate the tree by a depth-first search. Since the depth of the tree is bounded by $p(|x|)$ and the fanout of each node is also bounded by $q(|x|)$ for some polynomial $q$, the evaluation only needs space $O(p(n) \cdot q(n))$ in order to keep track of the current simulation path (cf. Theorem 3.17).  ∎

Are the converses of the above proposition true? That is, are all sets in $PSPACE$ $tt$-self-reducible and are all sets in $NP$ $d$-self-reducible? These are interesting open questions. In the following, we are going to show that a $tt$-self-reducible set cannot be sparse unless it is in $P$ (Theorem 7.22(b)). Therefore, unless all sets in $PSPACE - P$ are nonsparse, the answer to the first question is *no*. By a simple padding argument, it can be proved that if $EXPSPACE \neq EXP$ then there exists a tally set in $PSPACE - P$ (cf. Exercise 1.30). Thus, if $EXPSPACE \neq EXP$, then the answer to the first question is *no*.

Similarly, Theorem 7.22(a) shows that a $c$-self-reducible set cannot be sparse. So, it follows that the answer to the second question is also NO unless there is a co-sparse set in $NP - P$. Or, equivalently, if $NEXP \neq EXP$, then the answer to the second question is also em no.

Are all $PSPACE$-complete sets $tt$-self-reducible, and are all $NP$-complete sets $d$-self-reducible? We do not know the precise answer. Note that it is easy to show that $d$-self-reducibility is preserved by polynomial-time isomorphisms (see Exercise 7.6). Thus, if the Berman-Hartmanis conjecture and its $PSPACE$ version are true, then the answers to both of the above questions are yes. On the other hand, this property is, like paddability, easy to verify for *natural* problems but appears more difficult to prove for all complete problems by a general method.

We now prove our first main result of this section: $tt$- and $c$-self-reducible sets cannot be sparse, unless they are in $P$. It follows that $PSPACE$- and $coNP$-complete sets are not sparse unless $PSPACE$ or $NP$ collapses to $P$, respectively.

**Theorem 7.22** (a) *If $A$ is $c$-self-reducible and $A$ is $\leq_m^P$-reducible to a sparse set, then $A \in P$.*

(b) *If $A$ is $tt$-self-reducible and both $A$ and $\overline{A}$ are $\leq_m^P$-reducible to sparse sets, then $A \in P$.*

*Proof.* (a) The main idea of the proof is to perform a depth-first tree-pruning procedure on the self-reducing tree of the instance $x$. Note that for the $c$-self-reducible set $A$, the self-reducing tree of $x$ has the following properties:

(i) The depth of the tree is bounded by $p(|x|)$ for some polynomial $p$.

(ii) $x \notin A$ if and only if there exists a path from root $x$ to a leaf of which all nodes are not in $A$.

Suppose we perform a depth-first search of the tree. Then, we can tell whether $x \in A$ as follows: (1) if some node (e.g., a leaf) in the tree is found to be not in $A$, then the search halts and output NO, or (2) if all nodes in the tree (or, all leaves in the tree) are found to be in $A$, then output YES. This search procedure, however, does not work in polynomial time as the tree may have as many as $2^{p(n)}$ nodes. So, the tree-pruning procedure is to be applied.

Assume that $A \leq_m^P S$ for some sparse set $S$ via a function $g$. Also assume that $C_S(n) \leq q(n)$ for some polynomial $q$. For each node $y$ of the tree, let us attach the string $g(y)$ to the node (called the *label* of $y$). Then, we note that the tree has the following additional properties:

(iii) If two nodes $y$ and $z$ have the same label, then both are in $A$ or both are in $\overline{A}$.

(iv) If $x \in A$, then *many* nodes have the same label.

The property (iv) above comes from the observation that, if $|x| = n$ and $x \in A$, then there are $2^{p(n)}$ many nodes in the self-reducing tree of $x$, all in $A$, but only $q(p(n))$ many possible labels for them. By property (iii), we can prune the subtree whose root is $z$ if $g(z) = g(y)$ for some $y$ which has been visited before in the depth-first search. Then, a careful calculation shows that the total number of nodes to be visited in the depth-first search can be reduced to a polynomially bounded size.

We now describe the procedure more precisely. Let $M_1$ and $M_2$ be the TMs which witness that $A$ is $c$-self-reducible. The procedure includes a recursive procedure *search* and a main procedure *decide*.

> Procedure *decide*$(x)$;
>
> > Set *Label* $:= \emptyset$;
> > Call *search*$(x)$;
> > Output YES and halt.

> Procedure *search*$(y)$;
>
> > (1) If $y$ is a leaf (i.e., $y$ is a minimal element with respect to $\prec$), then run $M_2(y)$ to decide whether $y \in A$;
> >
> > > (1.1) If $M_2(y)$ rejects, then output NO and halt;
> > > (1.2) If $M_2(y)$ accepts, then let *Label* $:=$ *Label* $\cup \{g(y)\}$ and return to the calling procedure.
> >
> > (2) If $y$ is not a leaf and $g(y) \in$ *Label* then return.
> > (3) If $y$ is not a leaf and $g(y) \notin$ *Label* then do the following:
> >
> > > (3.1) Run $M_1(y)$ to get the $tt$-condition $\langle y_1, y_2, \ldots, y_k \rangle$;
> > > (3.2) For $i := 1$ to $k$, call *search*$(y_i)$;
> > > (3.3) Let *Label* $:=$ *Label* $\cup \{g(y)\}$ and return.

We claim that the procedure *decide*$(x)$ works correctly. First, it is clear that it works correctly when it outputs NO, because that happens only when it finds a leaf $y$ which is not in $A$. When it outputs YES, we can prove, by induction on the order of the depth-first search that all strings added to *Label* are in $S$ and all nodes that have been examined are in $A$ and, hence, the YES answer is also correct:

(1) If *search*$(y)$ returns to the calling procedure at step (1.2), then $y$ must be in $A$, as $M_2(y)$ accepts. It follows that $g(y) \in S$.

(2) If *search*$(y)$ returns to the calling procedure at step (2), then $g(y) \in$ *Label*. By the inductive hypothesis, *Label* $\subseteq S$ and, hence, $y \in A$.

(3) If *search*$(y)$ returns to the calling procedure at step (3.3), then, in step (3.2), $y_i$ returns to the procedure *search*$(y)$ for all $i = 1, \ldots, k$. By the inductive hypothesis, $y_i \in A$ for all $i = 1, \ldots, k$. Since $A$ is $c$-self-reducible, we know that $y \in A$ and $g(y) \in S$.

The only thing left to verify is that the procedure *decide* works in polynomial time. Assume that $|x| = n$. Recall that the self-reducing tree has height $\leq p(n)$ and $C_S(n) \leq q(n)$. Also assume that $|y| \leq r(n)$ for all nodes $y$ in the self-reducing tree of $x$. We claim that the above procedure visits at most $p(n) + p(n) \cdot q(r(n))$ interior nodes (of the pruned tree) and, hence, the procedure halts by visiting at most a polynomially bounded number of nodes.

To check this claim, suppose that $y$ and $z$ are two interior nodes of the pruned tree with the same label. Then, they must have the ancestor–descendant relation. (Otherwise, one node would be visited first and its label be put in the set *Label*; and the second node would become a leaf.) Thus, for each string $u \in S$, there are at most $p(n)$ interior nodes having $u$ as its label. Since there are at most $q(r(n))$ many strings in the set *Label*, there are at most $p(n) \cdot q(r(n))$ interior nodes which are in $A$. So, if $x \in A$, then the procedure *decide*$(x)$ visits at most $p(n) \cdot q(r(n))$ interior nodes. If $x \notin A$, then there is at most one more path (of $\leq p(n)$ interior nodes) which is to be searched before the procedure concludes with answer NO. This completes the proof of part (a).

(b) Assume that $A$ is *tt*-self-reducible via TMs $M_1$ and $M_2$, and that $A \leq_m^P S_1$ via $f_1$ and $\overline{A} \leq_m^P S_0$ via $f_0$ for some sparse sets $S_1$ and $S_0$. As in part (a), we label each node $y$ of the self-reducing tree of $x$ by $\langle f_0(y), f_1(y) \rangle$, and keep a set *Label*$_0$ for labels $f_0(y)$ with $y \in \overline{A}$ and a set *Label*$_1$ for labels $f_1(y)$ with $y \in A$. If a node $y$ is found to have $f_0(y) \in$ *Label*$_0$ or $f_1(y) \in$ *Label*$_1$, then we prune the subtree below node $y$. We leave the detail of the procedure and its correctness as an exercise (Exercise 7.8). ∎

**Corollary 7.23** (a) *A sparse set cannot be $\leq_m^P$-hard for PSPACE, unless $P = PSPACE$.*

(b) *A sparse set cannot be $\leq_m^P$-hard for coNP unless $P = NP$.*

(c) *A tally set cannot be $\leq_m^P$-hard for NP unless $P = NP$.*

*Proof.* (a) If a sparse set $S$ is $\leq_m^P$-hard for *PSPACE*, then QBF $\leq_m^P S$ and $\overline{\text{QBF}} \leq_m^P S$, as both QBF and $\overline{\text{QBF}}$ are in *PSPACE*. Thus, by Theorem 7.22(b), QBF $\in P$ and so $P = PSPACE$.

(b) If a sparse set $S$ is $\leq_m^P$-hard for *coNP*, then $\overline{\text{SAT}} \leq_m^P S$ and so $\overline{\text{SAT}} \in P$, which implies $P = NP$.

(c) If a tally set $T \subseteq \{1\}^*$ is $\leq_m^P$-hard for *NP*, then the set $\{1\}^* - T$ is a sparse $\leq_m^P$-hard set for *coNP*. ∎

Unfortunately, the results of Theorem 7.22 are not known to hold for $d$-self-reducible sets, because the depth-first pruning procedure does not work for $d$-self-reducing trees. However, we can still prove that, assuming $P \neq NP$, $NP$-complete sets are not sparse by a different pruning technique on the $d$-self-reducing trees of SAT.

**Theorem 7.24** *If $P \neq NP$, then a sparse set cannot be $\leq_m^P$-hard for NP.*

*Proof.* Consider the $NP$-complete set SAT. For each Boolean formula $F$, let $m_F$ denote the number of variables in $F$. For convenience, for any formula $F$ with variables $v_1, v_2, \ldots, v_{m_F}$, and for any $a_1, a_2, \ldots, a_k \in \{0, 1\}$, with $k \leq m_F$, we write $F|_{a_1 a_2 \cdots a_k}$ to denote $F|_{v_1 = a_1, v_2 = a_2, \ldots, v_k = a_k}$. Thus, for any $a \in \{0, 1\}^k$, with $k < m_F$, $F|_{a0}$ and $F|_{a1}$ are the two children of the node $F|_a$ in the self-reducing tree of $F$.

Let $\prec$ be the lexicographic ordering on binary strings. We define

$$L = \{\langle F, w \rangle : |w| = m_F, (\exists u \in \{0, 1\}^{m_F}) [w \preceq u \text{ and } F|_u = 1]\}.$$

It is clear that $L \in NP$ and so $L \leq_m^P$ SAT.

Now assume that SAT $\leq_m^P S$ for some sparse set $S$. Then, $L \leq_m^P S$ by a reduction function $f$. For any Boolean formula $F$, we consider its self-reducing tree. For each binary string $u$ with $|u| < m_F$, let $u'$ be the string $u$ padded with 0's so that it is of length $m_F$. We attach to each node $F|_u$ a label $f(\langle F, u' \rangle)$.

The idea of the proof is to perform a *breadth-first search* of the self-reducing tree and, along the search, prune the tree according to the labels $f(\langle F, u' \rangle)$. We say a node $F|_u$ is *smaller* than a node $F|_v$ if $|u| = |v|$ and $u \prec v$. We also assume that the tree is ordered so that a smaller node is always to the left of a bigger node; that is, for any $a \in \{0, 1\}^k$ with $k < m_F$, $F|_{a0}$ is to the left of $F|_{a1}$. Assume that $F \in$ SAT and let $w$ be the maximum string that satisfies $F$. Then, it is easy to see that each node that lies to the left of, or belong to, the path from $F$ to $F|_w$ is attached with a label in $S$. In other words, to the left of $w$, we only have polynomially many different labels. So, at each level of the tree, we only need to keep, starting from left, polynomially many nodes with different labels, and we can prune all the other nodes to the right. The following is the more precise procedure. Assume that $C_S(n) \leq q(n)$ and that $f(\langle F, u' \rangle)$ is computable in time $p(|F|)$ (if $|u'| = m_F$) and so each label in the self-reducing tree of $F$ is of length $\leq p(|F|)$.

Procedure $bsearch(F)$;

(1) Set $W_0 := \{F\}$.
(2) For $i := 1$ to $m_F - 1$, do the following:
    (2.1) Set $W_i := \{F|_{u0}, F|_{u1} : F|_u \in W_{i-1}\}$.
    (2.2) For each pair $F|_u$, $F|_v$ in $W_i$ that have the same label, delete the smaller one from $W_i$.
    (2.3) If $|W_i| > q(p(|F|))$ then delete the greatest $|W_i| - q(p(|F|))$ nodes from $W_i$ so that $|W_i| = q(p(|F|))$.
(3) Set $W_{m_F} := \{F|_{u0}, F|_{u1} : F|_u \in W_{m_F-1}\}$.
(4) If there is an $F|_u$ in $W_{m_F}$ which evaluates to 1 then answer YES else answer NO.

It is apparent that if the above procedure answers YES then $F \in$ SAT, because it has found a leaf node of the tree that is true. It is also clear that the above procedure always halts in polynomial time because each $W_i$ has size bounded by $q(p(|F|))$. It is only left to show that if $F \in$ SAT then the procedure will answer YES.

Let $w$ be the maximum string that satisfies $F$, and let $w^{(i)}$ be the string of the first $i$ bits of $w$. We claim that at each level $i$, the set $W_i$ must contain the node $F|_{w^{(i)}}$. Thus, at the end, $W_{m_F}$ contains $F|_w$ and the procedure must answer YES. To see this, we note that at the beginning, $W_0$ contains $F = F|_{w^{(0)}}$. Suppose that, at the end of the $(i-1)$st iteration of the loop, $W_{i-1}$ contains $F|_{w^{(i-1)}}$. Then, in the $i$th iteration, after step (2.1), $W_i$ must contain $F|_{w^{(i)}}$. Note that $F|_{w^{(i)}}$ has a label in $S$ and all bigger nodes in $W_i$ have labels in $\overline{S}$. Therefore, step (2.2) never deletes $F|_{w^{(i)}}$ from $W_i$. After step (2.2), $W_i$ contains only nodes with all distinct labels. Since all nodes $F|_u$ that are smaller than $F|_{w^{(i)}}$ have labels in $S$ and since there are at most $q(p(|F|))$ distinct labels in $S$ of length $\leq p(|F|)$, there are at most $q(p(|F|))$ many nodes in $W_i$ smaller than or equal to $F|_{w^{(i)}}$. So, node $F|_{w^{(i)}}$ survives step (2.3), which leaves the smallest $q(p(|F|))$ nodes in $W_i$ untouched. This completes the induction proof of the claim and, hence, the theorem. ∎

*Remark.* We showed, in Section 6.2, that a set is $\leq_T^P$-reducible to a sparse set if and only if it has polynomial-size circuits. It follows from Theorem 6.11 that a sparse set cannot be $\leq_T^P$-complete for $NP$ unless the polynomial-time hierarchy collapses to $\Sigma_2^P$. Theorem 7.24 may be viewed as an improvement over Theorem 6.11. Note that the proof technique here is quite different. This proof technique can be further applied to show that if $P \neq NP$, then sparse sets cannot be $\leq_{ktt}^P$-complete for $NP$, for any fixed $k \geq 1$. (See Exercise 7.15.)

## 7.4 Density of *EXP*-Complete Sets

In view of the difficulty of the Berman-Hartmanis isomorphism conjecture, researchers have studied the conjecture with respect to other complexity classes in order to gain more insight into this problem. One of the much studied direction is the exponential-time version of the Berman-Hartmanis conjecture. One of the advantages of studying the conjecture with respect to $EXP$ is that we can perform diagonalization in $EXP$ against all sets in $P$. Indeed, in this and the next sections, we will show some stronger results that reveal interesting structures of $EXP$-complete sets which are not known for $NP$-complete sets.

We begin with the notion of $p$-immune sets which is motivated by the notion of immune sets in recursion theory. Recall that a set $A$ is called *immune* if $A$ does not have an infinite r.e. subset. Thus, an r.e. set whose complement is immune (such a set is called a *simple* set) has, intuitively, very high density (since it intersects with every co-r.e. set) and, hence, cannot be $\leq_m$-complete for r.e. sets. Here, we argue that $P$-immune sets cannot be $\leq_m^P$-complete for $EXP$.

**Definition 7.25** *An infinite set $A$ is called $P$-immune if it does not have an infinite subset in $P$. A set $A$ is called $P$-bi-immune (or, simply, bi-immune) if both $A$ and $\overline{A}$ are $p$-immune.*

There is an interesting characterization of bi-immune sets in terms of $\leq_m^P$-reductions. We say a function $f : \Sigma^* \to \Gamma^*$ is *finite-to-one* if for every $y \in \Gamma^*$, the set $f^{-1}(y)$ is finite.

**Proposition 7.26** *A set $A$ is bi-immune if and only if every $\leq_m^P$-reduction from $A$ to another set $B$ is a finite-to-one function.*

*Proof.* Assume that $A \leq_m^P B$ via a function $f$ that is not finite-to-one. Then, there exists a string $y$ such that $f^{-1}(y)$ is infinite. If $y \in B$, then $f^{-1}(y)$ is a polynomial-time computable infinite subset of $A$; otherwise, $f^{-1}(y)$ is a polynomial-time computable infinite subset of $\overline{A}$. In either case, $A$ is not bi-immune.

Conversely, assume that $A$ is not bi-immune; for instance, $A$ has an infinite subset $C \in P$. Let $y_0$ be a fixed string in $A$. Then, the function

$$f(x) = \begin{cases} x & \text{if } x \notin C, \\ y_0 & \text{if } x \in C \end{cases}$$

is a $\leq_m^P$-reduction from $A$ to $B = (A - C) \cup \{y_0\}$, and $f^{-1}(y) = C$ is not finite. ∎

A generalization of the bi-immune set based on the above characterization is the *strongly bi-immune* set.

**Definition 7.27** *A set $A$ is strongly bi-immune if every $\leq_m^P$-reduction from $A$ to another set $B$ is almost one-to-one (i.e., one-to-one except for a finite number of inputs).*

One might suspect that the requirement of almost one-to-oneness is too strong for any set to have this property. Interestingly, there are even sparse sets which are strongly bi-immune. We divide the construction of such a set into two steps.

**Theorem 7.28** *There exists a set in EXP which is strongly bi-immune.*

*Proof.* The basic idea is to diagonalize against all polynomial-time computable functions $g$ which have the property that $(\forall n)(\exists x, y, |x| \geq n)g(x) = g(y)$. Let $\{M_i\}$ be an enumeration of polynomial-time TMs (see Section 1.5). The diagonalization needs to satisfy the requirements $R_i$, for $i = 1, 2, \ldots$ :

$R_i$: If $M_i$ computes a function $g$ with the above property then there exist $x$ and $y$ such that $x \in A$, $y \notin A$ and $M_i(x) = M_i(y)$.

The construction of set $A$ is by delayed diagonalization. Let $w_n$ denote the $n$th string in $\{0,1\}^*$ under the lexicographic ordering. At stage $n$, we determine whether the string $w_n$ is in $A$ or not. While each requirement $R_i$ will eventually be satisfied, we do not try to do it in stage $i$. Instead, in stage $n$, we try to satisfy $R_i$ for some $i \leq n$ for which there exists some $x < w_n$ such that the fact $M_i(x) = M_i(w_n)$ can be verified in exponential time. The following is a more precise description of this diagonalization algorithm.

Before stage 1, we let $A$ and $D$ be $\emptyset$, and $k = 0$.

*Stage n.*

(1) If $M_k(w_n)$ halts in $2^{|w_n|}$ moves then let $D := D \cup \{k\}$ and $k := k + 1$.

(2) Search for the smallest $i \in D$ such that for some $x < w_n$, $M_i(x) = M_i(w_n)$ (if no such $i$, then go to stage $n + 1$).

(3) Let $x$ be the smallest $x$ such that $M_i(x) = M_i(w_n)$.

(4) If $x \notin A$ then let $A := A \cup \{w_n\}$.

(5) Let $D := D - \{i\}$. (We say integer $i$ is *cancelled*.)

(6) Go to stage $n + 1$.

We need to verify the following properties of the above algorithm for $A$:

(a) Every integer $k > 0$ must eventually enter set $D$.

(b) Every integer $i$ having the property that $M_i$ computes a function $g$ which is not almost one-to-one (called a *bad* function) will eventually be cancelled from $D$.

(c) When $i$ is cancelled from $D$, the requirement $R_i$ is satisfied.

(d) $A \in EXP$.

Conditions (a) and (c) are obvious from the construction. We prove condition (b) by contradiction. Assume that some bad function is never cancelled. Then, there must be a least integer $i$ such that $M_i$ computes a bad function and $i$ is never cancelled from $D$. Then, by some stage $n$, the following conditions hold:

(i) All $j < i$ which are eventually to be cancelled are already cancelled by stage $n$;

(ii) All $j < i$ which are to stay in $D$ forever have the property that for all $y \geq w_n$, $M_j(y) \neq M_j(z)$ for all $z < y$; and

(iii) $i \in D$.

That is, starting from stage $n$, $i$ is the least index in $D$ such that $M_i$ could satisfy the condition of step (2). Let $m$ be the least integer greater than $n$ such that $M_i(w_m) = M_i(x)$ for some $x < w_m$. Then, by stage $m$, $i$ will be found in step (2) and cancelled in step (5). This is a contradiction.

Finally, for condition (d), we note that steps (1) and (2) of stage $n$ need $n \cdot 2^{O(|w_n|)}$ moves each. Since $|w_n| = \lfloor \log_2(n+1) \rfloor$, each stage $n$ takes about time $2^{O(|w_n|)}$. To determine whether $w_n$ is in $A$, we need only simulate the above algorithm from stage 1 to stage $n$, and the total runtime is bounded by $n \cdot 2^{O(|w_n|)} = 2^{O(|w_n|)}$. ∎

**Corollary 7.29** *There exists a sparse, strongly bi-immune set in EXP.*

*Proof.* We modify the algorithm of Theorem 7.28 as follows:
Stage $n$.
   (1) If $M_k(w_n)$ halts in $2^{|w_n|}$ moves then let $D := D \cup \{k\}$ and $k := k+1$.
   (2) Search for the smallest $i \in D$ such that for some $x, y$, $x < y \leq w_n$, $w_{n/2} < y$, $M_i(x) = M_i(y)$ (if no such $i$, then go to stage $n+1$).
   (3) Let $y$ be the smallest string such that there is a string $x$ satisfying the above condition. Let $x$ be the smallest such string $x$.
   (4) If $x \notin A$ then $A := A \cup \{y\}$.
   (5) Let $D := D - \{i\}$.
   (6) Go to stage $2n$.

We claim that condition (b) still holds. Assume otherwise that $i$ is the least integer such that $M_i$ computes a bad function and $i$ is never cancelled. Then, again, there exists a stage $n$ such that (i), (ii), and (iii) in the proof of Theorem 7.28 hold. Let $y$ be the least string $y > w_n$ such that for some $x < y$, $M_i(x) = M_i(y)$. Let $y = w_r$. Then, there exists an integer $m$, $m/2 < r \leq m$ (or, $r \leq m < 2r$), such that we will enter stage $m$, because we cannot jump from a stage less than $r$ to a stage $\geq 2r$. Then, in that stage, we must cancel $i$ and that gives us the contradiction.

We note that condition (c) holds but not as trivially as in the last proof. In other words, whenever we satisfy a requirement $R_i$ in stage $n$ by finding $x < y$ such that $M_i(x) = M_i(y)$ and making one of them in $A$ and the other in $\overline{A}$, we must make sure that the memberships of $x$ and $y$ in $A$ will not change later. This is true because we jump to stage $2n$ after that and will never change the membership of any string $\leq w_n$.

Condition (d) also holds, because we can determine whether a string $w_n$ is in $A$ by simulating the algorithm up to stage $2n$.

Finally, we note that, for any $n$,

$$|A \cap \{y : y \leq w_n\}| \leq \lceil \log_2 n \rceil + 1,$$

because we would reach a stage $m \geq 2^n$ after we put $n$ strings into $A$. It follows that $A$ is sparse. ∎

From the above theorem, we obtain a stronger version of Theorem 7.24 on *EXP*. First, we give some more definition.

**Definition 7.30** *A set $A$ is said to have* subexponential density *if for every polynomial function $q$, $C_A(q(n)) = o(2^n)$.*

Obviously a sparse set has subexponential density. A set of density $O(n^{\log n})$ also has subexponential density. We note that if a set $B$ has subexponential density and $A \leq_m^P B$ via an almost one-to-one reduction then $A$ also has subexponential density (Exercise 7.9). Thus, let $B$ be any *EXP*-complete set, and $A$ a strongly bi-immune set in *EXP*. We can see that $B$ does not have subexponential density because otherwise both $A$ and $\overline{A}$ have subexponential density, which is impossible. (Note that if $B$ is $\leq_m^P$-complete for *EXP*, then so is $\overline{B}$.)

**Corollary 7.31** *An EXP-complete set cannot have subexponential density and, hence, cannot be sparse.*

## 7.5   One-Way Functions and Isomorphism in *EXP*

In Section 7.3, we mentioned that one of the difficulties of proving the Berman-Hartmanis conjecture (for the class *NP*) is that it is a strong statement which implies $P \neq NP$. Another difficulty involves with the notion of one-way functions. Recall that a function $f : \Sigma^* \to \Gamma^*$ is a (weak) one-way function if it is one-to-one, polynomially honest, polynomial-time computable but not polynomial-time invertible (see Section 4.2). Suppose such a one-way function $f$ exists. Then, consider the set $A = f(\text{SAT})$. It is easy to see that $A$ is *NP*-complete. However, it is difficult to see how it is *p*-isomorphic to SAT: the isomorphism function, if exists, must be different from the function $f$, for otherwise we could use it to compute $f^{-1}$ in polynomial time. (If $f$ is a strong one-way function then the isomorphism function would be substantially different from $f$, as we cannot compute even a small fraction of $f^{-1}$.) Thus, if one-way functions exist, then it seems possible to construct a set $A$ that is *NP*-complete but not *p*-isomorphic to SAT. On the other hand, if one-way functions do not exist, then all one-to-one, polynomially honest reduction functions are polynomial-time invertible, and so the Berman-Hartmanis conjecture seems more plausible.

Based on this idea and the study of the notion of *k-creative* sets, Joseph and Young proposed a new conjecture:

**Joseph-Young Conjecture:** The Berman-Hartmanis conjecture is true if and only if one-way functions do not exist.

Although this conjecture has some interesting ideas behind it, there is not much progress in this direction yet (but see Exercises 7.12 and 7.13 for the study on $k$-creative sets). In this section, instead, we investigate the isomorphism conjecture on $EXP$-complete sets and its relation with the existence of one-way functions.

We first generalize the notion of paddable sets to *weakly paddable* sets, and use this notion to show that all $EXP$-complete sets are equivalent under the $\leq_{1,li}^P$-reduction.

**Definition 7.32** *A set $A \subseteq \Sigma^*$ is* weakly paddable *if $A \times \Sigma^* \leq_{1,li}^P A$.*

In Section 7.2, we showed that a paddable $NP$-complete set must be $p$-isomorphic to SAT. The following is its analogue for weakly paddable sets.

**Lemma 7.33** *Assume that $A \leq_m^P B$ for some weakly paddable set $B$. Then $A \leq_{1,li}^P B$.*

*Proof.* Let $B \times \Sigma^* \leq_{1,li}^P B$ via $f$ and $A \leq_m^P B$ via $g$. Define $h(x) = f(\langle g(x), x \rangle)$. Then, $h$ is one-to-one and length-increasing, and $A \leq_{1,li}^P B$ via $h$. ∎

Thus, it follows immediately that all $\leq_m^P$-complete sets in $EXP$ are equivalent under $\leq_{1,li}^P$, if all $EXP$-complete sets are known to be weakly paddable.

**Lemma 7.34** *Every $EXP$-complete set $A$ is weakly paddable.*

*Proof.* Let $\{M_i\}$ be an enumeration of all polynomial-time TMs and $\phi_i$ be the function computed by the $i$th machine $M_i$. Intuitively, we will construct a set $C$ in $EXP$ such that if $C \leq_m^P A$ via $\phi_i$, then $\phi_i$ on inputs of the form $\langle i, x, y0^{n_i} \rangle$ is one-to-one and length-increasing, where $n_i$ is an integer to be defined later. In addition, we make $\phi_i(\langle i, x, y0^{n_i} \rangle) \in C$ if and only if $x \in A$. Thus, the mapping $f(x,y) = \phi_i(\langle i, x, y0^{n_i} \rangle)$ is a reduction for $A \times \Sigma^* \leq_{1,li}^P A$. More precisely, we define set $C$ by the following algorithm:

*Algorithm $M_C$ for set $C$.*

Assume that input is $z = \langle i, x, y \rangle$ and that $|z| = n$.

(1) Simulate $M_i(\langle i, x, y \rangle)$ for $2^n$ moves. If $M_i(\langle i, x, y \rangle)$ does not halt in $2^n$ moves then go to step (4).

(2) Let $u := \phi_i(\langle i, x, y \rangle)$. If $|u| \leq |z|$ then accept $z$ if and only if $u \notin A$ and halt; else go to step (3).

(3) For each $\langle x', y' \rangle < \langle x, y \rangle$, simulate $M_i(\langle i, x', y' \rangle)$ for $2^n$ moves.

    (3.1) If, for any $\langle x', y' \rangle$, $M_i(\langle i, x', y' \rangle)$ does not halt in $2^n$ moves then go to step (4).

(3.2) If, for any $\langle x', y' \rangle$, $u = \phi_i(\langle i, x', y' \rangle)$ then accept $z$ if and only if $x' \notin A$ and halt.

(4) Accept $z$ if and only if $x \in A$.

It is clear that $C \in EXP$. Furthermore, we claim that if $C \leq_m^P A$ via $\phi_i$, then there exists an integer $n_i$ satisfying the following conditions:

(i) $|\phi_i(\langle i, x, y0^{n_i} \rangle)| > |\langle i, x, y0^{n_i} \rangle|$ (i.e., $\phi_i$ is length-increasing on strings of the form $\langle i, x, y0^{n_i} \rangle$).

(ii) For all $\langle x', y' \rangle \neq \langle x, y \rangle$, $\phi_i(\langle i, x', y'0^{n_i} \rangle) \neq \phi_i(\langle i, x, y0^{n_i} \rangle)$ (i.e., $\phi_i$ is one-to-one on strings of the form $\langle i, x, y0^{n_i} \rangle$).

(iii) For all $\langle x, y \rangle$, $\langle i, x, y0^{n_i} \rangle \in C$ if and only if $x \in A$.

To show the claim, let $n_i$ be a sufficiently large integer such that for all $n > n_i$, $M_i$ on inputs of length $\leq n$ must halt in $2^n$ moves. Then, the computation of the algorithm $M_C$ on any $z = \langle i, x, y0^{n_i} \rangle$ must go to step (2).

Now, step (2) guarantees that if $\phi_i(\langle i, x, y0^{n_i} \rangle)$ is not length-increasing, then it is not a reduction from $C$ to $A$. Therefore, $\phi_i$ satisfies condition (i).

Next, if $\phi_i(\langle i, x, y0^{n_i} \rangle) = \phi_i(\langle i, x', y' \rangle) = u$ for some $\langle x', y' \rangle < \langle x, y0^{n_i} \rangle$, then for the smallest such $\langle x', y' \rangle$, we have $x' \notin A$ if and only if $\langle i, x, y0^{n_i} \rangle \in C$. Note that for the smallest such $\langle x', y' \rangle$, the computation of $\phi_i(\langle i, x', y' \rangle)$ must either go from step (1) to step (4) or pass through steps (2), (3), and (4), and it will determine, in step (4), that $\langle i, x', y' \rangle \in C$ if and only if $x' \in A$. Thus, we have $\langle i, x', y' \rangle \in C$ if and only if $\langle i, x, y0^{n_i} \rangle \notin C$, but $\phi_i(\langle i, x', y' \rangle) = \phi_i(\langle i, x, y0^{n_i} \rangle)$. So, $\phi_i$ cannot be a reduction from $C$ to $A$. Therefore, $\phi_i$ must satisfy condition (ii).

Finally, from the above discussion, we see that the computation of $\phi_i(\langle i, x, y0^{n_i} \rangle)$ must determine the membership in $C$ in step (4), and so condition (iii) also holds.

Now assume that $C \leq_m^P A$ via $\phi_i$. Let $f(\langle x, y \rangle) = \phi_i(\langle i, x, y0^{n_i} \rangle)$. This is a one-one, length-increasing reduction from $A \times \Sigma^*$ to $A$.  ∎

Combining Lemmas 7.33 and 7.34, we get the following theorem.

**Theorem 7.35** *All EXP-complete sets are $\leq_{1,li}^P$-equivalent.*

From Theorem 7.4, we know that if two sets $A$ and $B$ are equivalent under length-increasing polynomial-time invertible reductions, then they are $p$-isomorphic. Observe that if one-way functions do not exist then a $\leq_{1,li}^P$-reduction is also a $\leq_{inv,li}^P$-reduction. So, we get the following corollary.

**Corollary 7.36** *If one-way functions do not exist, then all EXP-complete sets are polynomial-time isomorphic to each other.*

In the following, we show that the converse of the above corollary holds for sets that are $\leq_{2tt}^P$-complete for $EXP$.

In order to prove this result, we need one-way functions that are length-increasing. The following lemma shows that this requirement is not too strong.

**Lemma 7.37** *If one-way functions exist, then one-way functions that are length-increasing exist.*

*Proof.* Let $f$ be a one-way function. We simply pad enough 1's after $f(x)$ to create a new length-increasing one-way function. More precisely, define $g(x) = f(x)01^{|x|+1}$. It is clear that $g$ is one-to-one, length-increasing, and polynomial-time computable. To see that $g$ is a one-way function, we show that if $g^{-1}$ is polynomial-time computable, then we can compute $f^{-1}$ in polynomial time as follows: Assume that $p$ is a polynomial function such that $|f^{-1}(y)| \leq p(|y|)$ for all $y$ in the range of $f$ ($f$ is known to be polynomially honest). For a given string $y$, we let $y_i = y01^i$ and compute $x_i = g^{-1}(y_i)$ (if it exists), for $i = 1, \ldots, |p(y)| + 1$, If $f^{-1}(y)$ exists, it must by one of $x_i$; we can compute $f(x_i)$ to verify whether $f^{-1}(y) = x_i$ in polynomial time. ∎

**Theorem 7.38** *Assume that one-way functions exist. Then, there exist two sets $A$ and $B$ in EXP such that*
  (a) $A \equiv_{1,li}^P B$;
  (b) $A \not\equiv^P B$; and
  (c) *Both $A$ and $B$ are $\leq_{2tt}^P$-complete for EXP.*

*Proof.* For convenience, we will define sets $A$ and $B$ over the alphabet $\Sigma = \{0, 1, \$\}$. Before we construct sets $A$ and $B$, we first define two one-way functions $f_0$ and $f_1$ and discuss their properties.

Let $f : \{0,1\}^* \to \{0,1\}^*$ be a length-increasing one-way function. Extend $f$ to $\bar{f}$ on $\Sigma^*$ as follows:
  *Case 1.* If $x \in \{0,1\}^*$, then $\bar{f}(x) = f(x)$.
  *Case 2.* If $x = y\$z$ for some $y, z \in \{0,1\}^*$, then $\bar{f}(x) = f(y)\$z$.
  *Case 3.* If $x$ contains more than one \$'s, then $\bar{f}(x) = x$.
Now, define $f_0(x) = \bar{f}(x)0$ and $f_1(x) = \bar{f}(x)1$. It is clear that both $f_0$ and $f_1$ are length-increasing one-way functions on $\Sigma^*$.

Next, define, for each $x \in \Sigma^*$, four sets as follows:

$$A_1(x) = \{y \in \Sigma^* : (f_1 f_0)^k(x) = y \text{ or } (f_1 f_0)^k(y) = x \text{ for some } k \geq 0\},$$
$$A_0(x) = \{y \in \Sigma^* : (f_0 f_1)^k f_0(x) = y \text{ or } (f_1 f_0)^k f_1(y) = x \text{ for some } k \geq 0\},$$
$$B_1(x) = \{y \in \Sigma^* : (f_0 f_1)^k(x) = y \text{ or } (f_0 f_1)^k(y) = x \text{ for some } k \geq 0\},$$
$$B_0(x) = \{y \in \Sigma^* : (f_1 f_0)^k f_1(x) = y \text{ or } (f_0 f_1)^k f_0(y) = x \text{ for some } k \geq 0\}.$$

Each of these sets is called a *chain* (since $f_0$ and $f_1$ are length-increasing). We call $A_0(x) \cup A_1(x)$ the *A-chain* for $x$ and $B_0(x) \cup B_1(x)$ the *B-chain* for $x$. (Note that the *A*-chain for $x$ could be the *B*-chain for some other string $y$.) We list their basic properties below (see Figures 7.3–7.5):
  (a) For any $x \in \Sigma^*$, $(A_1(x) \cup A_0(x)) \cap (B_1(x) \cup B_0(x)) = \{x\}$.
  (b) If $x$ ends with 0, then $x$ is the smallest element in $A_1(x)$ and $f_0(x)$ is the smallest element of $A_0(x)$.
  (c) If $x$ ends with 1, then $x$ is the smallest element in $B_1(x)$ and $f_1(x)$ is the smallest element of $B_0(x)$.

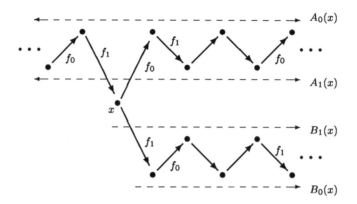

Figure 7.3: Chains for $x$ that ends with 1.

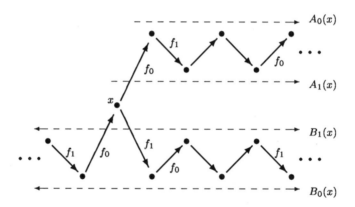

Figure 7.4: Chains for $x$ that ends with 0.

(d) If $x$ ends with \$, then $x$ is the smallest element of both $A_1(x)$ and $B_1(x)$, $f_0(x)$ is the smallest element of $A_0(x)$, and $f_1(x)$ is the smallest element of $B_0(x)$.

Let $\{M_i\}$ be an enumeration of polynomial-time TMs, $\phi_i$ the function computed by $M_i$, and $p_i$ a polynomial bound of the runtime of $M_i$. Then, the following holds:

(e) Let $C_1, C_2, \ldots, C_m$ be finitely many chains. For any $i, j \geq 0$, either

(i) $(\exists x \in \Sigma^*)$ $[\phi_i(\phi_j(x)) \neq x$ or $\phi_j(\phi_i(x)) \neq x]$, or

(ii) $(\exists x \in \{0, 1\}^*)$ $[x\$ \notin \bigcup_{k=1}^{m} C_k$ and $\phi_i(x\$) \notin A_0(x\$)]$.

*Proof of* (e). We show how to compute $f^{-1}$ in polynomial time if neither (i) nor (ii) is true for some $i, j \geq 0$.

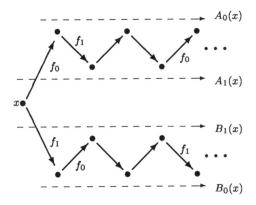

Figure 7.5: Chains for $x$ that ends with $.

*Algorithm for $f^{-1}(y)$.*

(1) For $k := 1$ to $m$ do the following:

   If the minimum element of $C_k$ is of the form $x\$$ for some $x \in \{0,1\}^*$ and $f(x) = y$ then output $x$ and halt.

(2) Let $z := y\$0$.

(3) Repeat the following while $|z| \leq p_i(|y\$0|)$:

   If $f_0(\phi_j(z)) = y\$0$ then output $x$ and halt, where $\phi_j(z) = x\$$; else let $z := f_0(f_1(z))$ and continue.

(4) Report that $y \notin f(\{0,1\}^*)$ and halt.

To see that the algorithm is correct, we first assume that $y = f(x)$ for some $x \in \{0,1\}^*$.

*Case 1.* $x\$ \in \bigcup_{k=1}^m C_k$. By property (d), $x\$$ must be the minimum element of some $C_k$. Thus, $x$ can be found by step (1).

*Case 2.* $x\$ \notin \bigcup_{k=1}^m C_k$. Since condition (ii) is not true, we know that $\phi_i(x\$) \in A_0(x\$)$. By property (d), $x\$$ is the minimum element of $A_1(x\$)$ and the minimum element of $A_0(x\$)$ is $f_0(x\$) = y\$0$. In addition, $|\phi_i(x\$)| \leq p_i(|x\$|) \leq p_i(|y\$0|)$, since $f_0$ is length-increasing. So, one of $z$ of step (3) must be equal to $\phi_i(x\$)$. Now, condition (i) being false implies that $\phi_j(\phi_i(x\$)) = x\$$ and so step (3) can find $x$ correctly.

From the above analysis, we can also see that if steps (1) and (3) fail to find $x$ such that $f(x) = y$ then $y \notin f(\{0,1\}^*)$.                    □

We now describe the construction of sets $A$ and $B$. The construction needs to satisfy three main goals.

(1) (*Chain Assignments*) We will constructs sets $A$ and $B$ such that $A \leq_m^P B$ via $f_0$ and $B \leq_m^P A$ via $f_1$ so that $A \equiv_{1,li}^P B$. To satisfy $A \leq_m^P B$ via $f_0$,

whenever a string $x$ is assigned to $A$ (or $\overline{A}$), we add all strings in $A_1(x)$ to $A$ (or, to $\overline{A}$), and all strings in $A_0(x)$ to $B$ (or, to $\overline{B}$, respectively). Likewise, to satisfy $B \leq_m^P A$ via $f_1$, we will assign chains $B_1(x)$ and $B_0(x)$ together with $x$.

(2) (*Diagonalization*) For the condition $A \not\leq_{inv}^P B$, we perform a diagonalization to satisfy the requirements $R_{i,j}$ for $i, j \geq 0$:

$R_{i,j}$: If $\phi_i$ and $\phi_j$ are total, one-to-one functions such that $\phi_i = \phi_j^{-1}$, then $A \not\leq_m^P B$ via $\phi_i$.

We satisfy requirement $R_{i,j}$ in stage $\langle i, j \rangle$ by trying to find a witness $w$ such that $\phi_i(w) \notin A_0(w)$ and make $w \in A$ if and only if $\phi_i(w) \notin B$. By property (e), such a witness $w$ exists if $\phi_i = \phi_j^{-1}$.

(3) (*Completeness*) To make $A$ and $B \leq_{2tt}^P$-complete for *EXP*, we let $E$ be an *EXP*-complete set and get $E \leq_{2tt}^P A$ by constructing $A$ so that for any $w$, $w \in E$ if and only if exactly one of $w\$\$$ or $w\$\$\$$ is in $A$. Note that $w\$\$$ and $w\$\$\$$ are not involved in the diagonalization or chain assignments and so can be used to satisfy the completeness condition.

We now show how to assign a string $x$ into $A$ or $\overline{A}$ while satisfying goals (1) and (3). We are going to use the following procedures:

Procedure $assign_A(x)$;

   $A := A \cup A_1(x);\ \ B := B \cup A_0(x).$

Procedure $assign_{\overline{A}}(x)$;

   $\overline{A} := \overline{A} \cup A_1(x);\ \ \overline{B} := \overline{B} \cup A_0(x).$

Procedure $assign_B(x)$;

   $B := B \cup B_1(x);\ \ A := A \cup B_0(x).$

Procedure $assign_{\overline{B}}(x)$;

   $\overline{B} := \overline{B} \cup B_1(x);\ \ \overline{A} := \overline{A} \cup B_0(x).$

Procedure $assign(x)$;

(1) If $x$ is not yet assigned to $A$ or $\overline{A}$, then consider the following cases.

   (1.1) $x = y\$\$$ for some $y \in \{0,1\}^*$.

   (a) $y\$\$\$$ is not yet assigned to $A$ or $\overline{A}$. Then, $assign_A(y\$\$\$)$, and go to case (b).

   (b) $y\$\$\$ \in A$. If $y \in E$ then $assign_{\overline{A}}(x)$ else $assign_A(x)$.

   (c) $y\$\$\$ \notin A$. If $y \in E$ then $assign_A(x)$ else $assign_{\overline{A}}(x)$.

   (1.2) $x = y\$\$\$$ for some $y \in \{0,1\}^*$. (The action is symmetric to Case (1.1) such that $y\$\$$ and $y\$\$\$$ will be assigned to $A$ or $\overline{A}$ such that $y \in E$ if and only if exactly one of them is in $A$.)

(1.3) $x$ is not $y$\$\$ or $y$\$\$\$ for any $y \in \{0,1\}^*$. Then, $assign_A(x)$.

(2) If $x$ is not assigned to $B$ or $\overline{B}$ yet, then $assign_B(x)$.

Stage $\langle i, j \rangle$ of the main procedure is described below using these procedures.

Stage $n = \langle i, j \rangle$.

(1) Set $x := \lambda$; $witness := 0$; $bound := 2^n$.

(2) Repeat the following while ($witness = 0$ or $|x| \leq bound$):

    (2.1) If $witness = 1$ then $assign(x)$ and go to step (2.5).

    (2.2) If $\phi_i(\phi_j(x)) \neq x$ or $\phi_j(\phi_i(x)) \neq x$, then let $witness := 1$, $assign(x)$ and go to step (2.5).

    (2.3) If ($x = w$\$ for some $w \in \{0,1\}^*$) and ($x$ is not in the chains $A_0(z)$ or $A_1(z)$ assigned so far) and ($\phi_i(x) \notin A_0(x)$), then do the following:

        (a) let $witness := 1$,

        (b) if $\phi_i(x) \in B$, then $assign_{\overline{A}}(x)$,

        (c) if $\phi_i(x) \in \overline{B}$ or if it is not assigned to $B$ or $\overline{B}$, then $assign_{\overline{B}}(\phi_i(x))$, $assign_A(x)$ and let $bound := \max\{bound, |\phi_i(x)| + 1\}$,

        (d) $assign(x)$,

        (e) go to step (2.5).

    (2.4) If not (2.3), then $assign(x)$ and go to step (2.5).

    (2.5) Let $x := x + 1$.

We now verify the correctness of the above algorithm. First, it is easy to see that each $x$ is eventually assigned to $A$ or $\overline{A}$ and to $B$ or $\overline{B}$. In addition, the procedures $assign_A$, $assign_{\overline{A}}$, $assign_B$, and $assign_{\overline{B}}$ guarantee that the chain assignments are consistent with goal (1). Thus, we get $A \leq^P_m B$ via $f_0$ and $B \leq^P_m A$ via $f_1$.

Next, we check that $A \not\leq^P_{inv} B$ by arguing that stage $n = \langle i, j \rangle$ satisfies $R_{i,j}$. If there is an $x$ satisfying condition (i) of property (e), then $R_{i,j}$ is obviously satisfied. Thus, we assume that $\phi_i = \phi_j^{-1}$. We further assume, by induction, that each previous stage $n_1 < n$ halts, and so, at the beginning of stage $n$, there are only finitely many chains $C_1, \ldots, C_m$ that have been assigned. Property (e) implies that there is a smallest $x \in \{0,1\}^*$ satisfying condition (ii) of property (e). When the algorithm considers $x' < x$\$ in stage $n$, it might create new chains other than $C_1, \ldots, C_m$. However, none of these chains includes $x$\$, since $x$\$ is the smallest element in $A_1(x$\$$)$. Therefore, $R_{i,j}$ will be satisfied when the loop step (2) considers $x$\$. Furthermore, when such an $x$ is found, $witness$ is set to 1, and $bound$ will not be reset later and so stage $n$ will halt.

Finally, we claim that $y \in E$ if and only if exactly one of $y$\$\$ and $y$\$\$\$ is in $A$, and so $E \leq^P_{2tt} A$. First observe that if the assignment of at least one of $y$\$\$

and $y$\$\$$ to $A$ or $\overline{A}$ is made by the procedure *assign*, then the claim is true. On the other hand, there is only one place in stage $n = \langle i, j \rangle$, where we may assign $y$\$\$$ and $y$\$\$\$$ to $A$ or $\overline{A}$ without calling procedure *assign*. That happens in case (c) of step (2.3), when $\phi_i(x)$ is not assigned to $B$ or $\overline{B}$ and $y$\$\$$ or $y$\$\$\$$ is in $B_0(\phi_i(x))$. (The procedures $assign_A(x)$ and $assign_{\overline{A}}(x)$ cannot assign $y$\$\$$ or $y$\$\$\$$ because $y$\$\$$, $y$\$\$\$$, and $x$ all end with \$.) In this case, only one of $y$\$\$$ or $y$\$\$\$$ is assigned to $\overline{A}$ since $y$\$\$$ and $y$\$\$\$$ cannot be in the same chain $B_0(\phi_i(x))$. (They are the smallest elements of two different chains.) After this assignment, one of the two strings $y$\$\$$ and $y$\$\$\$$ remains unassigned. In the meantime, the bound has been reset to $|\phi_i(x)| + 1$. Since one of $y$\$\$$ or $y$\$\$\$$ is the smallest element of the chain $B_0(\phi_i(x))$, we must have $|y$\$\$| \leq |\phi_i(x)|$. Therefore, the unassigned one of $y$\$\$$ and $y$\$\$\$$ will also be assigned in stage $n$ by procedure *assign* in step (2.1). (Again, $y$\$\$$ or $y$\$\$\$$ is the smallest element of chain $A_1(y$\$\$)$ or $A_1(y$\$\$\$)$, respectively, and will not be assigned to $A$ or $\overline{A}$ by other assignments.) So, we have proved $E \leq_{2tt}^P A$.

The only thing left is to verify that $A$ and $B$ are in the class *EXP*. For a given input $z$, we are going to simulate the stage construction of sets $A$ and $B$ to determine whether $z \in A$ and whether $z \in B$. We need to modify the stage construction and verify the following time bounds.

(a) We only need to simulate the construction from stage 1 to stage $m = \lceil \log |z| \rceil$ and stop the simulation as soon as the membership of $z$ in $A$ or $B$ is decided.

(b) During the simulation, we only need to store the initial segment of each chain up to strings of length $b(z) = 2^{O((\log |z|)^2)}$.

(c) The total size of the initial segments of all chains together is bounded by $2^{O(|z|)}$.

(d) The computation in each step can be done in time $2^{O(|z|)}$.

First, from the setting of the parameter *bound*, we know that by the end of stage $m = \lceil \log |z| \rceil$, we would have determined the memberships of all strings of length $\leq 2^m$. So, we only need to simulate up to stage $m$ to determine the membership of $z$. In addition, because we can stop simulation as soon as $z$ is assigned, we will never consider, in any stage, a string $x$ that is longer than $z$, because the assignment of $z$ is done before we reach $x$. This verifies claim (a) above.

For claim (b), we estimate how long the portion of a chain that we need to store. Since, in the iteration of the loop that considering string $x$, the longest string involved is either $x$ or $\phi_i(x)$, we need to store only the portion of the chains that contain strings of length bounded by

$$b(z) = \max(\{|\phi_i(x)| : i \leq m, |x| \leq |z|\} \cup \{|z|\}).$$

We may assume, as discussed in Section 1.5, that $M_i$'s are TMs attached with the $i$th clock $p_i(n)$ and $p_i(n) \leq i \cdot n^i + i$. Then,

$$b(z) \leq m|z|^m + m = 2^{O((\log |z|)^2)}.$$

In other words, in the simulation of the four procedures $assign_A(x)$, $assign_{\overline{A}}(x)$, $assign_B(x)$, $assign_{\overline{B}}(x)$, we only need to assign strings of length $\leq b(z)$ to sets $A$, $\overline{A}$, $B$, and $\overline{B}$.

For claim (c), we can estimate the number of chains that are assigned from stage 1 to stage $m$ as follows: Assume that the chains assigned are stored in a table $T$. In each iteration of the loop step (2), we add at most two $A$-chains and two $B$-chains to $T$. From claim (b), each chain is of size at most $b(z)$ and, from claim (a), we will simulate at most $\log |z|$ stages and, in each stage, we go through the loop step (2) for at most $3^{|z|}$ times. Furthermore, each string in table $T$ is of length at most $b(z)$. Therefore, the total size of $T$ is bounded by $4 \cdot (b(z))^2 \cdot 3^{|z|} \cdot \log |z| = 2^{O(|z|)}$.

For claim (d), we need to check that the following problems involved in the algorithm can be implemented in time $2^{O(|z|)}$:

(i) Determine whether a string $x$ or $\phi_i(x)$ has been assigned to $A$, $\overline{A}$, $B$, or $\overline{B}$.

(ii) Determine whether a string $y$ is in $E$ or not.

(iii) Determine whether $\phi_i(\phi_j(x)) = x$ and $\phi_j(\phi_i(x)) = x$.

Problems (ii) and (iii) are clearly computable in time $2^{O(|z|)}$. Problem (i), however, needs some careful analysis. Note that in order to know whether $x$ has been assigned to sets $A$ and $B$ or their complements, we cannot simply search table $T$ for $x$. The reason is as follows: When an $A$-chain or a $B$-chain is assigned, it can happen in two ways. For instance, consider the assignment of a $B$-chain before we consider string $x$. It can happen when we call $assign_B(w)$ or $assign_{\overline{B}}(w)$, where $w$ is the smallest element of the $B$-chain. In this case, the whole $B$-chain, up to strings of length $b(z)$ is put in table $T$. Thus, if $x$ is in this chain, we can simply search table $T$ to decide that. However, it can also happen when we call $assign_{\overline{B}}(\phi_i(w))$ to construct the $B$-chain, with $\phi_i(w)$ not the smallest element. In this case, we only stored those strings $y$ in the chain that have $|\phi_i(w)| \leq |y| \leq b(z)$ to table $T$. Strings $y$ in the chain whose lengths are less than $|\phi_i(w)|$ are not stored in $T$ yet. Therefore, in order to determine whether $x$ is in the $B$-chain, we need to generate the $B$-chain that contains $x$ and checks whether any string in the chain is in $T$. More precisely, $x$ is in $B$ if there exists a string $y \in B_1(x)$ such that the fact that $y$ is in $B$ can be found in table $T$, or if there exists a string $y \in B_0(x)$ such that the fact that $y$ is in $A$ can be found in table $T$. It is easy to see, from claims (b) and (c), that this verification can be done in time $2^{O(|z|)}$. This completes the proof of claim (d), as well as the theorem. ∎

Improving Theorem 7.38 is not an easy job. First, it is known that every $\leq_{1tt}^P$-complete set for *EXP* is $\leq_m^P$-complete for *EXP* (see Exercise 7.20). So, the $\leq_{2tt}^P$-completeness of Theorem 7.38 is probably the best one can get without settling the exponential-time version of the Joseph-Young conjecture.

In a different direction, we may consider the overall structure of $\leq_{2tt}^P$-complete sets and study, for instance, whether $\leq_{1,li}^P$-equivalent sets in this class must contain non-$p$-isomorphic sets. To be more precise, let us call the equivalence classes with respect to the reduction $\leq_r^P$ (including $\equiv^P$) an $r$-*degree*. It is obvious that an *iso*-degree (or, an *isomorphism type*) is always contained in a $(1,li)$-degree, which, in turn, must be contained in an $m$-degree. The questions here are whether an $m$-degree *collapses* to a $(1,li)$-degree (i.e., any two sets in the $m$-degree are $\leq_{1,li}^P$-equivalent) and whether a $(1,li)$-degree *collapses* to an isomorphism type (i.e., any two sets in the $(1,li)$-degree are $p$-isomorphic). Theorem 7.38 showed that one-way functions exist if and only if there is a $(1,li)$-degree that does not collapse to a single isomorphism type. Can we improve Theorem 7.38 so that if one-way functions exist then none of the $(1,li)$-degrees within the $\leq_{2tt}^P$-complete degree in *EXP* collapses to a single isomorphism type? The following result shows that this is not possible. Together with Theorem 7.38, it provides evidence that the structure of $\leq_{2tt}^P$-complete sets for *EXP* is not simple.

**Theorem 7.39** *There exists a set $A$ which is $\leq_{2tt}^P$-complete for EXP such that if $B \equiv_m^P A$ then $B \equiv_{iso}^P A$.*

*Proof.* See Exercise 7.21.                                                                ∎

## 7.6   Density of $P$-Complete Sets

Starting with Corollary 7.23, the questions of whether a sparse set can be complete for a complexity class have been studied for different types of completeness and for different complexity classes. In this section, we show one more result of this type, the polynomial-time version of Theorem 7.24. The interesting part here is that it uses yet another proof technique from the theory of linear algebra.

**Theorem 7.40** *If $P \neq NC^2$, then a sparse set cannot be $\leq_m^{log}$-hard for $P$.*

Recall that the following problem is complete for $P$ under the $\leq_m^{log}$-reduction:

> CIRCUIT VALUE PROBLEM (CVP): Given a circuit $C$ and an input $x$ to the circuit, determine whether the output of $C$ is 1 or 0.

Suppose, by way of contradiction, that a sparse set $S$ is $\leq_m^{log}$-hard for $P$. We will show that CVP $\in NC^2$. To do so, we need to introduce a variation of the problem CVP.

> PARITY-CVP: Given a circuit $C$ of $n$ gates $g_1, \ldots, g_n$, an input $x$ to the circuit, a subset $I \subseteq \{1, 2, \ldots, n\}$, and $b \in \{0, 1\}$, determine whether $\bigoplus_{i \in I} g_i(x) = b$, where $\oplus$ denotes the exclusive-or operation and $g_i(x)$ is the value of gate $g_i$ when $C$ receives the input $x$.

It is clear that PARITY-CVP $\in P$ and so PARITY-CVP $\leq_m^{log} S$ via a log-space computable function $f$. For any fixed $C$ and $x$, let

$$A_{C,x} = \{\langle C, x, I, b\rangle : I \subseteq \{1, 2, \ldots, n\}, b \in \{0, 1\}, \bigoplus_{i \in I} g_i(x) = b\}.$$

Since $S$ is sparse, there exists a polynomial $p$ such that

$$|f(A_{C,x})| \leq p(n).$$

Note that $|A_{C,x}| = 2^n$ and so *many* elements in $A_{C,x}$ are mapped by $f$ to the same string in $S$. For each pair of $\langle C, x, I, b\rangle$ and $\langle C, x, J, c\rangle$ with $f(\langle C, x, I, b\rangle) = f(\langle C, x, J, c\rangle)$, we obtain an equation

$$\bigoplus_{i \in I \triangle J} g_i(x) = b \oplus c, \tag{7.1}$$

where $I \triangle J = (I - J) \cup (J - I)$. (Note that this is true whether or not $\langle C, x, I, b\rangle \in A_{C,x}$.)

The idea of the proof is to use the above property to find many equations of the form (7.1) and then explore the local relations between the gates and these equations to determine the value of the output gate. More precisely, let us assume that we can find enough equations to form a system of linear equations of rank $n - t$, with $t = O(\log n)$. Then, we can find the values of the gates of $C$ on input $x$ as follows:

(1) Find $n - t$ many gates $g_{i_1}, g_{i_2}, \ldots, g_{i_{n-t}}$ such that their coefficients in the system of equations form a submatrix of rank $n - t$.

(2) Solve the system of equations to find the values of the gates $g_{i_1}, \ldots, g_{i_{n-t}}$.

(3) For each of the $2^t$ many possible Boolean assignments of the remaining gates, check whether the values of gates $g_1, \ldots, g_n$ are consistent with the local relations defined by $C$ (such as $g_i = g_j \wedge g_k$).

We remark that the problems of finding the rank of a given Boolean matrix and of solving a full-rank system of equations over $GF(2)$ are both known to be in $NC^2$ (see Mulmuley [1987] and Borodin et al. [1982]). In addition, checking the local relations of gate values can be done in $NC^1$, and the loop (3) of cycling through $2^t$ Boolean assignments can be made parallel. Therefore, the whole procedure above can be implemented in $NC^2$.

From the above analysis, all we need to do is to find polynomially many subsets $I$ of gates which are guaranteed to generate a system of equations of rank at least $n - O(\log n)$. For people who are familiar with the idea of randomization, it is not hard to see that a randomly selected collection of $O(n^2 p(n))$ subsets $I$ is sufficient. In the following, we use the theory of linear algebra to *derandomize* this selection process to get a deterministic selection of subsets $I$ which will generate the desired system of equations.

First, we note that each subset $I$ of gates can be represented by an $n$-dimensional vector over the Galois field $GF(2) = \mathbf{Z}_2$ such that each component corresponds to a gate and a component has value 1 if and only if the corresponding gate is in the subset. Let $\mathcal{F}$ denote the $n$-dimensional vector space over $GF(2)$. Denote $k = 1 + \lceil \log p(n) \rceil = O(\log n)$ and $m = k + \lceil \log n \rceil = O(\log n)$. Consider the Galois field $GF(2^m)$ as an $m$-dimensional vector space over $GF(2)$. Choose any basis $\{e_1, e_2, \ldots, e_m\}$. Then every vector $u$ in $GF(2^m)$ can be represented as $u = \sum_{i=1}^{m} u_i e_i$, with $u_i \in GF(2)$. We can also define an inner product

$$\langle\!\langle u, v \rangle\!\rangle = \sum_{i=1}^{m} u_i v_i$$

for $u = \sum_{i=1}^{m} u_i e_i$ and $v = \sum_{i=1}^{m} v_i e_i$, where all arithmetics are done in $GF(2)$.
    Now, for any $u, v \in GF(2^m)$, define a vector

$$I(u, v) = (\langle\!\langle 1, v \rangle\!\rangle, \langle\!\langle u, v \rangle\!\rangle, \ldots, \langle\!\langle u^{n-1}, v \rangle\!\rangle)$$

in $\mathcal{F}$, and let $D = \{I(u, v) : u, v \in GF(2^m)\}$. Clearly, $|D| \leq (2^m)^2 = n^{O(1)}$. An important property of $D$ is as follows.

**Lemma 7.41** *For any $(n - k)$-dimensional subspace $L$ of $\mathcal{F}$ and $p(n)$ vectors $y^{(1)}, y^{(2)}, \ldots, y^{(p(n))}$ in $\mathcal{F}$, $D$ cannot be covered by $\bigcup_{i=1}^{p(n)} (L + y^{(i)})$.*

*Proof.* Note that for any two $n$-dimensional vectors $y$ and $z$, either $L + y = L + z$ or $(L + y) \cap (L + z) = \emptyset$. In fact, if $a + y = b + z$ for some $a, b \in L$, then $y - z = b - a \in L$. Hence, $L + y = L + (y - z) + z = L + z$. Since the dimension of $L$ is $n - k$, $L$ contains exactly $2^{n-k}$ vectors over $GF(2)$. Thus, for every vector $y$, $L + y$ contains exactly $2^{n-k}$ elements. We call such a set an *affine subspace*. Note that the total number of $n$-dimensional vectors over $GF(2)$ is $2^n$. Therefore, there are totally $2^k$ affine subspaces in the form $L + y$. This conclusion can also be seen from another representation of the affine subspace $L + y$.
    Every affine subspace $L + y$ of dimension $n - k$ can be represented as the solution space of a system of equations

$$\sum_{j=1}^{n} a_{ij} x_j = b_i, \quad i = 1, 2, \ldots, k, \tag{7.2}$$

where vectors $(a_{i1}, a_{i2}, \ldots, a_{in})$, for $i = 1, 2, \ldots, k$, are linearly independent. Note that each $(b_1, b_2, \ldots, b_k)$ corresponds to one affine subspace $L + v$. There are exactly $2^k$ possible vectors $(b_1, b_2, \ldots, b_k)$, which represent the $2^k$ affine subspaces in the form $L + v$.
    Note that $2^k \geq 2p(n)$. To prove this lemma, it suffices to show that every affine subspace $L + y$ contains some element in $D$. Let $L + y$ be the solution space of the system of equations (7.2). Then, $I(u, v) \in L + y$ if and only if the following system of equations holds:

$$\sum_{j=1}^{n} a_{ij} \langle\!\langle u^{j-1}, v \rangle\!\rangle = \left\langle\!\!\left\langle \sum_{j=1}^{n} a_{ij} u^{j-1}, v \right\rangle\!\!\right\rangle = b_i, \quad i = 1, 2, \ldots, k. \tag{7.3}$$

Let $Q$ be the collection of the following polynomials (with respect to variable $X$):

$$Q = \left\{ \sum_{i=1}^{k} \beta_i \sum_{j=1}^{n} a_{ij} X^{j-1} : \beta_i \in GF(2), \text{but not all } 0 \right\}.$$

Since $(a_{i1}, a_{i2}, \ldots, a_{in})$ for $i = 1, 2, \ldots, k$ are linearly independent, $Q$ contains exactly $2^k - 1$ polynomials and none of them is zero. If $u \in GF(2^m)$ is not a root of any polynomial in $Q$, then the system of linear equations (7.3), with respect to variable $v$, has full row rank $k$. Thus, the system of equations (7.3) has a solution and, hence, $D \cap (L + y) \neq \emptyset$.

It remains to show that there exists an element $u \in GF(2^m)$ such that no polynomial in $Q$ has $u$ as a root. This can be proved by a simple counting argument: There are totally $2^k - 1$ polynomials, each of which has at most $n - 1$ roots. Thus, the total number of roots of all polynomials in $Q$ is at most $(n-1)(2^k - 1)$. There are totally $2^m > (n-1)(2^k - 1)$ possible $u$'s and so not every one of them is a root of some polynomial in $Q$. ∎

Let us identify a vector $I(u, v)$ in $D$ with the subset $I$ of gates represented by $I(u, v)$. We choose all subsets $I \in D$, compute $f(\langle C, x, I, 0 \rangle)$ and $f(\langle C, x, I, 1 \rangle)$, and form equations of the form (7.1). We claim that the equations obtained from those $\langle C, x, I, b \rangle$ in $A_{C,x}$ form a system of rank at least $n - k + 1$.

To see this, we attach a *label* $f(\langle C, x, I, b \rangle)$ to each subset $I \in D$, if $\langle C, x, I, b \rangle \in A_{C,x}$. Note that for each $I \in D$, exactly one of $\langle C, x, I, 0 \rangle$ and $\langle C, x, I, 1 \rangle$ is in $A_{C,x}$. Thus, each element in $D$ receives exactly one label. It follows that $D$ can be divided into $p(n)$ subsets $D_1, D_2, \ldots, D_{p(n)}$ according to their labels, each subset containing elements with a same label. For $i = 1, 2, \ldots, p(n)$, let $L_i$ be the subspace generated by the differences of the vectors in $D_i$. Note that each equation (7.1) obtained from two sets $I, J$ in a subset $D_i$ corresponds to the difference of two vectors $I, J$ in $D_i$. Thus, the rank of the system of equations cannot be smaller than the dimension of $L = L_1 + L_2 + \cdots + L_{p(n)}$. We check that, from Lemma 7.41, the dimension of $L$ is at least $n - k + 1$, since each $D_i$, $1 \leq i \leq p(n)$, can be covered by $L_i + v^{(i)}$ for any fixed $v^{(i)} \in D_i$ and, hence, $D$ can be covered by $\bigcup_{i=1}^{p(n)} (L + v^{(i)})$. This completes the proof of the claim.

Finally, we note that each $u \in GF(2^m)$ can be represented as an $m$-dimensional vector over $GF(2)$ of length $O(\log n)$. With this representation, the inner product $\langle\langle u, v \rangle\rangle$ of two elements $u, v \in GF(2^m)$ can be computed in space $O(\log n)$. The multiplication of $u$ and $v$ in the field $GF(2^m)$ can also be computed in space $O(\log n)$. Therefore, set $D$ can be constructed in logarithmic space. Since $LOGSPACE \subseteq NC^2$, the above algorithm for CVP is an $NC^2$ algorithm and Theorem 7.40 is proven.

Theorem 7.40 can actually be improved to the following form. Recall that an $\leq_m^{NC^1}$-reduction is a many-one reduction which is computable in $NC^1$ (see Exercise 6.22).

**Theorem 7.42** *If $P \neq NC^1$, then a sparse set cannot be $\leq_m^{NC^1}$-hard for $P$.*

The main idea of the proof of Theorem 7.42 remains the same. It, however, involves much more work on computational linear algebra and we omit it here.

## Exercises

**7.1** Prove that if $A \leq_{inv,li}^P B$ and $B \leq_{inv}^P A$, then $A \equiv^P B$.

**7.2** A set $A$ is called a *cylinder* if $A \equiv^P (B \times \Sigma^*)$ for some set $B$. Show that the following four conditions are equivalent:
(a) $A$ is a cylinder.
(b) For any $B$, $B \leq_m^P A \Rightarrow B \leq_{inv}^P A$.
(c) $A \times \Sigma^* \leq_{inv}^P A$.
(d) $A$ is paddable.

**7.3** Show that $A$ is a cylinder if and only if both $A \times \{\lambda\}$ and $(A \times \{\lambda\}) \cup (\Sigma^* \times (\Sigma^* \times \{\lambda\}))$ are cylinders.

**7.4** Show that for every subset $B$ of a paddable $\leq_m^P$-complete set $A$ in $NP$, if $B$ is in class $P$, then there exists a subset $C$ of $A$ such that $C \in P$, $B \subseteq C$, and $|C - B|$ is infinite.

**7.5** Show that every paddable $\leq_m^P$-complete set $A$ in $NP$ has an infinite $p$-immune subset $B$ in $EXP$.

**7.6** Show that if $A$ is $c$-, $d$- or $tt$-self-reducible and $A \equiv^P B$, then $B$ is also $c$-, $d$- or $tt$-self-reducible, respectively.

**7.7** A set $L$ is *$T$-self-reducible* if there is a polynomially related ordering $\prec$ on $\Sigma^*$ and a polynomial-time oracle TM $M$ such that $M^{L_x}(x) = \chi_L(x)$, where $L_x = \{y \in L : y \prec x\}$. Show that a $T$-self-reducible set must be in *PSPACE*.

**7.8** Complete the proof of Theorem 7.22(b).

**7.9** Show that if $B$ has subexponential density and $A \leq_m^P B$ via an almost one-to-one function $f$, then $A$ also has subexponential density.

In the following four exercises, we study the notion of *creative* sets. In these four exercises, we let $p_{k,i}$ denote the polynomial $p_{k,j}(n) = j \cdot n^k + j$, where $k, i > 0$.

**7.10** Let $\{M_i\}$ be a recursive enumeration of polynomial-time DTMs with the runtime of $M_i$ bounded by $p_{|i|^{1/3},|i|}(n)$. A set $A$ is called *$p$-creative for $P$* if there exists a polynomial-time computable function $f$ such that

$$(\forall i)[L(M_i) \subseteq \overline{A} \Rightarrow f(i) \in \overline{A} - L(M_i)].$$

(a) Show that a set $A \in EXP$ is $p$-creative for $P$ if and only if $A$ is $\leq_m^P$-complete for $EXP$.

(b) Show that $A$ is $p$-creative for $P$ if and only if there exists a polynomial-time computable function $f$ such that

$$(\forall i)[f(i) \in A \iff f(i) \in L(M_i)].$$

**7.11** Let $\{N_i\}$ be a recursive enumeration of polynomial-time NTMs with the runtime of $N_i$ bounded by $p_{|i|^{1/3}, |i|}(n)$. A set $A$ is called $p$-*creative for NP* if there exists a polynomial-time computable function $f$ such that

$$(\forall i)[L(N_i) \subseteq \overline{A} \Rightarrow f(i) \in \overline{A} - L(N_i)].$$

(a) Show that a set $A \in NEXP$ is $p$-creative for $NP$ if and only if $A$ is $\leq_m^P$-complete for $NEXP$.

(b) Show that a set $A$ is $p$-creative for $NP$ if and only if there exists a polynomial-time computable function $f$ such that

$$(\forall i)[f(i) \in A \iff f(i) \in L(N_i)].$$

**7.12** Let $\{N_i\}$ be a recursive enumeration of polynomial-time NTMs, and $k$ a positive integer. A set $A$ is called a $k$-*creative set* if there exists a polynomial-time computable function $f$ such that for all $i$, if $N_i$ has a time bound $p_{k,|i|}(|x|)$, then

$$f(i) \in A \Leftrightarrow f(i) \in L(N_i).$$

(a) Let $f$ be a polynomially honest, polynomial-time computable function. Define $K_f^k = \{f(i) : N_i \text{ accepts } f(i) \text{ in time } p_{k,|i|}(|f(i)|)\}$. Show that $K_f^k$ is a $k$-creative set for each $k \geq 1$.

(b) Show that for every $k \geq 1$, a $k$-creative set is $\leq_m^P$-hard for $NP$.

(c) Define $A^k = \{\langle i, x \rangle : M_i \text{ accepts } (i, x) \text{ within time } p_{k,|i|}(|\langle i, x \rangle|)\}$. Show that $A^k$ is a $(k-2)$-creative set in $NP$.

**7.13** Let $A, B \in NP$. Prove that if $A$ is $k$-creative and $A \leq_m^P B$ via $f$ which is computable in time $O(n^t)$ $(k \geq t > 0)$, then $B$ is $\lfloor k/t \rfloor$-creative.

**7.14** Prove that an $NP$-complete set $A$ is weakly paddable if and only if $A \equiv_{1,li}^P$ SAT.

**7.15** Prove that, for any $k \geq 1$, a sparse set cannot be $\leq_{ktt}^P$-complete for $NP$ unless $P = NP$.

**7.16** A set $A \subseteq \Sigma^*$ is $p$-*selective* if there is a polynomial-time computable function $f : \Sigma^* \times \Sigma^* \to \Sigma^*$ such that (i) for any $x, y \in \Sigma^*$, $f(x,y) \in \{x,y\}^*$, and (ii) $x \in A$ or $y \in A$ implies $f(x,y) \in A$.

(a) Prove that if $A$ is $p$-selective then $A \in P/poly$.

(b) Prove that a $p$-selective set cannot be $\leq_m^P$-hard for $NP$ if $P \neq NP$, and it cannot be $\leq_m^P$-hard for $PSPACE$ if $P \neq PSPACE$.

**7.17** We extend the notions of immunity and bi-immunity to general complexity classes. That is, a set $A$ is *immune* relative to the complexity class $\mathcal{C}$ if any subset $B \subseteq A$ which is in class $\mathcal{C}$ must be a finite set. A set $S$ is *bi-immune* relative to $\mathcal{C}$ if both $A$ and $\overline{A}$ are immune relative to $\mathcal{C}$.

(a) Show that if $t_1(n)$ is fully time-constructible and $t_2(n)\log(t_2(n)) = o(t_1(n))$, then there exists a set $A \in DTIME(t_1(n))$ that is bi-immune relative to $DTIME(t_2(n))$.

(b) Show that there exists a set $A \in NEXP$ which is immune relative to $NP$.

**7.18** We say a set $X$ is a *(polynomial-time) complexity core* of set $A$ if, for any DTM $M$ that accepts $A$ and any polynomial $p$, $M(x)$ does not halt in time $p(|x|)$ for all but a finite number of $x \in X$. For instance, if $A \subseteq \Sigma^*$ is bi-immune then $\Sigma^*$ is a complexity core of $A$.

(a) Show that for any $A \notin P$, $A$ has an infinite complexity core.

(b) We say $B \subseteq A$ is a *maximal subset of property $Q$* if $B$ has property $Q$ and, for all $C \subseteq A$ that have property $Q$, $C - B$ is finite. Let $A$ be an infinite recursive set. Prove that $A$ has a maximal subset $B$ in $P$ if and only if $A$ has a maximal complexity core $C \subseteq A$.

(c) Show that if $A \notin P$ is length-increasingly paddable, then $A$ does not have a maximal subset in $P$. (Such a set is called *P-levelable*.)

**7.19** Prove that if $EXP$ contains two non-$p$-isomorphic, $\leq_m^P$-complete sets $A$ and $B$, then there exist an infinite number of sets $A_1, A_2, \ldots$, all $\leq_m^P$-complete for $EXP$ but pairwisely non-$p$-isomorphic.

**7.20** (a) Show that if a set $A$ is $\leq_{1tt}$-complete for r.e. sets, then it is $\leq_m$-complete for r.e. sets.

(b) Show that if a set $A$ is $\leq_{1tt}^P$-complete for $EXP$, then it is $\leq_m^P$-complete for $EXP$.

**7.21** Prove Theorem 7.39.

**7.22** Show that there exists an infinite number of sets $A_1, A_2, \ldots$ such that all of them are $\leq_{2tt}^P$-complete for $EXP$ but are pairwisely nonequivalent under the $\leq_1^P$-reducibility.

**7.23** Discuss the polynomial-time isomorphism of *PSPACE*-complete sets. Check whether the results on collapsing degrees hold for the class *PSPACE*.

**7.24** Prove that there exists a bi-immune set $A$ in $EXP$ which is not strongly bi-immune.

**7.25** (a) Prove that there exists a $P$-immune set $A$ which is $\leq_{tt}^P$-complete for $EXP$. Therefore, there exists a set $B$ which is $\leq_{tt}^P$-complete for $EXP$ but not $\leq_m^P$-complete for $EXP$.

(b) Compare complete sets in $EXP$ under the following reducibility: $\leq_m^P$, $\leq_{btt}^P$, $\leq_{tt}^P$, and $\leq_T^P$. More precisely, show that the classes of complete sets for $EXP$ under these different notions of reducibility are all distinct.

**7.26** (a) Prove Theorem 7.42.

(b) Prove that if $P \neq NC^1$, then a sparse set cannot be $\leq_{ktt}^{NC^1}$-hard for $P$ for any $k \geq 1$.

## Historical Notes

It has been noticed by many people, including Simon [1975] and L. Berman [1977], that many reductions among natural $NP$-complete problems are actually much stronger than the requirement of a $\leq_m^P$-reduction. Berman and Hartmanis [1977] proved the polynomial-time version of the Bernstein-Schröder-Cantor Theorem, established the technique of paddability, and proposed the Berman-Hartmanis conjecture. The recursive version of this conjecture, that all $\leq_m$-complete r.e. sets are recursively isomorphic, is known to be true [Myhill, 1955]. Berman and Hartmanis [1977] also proposed the weaker density conjecture, that sparse sets cannot be $\leq_m^P$-complete for $NP$. P. Berman [1978] was the first to use the self-reducibility property to attack the density conjecture. He showed that if $P \neq NP$ then tally sets cannot be $\leq_m^P$-complete for $NP$. Meyer and Paterson [1979] and Ko [1983] further studied the general notion of self-reducibility. Meyer and Paterson [1979] and Fortune [1979] used the generalized self-reducibility to improve P. Berman's [1978] result. The density conjecture was proved to be equivalent to $P \neq NP$ by Mahaney [1982]. Our proof of Theorem 7.24 is based on the new technique of Ogiwara and Watanabe [1991]. The related result, that sparse sets cannot be $\leq_T^P$-complete for $NP$ if $PH \neq \Sigma_2^P$, was proved in Chapter 6. Selman [1979] introduced $p$-selective sets. Exercise 7.16 is from Selman [1979] and Ko [1983].

Berman and Hartmanis [1977] proved that the weaker density conjecture holds true for the class $EXP$. The notions of $P$-immune sets, bi-immune sets, and strongly bi-immune sets have been studied in L. Berman [1976], Ko and Moore [1981], and Balcázar and Schöning [1985]. Theorem 7.28 was first proved in L. Berman [1976] and strengthened by Balcázar and Schöning [1985]. Exercise 7.25 is from Ko and Moore [1981] and Watanabe [1987]. The notion of complexity cores was first defined in Lynch [1975]. It has been further studied in Du et al. [1984], Ko [1985b], Orponen and Schöning [1986], Orponen, Russo, and Schöning [1986] and Du and Book [1989]. Exercise 7.18 is from Du et al. [1984] and Orponen, Russo, and Schöning [1986].

Berman and Hartmanis [1977] also pointed out that all $\leq_m^P$-complete sets for $EXP$ are $\leq_{1,li}^P$-equivalent. Our proof of Theorem 7.35 is based on a simplified proof given by Watanabe [1985]. The relationship between one-way functions and the Berman-Hartmanis conjecture was first suggested by Joseph and Young [1985]. They constructed a special type of $NP$-complete sets, called *k-creative sets*, and conjectured that if one-way functions exist then $k$-creative sets are not polynomial-time isomorphic to SAT and, hence, the Berman-Hartmanis conjecture fails. Wang [1991, 1992, 1993] further studied the notion of creative sets and polynomial-time isomorphism. Selman [1992] surveyed the role of one-way functions in complexity theory. Theorem 7.38

was proved by Ko et al. [1986], and Theorem 7.39 was proved by Kurtz et al. [1988]. Kurtz et al. [1990] included a survey of results in this direction. Exercise 7.20 is from Homer et al. [1993].

Hartmanis [1978] conjectured that there is no sparse $P$-complete set under log-space many-one reduction. Ogihara [1995] showed that the Hartmanis conjecture is true if $P \neq NC^2$. Cai and Sivakumar [1995] improved this result by showing that the Hartmanis conjecture is true if $P \neq NC^1$. Cai and Ogihara [1997] and van Melkebeek and Ogihara [1997] provide surveys in this direction.

The Berman-Hartmanis conjecture has also been studied in the relativized form. It has been shown to hold relative to some oracles (see Kurtz [1983], Kurtz et al. [1989], and Hartmanis and Hemachandra [1991]) and also to fail relative to some other oracles (see Goldsmith and Joseph [1986] and Fenner et al. [1992]). Rogers [1995] found an oracle relative to which the Berman-Hartmanis conjecture fails but one-way functions exist. Wang and Belanger [1995] studied the isomorphism problem with respect to random instances.

# Part III

# Probabilistic Complexity

*It is true that you may fool all the people some of the time; you can even fool some of the people all the time; but you can't fool all of the people all the time.*

— Abraham Lincoln

# 8

# *Probabilistic Machines and Complexity Classes*

We develop, in this chapter, the theory of randomized algorithms. A randomized algorithm uses a random number generator to determine some parameters of the algorithm. It is considered as an efficient algorithm if it computes the correct solutions in polynomial time with high probability, with respect to the probability distribution of the random number generator. We introduce the probabilistic Turing machine as a model for randomized algorithms and study the relation between the class of languages solvable by polynomial-time randomized algorithms and the complexity classes $P$ and $NP$.

## 8.1 Randomized Algorithms

A randomized algorithm is a deterministic algorithm with the extra ability of making random choices during the computation that are independent of the input values. Using randomization, some worst-case scenarios may be hidden so that it only occurs with a small probability, and so the expected runtime is better than the worst-case runtime. We illustrate this idea by examples.

**Example 8.1** (*Quicksort*) We consider the following two versions of the well-known Quicksort algorithm.

*Deterministic Quicksort*
Input: a list $L = (a_1, \ldots, a_n)$ of integers;
if $n \leq 1$ then return $L$
else begin
1:   let $i := 1$;
     let $L_1$ be the sublist of $L$ whose elements are $< a_i$;
     let $L_2$ be the sublist of $L$ whose elements are $= a_i$;
     let $L_3$ be the sublist of $L$ whose elements are $> a_i$;
     recursively Quicksort $L_1$ and $L_3$;
     return $L$ as the concatenation of the lists $L_1$, $L_2$, and $L_3$
end.

*Randomized Quicksort*
{same as deterministic Quicksort, except line 1:}
1:   choose a random integer $i$, $1 \leq i \leq n$;

It is easy to see that both algorithms sort list $L$ correctly. We consider the runtime of the algorithms. Let $T_d(n)$ denote the maximum number of comparisons (of elements in $L$) made by the deterministic Quicksort algorithm on an input list $L$ of $n$ integers. It is clear that the worst case occurs when $L$ is reversely sorted and, hence, $a_1$ is always the maximum element in $L$. Thus, we get the following recurrence inequality:

$$T_d(n) \geq cn + T_d(n-1),$$

for some constant $c > 0$. This observation yields $T_d(n) = \Omega(n^2)$.

Let $T_r(n)$ be the expected number of comparisons made by the randomized Quicksort algorithm on an input list $L$ of $n$ integers. Let $s$ be the size of $L_1$. Since the integer $i$ is a random number in $\{1, \ldots, n\}$, the value $s$ is actually a random variable with the property $\Pr[s = j] = 1/n$ for all $j = 0, \ldots, n-1$ (assuming that all elements in $L$ are distinct). Thus, we obtain the following recurrence inequality:

$$T_r(n) \leq cn + \frac{1}{n} \sum_{j=0}^{n-1} [T_r(j) + T_r(n-1-j)],$$

for some constant $c > 0$. Solving this recurrence inequality, we get $T_r(n) = O(n \log n)$.

Using essentially the same proof, we can show that the deterministic Quicksort algorithm also has an expected runtime $O(n \log n)$, under the uniform distribution over all lists $L$ of $n$ integers. This result, however, is of quite a different nature from the result that $T_r(n) = O(n \log n)$. The expected runtime $O(n \log n)$ of the randomized Quicksort algorithm is the same for any input list $L$, since the probability distribution of the random number $i$ is independent of the input distribution. On the other hand, the expected runtime

$O(n \log n)$ of the deterministic Quicksort algorithm is only correct for the uniform distribution on the input domain. In general, it varies when the input distributions change. □

In the above, we have seen an example in which the randomized algorithm does not make mistakes and it has an expected runtime lower than the worst-case runtime. We achieved this by changing a deterministic parameter $i$ into a randomized parameter that transforms a worst-case input into an average-case input. (Indeed, all inputs are average-case inputs after the transformation.) In the following, we show another example of the randomized algorithms which run faster than the brute-force deterministic algorithm, but might make mistakes with a small probability. In this example, we avoid the brute-force exhaustive computation by a random search for witnesses (for the answer NO). A technical lemma shows that such witnesses are abundant and so the random search algorithm works efficiently.

**Example 8.2** (DETERMINANT OF A POLYNOMIAL MATRIX, or DPM). In this example, we consider matrices of multi-variable integer polynomials. An $n$-variable integer polynomial $Q(x_1, \ldots, x_n)$ is said to be of the *standard form* if it is written as the sum of terms of the form $c_k x_{i_1}^{d_1} x_{i_2}^{d_2} \ldots x_{i_k}^{d_k}$. For instance, the following polynomial is in the standard form:

$$Q(x_1, \ldots, x_5) = 3x_1 x_2^2 x_4 - 4x_2^5 x_3 x_5 + 3x_3 x_4 x_5^3.$$

A polynomial $Q$ in the standard form has degree $d$ if the maximum of the total degree of each term is $d$. For instance, the above polynomial $Q$ has degree 7.

The problem DPM is the following: Given an $m \times m$ matrix $\mathbf{Q} = [Q_{i,j}]$ of $n$-variable integer polynomials, and an $n$-variable integer polynomial $Q_0$, with $Q_0$ and all $Q_{i,j}$ in $\mathbf{Q}$ written in the standard form, determine whether $\det(\mathbf{Q}) = Q_0$, where $\det(\mathbf{Q})$ denotes the determinant of $\mathbf{Q}$. Even for small sizes of $m$ and $n$, the brute-force evaluation of the determinant of $\mathbf{Q}$ appears formidable. On the other hand, the determinant of an $m \times m$ matrix over integers in $\{-k, -k+1, \ldots, k-1, k\}$ can be computed in time polynomial in $(m + \log k)$. (In fact, as discussed in Chapter 6, this problem is solvable in $NC^2$.) In the following we present a randomized algorithm for DPM, based on the algorithm for evaluating determinants of integer matrices.[1]

> *Randomized Algorithm for* DPM
> Input: $\mathbf{Q} = [Q_{i,j}]$, $Q_0$, $\epsilon$;
> let $d' := \max\{\text{degree of } Q_{i,j} : 1 \leq i, j \leq m\}$;
> let $d_0 := \text{degree of } Q_0$;
> let $d := \max\{md', d_0\}$;
> let $k := \lceil -\log \epsilon \rceil$;

---

[1] In the rest of this example, $|u|$ denotes the absolute value of an integer $u$; do not confuse it with the length of the string $u$.

for $i := 1$ to $k$ do
begin
      choose $n$ random integers $u_1, \ldots, u_n$ in the range $\{-d, \ldots, d\}$;
      if $\det(\mathbf{Q}(u_1, \ldots, u_n)) \neq Q_0(u_1, \ldots, u_n)$ then output NO and halt
end;
output YES and halt.

Suppose the absolute values of the coefficients of the polynomials $Q_{i,j}$ and $Q_0$ are bounded by $r$. Then, the absolute value of each $Q_{i,j}(u_1, \ldots, u_n)$ is bounded by $d^{O(n+d)}r$, and so $\det(\mathbf{Q}(u_1, \ldots, u_n))$ can be computed in time polynomial in $(m + n + d + \log r)$. Thus this algorithm always halts in polynomial time. It, however, might make mistakes. We claim nevertheless that the error probability is smaller than $\epsilon$:

(1) If $\det(\mathbf{Q}) = Q_0$, then the above algorithm always outputs YES.

(2) If $\det(\mathbf{Q}) \neq Q_0$, then the above algorithm outputs NO with probability $\geq 1 - \epsilon$.

The claim (1) is obvious since the algorithm outputs NO only when a witness $(u_1, \ldots, u_n)$ is found such that $\det(\mathbf{Q}(u_1, \ldots, u_n)) \neq Q_0(u_1, \ldots, u_n)$. The claim (2) follows from the following lemma. Let $A_{n,m} = \{(u_1, \ldots, u_n) : |u_i| \leq m, 1 \leq i \leq n\}$.

**Lemma 8.3** *Let $Q$ be an $n$-variable integer polynomial of degree $d$ that is not identical to zero. Let $m \geq 0$. Then, $Q$ has at most $d \cdot (2m + 1)^{n-1}$ zeros $(u_1, \ldots, u_n)$ in $A_{n,m}$.*

*Proof.* The case $n = 1$ is trivial since a 1-variable nonzero polynomial $Q$ of degree $d$ has at most $d$ zeros. We show the general case $n > 1$ by induction. Write $Q$ as a polynomial of a single variable $x_1$, and let $d_1$ be the degree of $Q$ in $x_1$. Let $Q_1$ be the coefficient of $x_1^{d_1}$ in $Q$. Then $Q_1$ is an $(n - 1)$-variable polynomial of degree $\leq d - d_1$.

By induction, since $Q_1$ is not identical to zero, it has at most $(d - d_1)(2m + 1)^{n-2}$ zeros in $A_{n-1,m}$. For any $(v_2, \ldots, v_n)$ in $A_{n-1,m}$, if $(v_2, \ldots, v_n)$ is not a zero of $Q_1$, then $Q(x_1, v_2, \ldots, v_n)$ is a 1-variable polynomial not identical to zero and, hence, has at most $d_1$ zeros $u_1$, with $|u_1| \leq m$. If $(v_2, \ldots, v_n)$ is a zero of $Q_1$, then $Q(x_1, v_2, \ldots, v_n)$ might be identical to zero and might have $2m + 1$ zeros $u_1$, with $|u_1| \leq m$. Together, the polynomial $Q$ has at most

$$d_1 \cdot (2m + 1)^{n-1} + (2m + 1) \cdot (d - d_1) \cdot (2m + 1)^{n-2} = d \cdot (2m + 1)^{n-1}$$

zeros in $A_{n,m}$. ∎

The above lemma shows that if $\det(\mathbf{Q}) \neq Q_0$ then the probability that a random $(u_1, \ldots, u_n)$ in $A_{n,d}$ satisfies $\det(\mathbf{Q}(u_1, \ldots, u_n)) = Q_0(u_1, \ldots, u_n)$ is less than $1/2$. Thus, the probability that the algorithm fails to find a witness in $A_{n,d}$ for $k$ times is $\leq 2^{-k} \leq \epsilon$.

Our third example is the problem PRIME. We have seen in Section 4.1 that PRIME is in $NP \cap coNP$ but is not known to be in $P$. The following randomized algorithm solves the problem PRIME in polynomial time, with an arbitrarily small error probability $\epsilon$. Again, we search for a witness that asserts that the given input number is a composite number. A technical lemma shows that if the given number is a composite number then there are abundant witnesses and, hence, the error probability is small. The technical lemmas are purely number-theoretic results, and we omit the proofs.

**Example 8.4** (PRIMALITY TESTING, or PRIME) Recall that for any number $n > 0$, $Z_n^* = \{x \in \mathbf{N} : 0 < x < n,\ \gcd(x, n) = 1\}$ (see Example 4.1). In particular, if $p$ is a prime, then $Z_p^* = \{1, 2, \ldots, p - 1\}$. Note that $Z_n^*$ is a multiplicative group. For any $x \in Z_n^*$, $x$ is a *quadratic residue* modulo $n$ if $x \equiv y^2 \bmod n$ for some $y \in Z_n^*$. For any prime $p$ and $x \in Z_p^*$, the *Legendre symbol* $\left(\frac{x}{p}\right)$ is defined as

$$\left(\frac{x}{p}\right) = \begin{cases} 1 & \text{if } x \text{ is a quadratic residue modulo } p, \\ -1 & \text{otherwise.} \end{cases}$$

For $x > p$ with $\gcd(x, p) = 1$, we also write $\left(\frac{x}{p}\right)$ to denote $\left(\frac{x \bmod p}{p}\right)$. For any number $n = p_1 p_2 \cdots p_i$, where $p_1, \ldots, p_i$ are primes, not necessarily distinct, and for any $x > 0$ with $\gcd(x, n) = 1$, the *Legendre-Jacobi symbol* $\left(\frac{x}{n}\right)$ is defined as

$$\left(\frac{x}{n}\right) = \left(\frac{x}{p_1}\right) \left(\frac{x}{p_2}\right) \cdots \left(\frac{x}{p_i}\right).$$

Both Legendre symbols and Legendre-Jacobi symbols are polynomial-time computable, based on the following lemma in number theory.

**Lemma 8.5** *(a)* (Euler's Criterion) *For all primes $p > 2$ and all $x \in Z_p^*$, $\left(\frac{x}{p}\right) \equiv x^{(p-1)/2} \bmod p$.*

*(b)* (Law of Quadratic Reciprocity) *For all odd $m, n > 2$, if $\gcd(m, n) = 1$, then $\left(\frac{m}{n}\right) \cdot \left(\frac{n}{m}\right) = (-1)^{(m-1)(n-1)/4}$.*

Lemma 8.5(b) above shows that the Legendre-Jacobi symbols can be computed by a polynomial-time algorithm similar to the Euclidean algorithm for the greatest common divisors.

Now we are ready for the randomized algorithm for primality testing.

> *Randomized Algorithm for* PRIME
> Input: $n$, $\epsilon$;
> let $k := -\lceil \log \epsilon \rceil$;
> for $i := 1$ to $k$ do
> begin
>     choose a random $x$ in the range $\{1, \ldots, n - 1\}$;
>     if $\gcd(x, n) \neq 1$ then output NO and halt;
>     if $\left(\frac{x}{n}\right) \not\equiv x^{(n-1)/2} \bmod n$ then output NO and halt;
> end;
> Output YES and halt.

From the polynomial-time computability of the Legendre-Jacobi symbols, the above algorithm always halts in polynomial time. In addition, by Euler's criterion, if $n$ is a prime, then the above algorithm always outputs YES. Finally, the following lemma establishes the property that if $n$ is a composite number then the above algorithm outputs NO with probability $\geq 1 - \epsilon$.

**Lemma 8.6** *If $n$ is an odd, composite number, then set*

$$A = \left\{ x \in Z_n^* : \gcd(x, n) = 1 \text{ and } x^{(n-1)/2} \equiv \left(\frac{x}{n}\right) \bmod n \right\}$$

*is a proper subgroup of $Z_n^*$ and, hence, has size $|A| \leq (n-1)/2$.*

The above examples demonstrate the power of randomization in the design and analysis of algorithms. In the following sections, we present a formal computational model for randomized computation and discuss the related complexity problems.

## 8.2   Probabilistic Turing Machines

Our formal model for randomized computation is the probabilistic Turing machine. A multi-tape *probabilistic Turing machine* (PTM) $M$ consists of two components $(\tilde{M}, \phi)$, where $\tilde{M}$ is a regular multi-tape deterministic Turing machine and $\phi$ is a random bit generator. In addition to the input, output, and work tapes, the machine $M$ has a special *random-bit* tape. The computation of the machine $M$ on an input $x$ can be informally described as follows: At each move, the generator $\phi$ first writes a random bit $b = 0$ or $1$ on the square currently scanned by the tape head of the random-bit tape. Then the deterministic TM $\tilde{M}$ makes the next move according to the current state and the current symbols scanned by the tape heads of all (but output) tapes, including the random-bit tape. The tape head of the random-bit tape always moves to the right after each move.

For the sake of simplicity, we only define formally the computation of a 2-tape PTM, where the first tape is the input/output/work tape and the second tape is the random-bit tape. Also, we will use a fixed random bit generator $\phi$ which, when asked, outputs a bit $0$ or $1$ with equal probability. This generator, thus, defines a uniform probability distribution $U$ over infinite sequences in $\{0,1\}^\infty$.

With this fixed uniform random-bit generator $\phi$, a 2-tape PTM $M$ is formally defined by five components $(Q, \Sigma, q_0, F, \delta)$, which defines, as in Section 1.2, a 2-tape deterministic Turing machine $\tilde{M}$, with the following extra conditions:

(1) $\{0,1\} \subseteq \Sigma$.

(2) The function $\delta$ is defined from $Q \times \Sigma \times \{0,1\}$ to $Q \times \Sigma \times \{L, R\} \times \{R\}$. (That is, the random-bit tape is a read-only tape holding symbols in $\{0,1\}$ and always moves to the right.)

From the above conditions, we define a *configuration* of $M$ to be an element in $Q \times \Sigma^* \times \Sigma^* \times \{0,1\}^*$. Informally, a configuration $\gamma = (q, s, t, \alpha)$ indicates that the machine is currently in the state $q$, the current tape symbols in the first tape are $B^\infty s t B^\infty$, its tape head is currently scanning the first symbol of $t$, the random-bit tape contains symbols $\alpha$ that were generated by $\phi$ so far, and its tape head is scanning the cell to the right of the last bit of $\alpha$. (Note that the random bit tape symbols $\alpha$ do not affect the future configurations. We include them in the configuration only for the convenience of the analysis of the machine $M$.) Recall that $\vdash_{\tilde{M}}$ is the next-configuration function on the configurations of the machine $\tilde{M}$. Let $\gamma_1 = (q_1, s_1, t_1, \alpha_1)$ and $\gamma_2 = (q_2, s_2, t_2, \alpha_2)$ be two configurations of $M$. We write $\gamma_1 \vdash_M \gamma_2$, and say that $\gamma_2$ is a *successor* of $\gamma_1$, if there exists a bit $b \in \{0,1\}$ such that

(a) $\alpha_2 = \alpha_1 b$, and

(b) $(q_1, s_1, t_1, \lambda, b) \vdash_{\tilde{M}} (q_2, s_2, t_2, b, \lambda)$.

(Informally, $b$ is the random bit generated by $\phi$ at this move. Thus, the configuration $(q, s_1, t_1, \lambda, b)$ of the DTM $\tilde{M}$ shows, in addition to the configuration of the PTM $M$, the new random bit $b$ to be read by $\tilde{M}$.) We let $\vdash^*_M$ denote the reflexive, transitive closure of $\vdash_M$.

For each input $x$, let the initial configuration be $\gamma_x$; that is, $\gamma_x = (q_0, \lambda, x, \lambda)$. We say that a configuration $\gamma_1$ of $M$ is a *halting configuration* if there does not exist a configuration $\gamma_2$ such that $\gamma_1 \vdash_M \gamma_2$. In the following, we write $\alpha \prec \beta$ to mean that $\alpha$ is an initial segment of $\beta$.

**Definition 8.7** *Let $M$ be a PTM and $x$ an input string. Then the* halting probability $halt_M(x)$ *and the* accepting probability $accept_M(x)$ *of $M$ on input $x$ are defined as, respectively,*

$$halt_M(x) = \Pr\{\alpha_1 \in \{0,1\}^\infty : \text{there exists a halting configuration}$$
$$\gamma = (q, s, t, \alpha), \alpha \prec \alpha_1, \text{ such that } \gamma_x \vdash^*_M \gamma\},$$
$$accept_M(x) = \Pr\{\alpha_1 \in \{0,1\}^\infty : \text{there exists a halting configuration}$$
$$\gamma = (q, s, t, \alpha), q \in F, \alpha \prec \alpha_1, \text{ such that } \gamma_x \vdash^*_M \gamma\},$$

*where the probability is measured over the uniform distribution $U$ over the space $\{0,1\}^\infty$. The* rejecting probability *of $M$ on $x$ is*

$$reject_M(x) = halt_M(x) - accept_M(x).$$

The above definition involves the probability on the space of infinite sequences $\{0,1\}^\infty$ and is not convenient in the analysis of practical randomized algorithms. Note, however, that in each halting computation, the machine $M$ only uses finitely many random bits. Therefore, there is an equivalent definition involved only with the probability on finite strings in $\{0,1\}^*$. To be more precise, we treat $\tilde{M}$ as a 2-input DTM that satisfies the conditions (1) and (2) above. We assume that initially the first tape holds the first input $x \in \Sigma^*$, with all other squares having blank symbols $B$, and its tape head scans the first symbol of $x$. Also assume that the second tape initially contains the second input $\alpha \in \{0,1\}^*$, with all other squares having a symbol $0$,[2] and the tape head scans the first symbol of $\alpha$.

**Proposition 8.8** *Assume that $\tilde{M}$ has the above initial setting. Then,*

$$halt_M(x) = \sum\{2^{-|\alpha|} : \alpha \in \{0,1\}^*, \tilde{M}(x,\alpha) \text{ halts in exactly } |\alpha| \text{ moves}\},$$
$$accept_M(x) = \sum\{2^{-|\alpha|} : \alpha \in \{0,1\}^*, \tilde{M}(x,\alpha) \text{ halts in a state}$$
$$q \in F \text{ in exactly } |\alpha| \text{ moves}\}.$$

*Proof.* See Exercise 8.3.                                                              ∎

For notational purposes, we write $M(x) = 1$ to denote that $M$ accepts $x$, and $M(x) = 0$ to denote that $M$ rejects $x$. That is, $M(x)$ is a random variable with the probability $\Pr[M(x) = 1] = accept_M(x)$ and $\Pr[M(x) = 0] = reject_M(x)$.

**Definition 8.9** *Let $M$ be a PTM. We say that $M$ accepts an input $x$ if $\Pr[M(x) = 1] > 1/2$. We write $L(M)$ to denote the set of all inputs $x$ accepted by $M$. For any input $x$, the error probability $err_M(x)$ of $M$ on $x$ is defined to be $\Pr[M(x) \neq \chi_{L(M)}(x)]$.*

Our definition above appears arbitrary when, for instance, $\Pr[M(x) = 1] = \Pr[M(x) = 0] = 1/2$. An alternative, stronger definition for the set $L(M)$ requires that if $x \notin L(M)$, then the rejecting probability of $M(x)$ is greater than $1/2$; and, if both the accepting probability and the rejecting probability of $M(x)$ are $\leq 1/2$, then let $M(x)$ be undefined. For time-bounded probabilistic computation, however, to be shown in Proposition 8.15, these two definitions are essentially equivalent.

Finally, we point out some discrepancies between our model of probabilistic Turing machines and practical randomized algorithms. The first difference is that we have simplified the model so that the random-bit generator $\phi$ can only generate a bit 0 or 1. In practice, one might wish to get a random number $i$ in any given range. For instance, our model is too restrictive to generate a random number $i \in \{0,1,2\}$ with $\Pr[i = 0] = \Pr[i = 1] = \Pr[i = 2] = 1/3$. We might solve this problem by extending the model of probabilistic Turing machines to allow more general random number generators. Alternatively, we could just approximate the desired random number by our fixed random-bit generator $\phi$ (cf. Exercise 8.5).

The second difference is that in our formal model of probabilistic Turing machines, we defined the random-bit generator $\phi$ to generate a random bit at every step, no matter whether the deterministic machine $\tilde{M}$ needs it or not. In practice, we only randomize a few parameters when it is necessary. We could modify our definition of probabilistic Turing machines to reflect this property.

---

[2] The use of symbol 0 here is arbitrary. The machine $\tilde{M}$ is not supposed to read more than $|\alpha|$ bits on the second tape.

Exercise 8.4 shows that these two models are equivalent. In many applications, we note, however, that the model of Exercise 8.4 is more practical because the generation of a random bit could be much costlier than the runtime of the deterministic machine $\tilde{M}$.

## 8.3 Time Complexity of Probabilistic Turing Machines

The time complexity of a PTM can be defined in two ways: the worst-case time complexity and the expected time complexity. The worst-case time complexity of a PTM $M$ on an input $x$ measures the time used by the longest computation paths of $M(x)$. The expected time complexity of $M$ on $x$ is the average number of moves for $M(x)$ to halt with respect to the uniform distribution of the random bits used. In both cases, if $halt_M(x) < 1$ then the time complexity of $M$ on $x$ is $\infty$.

**Definition 8.10** *Let $M$ be a PTM, $x$ a string, and $n$ an integer. If $halt_M(x) < 1$, then we define $time_M(x) = \infty$; otherwise, we define $time_M(x)$ to be the the maximum value of the set $A_x = \{|\alpha| : \tilde{M}(x, \alpha)$ halts in exactly $|\alpha|$ moves$\}$. The time complexity function of $M$ is*

$$t_M(n) = \max\{time_M(x) : |x| = n\}.$$

**Definition 8.11** *Let $M$ be a PTM, $x$ a string, and $n$ an integer. If $halt_M(x) < 1$, then we define $expecttime_M(x) = \infty$; otherwise, we define $expecttime_M(x)$ to be the value*

$$\sum_{\alpha \in A_x} |\alpha| \cdot 2^{-|\alpha|}.$$

*The expected time complexity of $M$ is* [3]

$$\hat{t}_M(n) = \max\{expecttime_M(x) : |x| = n\}.$$

The expected time complexity is a natural complexity measure for randomized algorithms. People often use this complexity measure in the analysis of randomized algorithms. In the theory of polynomial-time computability, however, we will mostly work with the worst-case time complexity of probabilistic Turing machines. The main reason is that the worst-case time complexity is consistent with the time complexity measures of deterministic and nondeterministic machines, allowing us to study them in a unified model. In addition, for time- and error-bounded probabilistic computation, the worst-case time complexity is very close to the expected time complexity, as shown in the following proposition. We say two PTMs, $M_1$ and

---

[3]Note that the expected time of $M$ on $n$ is defined to be the maximum, rather than the average, of the expected time of $M$ on inputs of length $n$. The average of the expected time can be defined only dependent on the probability distribution of the input instances.

$M_2$, are *equivalent* if $L(M_1) = L(M_2)$, and they are *strongly equivalent* if $accept_{M_1}(x) = accept_{M_2}(x)$ for all inputs $x$.

**Proposition 8.12** *Assume that $M$ is a PTM with $accept_M(x) > 1/2 + \epsilon$ for all $x \in L(M)$ and $\hat{t}_M(n) \le t(n)$. Then, there exists an equivalent PTM $M_1$ with $t_{M_1}(n) \le t(n)/\epsilon$ such that for all $x$, $err_{M_1}(x) \le err_M(x) + \epsilon$.*

*Proof.* Let $M_1$ be the PTM that simulates $M$ on each input $x$ for at most $t(n)/\epsilon$ moves (it rejects if $M$ does not halt in $t(n)/\epsilon$ moves). For any $x$, let $\alpha$ be a binary string of length $t(|x|)/\epsilon$. Then, it is clear that the probability of $\tilde{M}(x, \alpha)$ not halting in $|\alpha|$ moves is at most $\epsilon$, which bounds the extra error probability $M_1$ might get on input $x$.                    ∎

Since we are going to work mostly with the worst-case time complexity, we restrict ourselves to PTMs with a uniform time complexity. A PTM $M$ has a *uniform time complexity* $t(n)$ if for all $x$, for all $\alpha$, $|\alpha| = t(|x|)$, $\tilde{M}(x, \alpha)$ halts in exactly $t(|x|)$ moves. The following proposition shows that we don't lose much by considering only PTMs with uniform time complexity.

**Proposition 8.13** *For any PTM $M$ with time complexity $t_M(n) \le t(n)$ for some time-constructible function $t$, there exists a strongly equivalent PTM $M_1$ that has a uniform time complexity $ct(n)$, for some constant $c > 0$.*

*Proof.* Since $t$ is time-constructible, we can attach a $t(n)$-time clock machine to $M$ to force it to halt in exactly $t(n)$ moves. We leave it as an exercise to fill in the details.                    ∎

We are ready to define the complexity classes of probabilistic computation.

**Definition 8.14** *For any function $t(n)$, we define $RTIME(t(n))$ to be the class of all sets accepted by a PTM with a uniform time complexity $t_M(n) \le t(n)$, and*

$$PP = \bigcup_{k>0} RTIME(n^k).$$

In Definition 8.9, the set $L(M)$ was defined to be the set of all strings $x$ with the accepting probability greater than $1/2$, and all other strings are defined to be in $\overline{L(M)}$. A slightly stronger definition requires that for all $x \in \overline{L(M)}$, its rejecting probability is greater than $1/2$. The following proposition shows that the class $PP$ is robust with respect to these two different definitions.

**Proposition 8.15** *Let $M$ be a PTM with the uniform time complexity $t(n)$. There exists an equivalent PTM $M_1$ with the uniform time complexity $t(n) + 2$ such that for all $x$, $\Pr[M_1(x) = \chi_{L(M)}(x)] > 1/2$.*

*Proof.* The idea of the proof is to increase the rejecting probability by $2^{-(t(n)+2)}$ for all strings of length $n$. Since all strings $x$ of length $n$ that are in $L(M)$ have the accepting probability $\ge 2^{-1} + 2^{-t(n)}$, this smaller amount of increase in the rejecting probability will not affect the language $L(M)$. Formally, we define $M_1$ by the following deterministic machine $\tilde{M}_1$. Recall that $0 \prec \alpha$ means that $\alpha$ begins with the symbol 0.

Deterministic Machine $\tilde{M}_1$
Input: $(x, y\alpha)$, $|y| = 2$ and $|\alpha| = t(n)$;
If $y \neq 00$ then accept $(x, y\alpha)$ if and only if $\tilde{M}$ accepts $(x, \alpha)$;
If $y = 00$ and $0 \prec \alpha$ and $\alpha \neq 0^n$ then accept $(x, y\alpha)$
else reject $(x, y\alpha)$.

Since machine $M$ has a uniform time complexity $t(n)$, we know that on an input $x$ of length $n$, it has an accepting probability $r \cdot 2^{-t(n)}$ for some integer $r$, $0 \leq r \leq 2^{t(n)}$. Then the accepting probability of $x$ by the new machine $M_1$ is equal to $(3r + 2^{t(n)-1} - 1) \cdot 2^{-(t(n)+2)}$. This accepting probability is greater than $1/2$ if $r > 2^{t(n)-1}$, and is less than $1/2$ if $r \leq 2^{t(n)-1}$. ∎

We defined the class $PP$ as the class of sets accepted by some polynomial-time PTMs. Our goal is to define the class of sets feasibly computable by randomized algorithms. Does the class $PP$ really consist of probabilistically feasible sets? The answer is NO; the class $PP$ appears too general, as demonstrated by the following relations.

**Theorem 8.16** $NP \subseteq PP \subseteq PSPACE$.

*Proof.* The fact that $PP$ is a subclass of $PSPACE$ is easy to see. Let $M$ be a PTM with a uniform time complexity $p(n)$. We consider the following deterministic machine $M_1$: On each $x$ of length $n$, $M_1$ enumerates all strings $\alpha$ of length $p(n)$ and verifies whether $\tilde{M}(x, \alpha)$ accepts; $M_1$ accepts $x$ if and only if $\tilde{M}(x, \alpha)$ accepts for more than $2^{p(n)-1}$ many $\alpha$. Since each $\alpha$ is only used in one simulation, the total amount of space used by $M_1$ on input $x$ is only $O(p(n))$.

For the relation $NP \subseteq PP$, assume that $A \in NP$. By the existential-quantifier characterization of $NP$, there exist a polynomial function $p$ and a polynomial-time computable predicate $R$ such that for all $x$,

$$x \in A \iff (\exists y, |y| = p(|x|))\ R(x, y).$$

Consider the following PTM $M$ for $A$ that has a uniform time complexity $p(n) + 1$:

$\tilde{M}$ accepts input $(x, by)$, with $|b| = 1$ and $|y| = p(|x|)$, if and only if $R(x, y)$ or $b = 1$.

If $x \in A$, then there exists at least one string $y$ of length $p(n)$ such that $R(x, y)$. Thus, the accepting probability of $x$ by $M$ is at least $1/2 + 2^{-(p(n)+1)}$. On the other hand, if $x \notin A$, then the accepting probability of $x$ by $M$ is exactly $1/2$. (Applying Proposition 8.15, we can reduce the accepting probability of $x \notin A$ to strictly below $1/2$.) ∎

The above theorem shows that each *NP*-complete problem has a polynomial-time randomized algorithm. A closer look at the above algorithm, however, reveals that this algorithm consists mainly of random guesses. Indeed, the error probability of the above algorithm is arbitrarily close to $1/2$ and so we do not have much confidence in the answer given by the algorithm. Thus, it suggests that *PP* is too loose as a candiate for the class of probabilistically feasible sets. Such a class must be defined by PTMs with stronger accepting requirements. We will study these sets in the next section.

The class *PP*, though not representing probabilistically feasible sets, is nevertheless an interesting complexity class. We study its computational power in Chapter 9. Some interesting properties of the class *PP* can be found in Exercises 8.6–8.8.

Finally we remark that the space complexity of a PTM, and the related space-bounded probabilistic complexity classes can be defined in a similar way. However, the space bounded probabilistic complexity classes do not appear to be natural, and are not directly related to practical randomized algorithms. We will not study them formally, but only include some interesting questions regarding space complexity of PTMs in the exercises (Exercise 8.10).

## 8.4  Probabilistic Machines with Bounded Errors

As we have seen in the last section, since the error probability of a probabilistic algorithm for a problem in *PP* could be arbitrarily close to $1/2$, the class *PP* is not a good candidate for this class. Instead, we need to consider stronger PTMs whose error probability is bounded away from $1/2$. We say a PTM $M$ has the error probability *bounded* by $c$ if $err_M(x) \leq c$ for all $x$. A PTM $M$ is a *polynomial-time* machine if it has a uniform time complexity $t_M(n) \leq p(n)$ for some polynomial $p$.

**Definition 8.17** *The class BPP (meaning* bounded-error probabilistic polynomial-time*) is the class of sets computable by polynomial-time PTMs that have the error probability bounded by a constant $c < 1/2$.*

For any $c$ such that $0 < c < 1/2$, let $BPP_c$ denote the class of sets defined by polynomial-time PTMs with the error probability $\leq c$. Then, *BPP* is the union of all $BPP_c$, $0 \leq c < 1/2$. It appears that the subclasses $BPP_c$ define a hierarchy of probabilistic classes between the classes $P$ and $PP$, and it is not clear where the correct cut point is for the class of probabilistically feasible problems. It turns out that there exist simple simulations among these classes so that they actually define a unique class. In other words, for all $c$, $0 < c < 1/2$, $BPP_c = BPP$. Furthermore, this simulation technique even allows us to reduce the error probability to $O(2^{-n})$, where $n$ is the input size. To prove this result, we first observe the following fact in probability theory.

**Lemma 8.18** *Let E be an event to occur with probability at least $1/2 + \epsilon$, $0 < \epsilon < 1/2$. Then, with probability at least $q = 1 - (1/2) \cdot (1 - 4\epsilon^2)^{k/2}$, E occurs more than $k/2$ times in $k$ independent trials.*

*Proof.* Note that $q$ increases as $\epsilon$ increases. Therefore, it suffices to show the lemma for the case of $\Pr[E] = 1/2 + \epsilon$. For each $i$, $0 \leq i \leq k$, let $q_i$ be the probability that $E$ occurs exactly $i$ times in $k$ independent trials. Then, for $i \leq k/2$, we have

$$q_i = \binom{k}{i} \cdot \left(\frac{1}{2} + \epsilon\right)^i \cdot \left(\frac{1}{2} - \epsilon\right)^{k-i}$$

$$\leq \binom{k}{i} \cdot \left(\frac{1}{2} + \epsilon\right)^i \cdot \left(\frac{1}{2} - \epsilon\right)^{k-i} \cdot \left(\frac{1/2 + \epsilon}{1/2 - \epsilon}\right)^{k/2-i} = \binom{k}{i} \cdot \left(\frac{1}{4} - \epsilon^2\right)^{k/2}.$$

Thus, the probability that $E$ occurs more than $k/2$ times is

$$1 - \sum_{i=0}^{\lfloor k/2 \rfloor} q_i \geq 1 - \sum_{i=0}^{\lfloor k/2 \rfloor} \binom{k}{i} \cdot \left(\frac{1}{4} - \epsilon^2\right)^{k/2}$$

$$\geq 1 - 2^{k-1} \cdot \left(\frac{1}{4} - \epsilon^2\right)^{k/2} = 1 - \frac{1}{2} \cdot \left(1 - 4\epsilon^2\right)^{k/2}. \qquad \blacksquare$$

**Theorem 8.19** *Let A be a set of strings. The following are equivalent:*

*(a) A is in BPP; that is, there exist a constant $c$, $0 < c < 1/2$, and a polynomial-time PTM M, with $L(M) = A$, such that for every $x$, $err_M(x) \leq c$.*

*(b) For every constant $c$, $0 < c < 1/2$, there exists a polynomial-time PTM M, with $L(M) = A$, such that for every $x$, $err_M(x) \leq c$.*

*(c) For every polynomial $q$, there exists a polynomial-time PTM M, with $L(M) = A$, such that for every $x$, $err_M(x) \leq 2^{-q(|x|)}$.*

*Proof.* The main idea is to simulate the same machine $M$ in (a) many times and take the majority answer. The relation between the number of simulations and the error probability follows easily from Lemma 8.18. This proof technique is usually termed as the *majority vote* technique or the *amplification of accepting probability*.

More precisely, assume that $A$ is computable by a PTM $M$ with runtime $p(n)$ and the error probability $\leq c$, where $p$ is a polynomial function and $0 < c < 1/2$. Let $r$ be a polynomial function. We consider the following PTM $M_r$, which has the uniform time complexity $O(p(n)r(n))$:

> On input $(x, \alpha_1 \cdots \alpha_{r(n)})$, where $|x| = n$ and $|\alpha_i| = p(n)$ for $1 \leq i \leq r(n)$, the deterministic machine $\tilde{M}_r$ accepts if and only if $\tilde{M}(x, \alpha_i)$ accepts for more than $r(n)/2$ many $i \in \{1, \ldots, r(n)\}$.

From Lemma 8.18, the error probability of $M_r$ is bounded by $(1/2) \cdot (1 - 4(1/2 - c)^2)^{r(n)/2}$. This error probability is bounded by $2^{-q(n)}$ if $r(n) =$

$m \cdot q(n)$, where

$$m = \left\lceil \frac{2}{-\log_2(1 - 4(1/2 - c)^2)} \right\rceil.$$

This shows that we can reduce the error probability from $c$ to $2^{-q(n)}$ with the time complexity increased by a factor of $m \cdot q(n)$.                                   ∎

The above theorem shows that $BPP$ is robust against the change of error probability, and is a good candidate for the class of probabilistically feasible sets. In addition, it is easy to see that $BPP$ is closed under union, intersection, and complementation.

The class $BPP$ contains sets computed by polynomial-time PTMs that make *two-sided* errors. Many randomized algorithms, however, have been found to only make *one-sided* errors in the sense that the algorithms do not make mistakes when the inputs are not in the sets accepted by the algorithms (or, equivalently, the answer YES is always correct). We formally define the complexity class $RP$ to contain the sets computed by feasible, randomized algorithms making one-sided errors.

**Definition 8.20** *A set $A$ is in $RP$ (meaning randomized polynomial-time) if there exists a polynomial-time PTM $M$ that accepts set $A$ and has $\mathrm{err}_M(x) = 0$ for every $x \notin A$.*

It is clear from the definition of $BPP$ that $BPP$ is closed under complementation. Since the machine for sets in $RP$ is allowed to make only one-sided errors, it is not known whether $RP$ is closed under complementation. We write $coRP$ to denote the class of all sets $A$ whose complements $\overline{A}$ are in $RP$. The problems DPM and PRIME in Section 8.1 are examples of problems in $coRP$: the answer NO of both algorithms of Examples 8.2 and 8.4 are always correct.

Similar to the class $BPP$, the classes $RP$ and $coRP$ are robust against the change of the error probability. The proof is essentially the same as Proposition 8.19. We leave it as an exercise.

**Theorem 8.21** *Let $A$ be a set of strings. The following are equivalent:*

*(a) $A \in RP$.*

*(b) For any constant $0 < c < 1$, there exists a polynomial-time PTM $M$ such that for all $x \in A$, $\Pr[M(x) = 1] \geq c$ and for all $x \notin A$, $\Pr[M(x) = 0] = 1$. (Note that $c$ does not have to be greater than $1/2$.)*

*(c) For any polynomial $q$, there exists a polynomial-time PTM $M$ such that $\mathrm{err}_M(x) \leq 2^{-q(n)}$ for all $x \in A$, and $\mathrm{err}_M(x) = 0$ for all $x \notin A$.*

Since $RP$ is not known to be closed under complementation, the intersection class $RP \cap coRP$ is not known to be equal to $RP$. We let $ZPP$ denote this class. The name $ZPP$ means *zero-error probabilistic (expected) polynomial-time*, which is justified by the following observation.

**Theorem 8.22** *A set A is in ZPP if and only if there exists a PTM M such that M computes A with zero error probability in the expected time p(n) for some polynomial p.*

*Proof.* First we prove the backward direction. Let $M$ be a PTM that has a zero error probability and has the expected time $p(n)$. Let $A = L(M)$. We define a new PTM $M_1$ that simulates $M$ on each input $x$ for at most $4p(n)$ moves. If a computation of $M$ halts after $4p(n)$ moves, then $M_1$ accepts if and only if $M$ accepts; otherwise, $M_1$ rejects the input $x$.

From the same argument as the proof of Proposition 8.12, the error probability of $M_1$ is bounded by $1/4$. Furthermore, consider any $x \notin A$. Since $M$ does not make mistakes, all its halting computations reject $x$. In addition, when $M$ does not halt on $x$ in $4p(|x|)$ moves, $M_1$ correctly rejects $x$. Thus, the error probability of $M_1$ on $x$ is zero. This shows that $A \in RP$. The proof for $A \in coRP$ is symmetric.

Conversely, assume that $A \in RP \cap coRP$. Then there exist two polynomial-time PTMs, $M_0$ and $M_1$, such that

$$x \in A \;\Rightarrow\; \Pr[M_0(x) = 1] = 1 \text{ and } \Pr[M_1(x) = 1] > 1/2,$$
$$x \notin A \;\Rightarrow\; \Pr[M_0(x) = 0] > 1/2 \text{ and } \Pr[M_1(x) = 0] = 1.$$

Now, we define a new PTM $M$ that on each $x$ alternatingly simulates $M_0(x)$ and $M_1(x)$ until it gets a definite answer (i.e., either when $M_1$ accepts $x$ or when $M_0$ rejects $x$). More precisely, let $p_0$ and $p_1$ be two polynomial functions that bound the runtime of $M_0$ and $M_1$, respectively. For any $x$ of length $n$ and any infinte sequence $\alpha \in \{0,1\}^\infty$, write $\alpha = \alpha_0 \alpha_1 \cdots$, with $|\alpha_{2i}| = p_0(n)$ and $|\alpha_{2i+1}| = p_1(n)$, for all $i = 0,1,2,\ldots$. Then, the deterministic machine $\tilde{M}(x, \alpha)$ accepts if $\tilde{M}_1(x, \alpha_{2i+1})$ accepts for some $i \geq 0$, and it rejects if $\tilde{M}_0(x, \alpha_{2i})$ rejects for some $i \geq 0$. It is clear that $M$ never makes mistakes. Let $|x| = n$. The expected runtime of $M$ on $x$ is at most

$$(p_0(n) + p_1(n)) \sum_{i=1}^\infty \frac{i}{2^i} = 2(p_0(n) + p_1(n)). \qquad \blacksquare$$

## 8.5 *BPP* and *P*

We now study the relationship between the probabilistic complexity classes and other complexity classes such as *P*, *NP*, and the polynomial-time hierarchy. We first observe that all the probabilistic complexity classes are defined by PTMs with polynomial uniform time complexity. This type of probabilistic machine is very similar to a polynomial-time nondeterministic machine. Indeed, let $M$ be a $p(n)$-time nondeterministic Turing machine in which any nondeterministic move consists of at most two nondeterministic choices. For the sake of simplicity, we may assume that $M$ has exactly two nondeterministic choices at each step (which are allowed to be the same choice). Then, for each input $x$ of length $n$, the computation of $M(x)$ consists of $2^{p(n)}$ computation paths each of length $p(n)$. We defined that $M$ accepts $x$ if and only if one of these computation paths accepts $x$. We may treat this nondeterministic machine $M$ as a probabilistic TM. The only difference is that we say the probabilistic TM $M$ accepts $x$ if and only if $M(x)$ has more than $2^{p(n)-1}$ accepting computation paths. By different requirements on the number of accepting and rejecting computation paths that $M(x)$ must have, the classes $BPP$ and $RP$ can be defined likewise. This view is very useful for the comparison between the probabilistic complexity classes with other complexity classes, as well as for the error analysis of PTMs.

From the above observation and other trivial simulations, we obtain the following relations.

**Proposition 8.23** *(a)* $P \subseteq ZPP \subseteq RP \subseteq BPP \subseteq PP$.
*(b)* $RP \subseteq NP$.

The question of whether $ZPP$, $RP$, or $BPP$ is equal to $P$ is an interesting one. Although these probabilistic classes appear intuitively stronger than the class $P$, we do not know many candidates of sets that are in these probabilistic classes and yet not known to be in $P$. We have seen in Section 8.1 that the problems DPM and PRIME are in $coRP$ but are not known to be in $P$. It is actually provable that the problem PRIME is also in $RP$ and, hence, it is a candidate for a problem in $ZPP - P$. The proof of this result involves complicated number-theoretic arguments and is beyond our scope. In the following, we include a simpler problem that is provably in $ZPP$ but not known to be in $P$.

**Example 8.24** (SQUARE ROOT MODULO A PRIME) Recall that a number $x$ is a quadratic residue modulo an integer $p$ ($x \in QR_p$) if there exists a number $y$ such that $y^2 \equiv x \bmod p$. The problem to be considered here is to find, for two given integers $x$ and $p$, where $p$ is a prime and $x \in QR_p$, the square root $y$ such that $y^2 \equiv x \bmod p$. This problem is known to be as hard as factoring the integer $p$ if $p$ is not known to be a prime.

Recall from Example 8.4 that for a prime number $p$, $Z_p^* = \{1, 2, \ldots, p-1\}$. For any $b \in Z_p^*$, $(\frac{b}{p})$ is the Legendre symbol of $b$ which is equal to 1 if $b \in QR_p$ and equal to $-1$ if $b \notin QR_p$. The randomized algorithm for the square roots modulo a prime number $p$ is based on the following number-theoretic lemmas.

**Lemma 8.25** *Let $p$ be an odd prime. Then there are exactly $(p-1)/2$ numbers in $Z_p^*$ that are quadratic residue modulo $p$.*

*Proof.* Observe that for any $x \in Z_p^*$, $x^2 \equiv (p - x)^2 \bmod p$. In addition, for $x, y \in Z_p^*$, $x^2 \equiv y^2 \bmod p$ implies $x = y$ or $x + y = p$. ∎

**Lemma 8.26** *There exists a deterministic polynomial-time algorithm that, given a prime $p$, a number $x \in QR_p$, and a number $u \in Z_p^*$ such that $u \notin QR_p$, finds a square root $y$ of $x$ modulo $p$.*

*Proof.* Let $p - 1 = 2^e q$ for some odd number $q$. Define inductively a sequence $x_1, x_2, \ldots$ of quadratic residues modulo $p$ and a sequence $e = k(0) > k(1) > k(2) > \cdots$ of strictly decreasing integers as follows:

$$x_1 = x,$$
$$k(n) = \min\{h : h \geq 0, x_n^{2^h q} \equiv 1 \bmod p\},$$
$$x_{n+1} = x_n \cdot u^{2^{e-k(n)}} \bmod p.$$

We claim that if $k(n) > 0$, then $x_{n+1} \in QR_p$ and $k(n+1) < k(n)$.

We prove this claim by induction. First, since $x = x_1 \in QR_p$, $x_1^{2^{e-1}q} \equiv 1 \bmod p$ (by Euler's criterion) and, hence, $k(1) < e = k(0)$. Next, assume that $k(n) > 0$. Then

$$x_{n+1}^{2^{k(n)-1}q} \equiv (x_n \cdot u^{2^{e-k(n)}})^{2^{k(n)-1}q}$$
$$\equiv x_n^{2^{k(n)-1}q} \cdot u^{2^{e-1}q} \equiv (-1)(-1) \equiv 1 \pmod{p},$$

where the last congruence uses the minimality of $k(n)$ and Euler's criterion on $u$ which is not a quadratic residue modulo $p$. Thus, we have $k(n+1) < k(n)$. We also have $x_{n+1}^{2^{e-1}q} \equiv 1 \bmod p$ and, hence, $x_{n+1} \in QR_p$.

The above claim implies that there exists an $m > 0$ such that $k(m) = 0$. For this $m$, we have $[x_m^q \equiv 1 \bmod p]$ and, hence, $y_m = (x_m^{(q+1)/2} \bmod p)$ is a square root of $x_m$. Inductively, for $n = m - 1, m - 2, \ldots, 1$, we can find the square root of $x_n$ by

$$y_n = y_{n+1} \cdot \left(u^{2^{e-k(n)-1}}\right)^{-1} \bmod p.$$

Note that $m \leq e \leq \log_2 p$, and so the above algorithm finds $y_1$, the square root of $x$ modulo $p$, in polynomial time. ∎

It is now clear how to find a square root of $x$ modulo $p$:

> *Randomized Algorithm for* SQUARE ROOT MODULO A PRIME.
> Input: a prime $p$, a number $x \in QR_p$.
> 1: Select a random number $u \in \{1, 2, \ldots, p-1\}$;
> If $\left(\frac{u}{p}\right) = -1$ then compute square root of $x$ modulo $p$ as above
> else goto 1.

The above algorithm never outputs a wrong number. Furthermore, the expected time for the algorithm to hold is only about two times the time for the deterministic algorithm of Lemma 8.26.

The lack of natural problems in *BPP*, *RP*, and *ZPP* that are not known to be in *P* makes it difficult to distinguish these probabilistic classes from the class *P*. To make it even more difficult, these few candidates are not known to be complete for these probabilistic complexity classes. In fact, the probabilistic complexity classes *BPP*, *RP*, and *ZPP*, like *NP* ∩ *coNP*, do not seem to

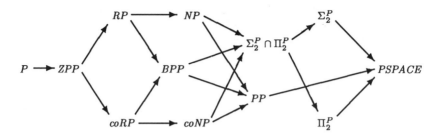

Figure 8.1: Inclusive relations on probabilistic complexity classes. $A \to B$ means $A \subseteq B$.

possess complete sets at all. From the proof of the first, generic $NP$-complete problem BHP, we see that a complexity class $\mathcal{C}$ has a complete set if the underlying machines defining class $\mathcal{C}$ are effectively enumerable. For a class $\mathcal{C}$ without this property, it appears difficult to even find the first complete set for $\mathcal{C}$. While polynomial time-bounded PTMs are effectively enumerable, those with bounded error probabilities are not likely to be effectively enumerable. In other words, there does not seem to exist a *universal BPP* machine that can simulate, from the codes of PTMs, all *BPP* machines and keep a bounded error probability. Without the universal *BPP* machine, no other method is known to establish the first *BPP*-complete problem.

## 8.6 *BPP* and *NP*

The relationship between the class *BPP* and class *NP* and the polynomial-time hierarchy *PH* is rather complex. We do not know whether *BPP* is contained in *NP* or $\Delta_2^P$. The best we know is that *BPP* is contained in $\Sigma_2^P \cap \Pi_2^P$. We also have some evidence that *BPP* does not contain *NP*. Figure 8.1 shows the inclusive relations between these classes that are known so far.

To study these relations, we first prove that *BPP* has polynomial-size circuits. Recall that *P/poly* denotes the class of sets solvable by polynomial-size circuits, or, equivalently, a set $A$ is in *P/poly* if there exist a set $B \in P$ and a polynomial length-bounded function $h : \mathbf{N} \to \{0,1\}^*$ such that for all $x$, $x \in A$ if and only if $\langle x, h(|x|) \rangle \in B$. From Theorem 6.11, we have seen that it is unlikely that $NP \subseteq P/poly$ unless the polynomial-time hierarchy collapses. Thus, the following result shows that it is unlikely that $NP \subseteq BPP$.

**Theorem 8.27** $BPP \subseteq P/poly$.

*Proof.* Assume that $A \in BPP$. Then, from Theorem 8.19, there exists a polynomial-time PTM $M$ such that for each $x$ of length $n$, $err_M(x) \leq 2^{-2n}$. Equivalently, there exists a polynomial function $p$ such that for every $x$ of length $n$,

$$|\{\alpha : |\alpha| = p(n), \tilde{M}(x, \alpha) \neq \chi_A(x)\}| \leq 2^{p(n)-2n}.$$

Consider all strings $x$ of length $n$. There are $2^n$ such strings $x$. This implies that there are at least

$$2^{p(n)} - 2^n \cdot 2^{p(n)-2n} = 2^{p(n)} \cdot (1 - 2^{-n}) > 1$$

strings $\alpha \in \{0, 1\}^*$, $|\alpha| = p(n)$, for which $\tilde{M}(x, \alpha) = \chi_A(x)$ for all $x$ of length $n$. Choose a fixed such $\alpha$ and call it $h(n)$. Then we can see that $A \in P/poly$ since $x \in A$ if and only if $\tilde{M}(x, h(|x|))$ accepts. ∎

*Remark.* An easy way to visualize the argument in the above proof is as follows: First define a $2^n \times 2^{p(n)}$ binary matrix $E$, and associate each row with a binary string $x$ of length $n$ and each column with a binary string $\alpha$ of length $p(n)$. Each element $E(x, \alpha)$ has the value 1 if $\tilde{M}(x, \alpha) = \chi_A(x)$ and has the value 0 otherwise. Then, the assumption that $err_M(x) \leq 2^{-2n}$ for all $x$ of length $n$ means that each row of $E$ has at most $2^{p(n)-2n}$ zeros. Since there are only $2^n$ rows, $E$ has totally at most $2^{p(n)-n}$ zeros which cannot fill out one single row. Therefore, there must be at least one column $\alpha$ on which $E$ has all 1's.

**Corollary 8.28** *If $NP \subseteq BPP$, then the polynomial-time hierarchy collapses to the second level $\Sigma_2^P$.*

*Proof.* The corollary follows from Theorems 8.27 and 6.11. ∎

The above result also has an interesting application to the relation between $P$ and $BPP$. Recall that $EXP$ and $EXPSPACE$ denote the classes of sets computable by DTMs in time $2^{O(n)}$ and in space $2^{O(n)}$, respectively.

**Corollary 8.29** *If $P \neq BPP$ then $EXP \neq EXPSPACE$.*

*Proof.* First we note that by a translation between binary encoding and unary encoding, we know that $EXP \neq EXPSPACE$ if and only if there exists a tally set (a set over a singleton alphabet $\{0\}$) in $PSPACE - P$ (cf. Exercise 1.30).

Assume that $A \in BPP - P$. Then, by Theorem 8.27, there exist a PTM $M$ and a polynomial $p$ having the following property: For each $n$ there is a string $\alpha$ of length $p(n)$ such that for all $x$ of length $n$, $x \in A$ if and only if $\tilde{M}(x, \alpha)$ accepts. Let $w_n$ be the first string of length $p(n)$ having this property.

Define $C = \{0^{\langle n,k \rangle} : \text{the } k\text{th bit of } w_n \text{ is } 1\}$. We claim that $C$ is a tally set in $PSPACE - P$. First, set $A$ is computable in polynomial time with oracle $C$: for each $x$ of length $n$, first query the oracle $C$ to get $w_n$ and then verify whether $\tilde{M}(x, w_n)$ accepts. This implies that $C \notin P$ because $A \notin P$. Also, $C \in PSPACE$ because we can find the least string $w_n$ having the property that for all $x$ of length $n$, $[x \in A \iff \tilde{M}(x, w_n)$ accepts] by checking through every string of length $p(n)$. ∎

Another interesting observation on the relation between $NP$ and $BPP$ is that if problems in $NP$ are polynomial-time computable by PTMs with two-sided bounded errors, then it actually can be computed in polynomial time by PTMs with one-sided bounded errors. Intuitively, this is so because the class $NP$ itself is defined by the notion of one-sided errors.

**Theorem 8.30** *If $NP \subseteq BPP$ then $NP = RP$.*

*Proof.* Since $RP \subseteq NP$, it suffices to show that if $NP \subseteq BPP$ then $NP \subseteq RP$. Next observe that $RP$ is closed under $\leq_m^P$-reducibility. Therefore, it suffices to show that if the $NP$-complete set SAT is in $BPP$ then SAT is in $RP$.

Recall that SAT has the $c$-self-reducibility property (see Section 7.3). That is, for any Boolean formula $F(x_1, \ldots, x_n)$, with $n \geq 1$, $F \in$ SAT if and only if $F|_{x_1=0} \in$ SAT or $F|_{x=1} \in$ SAT. We assume, by a simple padding technique, that for each formula $F$ and any Boolean values $b_1, \ldots, b_i$, $1 \leq i \leq n$, the induced formula $F|_{x_1=b_1, \ldots, x_i=b_i}$ has the same length as $F$. Assume that SAT $\in BPP$ such that there exists a polynomial-time PTM $M$ computing SAT with the error probability of $M(F)$ bounded by $2^{-|F|}$ (by Theorem 8.19). Now for each $F \in$ SAT, we can use the self-reducibility of SAT to produce a truth assignment for $F$ as follows.

> *Randomized Algorithm $M_1$ for* SAT:
>
> Input: A Boolean formula $F$ with $n$ variables.
> If $M(F) = 0$ then reject $F$;
> For $i := 1$ to $n$ do
>     If $M(F|_{x_1=a_1, \ldots, x_{i-1}=a_{i-1}, x_i=0}) = 1$ then let $a_i := 0$
>     else if $M(F|_{x_1=a_1, \ldots, x_{i-1}=a_{i-1}, x_i=1}) = 1$ then let $a_i := 1$
>                               else reject $F$ and halt;
> If $F|_{x_1=a_1, \ldots, x_n=a_n} = 1$ then accept $F$ else reject $F$.

Observe that $M_1$ accepts $F$ only if a truth assignment $t(x_i) = a_i$ that satisfies $F$ is found. Therefore, $M_1$ never makes mistakes if $F \notin$ SAT. On the other hand, if $F \in$ SAT, then the probability of $M$ rejecting $F$ in each iteration of the loop is only $2^{-|F|}$. So, the probability of $M_1$ accepting $x$ is $(1 - 2^{-|F|})^n$, which is greater than $1/2$ if $|F| \geq n > 1$. ∎

## 8.7 *BPP* and the Polynomial-Time Hierarchy

In this section, we prove that $BPP$ is contained in the second level of the polynomial-time hierarchy. In order to prove this result, we need to further explore the similarity between probabilistic and nondeterministic Turing machines and to establish a polynomial quantifier characterization of the class $BPP$ similar to that for $PH$ in Theorem 3.8.

For any predicate $R(y)$ and any rational number $r \in (0,1)$, we write $(\exists_r^+ y, |y| = m) \, R(y)$ to denote the following predicate: There exist at least

$r \cdot 2^m$ many strings $y$ of length $m$ for which $R(y)$ holds. To simplify the notation, we drop the subscript $r$ when $r = 3/4$. We call $\exists_r^+$ a *probabilistic quantifier*, and $\exists^+$ a *majority quantifier*. Using the majority quantifier, we can characterize complexity classes *BPP* and *RP* in the following form: A set $A$ is in *BPP* if and only if there exist a predicate $R \in P$ and a polynomial $p$ such that for all $x$,

$$x \in A \;\Rightarrow\; (\exists^+ y, |y| = p(|x|))\; R(x, y),$$
$$x \notin A \;\Rightarrow\; (\exists^+ y, |y| = p(|x|))\; \neg R(x, y);$$

A set $A$ is in *RP* if and only if there exist a predicate $R \in P$ and a polynomial $p$ such that for all $x$,

$$x \in A \;\Rightarrow\; (\exists^+ y, |y| = p(|x|))\; R(x, y),$$
$$x \notin A \;\Rightarrow\; (\forall y, |y| = p(|x|))\; \neg R(x, y).$$

The above characterization of *RP* is *decisive*, in the sense that if we replace the quantifier $\exists^+$ by the weaker quantifier $\exists$, then the two predicates $(\exists y, |y| = p(|x|))R(x, y)$ and $(\forall y, |y| = p(|x|))\neg R(x, y)$ are still complementary to each other, and for any $x$, exactly one of these predicates holds for $x$. (Two predicates $R_1$ and $R_0$ are *complementary* if $R_0 \Rightarrow \neg R_1$.) Furthermore, when the quantifier $\exists^+$ is replaced by $\exists$, the two predicates characterize the class *NP* and, hence, $RP \subseteq NP$. On the other hand, the above characterization of *BPP* is not *decisive*, since if we replace the quantifier $\exists^+$ by the quantifier $\exists$, the two resulting predicates are not complementary and do not define a complexity class. What we need is a decisive quantifier characterization for *BPP* such that when we replace $\exists^+$ by $\exists$, we get a characterization of a set in the polynomial-time hierarchy. To do this, we first show that the quantifiers $\exists^+$ and $\forall$ can be *swapped* under certain circumstances.

**Lemma 8.31** (Swapping Lemma) *Let $R(x, y, z)$ be a predicate that holds only if $|y| = |z| = p(|x|)$ for some polynomial $p$. Let $|x| = n$ for some sufficiently large $n$. In the following, we write $\vec{w}$ to denote $\langle w_1, \ldots, w_{p(n)} \rangle$, with each $w_i$ of length $p(n)$, $i = 1, \ldots, p(n)$. Also, $z \in \vec{w}$ means $z = w_i$ for some $i = 1, \ldots, p(n)$.*

*(a)* $(\forall y, |y| = p(n))\, (\exists^+ z, |z| = p(n))\; R(x, y, z)$
$$\Rightarrow (\exists^+ \vec{w})\, (\forall y, |y| = p(n))\; [\exists z \in \vec{w}\; R(x, y, z)].$$
*(b)* $(\forall z, |z| = p(n))\, (\exists^+_{1-2^{-n}} y, |y| = p(n))\; R(x, y, z)$
$$\Rightarrow (\forall \vec{w})\, (\exists^+ y, |y| = p(n))\; [\forall z \in \vec{w}\; R(x, y, z)].$$

*Remark.* Again, we can view the above lemma in terms of a binary matrix $E$ of size $2^{p(n)} \times 2^{p(n)}$, with $E(y, z) = 1$ if and only if $R(x, y, z)$. Part (a) states that if every row of $E$ has more than $(3/4) \cdot 2^{p(n)}$ many 1's, then for the majority of the choices $\vec{w}$ of $p(n)$ many columns, every row of $E$ contains at least one 1 in these columns. Part (b) states that if every column of $E$ has at most $2^{p(n)-n}$ many 0's, then for all choices $\vec{w}$ of $p(n)$ many columns, most rows have 1's in all of these columns.

*Proof.* (a) Let $\vec{w}$ be a list of $p(n)$ strings randomly chosen from $\{0,1\}^{p(n)}$. Then,

$$\Pr[(\exists y, |y| = p(n)) \, (\forall z \in \vec{w}) \, \neg R(x,y,z)]$$
$$\leq \sum_{|y|=p(n)} \Pr[(\forall z \in \vec{w}) \, \neg R(x,y,z)]$$
$$= \sum_{|y|=p(n)} \prod_{z \in \vec{w}} \Pr[\neg R(x,y,z)] \leq \sum_{|y|=p(n)} \prod_{z \in \vec{w}} \frac{1}{4} \leq \frac{1}{4} \, .$$

(b) Assume that $n$ is sufficiently large so that $2^n \geq 4p(n)$. Let $y$ be a string randomly chosen from $\{0,1\}^{p(n)}$. For any $z \in \{0,1\}^{p(n)}$, we have $\Pr[\neg R(x,y,z)] \leq 2^{-n}$. So, for any $\vec{w}$ of $p(n)$ strings,

$$\Pr[(\exists z \in \vec{w}) \, \neg R(x,y,z)] \leq \sum_{z \in \vec{w}} \Pr[\neg R(x,y,z)] \leq p(n) \cdot 2^{-n} \leq \frac{1}{4} \, . \qquad \blacksquare$$

From this Swapping Lemma we obtain the following decisive characterization of *BPP*.

**Theorem 8.32** (BPP Theorem) *The following are equivalent:*
    *(a) $A \in BPP$.*

   *(b) There exist a polynomial-time predicate $R$ and a polynomial function $p$ such that for all $x$ of length $n$,*

$$x \in A \Rightarrow (\exists^+ y, |y| = p(n)) \, (\forall z, |z| = p(n)) \, R(x,y,z),$$
$$x \notin A \Rightarrow (\forall y, |y| = p(n)) \, (\exists^+ z, |z| = p(n)) \, \neg R(x,y,z). \qquad (8.1)$$

   *(c) There exist a polynomial-time predicate $R$ and a polynomial function $p$ such that for all $x$ of length $n$,*

$$x \in A \Rightarrow (\forall y, |y| = p(n)) \, (\exists^+ z, |z| = p(n)) \, R(x,y,z),$$
$$x \notin A \Rightarrow (\exists^+ y, |y| = p(n)) \, (\forall z, |z| = p(n)) \, \neg R(x,y,z). \qquad (8.2)$$

*Proof.* Let $\oplus$ denote the exclusive-or operator on two strings of the equal length; that is, $0 \oplus 0 = 1 \oplus 1 = 0$, $1 \oplus 0 = 0 \oplus 1 = 1$, and if $|y| = |z|$ and $|a| = |b| = 1$, then $(ya \oplus zb) = (y \oplus z)(a \oplus b)$.

We first prove (a) $\Rightarrow$ (b). Let $A \in BPP$. Then, by Theorem 8.19, there exist a polynomial-time predicate $Q$ and a polynomial function $q$ such that for all $x$ of length $n$,

$$x \in A \Rightarrow (\exists^+_r y, |y| = q(n)) \, Q(x,y),$$
$$x \notin A \Rightarrow (\exists^+_r y, |y| = q(n)) \, \neg Q(x,y), \qquad (8.3)$$

where $r = 1 - 2^{-n}$. Now, by part (a) of the Swapping Lemma, we have, for all $x$ of length $n$,

$$x \in A \;\Rightarrow\; (\exists_r^+ z, |z| = q(n)) \; Q(x, z)$$
$$\Rightarrow\; (\forall y, |y| = q(n)) \, (\exists_r^+ z, |z| = q(n)) \; Q(x, y \oplus z)$$
$$\Rightarrow\; (\exists^+ \vec{w}) \, (\forall y, |y| = q(n)) \, [\exists z \in \vec{w} \; Q(x, y \oplus z)],$$

where $\vec{w}$ denotes a list of $q(n)$ strings each of length $q(n)$. In addition, by part (b) of the Swapping Lemma, we have

$$x \notin A \;\Rightarrow\; (\exists_r^+ y, |y| = q(n)) \; \neg Q(x, y)$$
$$\Rightarrow\; (\forall z, |z| = q(n)) \, (\exists_r^+ y, |y| = q(n)) \; \neg Q(x, y \oplus z)$$
$$\Rightarrow\; (\forall \vec{w}) \, (\exists^+ y, |y| = q(n)) \, [\forall z \in \vec{w} \; \neg Q(x, y \oplus z)].$$

Note that $R(x, \vec{w}, y) \equiv [(\exists z \in \vec{w}) \; Q(x, y \oplus z)]$ is a polynomial-time predicate. This proves the forward direction.

Conversely, for (b) $\Rightarrow$ (a), assume that there exist a polynomial-time predicate $R$ and a polynomial function $p$ such that $A$ satisfies (8.1). For each string $w$ of length $2p(n)$, we write $w_1$ as the first half of $w$, and $w_2$ the second half of $w$; that is, $|w_1| = |w_2| = p(n)$ and $w = w_1 w_2$. Then, for each string $x$ of length $n$,

$$x \in A \;\Rightarrow\; (\exists^+ y, |y| = p(n)) \, (\forall z, |z| = p(n)) \; R(x, y, z)$$
$$\Rightarrow\; (\exists^+ w, |w| = 2p(n)) \; R(x, w_1, w_2),$$
$$x \notin A \;\Rightarrow\; (\forall y, |y| = p(n)) \, (\exists^+ z, |z| = p(n)) \; \neg R(x, y, z)$$
$$\Rightarrow\; (\exists^+ w, |w| = 2p(n)) \; \neg R(x, w_1, w_2).$$

This shows that $A$ is in *BPP*.

The equivalence of (a) and (c) follows from the fact that *BPP* is closed under complementation. ∎

By changing in the above the quantifier $\exists^+$ to $\exists$, we obtain immediately the following relation:

**Corollary 8.33** *BPP* $\subseteq \Sigma_2^P \cap \Pi_2^P$.

Observe that in the forward direction of Theorem 8.32, the proof did not use the fact that $Q$ is polynomial-time computable. Thus, it actually shows that if a set $A$ satisfies (8.3) with respect to any predicate $Q$ and any polynomial function $q$, then it can be characterized as in (8.1), with the predicate $R$ polynomial-time computable relative to $Q$. This observation will be useful in Chapter 10, when we study other complexity classes defined by alternating $\exists$ and $\exists^+$ quantifiers. We state this more general result as the *generalized BPP Theorem*.

**Theorem 8.34** (Generalized BPP Theorem) *Assume that $R_1$ and $R_0$ are two complementary predicates, and $q$ is a polynomial. Let*

$$S_i(x) \equiv (\exists_r^+ y, |y| = q(|x|)) \; R_i(x, y),$$

*for $i = 0, 1$, where $r = 1 - 2^{-|x|}$. Then, for all $x$, $|x| = n$,*

$$S_1(x) \;\Rightarrow\; (\forall \vec{w})(\exists^+ y, |y| = q(n)) \, [(\forall i \leq q(n)) R_1(x, y \oplus w_i)],$$
$$S_0(x) \;\Rightarrow\; (\exists^+ \vec{w})(\forall y, |y| = q(n)) \, [(\exists i \leq q(n)) R_0(x, y \oplus w_i)],$$

*where $\vec{w} = \langle w_1, \ldots, w_{q(n)} \rangle$ and each $w_i$ in $\vec{w}$ has length $q(n)$.*

## 8.8   Relativized Probabilistic Complexity Classes

In the last section, we proved that $BPP$ is contained in the second level of the polynomial-time hierarchy, but we do not know whether $BPP$ is contained in $NP$ or whether $NP$ is contained in $BPP$. In this section, we study this question in the relativized form. First, we need to define the oracle PTMs and to extend the class $BPP$ to the relativized class $BPP^A$ relative to a set $A$.

The *oracle probabilistic Turing machine* $M$ can be easily defined as an oracle DTM $\tilde{M}$ together with a random bit generator $\phi$. For any fixed $\alpha \in \{0, 1\}^\infty$, the computation of $\tilde{M}^A(x, \alpha)$ is just the same as an oracle DTM. The concepts of accepting probability, error probability, and time complexity of $M$ can be defined exactly as those of an ordinary PTM. Let $A$ be an arbitrary set. The class $PP^A$ is defined to be the class of sets accepted by polynomial-time oracle PTMs $M$, relative to the oracle $A$. The class $BPP^A$ is the class of all sets accepted by polynomial-time oracle PTMs $M$ relative to $A$ such that the error probability of $M^A(x)$ is bounded by some constant $c < 1/2$.[4] The class $RP^A$ is defined, in a similar way, to be the class of sets accepted by polynomial-time oracle PTMs $M$ relative to $A$ such that the error probability of $M^A(x)$ is bounded by some constant $c < 1/2$ and the error probability of $M^A(x)$ for $x \notin L(M^A)$ is 0. It is easy to verify that all the properties of $BPP$ and $RP$ proved in the previous sections also hold with respect to $BPP^A$ and $RP^A$, respectively, for all sets $A$.

To allow simulation and diagonalization, we need an effective enumeration of polynomial-time oracle PTMs. Such an enumeration exists because the class of all polynomial-time deterministic oracle machines $\{\tilde{M}_j\}$ satisfying properties (1) and (2) of Section 8.2 is enumerable.

We now consider the relation between $BPP^A$ and $P^A$ and $NP^A$. First, we show that relative to a random oracle $A$, $P^A = BPP^A$. That is, given a random oracle $A$, a deterministic machine is able to use oracle $A$ to simulate the random bit generator of a randomized machine. For the notion of random oracles, refer to Section 4.7.

**Theorem 8.35** $P^A = BPP^A$ *relative to a random oracle $A$.*

---

[4]Note that we only require that the machine $M$ has a bounded error probability relative to oracle $A$. The machine $M$ does not have to have a bounded error probability relative to other oracles $X$.

*Proof.* It is sufficient to prove that the class of sets $A$ such that $P^A \neq BPP^A$ has probability less than $\delta$ for any $\delta < 1$. The idea of the proof is that a deterministic oracle machine can query the random oracle $A$ to find a sequence of *pseudorandom* bits that can be used to simulate an oracle PTM. Since a problem in $BPP^A$ has an arbitrarily small error probability, the probability of getting *bad* pseudorandom bits from set $A$ can be made to be arbitrarily small.

Let $\{M_i\}$ be an enumeration of polynomial-time oracle PTMs such that each $M_i$ has a uniform time complexity $p_i(n)$ for some polynomial function $p_i$. We say machine $M_i$ is a *BPP-type* machine relative to oracle $A$ if for all $x$, the error probability $err^A_{M_i}(x)$ is bounded by $1/4$. Assume that $M_i$ is a *BPP*-type machine relative to oracle $A$. Then, by a generalization of Theorem 8.19, the error probability can be reduced to $2^{-(i+2n+(-\log \delta)+2)}$ by repeatedly simulating $M^A_i$ for $m' = m \cdot \lceil i + 2n + (-\log \delta) + 2 \rceil$ times and taking the majority answer, where $m$ is an absolute constant. Following this observation, we consider the following *deterministic* oracle machine $D_{f(i)}$: [5]

*Deterministic oracle machine $D_{f(i)}$:*

Input: $x$, $|x| = n$. {Assume that $m' \cdot p_i(n) < 2^{p_i(n)+1}$.}
Oracle: $A$.

  (1) Query oracle $A$ on the first $m' \cdot p_i(n)$ strings of length $p_i(n) + 1$ to form $m'$ binary strings $\alpha_1, \ldots, \alpha_{m'}$, each of length $p_i(n)$; that is, the $j$th bit of $\alpha_k$ is 1 if and only if the $((k-1)p_i(n) + j)$th string of length $p_i(n)$ is in $A$.
  (2) For each $k := 1, \ldots, m'$, simulate $\tilde{M}^A_i(x, \alpha_k)$.
  (3) Accept $x$ if and only if the majority of simulations of step (2) above accept.

For any number $i > 0$ and any string $x$ of length $n$ satisfying $2^{p_i(n)+1} > m' \cdot p_i(n)$, let

$$E_{i,x} = \left\{ A : err^A_{M_i}(x) \leq \frac{1}{4} \text{ and } D^A_{f(i)}(x) \neq M^A_i(x) \right\}.$$

Note that the simulation of $\tilde{M}^A_i(x, \alpha_k)$ by $D^A_{f(i)}$ lasts only $p_i(n)$ moves and so it never queries about strings of length $p_i(n) + 1$. Therefore, the measure of $E_{i,x}$ is the same as the probability of $\alpha_1, \ldots, \alpha_{m'}$ that make the simulation result of $D^A_{f(i)}(x)$ different from the result of $M^A_i(x)$. That is,

$$\mu(E_{i,x}) \leq 2^{-(i+2n+(-\log \delta)+2)}.$$

---

[5]The index $f(i)$ indicates that this machine depends on $M_i$. It is actually true that for any acceptable indexing of oracle machines, $f$ is computable, though this fact is not used in the above proof.

Now, assume that $P^A \neq BPP^A$ for some set $A$. Then, there must exist an integer $i$ such that $M_i$ is a $BPP$-type machine relative to $A$ but the deterministic machine $D_{f(i)}$ with oracle $A$ disagrees with $M_i$ with oracle $A$ on infinitely many strings $x$. That is, $A$ must belong to some $E_{i,x}$ with $|x| = n$ satisfying $2^{p_i(n)+1} > m' \cdot p_i(n)$. Let $B_i$ be the set of integers $n$ satisfying the above inequality. Then, the probability of $P^A \neq BPP^A$ is bounded by

$$\sum_{i=1}^{\infty} \sum_{n \in B_i} \sum_{|x|=n} \mu(E_{i,x}) \leq \delta.$$

This completes the proof.                                                    ∎

Next we consider the relation between $BPP$ and $NP$. The following result follows immediately from the above result and Theorem 4.24.

**Corollary 8.36** $BPP^A \subsetneq NP^A$ *relative to a random oracle* $A$.

In the following, we show that some other possible relations between the classes $P$, $RP$, $coRP$, $BPP$, and $NP$ are possible. First we show separation results.

**Theorem 8.37** *(a) There exists a set $A$ such that $P^A \neq RP^A$.*
*(b) There exists a set $A$ such that $RP^A \neq coRP^A$.*
*(c) There exists a set $A$ such that $RP^A \neq NP^A$.*
*(d) There exists a set $A$ such that $BPP^A \not\subseteq \Delta_2^{P,A}$.*

*Proof.* We only show parts (b) and (d). Note that part (b) implies part (a), and part (c) follows from Corollary 8.36. We leave the constructive proofs for parts (a) and (c) as exercises.

(b): We are going to construct an oracle set $A$ by the standard stage construction. In each stage, we will satisfy a single diagonalization condition. When all stages are done, all diagonalization conditions are satisfied, and the separation between the relativized complexity classes is achieved (cf. Section 4.8).

Let $X_A = \{0^n : (\exists y, |y| = n)\ 0y \notin A\}$. Also let $\{M_i\}$ be an enumeration of all polynomial-time oracle PTMs. We say a machine $M_i$ is an $RP$-type machine relative to an oracle $B$ if for any input $x$, $\Pr[M_i^B(x) = 1]$ is either greater than $1/2$ or equal to 0. In the construction of set $A$, we diagonalize against all $RP$-type machines $M_i$ such that $0^n \in X_A$ if and only if $M_i^A$ accepts $0^n$ for some $n$ and, hence, $M_i^A$ does not recognize the complement of $X_A$. That is, we need to satisfy the following requirements for all $i > 0$ and $j > 0$:

$R_{0,i}$: $(\exists n = n_i)\ [0^n \in X_A$ if and only if $\Pr[M_i^A(0^n) = 1] > 0$.

$R_{1,j}$: if $0^j \in X_A$ then the number of $y$ such that $|y| = j$ and $0y \notin A$ is more than $2^{j-1}$.

In the above, if requirements $R_{1,j}$ are satisfied for all $j > 0$, then the set $X_A$ is in $RP^A$. If requirement $R_{0,i}$ is satisfied, then either machine $M_i$ is not an $RP$-type oracle machine relative to $A$, or $M_i$ does not recognize $\overline{X}_A$.

We construct set $A$ in stages. At each stage $i$, we satisfy requirement $R_{0,i}$ by witness $0^{n_i}$. In the meantime, we never add $0y$ to $A$ unless $|y| = n_i$ for some $i > 0$. In addition, we add either all $0y$, with $|y| = n_i$, to $A$ or add less than $2^{n_i-1}$ $0y$, with $|y| = n_i$, to $A$ such that $R_{1,j}$ is always satisfied. Let $m_0 = 0$ and $A(0) = \emptyset$. We assume that $m_{i-1}$ has been defined in such a way that no string of length greater than $m_{i-1}$ has been queried in earlier stages, and that all strings in $A(i-1)$ are of length $\leq m_{i-1}$.

> *Stage $i$.* We choose an integer $n = n_i$ that is greater than $m_{i-1}$ and satisfies $p_i(n) < 2^{n-1}$. Consider the computation of $\tilde{M}_i^{A(i-1)}(0^n, \alpha)$ for all $\alpha$ of length $p_i(n)$. For each $\alpha$, we simulate the computation of $\tilde{M}_i^{A(i-1)}(0^n, \alpha)$ as follows:
>
> (1) If a query $u$ is made and $u \notin \{0y : |y| = n\}$, then answer YES if and only if $u \in A(i-1)$.
>
> (2) If a query $u = 0y$, $|y| = n$, is made, then simulate both answers YES and NO.
>
> Then we obtain, for each $\alpha$, a computation tree of depth $\leq p_i(n)$. Now consider two cases:
>
> *Case 1.* There exists an $\alpha$ such that its computation tree contains at least one accepting path. Then, we let $Y = \{0y : |y| = n$, $0y$ was queried and answered YES in this accepting path$\}$. We let $A(i) := A(i-1) \cup Y$.
>
> *Case 2.* Not Case 1. Then, we let $A(i) := A(i-1) \cup \{0y : |y| = n\}$. In addition, we let $m_i = p_i(n) + 1$.
>
> *End of Stage $i$.*

We let $A := \bigcup_{i=1}^{\infty} A(i)$. First, we note that $m_i$ clearly satisfies the required property. Next, we claim that $A$ satisfies all requirements $R_{0,i}$ and $R_{1,j}$.

First, we check requirements $R_{1,j}$. Note that for any $j$ which is not equal to $n_i$ for any $i$, then $0y \notin A$ for all $y$ of length $j$, and so $R_{1,j}$ is satisfied. If $j = n_i$ for some $i > 0$, and if Case 1 holds in stage $i$, then we have added to $A$ at most $p_i(n) < 2^{n-1}$ many $0y$, $|y| = n_i$. This keeps the total number of $y$ with $|y| = n_i$ and $0y \notin A$ greater than $2^{n-1}$. Otherwise, if Case 2 occurs in stage $i$, then we added all $0y$, with $|y| = n$, to $A$, and so $R_{1,j}$ is still satisfied.

Next, for requirement $R_{0,i}$, we note that in Case 1 of Stage $i$, we left $0^n \in X_A$ but made $\Pr[M_i^{A(i)}(0^n) = 1] > 0$. On the other hand, if Case 2 occurs in Stage $i$, then for all $\alpha$, all computation paths of $\tilde{M}_i^{A(i)}(0^n, \alpha)$ rejects $0^n$. Since we do not add, in the later stages, any strings of length $\leq m_i = p_i(n) + 1$ to $A$, we must have $M_i^{A(i)}(0^n) = M_i^A(0^n)$, and the requirement $R_{0,i}$ is satisfied.

(d): We follow the setup of the stage construction above. Let $Y_A = \{0^n : (\exists^+ y, |y| = n)\ 1y \in A\}$. Also let $\{D_i\}$ be an enumeration of polynomial-time oracle DTMs, and $\{N_j\}$ an enumeration of oracle NTMs. (We assume that each move of $N_j$ has at most two nondeterministic choices.) We recall that

$$K(A) = \{\langle j, x, 0^k \rangle : N_j^A \text{ accepts } x \text{ in } k \text{ moves}\}$$

is the generic complete set for $NP^A$ (it was called $K_A$ in Chapter 4). Our requirements are

$R_{0,i}$: $(\exists n = n_i)$ $[0^n \in Y_A$ if and only if $D_i^{K(A)}$ rejects $0^n$.

$R_{1,j}$: if $0^j \notin Y_A$, then $(\exists^+ y, |y| = n)$ $1y \notin A$.

In the above, all requirements $R_{1,j}$ together imply that $Y_A \in BPP^A$. Requirement $R_{0,i}$ implies that the $i$th oracle machine $D_i$ does not recognize set $Y_A$ using $K(A)$ as the oracle. Together, they imply that $Y_A \notin P^{K(A)} = \Delta_2^{P,A}$.

> *Stage i.* We choose an integer $n = n_i$ that is greater than $m_{i-1}$ and satisfies $(p_i(n))^2 \leq 2^{n-2}$. We first let $A(i) := A(i-1)$ and $B = \emptyset$. Next we simulate the computation of $D_i^{K(A)}(0^n)$ and update the sets $A(i)$ and $B$. When a query $\langle j, x, 0^k \rangle \in ?K(A)$ is made, we decide the answer as follows:
>
> For all strings $z$ of length $k$, perform a simulation of the computation of $N_j^A$ on input $x$ for $k$ moves using $z$ as the *witness* (i.e., using the $\ell$th bit of $z$ to determine the $\ell$th nondeterministic move of $N_j$). For each simulation of $N_j$ on $x$ with the witness string $z$, we answer the queries made by $N_j$ to $A$ as follows:
>
> (1) If the query $u$ is not in $\{1y : |y| = n\}$, then we answer YES if and only if $u \in A(i)$.
>
> (2) If the query $u$ is in $\{1y : |y| = n\}$, then answer YES if $u \in A(i)$, answer NO if $u \in B$, and simulate both answers otherwise.
>
> Thus, for each $z$, we have a computation tree of depth $\leq k$. If there exists at least one $z$ such that its computation tree contains at least one accepting path, then we let $Z_1 = \{1y : |y| = n$, $1y$ was queried and answered YES in this accepting path$\}$; and $Z_0 = \{1y : |y| = n$, $1y$ was queried and answered NO in this accepting path$\}$. We add $Z_1$ to $A(i)$, and $Z_0$ to $B$ and answer YES to the query $\langle j, x, 0^k \rangle$. Otherwise, we don't change $A(i)$ or $B$ and answer NO to the query.
>
> Assume that the above simulation of $D_i^{K(A)}(0^n)$ accepts. Then, we keep the set $A(i)$ and let $m_i = p_i(n)$. (Note that $D_i$ can only query strings $\langle j, x, 0^k \rangle$ of length $\leq p_i(n)$, and for each such query, the simulation of $N_j$ on $x$ only lasts $k \leq p_i(n)$ moves.) Otherwise, if the simulation of $D_i^{K(A)}(0^n)$ rejects, then we add $(3/4) \cdot 2^n$ strings $1y \in \{1y : |y| = n, 1y \notin B\}$ to $A(i)$. (We will show below in (2) that this can be done.) Also let $m_i = p_i(n)$.
> *End of Stage i.*

We let $A := \bigcup_{i=1}^{\infty} A(i)$. To see that $A$ satisfies all requirements, we observe the following properties of the above stage construction.

(1) At stage $i$, during the simulation of $D_i^{K(A)}(0^n)$, we added at most $p_i(n)$ strings to either $A(i)$ or $B$ for each query to $K(A)$. The total number of strings of the form $1y$, $|y| = n$, added to $A(i)$ or to $B$ is bounded by $(p_i(n))^2 \leq 2^{n-2}$.

(2) From (1) above, after the simulation of $D_i^{K(A)}(0^n)$ is done in Stage $i$, there exist at least $(3/4) \cdot 2^n$ strings of the form $1y$, $|y| = n$, that are not in $B$. So, Stage $i$ is well defined.

(3) Since the simulation of each query $\langle j, x, 0^k \rangle \in ?K(A)$ made by $D_i^{K(A)}(0^n)$ chose the answer YES whenever possible, and since each answer YES was supported by adding $Z_0$ to $B$ and $Z_1$ to $A(i)$, the simulation is correct with respect to the final $A$.

From the above observations, it is easy to see that all requirements are satisfied. More specifically, observation (3) above shows that the simulation $D_i^{K(A)}(0^n)$ is correct. In addition, if the simulation accepts, then the final $A(i)$ contains at most $2^{n-2}$ strings in $\{1y : |y| = n\}$; and if it rejects, then the final $A(i)$ contains at least $(3/4) \cdot 2^{n-2}$ strings in $\{1y : |y| = n\}$. Therefore, requirement $R_{0,i}$ is satisfied at Stage $i$. Requirements $R_{1,j}$ are clearly satisfied by observations (1) and (2) above.  ∎

**Corollary 8.38** *There exists a set $A$ such that $P^A \neq coRP^A \neq RP^A \neq NP^A \nsubseteq BPP^A$ and $BPP^A \nsubseteq \Delta_2^{P,A}$.*

*Proof.* Each stage of the constructions in Theorem 8.37(a)–(d) involves the construction of the oracle in a local *window* area. Therefore, all the requirements can be satisfied by a single construction interleaving these stages.  ∎

Finally, we remark that it is possible that the class $NP^A$ collapses to the class $RP^A$ but not to $P^A$ (see Exercise 8.19), and it is possible that $NP^A$ is different from $P^A$ and yet $RP^A$ collapses to $P^A$ (by Theorems 4.24 and 8.35).

## Exercises

**8.1** Based on Lemma 8.5, find a polynomial-time algorithm that computes the Legendre-Jacobi symbols $(\frac{n}{m})$.

**8.2** The problem STRING MATCHING asks, for given strings $w, s \in \{0,1\}^*$, whether $w$ occurs in $s$ as a substring. Assume that $|w| = n$, $|s| = m$ and for each $1 \leq j \leq m$, let $s_j$ be the $j$th bit of $s$. The brute-force deterministic algorithm B compares $w$ with each substring $s(j) = s_j s_{j+1} \ldots s_{j+n-1}$, character by character, and needs deterministic time $O(nm)$ in the worst case.

For each string $w \in \{0,1\}^*$, let $i_w$ be the integer such that $w$ is its binary expansion (with leading zeros allowed). For each prime $p > 1$, let $H_p(w) = i_w \bmod p$. Now consider the following two randomized algorithms $R_1$ and $R_2$:

*Algorithm $R_1$*
Let $p$ be a random prime in the range $2 \le p < m^2 n$.
For $j := 1$ to $m - n + 1$ do
   If $H_p(w) = H_p(s(j))$ then if $w = s(j)$ then accept and halt;
Reject.

*Algorithm $R_2$*
For $j := 1$ to $m - n + 1$ do
Begin
   Choose a random prime $p$ in the range $2 \le p < m^2 n$
   If $H_p(w) = H_p(s(j))$ then if $w = s(j)$ then accept and halt
End;
Reject.

(a) Show that the probability of the algorithms $R_1$ and $R_2$ finding a mismatch (i.e., $w \ne s(j)$, but $H_p(w) = H_p(s(j))$ for some $j$) is bounded by $O(1/m)$.

(b) Analyze the expected time complexity of the algorithms $R_1$ and $R_2$. (Note that $H_p(s(j+1))$ can be computed from $H_p(s(j))$ in time $O(\log p)$ as follows: $H_p(s(j+1)) = (2H_p(s(j)) - 2^n s_j + s_{j+1}) \bmod p$. Also, to generate a random prime $p$, we can use the randomized algorithm for PRIME of Example 8.4 to test whether a random integer $p$ is a prime.)

**8.3** Prove Proposition 8.8. Note that if $M(x, \alpha)$ halts in $|\alpha|$ moves and $M(x, \beta)$ halts in $|\beta|$ moves with $\alpha \ne \beta$, then neither $\alpha \preceq \beta$ nor $\beta \preceq \alpha$.

**8.4** Consider the following alternative model of probabilistic Turing machines. A probabilistic Turing machine $M$ is a DTM equipped with a random-bit generator and three distinguished states: the *random state*, the *zero state*, and the *one state*. When $M$ is not in the random state, it behaves exactly the same as a deterministic machine. When $M$ is in the random state, the random-bit generator generates a bit $b \in \{0, 1\}$ and then $M$ enters the zero state or the one state according to the bit $b$. In other words, machine $M$ requests a random bit only when it needs it.

(a) Give the formal definition of the concepts of *halting probability* and *accepting probability* of such a PTM.

(b) Prove that this model of PTM is equivalent to the model defined in Section 8.2; that is, prove that for each PTM $M$ defined in Section 8.2, there exists a PTM $M'$ defined as above such that their halting and accepting probabilities on each input $x$ are equal, and vice versa.

**8.5** The above PTMs use a fixed random-bit generator $\phi$. We extend it to allow the random number generator to generate a random number in any given range. Let $M$ be a PTM defined as in Exercise 8.4 that is equipped with a random number generator $\psi$ that on any input $k > 0$ outputs a number $i \in \{0, 1, \ldots, k\}$ with $\Pr[i = j] = 1/(k+1)$ for all $j$, $0 \le j \le k$. Show that for any

$\epsilon > 0$, there exists a PTM $M_1$ with the standard random-bit generator $\phi$ such that $t_{M_1}(n) = O(t_M(n))$, and for all $x$ of length $n$, $err_{M_1}(x) \leq err_M(x) + \epsilon$.

**8.6** For any polynomial-time PTM $M$, let $f_M(x)$ be the number of accepting paths of $M(x)$ subtracting the number of rejecting paths of $M(x)$. For any two polynomial-time PTMs $M_1$ and $M_2$, show that there exist polynomial-time PTMs $M_3$ and $M_4$ such that for all $x$,

(a) $f_{M_3}(x) = f_{M_1}(x) + f_{M_2}(x)$, and

(b) $f_{M_4}(x) = f_{M_1}(x) \cdot f_{M_2}(x)$.

**8.7** Show that the class $PP$ is closed under union, intersection, and complementation. [*Hint:* For intersection, we need to find a PTM $M$ such that $f_M(x) > 0$ if and only if $f_{M_1}(x) > 0$ and $f_{M_2}(x) > 0$. Such a function $f_M(x)$ cannot be realized as a polynomial on $f_{M_1}(x)$ and $f_{M_2}(x)$. However, if we treat $f_{M_1}$ and $f_{M_2}$ as continuous functions, then the function $f_M(x)$ can be approximated by a rational function on $f_{M_1}(x)$ and $f_{M_2}(x)$. (This is well known in numerical analysis.) Now apply Exercise 8.6 to show that such a rational function can be defined as $f_M$ for some polynomial-time PTM $M$.]

**8.8** A polynomial-time oracle DTM $M$ is called a *log-query machine* if for any input $x$ of length and for any oracle $A$, it makes at most $c \log n$ queries, where $c$ is a constant. Let $\Theta_2^P$ denote the class of sets computable by polynomial-time log-query oracle machines with an oracle in $NP$. Prove that $\Theta_2^P \subseteq PP$.

**8.9** Compare the class $\Delta_2^P$ with $PP$. That is, prove that $\Delta_2^P \subseteq PP$ or find an oracle $A$ such that $\Delta_2^{P,A} \not\subseteq PP^A$.

**8.10** Formally define the notion of space complexity of a PTM. Let $RSPACE(s(n))$ be the class of sets accepted by PTMs with space bound $s(n)$.

(a) Prove that $NSPACE(s(n)) \subseteq RSPACE(s(n))$.

(b) Prove that $RSPACE(s(n)) \subseteq DSPACE(s(n)^{3/2})$.

(c) Prove that $RSPACE(s(n))$ is closed under complementation.

**8.11** Prove the following second form of the Swapping Lemma: Let $R(x, y, z)$ be a predicate that holds only if $|y| = p(|x|)$ and $|z| = q(|x|)$ for some polynomials $p$ and $q$. Assume that for some $x$ of length $n$, it holds that $(\forall y, |y| = p(n))\ (\exists^+ z, |z| = q(n))\ R(x, y, z)$. Then, for some polynomial $r$ it holds that $(\exists^+ \vec{w})\ (\forall y, |y| = p(n))\ [(Mz \in \vec{w})\ R(x, y, z)]$, where $\vec{w} = \langle w_1, \ldots, w_{r(n)} \rangle$ with each $w_i$ of length $q(n)$, and $(Mz \in \vec{w})$ means "for the majority of $z \in \{w_1, \ldots, w_{r(n)}\}$".

**8.12** Let $BPP^C$ denote the union of all classes $BPP^A$ for all $A \in C$. Prove that $BPP^{BPP} = BPP$. Is it true that $RP^{RP} = RP$? Why or why not?

**8.13** Let $C_1$ be the class of sets $A$ for which there exist a polynomial-time predicate $R$ and a polynomial $p$ such that for all $x$,

$$x \in A \;\Rightarrow\; (\exists y, |y| = p(n))\, (\forall z, |z| = p(n))\, R(x, y, z),$$
$$x \notin A \;\Rightarrow\; (\forall y, |y| = p(n))\, (\exists^+ z, |z| = p(n))\, \neg R(x, y, z).$$

Let $C_2$ be the class of sets $A$ for which there exist a polynomial-time predicate $R$ and a polynomial $p$ such that for all $x$,

$$x \in A \;\Rightarrow\; (\forall y, |y| = p(n))\, (\exists z, |z| = p(n))\, R(x, y, z),$$
$$x \notin A \;\Rightarrow\; (\exists^+ y, |y| = p(n))\, (\forall z, |z| = p(n))\, \neg R(x, y, z).$$

(a) Prove that $C_1 \subseteq C_2$.

(b) Prove that $NP^{BPP} \subseteq C_1$ and $C_2 \subseteq BPP^{NP}$.

(c) Prove that if $NP \subseteq BPP$ then the polynomial-time hierarchy collapses to $BPP$.

**8.14** Recall the notion of pseudorandom generators introduced in Exercise 4.8. We say a pseudorandom generator $f$ is an $n^c$-*generator*, where $c > 0$ is a constant, if $|f(x)| = |x|^c$, and an $n^c$-pseudorandom generator $f$ is *strongly unpredictable* if the range of $f$ is strongly unpredictable as defined in Definition 4.17.

(a) Prove that if, for every $c > 0$, there is a strongly unpredictable $n^c$-pseudorandom generator $f$, then $BPP \subseteq DTIME(2^{n^\epsilon})$ for every $\epsilon > 0$.

(b) We say that a collection $\{C_n\}$ of circuits *approximates* a set $A \subseteq \{0,1\}^*$ if for every polynomial function $p$, $\mathrm{Pr}_{|x|=n}[C_n(x) \neq \chi_A(x)] \leq 1/p(n)$ for almost all $n \geq 1$. Prove that if there exists a set $A \in EXPTIME$ that is not approximable by any polynomial-size circuits $\{C_n\}$, then for every $c > 0$, there exists a strongly unpredictable $n^c$-pseudorandom generator $f$ and, hence, $BPP \subseteq DTIME(2^{n^\epsilon})$ for every $\epsilon > 0$.

**8.15** For each $n \geq 1$, let $\Sigma_n^{BPP} = \bigcup_{A \in BPP} \Sigma_n^{P,A}$. Prove that $\Sigma_2^{BPP} = \Sigma_2^P$.

**8.16** Let *almost-P* be the class of sets that are computable in deterministic polynomial time with respect to a random oracle; that is, $almost\text{-}P = \{A : \mathrm{Pr}_X[A \in P^X] = 1\}$. Prove that $BPP = almost\text{-}P$.

**8.17** (a) Prove Theorem 8.37(c) by constructing the oracle directly.

(b) Construct directly an oracle $B$ such that $P^B = RP^B \neq NP^B$. [*Hint*: cf. Theorem 4.26.]

**8.18** Prove that there exists an oracle $A$ such that $P^A \neq ZPP^A$.

**8.19** Prove that there exists an oracle $A$ such that $P^A \neq RP^A = NP^A$.

## Historical Notes

Randomized algorithms can probably be traced back to the early days of programming. Two seminal papers on randomized algorithms for primality testing are Rabin [1976, 1980] and Solovay and Strassen [1977]. Our Example 8.4 is based on Solovay and Strassen's algorithm. Adleman and Huang [1987] gave a proof that the problem PRIME is in *ZPP*. Rabin [1976] also includes randomized algorithms for problems in computational geometry. Our Example 8.2 is based on the algorithms given in Schwartz [1980], and Example 8.24 is from Adleman, Manders and Miller [1977]. For a more complete treatment of randomized algorithms, see Motwani and Raghavan [1995]. Probabilistic Turing machines were first formally defined and studied in Gill [1977]. The complexity classes *PP*, *BPP*, *RP*, and *ZPP* and their basic properties were all defined there. Theorem 8.22 was first observed in Rabin [1976]. The fact that *RP* has polynomial-size circuits was first proved by Adleman [1978]. Theorem 8.27 is based on the same idea and was first pointed out by Bennett and Gill [1981]. Theorem 8.30 is from Ko [1982] and Zachos [1983]. The result that *BPP* is contained in the polynomial-time hierarchy was first proved by Sipser and Gacs (see Sipser [1983b]). A simpler version was later published in Lautemann [1983]. Our proof using the Swapping Lemma is from Zachos and Heller [1986]. Zachos [1986] contains a survey on the relations between probabilistic complexity classes.

The algorithm for the string matching problem in Exercise 8.2 is due to Karp and Rabin [1987]. Beigel, Reingold and Spielman [1991] showed that *PP* is closed under intersection (Exercise 8.7). Exercise 8.8 is from Beigel, Hemachandra and Wechsung [1991]. The space-bounded probabilistic classes have been studied in Gill [1977], Simon et al. [1980], Simon [1981], Ruzzo et al. [1984], and Saks and Zhou [1995]. Exercise 8.10 is from these papers. Yao [1982] and Nisan and Wigderson [1994] studied the existence of pseudorandom generators and the collapsing of complexity classes. Exercise 8.14 is from these papers. The notion *almost-P* was first defined by Bennett and Gill [1981]. Exercise 8.16 is from there. This notion can be generalized to *almost-C* for any complexity class $C$. See, for instance, Nisan and Wigderson [1994]. The relativized separation of *BPP* and *RP* from *P* was first studied by Rackoff [1982]. Exercises 8.17(b) and 8.19 are from there. Rackoff [1982] also gave a constructive proof for the result that $(\exists A)\ P^A = BPP^A \neq NP^A$. Our proofs are based on the random oracle technique developed in Bennett and Gill [1981]. Ko [1990] contains more results on relativized probabilistic complexity classes.

# 9

# *Complexity of Counting*

*How do I love thee? Let me count the ways.*
*I love thee to the depth and breadth and height*
*My soul can reach.*
— Elizabeth Barrett Browning

Informally, the complexity class *NP* contains the *decision* problems in which an input instance is accepted if and only if there exists a *good* witness. The complexity classes *PP* and *BPP* contain the problems in which an input instance is accepted if and only if the majority of the witnesses are *good*. All these problems are just the restricted forms of a more general counting problem: for any input instance, count the number of good witnesses for it. Such counting problems occur very often in practical problems. For instance, for the HAMILTONIAN CIRCUIT problem, one is often interested not only in the existence of a Hamiltonian circuit in a graph, but also interested in the total number of different Hamiltonian circuits. In this chapter, we study the complexity of these counting problems and its relation to the complexity of nondeterministic and probabilistic computation. We define two complexity classes, $\#P$ and $\oplus P$, of counting problems. The main result of this chapter shows that decision problems in the polynomial-time hierarchy are polynomial-time reducible to counting problems in these complexity classes. We also investigate the relation between the relativized $\#P$ and the relativized polynomial-time hierarchy.

## 9.1   Counting Class #*P*

Almost all problems we studied so far are decision problems that for an input $x$ ask for answers of a single bit (i.e., YES or NO). In practice, however, we often encounter search problems that for an input $x$ ask for an output $y$ that may be much longer than one bit. We have seen in Chapter 2 how to reduce some search problems, particularly the optimization problems, to the corresponding decision problems. Still, complexity classes for decision problems, such as *NP* and *BPP*, are not adequate to characterize precisely the complexity of search problems. A particularly interesting class of search problems is the class of counting problems related to nondeterministic computation. A decision problem in *NP* asks, for each input $x$, whether there exists a *witness* (or a *solution*) to $x$. The corresponding counting problem asks, for any input $x$, the *number* of witnesses for $x$.

**Example 9.1** In the form of a decision problem, the problem SAT asks whether a given Boolean formula $F$ has a satisfying Boolean assignment. The corresponding counting problem is as follows:

> #SAT: Given a Boolean formula $F$, find the total number of different Boolean assignments that satisfy $F$.

It is clear that #SAT is at least as difficult as the decision problem SAT and is intractable if $P \neq NP$.                                                                                   □

The above example can be extended to other *NP*-complete problems such as VERTEX COVER and HAMILTONIAN CIRCUIT. In general, for each *NP*-complete decision problem, there is a corresponding counting problem that is not polynomial-time solvable if $P \neq NP$.

**Example 9.2** Recall the problem PERFECT MATCHING (PM) defined in Chapter 6: For a given bipartite graph $G$, determine whether there exists a perfect matching for $G$. We define the corresponding counting problem as follows:

> #PM: Given a bipartite graph $G$, determine the number of perfect matchings there are for $G$.

It is well known that the decision problem PM is solvable in polynomial time and it was shown in Theorem 6.49 that PM is in *RNC*. However, these algorithms for the decision problem of bipartite matching do not seem applicable to the corresponding counting problem #PM. Indeed, in the next section, we will show that #PM is as hard as #SAT and is not solvable in polynomial time if $P \neq NP$. Therefore, in general, the complexity of a counting problem cannot be characterized precisely from the complexity of the corresponding decision problem.                                                                               □

Formally, a counting problem is a function $f$ mapping an input $x \in \Sigma^*$ to an integer $f(x) \in \mathbf{N}$. The computational model for counting problems is the deterministic Turing machine transducer. A counting problem $f : \Sigma^* \rightarrow \mathbf{N}$ is *computable* if there exists a DTM transducer $M$ such that for each input $x \in \Sigma^*$, $M(x)$ outputs integer $f(x)$ in the binary form; it is *polynomial-time computable* if it is computable by a polynomial-time DTM transducer. In other words, a counting problem $f$ is polynomial-time computable if it is in *FP*. Note that a function $f \in FP$ must be polynomially length-bounded; that is, for all $x$, $|f(x)| \leq p(|x|)$ for some polynomial $p$. Thus, if a counting problem $f$ is in *FP*, then $0 \leq f(x) \leq 2^{p(|x|)}$ for some polynomial $p$.

Many counting problems such as #SAT and #PM can easily be defined in terms of nondeterministic Turing machines. For any polynomial-time NTM $M$, we define a counting problem $\varphi_M$ as follows: For any $x \in \Sigma^*$, $\varphi_M(x)$ is the number of accepting computation paths of $M(x)$. Since each computation path of $M(x)$, with $|x| = n$, is of length $\leq p(n)$ for some polynomial $p$, we have the bounds $0 \leq \varphi_M(x) \leq c^{p(n)}$, where $c$ is a constant bounding the number of nondeterministic choices of $M$ in a nondeterministic move. Thus, all functions $\varphi_M$ are polynomial length-bounded. Let $\mathcal{M}$ denote the class of all polynomial-time NTMs.

**Definition 9.3** *The complexity class #P of counting functions is defined as* $\#P = \{\varphi_M : M \in \mathcal{M}\}$.

It is easy to see that each counting function in *FP* is also in #P. The fundamental question concerning counting problems is whether $\#P \subseteq FP$.

**Proposition 9.4** *If a counting function $f$ is in FP, then $f \in \#P$.*

*Proof.* Assume that $f : \Sigma^* \rightarrow \mathbf{N}$ is in *FP* with $f(x) \leq 2^{p(|x|)}$ for some polynomial $p$. Let $M$ be the NTM that, on input $x$, nondeterministically generates a string $y \in \{0, 1\}^*$ of length $p(|x|)$ and accepts if and only if $y$ is among the first $f(x)$ strings of length $p(|x|)$. Then, $f = \varphi_M \in \#P$. ∎

To understand the power of the counting class #P, we first consider its relation with class *NP*. From the definition of $\varphi_M$, it is clear that $M$ accepts $x$ if and only if $\varphi_M(x) > 0$. Thus, it follows that #P is at least as powerful as *NP*. Next we consider the relation between #P and *PP*. Assume that $A \in PP$ is computed by a polynomial-time PTM $M$ in uniform time $t(n)$; that is, every branch of the computation of $M(x)$ halts in exactly $t(|x|)$ moves. Define an NTM $M_1$ that on input $x$ guesses a string $y$ of length $t(|x|)$ and accepts $x$ if and only if $\tilde{M}(x, y)$ accepts. Then, it follows that $x \in A$ if and only if $\varphi_{M_1}(x) > 2^{t(|x|)-1}$. Thus, #P is at least as powerful as *PP*.

In the following, we formalize these observations in terms of polynomial-time reducibilities to show that the classes #P and *PP* are equivalent under the polynomial-time Turing reducibility. The notions of function-oracle Turing machines and polynomial-time Turing reducibility on functions have been defined in Chapter 2. Recall that a set $A$ is polynomial-time Turing reducible to a function $f$, written $A \leq_T^P f$ or $A \in P^f$, if there is a polynomial-time function-oracle Turing machine that computes the characteristic function $\chi_A$

of $A$ when the function $f$ is used as an oracle. Let $\mathcal{F}$ be a class of functions. We let $P^{\mathcal{F}}$ denote the class of all sets $A$ that are $\leq^P_T$-reducible to some functions $f \in \mathcal{F}$.

To prepare for this result, we first show that we can restrict $\#P$ to contain only functions $\varphi_M$ that correspond to NTMs that run in polynomial time in *uniform* time complexity.

**Proposition 9.5** *Let $f$ be a counting function. The followings are equivalent:*

*(a) $f \in \#P$.*

*(b) There exist an NTM $M$ and a polynomial $p$ such that (i) $M$ has at most two nondeterministic choices in each move; (ii) for each input $x$, all computation paths of $M(x)$ halts in exactly $p(n)$ moves; and (iii) $f = \varphi_M$.*

*(c) There exist a polynomial-time computable predicate $R$ and a polynomial $p$ such that for all inputs $x$ of length $n$, $f(x)$ is equal to the number of strings $y$ of length $p(n)$ such that $R(x, y)$.*

*Proof.* It is obvious that (b) $\Longleftrightarrow$ (c), and (b) $\Rightarrow$ (a). We only show (a) $\Rightarrow$ (b): Let $M$ be an NTM with time complexity $p(n)$ for some polynomial $p$. Assume that $M$ has at most $c$ nondeterministic choices in each move. Let $M_1$ be the NTM obtained from $M$ with the following modification:

(1) At each nondeterministic move of $M$, $M_1$ makes $\lceil \log c \rceil$ many nondeterministic moves, creating two computation paths in each move. Thus, each nondeterministic move of $M$ is simulated in $M_1$ by $2^{\lceil \log c \rceil}$ computation paths. Among them, let the first $c$ paths simulate the first $c$ paths of $M$ and the rest reject.

(2) Let $p_1(n)$ be a fully time constructible polynomial function such that $p_1(n) \geq \lceil \log c \rceil \cdot p(n)$. On any input $x$ of length $n$, $M_1$ runs exactly $p_1(n)$ moves before it halts. (If a computation path enters a halting state of $M$ before that, $M_1$ keeps the same output for the following moves until it runs exactly $p_1(n)$ moves.)

It is clear that $M_1(x)$ has exactly the same number of accepting paths as $M(x)$. ∎

**Theorem 9.6** $P^{\#P} = P^{PP}$.

*Proof.* It is immediate from the definition that a problem $A$ in $PP$ can be easily computed using an oracle function $f$ that counts the number of accepting computations of the PTM for $A$ and then accepts the input if and only if this number is greater than one half of all computations. It follows that $P^{PP} \subseteq P^{\#P}$.

Conversely, assume that $R$ is a polynomial-time computable relation such that for each $x$ of length $n$, $R(x, y)$ only if $|y| = p(|x|)$ for some polynomial $p$, and that $f(x)$ counts the number of strings $y$ such that $R(x, y)$. For any integer $m \geq 0$, let $w_m$ denote its binary expansion. We design in the following a PTM $M$ that accepts $(x, w_m)$ if and only if $f(x) > m$. Thus, the function $f(x)$ can be computed by a binary search using $L(M) \in PP$ as the oracle. ·

> *Probabilistic Machine M.* On input $(x, w_m)$, with $|x| = n$, flip a coin to get a random string $y \in \{0, 1\}^*$ of length $p(n) + 1$. Let $y = y_1 z$ with $|y_1| = 1$ and $|z| = p(n)$. Accept the input if and only if $[y_1 = 0$ and $z$ is among the first $2^{p(n)} - m$ strings of length $p(n)]$ or if $[y_1 = 1$ and $R(x, z)]$.

Note that $\Pr[M(x) = 1] = (2^{p(n)} - m + f(x))2^{-(p(n)+1)}$, and so $\Pr[M(x) = 1] > 1/2$ if and only if $f(x) > m$. ∎

From the above theorem and Theorem 8.16, we have immediately the following relation between $\#P$ and the polynomial-time hierarchy.

**Corollary 9.7** $\Delta_2^P \subseteq P^{\#P}$.

In Section 9.4, we will extend this result to show that the whole polynomial-time hierarchy is contained in $P^{\#P}$.

## 9.2 #P-Complete Problems

Similar to the notion of $NP$-completeness, the notion of $\#P$-completeness can be used to characterize precisely the complexity of certain counting problems. In this section, we present some $\#P$-complete problems.

**Definition 9.8** *A function* $f : \Sigma^* \to \mathbf{N}$ *is* $\#P$-*complete if (a)* $f \in \#P$*, and (b) for all* $g \in \#P$*,* $g \leq_T^P f$.

We establish the first $\#P$-complete problem $\#$SAT from the generic reduction of Cook's Theorem.

**Theorem 9.9** *The function* $\#$SAT *is* $\#P$-*complete.*

*Proof.* (Sketch) Recall the proof of Cook's Theorem. For each polynomial-time NTM $M$ and each input $w$, the proof constructed a Boolean formula $F$ such that each accepting computation path of $M(w)$ induces a Boolean assignment for variables in $F$ that satisfies $F$. It is easy to verify that no two accepting paths induce the same Boolean assignment, and that the induced assignment from an accepting path is unique. That is, the number of accepting paths of $M(w)$ is the same as the number of Boolean assignments satisfying $F$. ∎

Let $\#$3-SAT be the problem $\#$SAT restricted to 3-CNF Boolean formulas.

**Corollary 9.10** $\#$3-SAT *is* $\#P$-*complete.*

*Proof.* Note that the Boolean formula $F$ constructed in Cook's Theorem can be written in CNF. Moreover, it can be checked that the reduction from SAT to 3-SAT given in Corollary 2.13 preserves the number of satisfying assignments. Therefore, $\#$3-SAT is $\#P$-complete. ∎

We observe that most reductions used to prove $NP$-completeness in Section 2.3 either preserve a simple relation between the number of solutions of the two problems or can be easily modified to have this property. Thus, these reductions can be extended into reductions for the corresponding counting problems. In the following, we show that the counting problem for VERTEX COVER is $\#P$-complete. Other $\#P$-complete problems are included in exercises. We let $\#$VC denote the problem of counting the number of vertex covers of a given size $K$ for a given graph $G$.

**Theorem 9.11** *The problem $\#$VC is $\#P$-complete.*

*Proof.* Recall the reduction from SAT to VC given in the proof of Theorem 2.14. For a 3-CNF formula $F$ of $n$ variables and $m$ clauses, we constructed a graph $G_F$ of $2n + 3m$ vertices such that $F$ is satisfiable if and only if $G_F$ has a vertex cover of size $n + 3m$. This reduction does not preserve the number of witnesses. In particular, for each Boolean assignment $\tau$ for $F$, there is a unique $n$-vertex cover $X_\tau$ for $x_1, \bar{x}_1, \ldots, x_n, \bar{x}_n$. If $\tau$ satisfies one out of three literals of a clause $C_j$, then, with respect to $X_\tau$, there is a unique way to choose two vertices to cover the triangle $T_j = \{c_{j,1}, c_{j,2}, c_{j,3}\}$. If $\tau$ satisfies two out of three literals of $C_j$, then there are two ways to cover $T_j$ by two vertices. If $\tau$ satisfies all three literals of $C_j$, then there are three ways to cover $T_j$ by two vertices. That is, the number of vertex covers for graph $G_F$ depends not only on the number of satisfying assignments for $F$ but also on the number of literals satisfied in each clause. Therefore, there does not have a simple relation between the numbers of witnesses.

In the following, we make a simple modification of the graph $G_F$. Our new graph $G'_F$ has the property that the number of vertex covers for $G'_F$ is equal to $4^m$ times the number of satisfying assignments for $F$. $G'_F$ contains $G_F$ as a subgraph, and contains, for each clause $C_j$, $1 \leq j \leq m$, seven additional vertices: $c_{j,4}$, and $d_{j,k}, e_{j,k}$, $k = 1, 2, 3$. Assume that the three literals of $C_j$ are $y_{j_1}, y_{j_2}, y_{j_3}$. For each $j$, $1 \leq j \leq m$, it also contains the following edges:

(i) $\{c_{j,4}, c_{j,k}\}$, $k = 1, 2, 3$ (so $c_{j,\ell}$'s, with $1 \leq \ell \leq 4$, form a clique of size 4).

(ii) $\{c_{j,4}, y_{j_k}\}$, $k = 1, 2, 3$.

(iii) $\{d_{j,k}, \bar{y}_{j_k}\}$, $\{d_{j,k}, e_{j,k}\}$, $k = 1, 2, 3$.

We let $G_j$ denote the subgraph of $G'_F$ that includes vertices $c_{j,k}, d_{j,k'}, e_{j,k'}$, $k = 1, \ldots, 4$, $k' = 1, 2, 3$, and all edges incident on them. Figure 9.1 shows $G_j$ corresponding to the clause $C_j = (x_1 + \bar{x}_2 + x_3)$.

It is easy to see that for each $G_j$, we need at least six vertices to cover it (not including the edges $\{x_i, \bar{x}_i\}$). In addition, we verify that for each Boolean assignment $\tau$ that satisfies a clause $C_j$, there are exactly four ways to cover $G_j$.

*Case 1.* $\tau$ satisfies only one literal of $C_j$, say, $y_{j_1}$. Then, the minimum vertex cover must contain $c_{j,2}, c_{j,3}, c_{j,4}$, since each of them is connected to at least one of the unsatisfied literals $y_{j_2}$ or $y_{j_3}$. In addition, it must contain $d_{j,1}$, since it is connected to $\bar{y}_{j_1}$. For the other two vertices, we could choose

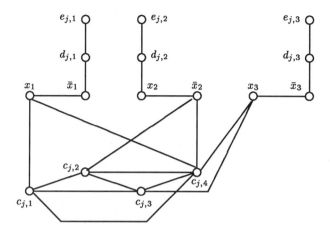

Figure 9.1: The subgraph $G_j$.

one from $d_{j,2}$ and $e_{j,2}$ and one from $d_{j,3}$ and $e_{j,3}$. So, there are four different ways to cover $G_j$.

*Case* 2. $\tau$ satisfies two literals of $C_j$, say, $y_{j_1}$ and $y_{j_2}$. Using a similar argument, we can see that the minimum cover must contain $c_{j,3}$, $c_{j,4}$, $d_{j,1}$, and $d_{j,2}$, and may contain one from $c_{j,1}$ and $c_{j,2}$ and one from $d_{j,3}$ and $e_{j,3}$.

*Case* 3. $\tau$ satisfies all three literals of $C_j$. Then, the minimum cover must contain $d_{j,1}, d_{j,2}$, and $d_{j,3}$. It may choose any three out of four vertices $c_{j,k}$, $1 \leq k \leq 4$.

Thus, for each satisfying assignment for $F$, there are exactly four ways to cover each $G_j$ and, hence, $4^m$ ways to cover the graph $G'_F$. This simple relation between the numbers of witnesses of $F$ and $G'_F$ shows that #VC is #P-complete.  ∎

From the above theorem and other similar results (see Exercise 9.1), one might suspect that #P-completeness is just a routine extension from NP-complete decision problems to their corresponding counting problems. The following result shows that this is not true: there exist some decision problems in $P$ whose corresponding counting problems are nevertheless #P-complete.

We first show that the problem PERM, of computing the permanent of an integer matrix, is #P-complete. The permanent of an $n \times n$ integer matrix $A$ is

$$perm(A) = \sum_{\sigma \in S_n} \prod_{i=1}^{n} A[i, \sigma(i)],$$

where $S_n$ is the set of all permutations on $\{1, 2, \ldots, n\}$. Note that the determinant of a matrix $A$ has a similar form except that a factor of $(-1)$ is multiplied to each *odd* permutation. It is well known that determinants of integer matrices are computable in polynomial time (indeed in $NC^2$). However, no polynomial-time algorithms for permanents are known, despite the similarity between the two problems. The #P-completeness of the function *perm* shows that the two problems are really quite different as far as their complexity is concerned (assuming $\#P \not\subseteq FP$).

The problem PERM has a number of equivalent formulation. One of them is the problem #PM of counting the number of perfect matchings (Example 9.2). Another one is the cycle covering problem. Let $G = (V, E)$ be a directed graph and $w : E \rightarrow \mathbf{Z}$ a weight function on the edges of $G$, where $\mathbf{Z}$ denotes the set of integers. A collection $C$ of node-disjoint cycles that cover all nodes in $G$ is called a *cycle cover* of $G$. The weight of a cycle cover $C$ is the product of the weights of all edges in $C$. The cycle covering problem is the following:

> #CC: Given a directed graph $G$ with a weight function $w$ on edges of $G$, find the sum of the weights of all cycle covers of $G$.

For any $n \times n$ integer matrix $A$, define $G_A$ to be the directed graph that has $n$ vertices $v_1, \ldots, v_n$ and has a weight function $w(v_i, v_j) = A[i, j]$; that is, the adjacency matrix of $G_A$ is $A$. The following lemma shows a simple relation between the problems #CC and PERM.

**Lemma 9.12** *For any integer matrix $A$, perm$(A)$ is equal to the sum of the weights of all cycle covers of $G_A$.*

*Proof.* Assume that $A$ is an $n \times n$ matrix. For each cycle cover $C$ of $G_A$, define a permutation $\sigma_C$ on $\{1, 2, \ldots, n\}$ as follows: $\sigma_C(i) = j$ if $(v_i, v_j) \in C$. Then, the weight of $C$ is just the product of $A[i, \sigma_C(i)]$, $i = 1, 2, \ldots, n$.  ∎

**Theorem 9.13** PERM *and* #CC *are* $\leq_T^P$*-equivalent.*

*Proof.* Lemma 9.12 above shows that PERM $\leq_T^P$ #CC. Conversely, for any edge-weighted directed graph $G$, let $A$ be its adjacency matrix; i.e., $A[i, j] =$ weight of the edge $(v_i, v_j)$. (If there is no edge $(v_i, v_j)$, then $A[i, j] = 0$.) Then, clearly, $G_A = G$. Thus, $\#CC(G) = perm(A)$ and so #CC $\leq_T^P$ PERM.  ∎

**Theorem 9.14** *The counting problems* PERM *and* #CC *are* #P*-complete.*

*Proof.* It is easy to see that PERM and #CC are in #P. From the above theorem, it suffices to construct a reduction from #SAT to #CC. For each Boolean formula $F$ as an instance of #SAT, we will construct an edge-weighted directed graph $G$ such that the sum of the weights of all cycle covers of $G$ is equal to $4^{3c(F)} \cdot s(F)$, where $c(F)$ is the number of clauses in $F$ and $s(F)$ is the number of Boolean assignments that satisfy $F$. This gives a Turing reduction from #SAT to #CC.

A basic component of the graph $G$ is a *junction* $J$, which is a 4-node directed graph with the following adjacency matrix:

$$X = \begin{bmatrix} 0 & 1 & -1 & -1 \\ 1 & -1 & 1 & 1 \\ 0 & 1 & 1 & 2 \\ 0 & 1 & 3 & 0 \end{bmatrix}.$$

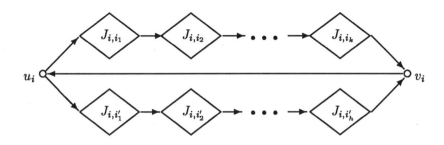

Figure 9.2: The track $T_i$.

For each junction $J$, we let $J[i]$, $i = 1, 2, 3, 4$, denote the $i$th node in $J$.

Let $X(a_1, \ldots, a_k; b_1, \ldots, b_\ell)$ denote the matrix obtained from $X$, removing rows $a_1, \ldots, a_k$ and columns $b_1, \ldots, b_\ell$. Then, the following properties of $X$ are easy to verify:

(i) $perm(X) = perm(X(1;1)) = perm(X(4;4)) = perm(X(1,4;1,4)) = 0$;

(ii) $perm(X(1;4)) = perm(X(4;1)) = 4$.

These properties on matrix $X$ imply certain interesting properties about cycle covers of $J$, which will be used later.

We now begin our construction. Let $F$ be a 3-CNF Boolean formula with clauses $C_1$, ..., $C_n$ and variables $x_1$, ..., $x_m$. The graph $G$ contains the following components:

(1) For each clause $C_j$ and each variable $x_i$ that occurs in $C_j$ (either positively or negatively), define a *junction* $J_{i,j}$ that is isomorphic to $J$ defined above.

(2) For each variable $x_i$, define a *track* $T_i$ that connects all $J_{i,j}$ together. Assume that $x_i$ occurs positively in clauses $C_{i_1}, \ldots, C_{i_k}$, and negatively in clauses $C_{i'_1}, \ldots, C_{i'_h}$. Then, $T_i$ contains two nodes $u_i$ and $v_i$, junctions $J_{i,i_r}$, $J_{i,i'_s}$, with $1 \leq r \leq k$ and $1 \leq s \leq h$, and the following edges:

$$(v_i, u_i), \ (u_i, J_{i,i_1}[1]), \quad (u_i, J_{i,i'_1}[1]), \ (J_{i,i_k}[4], v_i), \ (J_{i,i'_h}[4], v_i),$$
$$(J_{i,i_r}[4], J_{i,i_{r+1}}[1]), \quad r = 1, 2, \ldots, k-1,$$
$$(J_{i,i'_s}[4], J_{i,i'_{s+1}}[1]), \quad s = 1, 2, \ldots, h-1.$$

Note that, in the above, all edges enter a junction $J_{i,j}$ from node 1 and leave a junction from node 4. The track $T_i$ is shown in Figure 9.2.

(3) For each clause $C_j$, define an *interchange* $R_j$ that connects to some tracks $T_i$ through junctions $J_{i,j}$. Assume that the three variables occurring in $C_j$ are $x_{j_1}$, $x_{j_2}$, and $x_{j_3}$. Then the interchange $R_j$ contains nine nodes: $p_j$,

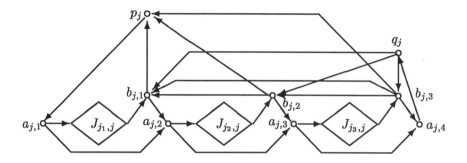

Figure 9.3: The interchange $R_j$.

$q_j$, $a_{j,k}$, $k = 1, 2, 3, 4$, $b_{j,h}$, $h = 1, 2, 3$; three junctions: $J_{j_k,j}$, $k = 1, 2, 3$; and the following edges between them:

$$(p_j, a_{j,1}), \ (a_{j,4}, q_j),$$
$$(a_{j,k}, J_{j_k,j}[4]), \ (J_{j_k,j}[1], b_{j,k}), \qquad k = 1, 2, 3,$$
$$(b_{j,k}, a_{j,k+1}), (a_{j,k}, a_{j,k+1}), \qquad k = 1, 2, 3,$$
$$(b_{j,3}, b_{j,2}), (b_{j,3}, b_{j,1}), (b_{j,2}, b_{j,1}),$$
$$(q_j, b_{j,k}), \ (b_{j,k}, p_j), \qquad\qquad k = 1, 2, 3.$$

Note that all edges in $R_j$ enter a junction at node 4 and leave from node 1. We show the interchange $R_j$ in Figure 9.3. In Figure 9.3, all junctions are shown to have node 1 to the right and node 4 to the left.

(4) In the above design, the weight of any edge not in a junction is defined to be 1.

We now observe the following properties of the graph $G$: First, define a *route* to be the collection of all cycle covers that have the same set of edges outside the junctions. A route is *good* if for each junction $J$ in $G$ it enters $J$ exactly once from node 1 or node 4 and leaves $J$ from node 4 or, respectively, node 1. The main idea of the design of matrix $X$ above is to ensure that all *bad* routes together contribute weight 0 to the total weight. Let $Q$ be a route. We write $w(Q)$ to denote the sum of weights of all cycle covers in $Q$.

*Claim* 1. A route $Q$ that is not good has weight 0.

*Proof.* A route $Q$ is not good if one of the following holds:

(a) It never enters a junction $J$;

(b) It enters a junction $J$ exactly once from node 1 and leaves from node 1;

(c) It enters a junction $J$ exactly once from node 4 and leaves from node 4; or

(d) It enters a junction $J$ twice.

In each case, the corresponding local cycles that cover the junction $J$ contribute weight 0 to the weight of $Q$, as shown in property (i) of matrix $X$. For instance, suppose that $Q$ is bad because of (b) above. Then all cycle covers in $Q$ leave nodes 2, 3, 4 of junction $J$ to be covered by local cycles. Let $Q_1$ be the set of cycle covers of nodes 2, 3, 4 of $J$. Let $G_1$ be the graph $G$ with nodes 2, 3, and 4 of junction $J$ removed. For any cycle cover $D_1$ of $G_1$, we have

$$\sum_{D \in Q, D_1 \subseteq D} \prod_{e \in D} w(e) = w(D_1) \cdot \sum_{D_2 \in Q_1} \prod_{e \in D_2} w(e)$$

$$= w(D_1) \cdot perm(X(1;1)) = 0.$$

Since each cycle cover $D$ in $Q$ must contain a cycle cover $D_1$ for $G_1$, it follows that the weight of $Q$ is 0. Other cases can be verified similarly.

*Claim 2.* A good route $Q$ has weight $4^{3c(F)}$.

*Proof.* The graph $G$ has totally $3c(F)$ many junctions. For each junction $J$, the route $Q$ either enters $J$ from node 1 and leaves from node 4 (we call $J$ a type 1 junction) or enters from node 4 and leaves from node 1 (a type 2 junction). For each type 1 junction $J$, there are six ways for a cycle cover in route $Q$ to cover junction $J$, corresponding to six permutations $\sigma$ over $\{1, 2, 3, 4\}$ with $\sigma(4) = 1$. ($\sigma(i) = j$, with $i \in \{1, 2, 3\}$ and $j \in \{2, 3, 4\}$, means the cycle includes the edge $(i, j)$ of $J$.) Therefore, a type 1 junction $J$ contributes a factor $perm(X(4;1)) = 4$ to the total weight of the route $Q$. Similarly, a type 2 junction $J$ contributes a factor $perm(X(1;4)) = 4$ to the total weight of the route $Q$. Together, the weight of $Q$ is $4^{3c(F)}$.

*Claim 3.* Each interchange $R_j$ has the following properties:
(1) There is no good route for $R_j$.
(2) Let $R_j'$ be $R_j$ with at least one of $J_{j_k, j}$ removed, $k = 1, 2, 3$. Then, there is a unique good route for $R_j'$.

*Proof.* We observe the following properties of $R_j$ by inspection: Any good route for $R_j$ or $R_j'$ must contain a big cycle that includes the following nodes: $p_j$, $a_{j,1}$, $a_{j,2}$, $a_{j,3}$, $a_{j,4}$, $q_j$. Also, if this big cycle covers a track $J_{j_k, j}$ in $R_j$, then it must cover it by the path from $a_{j,k}$ to $J_{j_k, j}$ to $b_{j,k}$. From $q_j$ to $p_j$, this cycle must go through at least one $b_{j,k}$, $k = 1, 2, 3$. However, the path from $a_{j,1}$ to $a_{j,4}$ must cover all three tracks $J_{j_k, j}$, $k = 1, 2, 3$ and, hence, all $b_{j,k}$, $k = 1, 2, 3$. It follows that the big cycle breaks down from $q_j$ to $p_j$, and $R_j$ does not have a good route.

On the other hand, each $R_j'$ has a good route because at least one $b_{j,k}$ is left uncovered in the path from $a_{j,1}$ to $a_{j,4}$, and it could be used to create a path from $q_j$ to $p_j$. Finally, the good route for each $R_j'$ is uniquely determined by the big cycle.

The following claim finishes the proof of the theorem.

*Claim 4.* There are exactly $s(F)$ good routes for $G$.

*Proof.* Each good route must cover each track $T_i$ either in a *positive cycle*, which goes from $u_i$ to $J_{i,i_1}, \ldots, J_{i,i_k}$ to $v_i$ and then comes back to $u_i$, or in a *negative cycle*, which goes from $u_i$ to $J_{i,i_1'}, \ldots, J_{i,i_k'}$ to $v_i$ and then comes back to $u_i$. (Note that this cycle enters each $J_{i,i_r}$ from node 1 and must leave from node 4.) Therefore it corresponds to exactly one Boolean assignment to variables $x_i$. From Claim 3, any good route for $G$ must cover at least one junction $J_{j_k,j}$ in each interchange $R_j$ by the cycle for track $T_{j_k}$. In other words, the corresponding Boolean assignment must make at least one literal true for each clause $C_j$ and vice versa.                                      ∎

Now we apply the above result to show that the perfect matching problem is #$P$-complete. Since the problem of determining whether a given bipartite graph $G$ has a perfect matching is polynomial-time solvable, this result demonstrates the first #$P$-complete problem whose corresponding decision problem is in $P$.

Let $G = (A, B, E)$ be a bipartite graph in which $A = B = \{1, 2, \ldots, n\}$. Define an $n \times n$ $(0, 1)$-matrix $M_G$ as follows:

$$M_G[i, j] = \begin{cases} 1 & \text{if } \{i, j\} \in E, \\ 0 & \text{otherwise.} \end{cases}$$

Then, it is clear that $perm(M_G)$ is exactly the number of perfect matchings in $G$. Thus, it suffices to prove that the problem of computing the permanents of $(0, 1)$-matrices is #$P$-complete.

**Theorem 9.15** *The problem of computing the permanent of a $(0, 1)$-matrix is #$P$-complete.*

*Proof.* We first prove that the problem of computing $perm(A)$ mod $p$ for any given $n \times n$ integer matrix $A$ and any prime $p < n^2$ is #$P$-complete. In Theorem 9.14, we have proved that the problem of computing $perm(A)$ is #$P$-complete if $A$ is an integer matrix with values in $\{-1, 0, 1, 2, 3\}$. Note that for such matrices $A$, $|perm(A)| \leq 3^n (n!)$. From the prime number theorem, for sufficiently large $n$, the product of all primes less than $n^2$ is greater than $3^n (n!)$. By the Chinese remainder theorem, $perm(A)$ could be found, in polynomial time, from $perm(A)$ mod $p_i$, for primes $p_i < n^2$ (see, e.g., Knuth [1981], Section 4.3.2). Therefore, the problem of computing $perm(A)$ is reduced to the problem of computing $perm(A)$ mod $p$, $p < n^2$.

Next, we reduce the problem of computing $perm(A)$ mod $p$ to the problem of computing $perm(B)$ for $(0, 1)$-matrices $B$. We recall the translation of Theorem 9.13 between the permanent problem and the weighted cycle covering problem. Let $G_A$ be the directed graph whose adjacency matrix is equal to $A$. Since we are only interested in $perm(A)$ mod $p$, we may assume that all weights in $G_A$ are in $\{0, 1 \ldots, p - 1\}$. For each edge $(u, v)$ in $G_A$ that has a weight $k \notin \{0, 1\}$, we replace it by a graph $H_k$ as follows:

The graph $H_k$ has $2k + 7$ nodes:

$$u, \ v, \ u_1, \ v_1, \ w_i, \ z_j, \qquad 0 \leq i \leq k + 1, \ 0 \leq j \leq k;$$

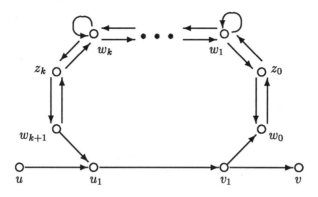

Figure 9.4: Graph $H_k$.

and $5k + 9$ edges:

$$(u, u_1),\ (u_1, v_1),\ (v_1, v),\ (v_1, w_0),\ (w_{k+1}, u_1),$$
$$(w_i, w_i), \qquad\qquad\qquad\qquad\qquad\qquad i = 1, \ldots, k,$$
$$(w_j, z_j),\ (z_j, w_j),\ (z_j, w_{j+1}),\ (w_{j+1}, z_j), \qquad j = 0, \ldots, k.$$

Each edge of $H_k$ has the weight 1. We show the graph $H_k$ in Figure 9.4.

Assume that in a graph $G$, an edge $(u, v)$ of weight $k$ is replaced by $H_k$ to form a new graph $G'$. Then, for any cycle cover $C$ of $G$ that does not include $(u, v)$, there is exactly one cycle cover $C'$ of $G'$ that has all the edges of $C$ plus a unique cycle cover for $H_k - \{u, v\}$. For any cycle cover $C$ of $G$ that includes $(u, v)$, there are $k$ cycle covers $C_1, \ldots, C_k$ of $G'$ that contain $C - \{(u, v)\}$ plus a cover of $H_k$. Thus, $\#CC(G) = \#CC(G')$.

Therefore, replacing each edge of weight $k$ in $G$ by a graph $H_k$ creates a new graph $G_1$ such that $G_1$ contains only edges of weight 1 or 0 and $\#CC(G) = \#CC(G_1)$. This completes the proof.                                                        ∎

**Corollary 9.16** *The problem* #PM *is* #*P-complete.*

For any function $f$ in $\#P$, we may define a set $A_f = \{\langle x, n \rangle : n \le f(x)\}$. Then, $A_f \equiv_T^P f$. By Theorem 9.6, we know that for all functions $f$, $f$ is #P-complete if and only if $A_f$ is complete for $PP$ under the $\le_T^P$-reducibility. Therefore, the class $PP$ has many $\le_T^P$-complete sets. In the following, we show that some of these sets are actually complete for $PP$ under the stronger $\le_m^P$-reducibility.

**Theorem 9.17** *Let* MAJSAT *be the set of all Boolean formulas $F$ that has $n$ variables and has more than $2^{n-1}$ truth assignments satisfying it. Then,* MAJSAT *is* $\le_m^P$*-complete for PP.*

*Proof.* Assume that a set $A$ in $PP$ is defined by an NTM $M$ with a polynomial uniform time complexity $t(n)$ such that $A = \{x : \varphi_M(x) > 2^{t(|x|)-1}\}$. (Note that we consider $M$ as an NTM and so $A \subseteq L(M)$ but they may not be equal.)

As discussed in the proof of Theorem 9.9, the reduction from $L(M)$ to SAT given in Cook's Theorem preserves the number of witnesses and so for each $x$, $|x| = n$, we have a Boolean formula $F_x$ of $m$ variables $v_1, v_2, \ldots, v_m$ such that $F_x$ has exactly $\varphi_M(x)$ truth assignments. Note that $m$ is usually greater than $t(n)$.

Now, we define a new formula $F_x'$ over variables $v_1, v_2, \ldots, v_m$ and $z$:

$$F_x' = z \cdot (v_1 + v_2 + \cdots + v_{m-t(n)+1}) + \bar{z} \cdot F_x.$$

Then, $F_x'|_{z=1}$ has $(2^{m-t(n)+1} - 1)2^{t(n)-1} = 2^m - 2^{t(n)-1}$ truth assignments, and $F_x'|_{z=0}$ has $\varphi_M(x)$ truth assignments. Together, $F_x'$ has $2^m - 2^{t(n)-1} + \varphi_M(x)$ truth assignments, which is greater than $2^m$ if and only if $\varphi_M(x) > 2^{t(n)-1}$. Thus, $x \in A$ if and only if $F_x'$ has more than $2^m$ truth assignments. (Note that $F_x'$ has $m + 1$ variables.) ∎

## 9.3　⊕$P$ and the Polynomial-Time Hierarchy

An interesting subclass of counting problems is the class of parity problems. For any set $A$ and any polynomial $p$, define $parity_A(x)$ to be 1 if the number of elements $\langle x, y \rangle$ in $A$, with $|y| = p(|x|)$, is odd, and to be 0 otherwise. Then $parity_A$ is a simplified counting problem. It is obvious that the parity problem $parity_A$ associated with a problem $A \in P$ is reducible to a problem in $\#P$. Although it intuitively seems a weaker problem than the $\#P$-complete counting problems, we show in this section that the parity problem is at least as difficult as any problem in the polynomial-time hierarchy. In the next section we will apply this result to show that $P^{\#P}$ contains the whole polynomial-time hierarchy.

We first formally define the complexity classes of parity problems. For each polynomial-time NTM $M$, we define a polynomial-time *parity machine* $M'$ as follows: on input $x$, $M'$ accepts $x$ if and only if the number of accepting computations of $M(x)$ is odd. The complexity class ⊕$P$ is the class of all sets $B$ that are accepted by polynomial-time parity machines.

Equivalently, we may define the *parity operator* ⊕ as follows:

**Definition 9.18** *Let $A$ be any set and $C$ be any complexity class. We say a set $B$ is in* ⊕$(A)$ *if there exists a polynomial $p$ such that for all $x$,*

$$x \in B \iff |\{y : |y| = p(|x|), \langle x, y \rangle \in A\}| \text{ is odd.}$$

*We say $B \in$ ⊕$(C)$ if $B \in$ ⊕$(A)$ for some $A \in C$.*

It is easy to see that ⊕$(P)$ is equivalent to the class ⊕$P$ defined by parity machines. We leave its proof to the reader.

**Proposition 9.19** $\oplus(P) = \oplus P$.

The following lemma on $\oplus P$ is useful for our main results.

**Lemma 9.20** *(a)* $\oplus(\oplus P) = \oplus P$.

*(b) Let $A \in \oplus P$ and $k > 0$. Then the set*

$$A_k = \{\langle x_1, \ldots, x_k \rangle : |x_1| = \cdots = |x_k|, (\exists i, 1 \leq i \leq k) \; x_i \in A\}$$

*is also in $\oplus P$.*

*(c) Let $A \in \oplus P$ and $q$ be a polynomial. Then, the set*

$$B = \{\langle x_1, \ldots, x_{q(n)} \rangle : |x_1| = \cdots = |x_{q(n)}| = n, (\exists i, 1 \leq i \leq q(n)) \; x_i \in A\}$$

*is also in $\oplus P$.*

*(d) Let $A \in \oplus P$ and $q$ be a polynomial. Then, the set*

$$C = \{\langle x_1, \ldots, x_{q(n)} \rangle : \quad |x_1| = \cdots = |x_{q(n)}| = n,$$
$$\text{the majority of } x_1, \ldots, x_{q(n)} \text{ are in } A\}$$

*is also in $\oplus P$.*

*Proof.* (a) Assume that $A \in P$, $B \in \oplus(A)$ and $C \in \oplus(B)$. From the definition, there exist two polynomial functions $q_1$ and $q_2$ such that for all $x$, with $|x| = n$,

$$x \in C \iff |\{y : |y| = q_1(n), \langle x, y \rangle \in B\}| \text{ is odd,}$$

and for all $x$, $|x| = n$, and all $y$, $|y| = q_1(n)$,

$$\langle x, y \rangle \in B \iff |\{z : |z| = q_2(n), \langle \langle x, y \rangle, z \rangle \in A\}| \text{ is odd.}$$

Define $A_1 = \{\langle x, y, z \rangle : \langle \langle x, y \rangle, z \rangle \in A, |y| = q_1(|x|), |z| = q_2(|x|)\}$, and consider a string $x$ of length $n$. First, if $x \in C$ then the number of pairs $\langle y, z \rangle$ such that $|y| = q_1(n)$, $|z| = q_2(n)$, $\langle x, y \rangle \in B$ and $\langle x, y, z \rangle \in A_1$ is odd because the sum of an odd number of odd numbers is odd. Also, the number of pairs $\langle y, z \rangle$ such that $|y| = q_1(n)$, $|z| = q_2(n)$, $\langle x, y \rangle \notin B$, and $\langle x, y, z \rangle \in A_1$ is even, because the sum of an odd number of even numbers is even. Together, we see that if $x \in C$ then the number of $\langle y, z \rangle$ such that $\langle x, y, z \rangle \in A_1$ is odd. Similarly, we can verify that if $x \notin C$ then the number of $\langle y, z \rangle$ such that $\langle x, y, z \rangle \in A_1$ is even.

(b) By a simple modification, we can see that the class $\oplus P$ remains the same if we use the *even* parity to define it in Definition 9.18. So, we assume that there exist a set $C \in P$ and a polynomial $p$ such that for all $x$, with $|x| = n$,

$$x \in A \iff |\{y : |y| = p(n), \langle x, y \rangle \in C\}| \text{ is even.}$$

Let

$$C_k = \{\langle x_1, \ldots, x_k, y_1, \ldots, y_k \rangle : |x_1| = |x_2| = \cdots = |x_k| = n,$$
$$|y_1| = |y_2| = \cdots = |y_k| = p(n), (\exists i, 1 \leq i \leq k) \; \langle x_i, y_i \rangle \in C\}.$$

It is clear that $C_k \in P$. We can see that for any $x_1, \ldots, x_k$, all of length $n$,

$$
\begin{aligned}
\langle x_1, \ldots, x_k \rangle \in A_k \iff & (\exists i, 1 \le i \le k)\, x_i \in A \\
\iff & (\exists i, 1 \le i \le k)\, |\{y_i : |y_i| = p(n), \langle x_i, y_i \rangle \in C\}| \text{ is even} \\
\iff & |\{\langle y_1, \ldots, y_k \rangle : \langle x_1, \ldots, x_k; y_1, \ldots, y_k \rangle \in C_k\}| \text{ is even}.
\end{aligned}
$$

The last two predicates are equivalent because the product of $k$ integers is even if and only if one of them is even.

(c) The proof is essentially the same as that of (b).

(d) Assume that there exist a set $D \in P$ and a polynomial $p$ such that for all $x$, $|x| = n$,

$$
x \in A \iff |\{y : |y| = p(n), \langle x, y \rangle \in D\}| \text{ is odd}.
$$

Now, let $E$ be the set of all strings of the form $\langle x_1, \ldots, x_{q(n)}, z_1, \ldots, z_{q(n)}, i_1, \ldots, i_m \rangle$ with the following properties:

 (i) $|x_1| = |x_2| = \cdots = |x_{q(n)}| = n$;

 (ii) $|z_1| = |z_2| = \cdots = |z_{q(n)}| = p(n) + 1$;

 (iii) $1 \le i_1 < i_2 < \cdots < i_m \le n$;

 (iv) $q(n)/2 < m \le q(n)$; and

 (v) $(\forall i, 1 \le i \le q(n))$ either $[z_i = 1y_i$ and $\langle x_i, y_i \rangle \in D]$ or $[z_i = 0^{p(n)+1}$ and $i \notin \{i_1, \ldots, i_m\}]$.

It is clear that $E \in P$. For any $x_1, \ldots, x_{q(n)}$ of length $n$, we claim that $\langle x_1, \ldots, x_{q(n)} \rangle \in C$ if and only if there are an odd number of witnesses of the form $\langle z_1, \ldots, z_{q(n)}, i_1, \ldots, i_m \rangle$ such that $\langle x_1, \ldots, x_{q(n)}, z_1, \ldots, z_{q(n)}, i_1, \ldots, i_m \rangle \in E$. Let $J = \{j : 1 \le j \le q(n), x_j \in A\}$, and for each $j$, $1 \le j \le q(n)$, let $t_j = |\{y : |y| = p(n), \langle x_j, y \rangle \in D\}|$. We note that $t_j$ is odd if and only if $j \in J$. Now, for any fixed $I = \{i_1, \ldots, i_m\}$, with $m > q(n)/2$ and $1 \le i_1 < i_2 < \cdots < i_m \le q(n)$, the number of witnesses $\langle z_1, \ldots, z_{q(n)}, i_1, \ldots, i_m \rangle$ for $\langle x_1, \ldots, x_{q(n)} \rangle$ with the fixed $i_1, \ldots, i_m$ is

$$
\prod_{j \in I} t_j \cdot \prod_{j \notin I} (t_j + 1).
$$

This number is odd if and only if $I = J$. So, the total number of witnesses for $\langle x_1, \ldots, x_{q(n)} \rangle$, ranging over all possible $I$'s, is odd if and only if one of $I$'s is equal to $J$ or, equivalently, if and only if $|J| > q(n)/2$.  ∎

Next, we define the $BP$-operator, or, the notion of randomized reduction based on the $BPP$-type probabilistic Turing machines. Intuitively, a set $B$ is randomly reducible to a set $A$ if there is a polynomial-time PTM transducer $M$ that on each input $x$ computes a string $y$ such that with a high probability, $x \in B$ if and only if $y \in A$. The following formal definition uses the $\exists^+$ quantifier to simplify the notation. Recall that $(\exists^+ y, |y| = m)$ means "for at least $(3/4) \cdot 2^m$ strings $y$ of length $m$."

**Definition 9.21** *Let $A$ be any set and $C$ be any complexity class. We say set $B$ is reducible to $A$ by a* randomized reduction, *and write $B \in BP(A)$, if there exists a polynomial $p$ such that for all $x$ of length $n$,*

$$x \in B \; \Rightarrow \; (\exists^{+}y, |y| = p(n)) \; \langle x, y \rangle \in A$$
$$x \notin B \; \Rightarrow \; (\exists^{+}y, |y| = p(n)) \; \langle x, y \rangle \notin A.$$

*We say $B \in BP(C)$ if $B \in BP(A)$ for some set $A \in C$.*

The $BP$-operator is, like the set $BPP$, robust with respect to polynomial-time computability.

**Lemma 9.22** *(a) If $B \in BP(A)$, then for each polynomial function $q$, there exist a set $A_1 \in P(A)$ and a polynomial $p$ such that for all $x$, with $|x| = n$,*

$$x \in B \; \Rightarrow \; (\exists_r^{+}y, |y| = p(n)) \; \langle x, y \rangle \in A_1$$
$$x \notin B \; \Rightarrow \; (\exists_r^{+}y, |y| = p(n)) \; \langle x, y \rangle \notin A_1,$$

*where $r = 1 - 2^{-q(n)}$.*

*(b) For any set $A$, $BP(BP(A)) \subseteq BP(A_1)$ for some $A_1 \in P(A)$.*

*Proof.* (a) The proof is almost identical to Theorem 8.19.

(b) The proof is similar to Exercise 8.12. ∎

One of the important properties about the $BP$-operator and the $\oplus$-operator is that they can be *swapped*.

**Lemma 9.23** $\oplus(BP(\oplus P)) = BP(\oplus P)$.

*Proof.* Assume that $L \in \oplus(BP(\oplus P))$. Then, there exist a set $A \in BP(\oplus P)$ and a polynomial $p$ such that for all $x$,

$$x \in L \; \Longleftrightarrow \; |\{y : |y| = p(|x|), \langle x, y \rangle \in A\}| \text{ is odd.}$$

First we claim that there exist a set $B \in \oplus P$ and a polynomial $r$ such that for all $x$ of length $n$, and for a random $z$ of length $r(n)$,

$$\Pr_z[(\forall y, |y| = p(n)) \; [\langle x, y, z \rangle \in B \; \Longleftrightarrow \; \langle x, y \rangle \in A]] \geq 3/4. \qquad (9.1)$$

*Proof of Claim.* From the fact that $A \in BP(\oplus P)$, we know that there exist a set $B_1 \in \oplus P$ and a polynomial $q$ such that $q(n) \geq p(n)$ and for all $\langle x, y \rangle$, with $|x| = n$, $|y| = p(n)$, and for a random $z$ of length $q(n)$,

$$\Pr_z[\langle x, y, z \rangle \in B_1 \; \Longleftrightarrow \; \langle x, y \rangle \in A] \geq 3/4. \qquad (9.2)$$

Now, applying the second Swapping Lemma (Exercise 8.11) to the above predicate $[\langle x, y, z \rangle \in B_1 \; \Longleftrightarrow \; \langle x, y \rangle \in A]$, we have, for a polynomial $q_1$, and a random $\vec{w} = \langle z_1, \ldots, z_{q_1(n)} \rangle$, each $|z_i| = q(n)$,

$$\Pr_{\vec{w}}[(\forall y, |y| = p(|x|)) \; [(\mathsf{M}z \in \vec{w}) \langle x, y, z \rangle \in B_1] \; \Longleftrightarrow \; \langle x, y \rangle \in A] \geq 3/4,$$

where $(Mz \in \vec{w})$ means "for the majority of $z \in \{z_1, \ldots, z_{q(n)}\}$". Let $B = \{\langle x, y, \vec{w}\rangle : (Mz \in \vec{w})\langle x, y, z\rangle \in B_1\}$. By Lemma 9.20(d), we know that $B$ is in $\oplus P$ since $B_1$ is in $\oplus P$. This completes the proof of the claim.

From the above claim, we assume the existence of set $B \in \oplus P$ and polynomial $r$ such that (9.1) holds for all $x$ of length $n$ and for a random $z$ of length $r(n)$. Define

$$C = \{\langle x, z\rangle : |z| = r(|x|), |\{y : |y| = p(|x|), \langle x, y, z\rangle \in B\}| \text{ is odd}\}.$$

By Lemma 9.20(a), $C \in \oplus(B) \subseteq \oplus P$. Now, if $x \in L$, then $|\{y : |y| = p(|x|), \langle x, y\rangle \in A\}|$ is odd and so, for a random $z$ of length $q(|x|)$,

$$
\begin{aligned}
\Pr_z[\langle x, z\rangle \in C] &= \Pr_z[|\{y : |y| = p(|x|), \langle x, y, z\rangle \in B\}| \text{ is odd}] \\
&\geq \Pr_z[(\forall y, |y| = p(n))\,[\langle x, y, z\rangle \in B \iff \langle x, y\rangle \in A]] \\
&\geq 3/4.
\end{aligned}
$$

Conversely, if $x \notin L$ then, for a random $z$ of length $q(|x|)$,

$$
\begin{aligned}
\Pr_z[\langle x, z\rangle \in C] &= \Pr_z[|\{y : |y| = p(|x|), \langle x, y, z\rangle \in B\}| \text{ is odd}] \\
&\leq \Pr_z[(\exists y, |y| = p(n))\,\neg[\langle x, y, z\rangle \in B \iff \langle x, y\rangle \in A]] \\
&< 1/4.
\end{aligned}
$$

Thus, $L \in BP(C) \subseteq BP(\oplus P)$. ∎

We are ready to prove the main results of this section. We first prove that $\oplus P$ is hard for the class *NP* under the randomized reduction.

**Theorem 9.24** $NP \subseteq BP(\oplus P)$.

*Proof.* We are going to apply the Isolation Lemma (Lemma 6.48) to create a random reduction from SAT to a set $B \in \oplus P$. For any Boolean formula $F$ with variables $x_1, \ldots, x_n$, assign weight $w_i$ to each $x_i$ for $i = 1, \ldots, n$. We choose the weights $w_i$ independently and randomly from the range $\{1, \ldots, 2n\}$. Let $\mathbf{w}$ denote $\langle w_1, \ldots, w_n\rangle$. For any Boolean assignment $\tau$ on $\{x_1, \ldots, x_n\}$, we let weight $\mathbf{w}(\tau)$ of $\tau$ be $\mathbf{w}(\tau) = \sum_{\tau(x_i)=1} w_i$. Then, by the Isolation Lemma, if $F \in$ SAT then, with probability at least $1/2$, there is a unique minimum-weight $\tau$ that satisfies $F$ (i.e., $F|_\tau = 1$). On the other hand, if $F \notin$ SAT, then there is no assignment $\tau$ satisfying $F$. Therefore, this is a random reduction from $F$ to the parity of the minimum-weight truth assignments to $F$. One technical problem here is that the question of whether $\mathbf{w}(\tau)$ is the minimum among all satisfying assignments $\tau$ such that $F|_\tau = 1$ is not decidable in polynomial time since there are $2^n$ assignments $\tau$ to be checked. In the following proof, we use a random guess of the minimum weight to avoid this problem.

Let $u$ be an integer in $\{1, \ldots, 2n^2\}$ and $m = 4n^2$. Define

$$A = \{\langle F, \mathbf{w}, u\rangle : |\{\tau \in \{0,1\}^n : F|_\tau = 1, \mathbf{w}(\tau) \leq u\}| \text{ is odd}\}.$$

Then, it is clear that $A \in \oplus P$. Next, define

$$B = \{ \langle F, \mathbf{w}^{(1)}, u_1, \mathbf{w}^{(2)}, u_2, \ldots, \mathbf{w}^{(m)}, u_m \rangle :$$
$$(\exists j, 1 \leq j \leq m) |\{ \tau \in \{0,1\}^n : F|_\tau = 1, \mathbf{w}^{(j)}(\tau) \leq u_j \}| \text{ is odd} \}.$$

From Lemma 9.20(c), we see that $B \in \oplus P$.

We define a random reduction $f$ from SAT to $B$ by

$$f(F) = \langle F, \mathbf{w}^{(1)}, u_1, \ldots, \mathbf{w}^{(m)}, u_m \rangle,$$

where each $w_i^{(j)}$ and each $u_j$, for $1 \leq i \leq n$ and $1 \leq j \leq m$, are chosen independently and randomly from $\{1, \ldots, 2n\}$ and $\{1, \ldots, 2n^2\}$, respectively. We observe that if $F \notin$ SAT, then for any $\mathbf{w}^{(j)}$ and any $u_j$, the set $\{ \tau \in \{0,1\}^n : F|_\tau = 1, \mathbf{w}^{(j)}(\tau) \leq u_j \}$ is empty and so, with probability 1, $f(F) \notin B$.

If $F \in$ SAT then, with probability at least $1/2$, there is a unique assignment $\tau$ that satisfies $F$ and has the minimum weight $\mathbf{w}^{(j)}(\tau)$ among all satisfying assignments. Also, with probability $1/(2n^2)$, the value $u_j$ is equal to the minimum weight $\mathbf{w}^{(j)}(\tau)$. Since the choice of $u_j$ is independent of the choice of $\mathbf{w}^{(j)}$, the probability is at least $1/m$ such that there is a unique $\tau$ such that $F|_\tau = 1$ and $\mathbf{w}^{(j)}(\tau) \leq u_j$. That is,

$$\Pr[\langle F, \mathbf{w}^{(j)}, u_j \rangle \in A] \geq \frac{1}{m} .$$

Since the choice of $\langle \mathbf{w}^{(j)}, u_j \rangle$ is independent of the choice of $\langle \mathbf{w}^{(k)}, u_k \rangle$ for $1 \leq j < k \leq m$, we get

$$\Pr[\langle F, \mathbf{w}^{(1)}, u_1, \ldots, \mathbf{w}^{(m)}, u_m \rangle \in B] \geq 1 - \left(1 - \frac{1}{m}\right)^m \geq 1 - \frac{1}{e} > \frac{1}{2} .$$

This shows that $f$ is a random reduction from SAT to set $B \in \oplus P$ and the proof of the theorem is complete. ∎

The above proof relativizes; in particular, it holds relative to an oracle in $\Pi_{k-1}^P$.

**Corollary 9.25** *For all $k \geq 1$, $\Sigma_k^P \subseteq BP(\oplus(\Pi_{k-1}^P))$.*

*Proof.* Recall the $\Sigma_k^P$-complete set $\text{SAT}_k$, which is the set of all Boolean formulas $F(X_1, \ldots, X_k)$ in which Boolean variables are partitioned into $k$ groups $X_1, \ldots, X_k$ such that

$$(\exists \tau_1)(\forall \tau_2) \cdots (Q_k \tau_k) \, F|_{\tau_1, \ldots, \tau_k} = 1,$$

where each $\tau_i$ ranges over all Boolean assignments on variables in $X_i$. To prove the corollary, replace, in the above proof of Theorem 9.24, set SAT by $\text{SAT}_k$, and the predicate $F|_\tau = 1$ by the following $\Pi_{k-1}^P$-predicate:

$$(\forall \tau_2) \cdots (Q_k \tau_k) \, F|_{\tau, \tau_2, \ldots, \tau_k} = 1.$$

It is clear that the resulting set $B$ is in $\oplus(\Pi_{k-1}^P)$. Thus, the same argument concludes that $\text{SAT}_k \in BP(A) \subseteq BP(\oplus(\Pi_{k-1}^P))$. ∎

Now we apply Lemma 9.23 to show that $\oplus P$ is hard for the polynomial-time hierarchy under the randomized reduction.

**Theorem 9.26** *For all $k \geq 1$, $\Sigma_k^P \subseteq BP(\oplus P)$.*

*Proof.* We prove the theorem by induction on $k$. Theorem 9.24 established the case $k = 1$. For $k > 1$, we have, from the inductive hypothesis and Corollary 9.25, $\Sigma_k^P \in BP(\oplus(BP(\oplus(P))))$. From Lemmas 9.22(b) and 9.23, the $BP$-operator and the $\oplus$-operator can be swapped and combined. We conclude that $\Sigma_k^P \subseteq BP(\oplus P)$. ∎

**Corollary 9.27** *(a) If $\oplus P \subseteq BPP$ then the polynomial-time hierarchy collapses to $BPP$ and, hence, to $\Sigma_2^P \cap \Pi_2^P$.*

*(b) For any $k \geq 1$, if $\oplus P \subseteq \Sigma_k^P$, then the polynomial-time hierarchy collapses to $\Sigma_{k+1}^P \cap \Pi_{k+1}^P$.*

*Proof.* Part (a) follows immediately from the above theorem and Corollary 8.33. Part (b) follows from the generalized $BPP$ theorem (Theorem 8.34), which implies that $BP(\Sigma_k^P) \subseteq \Sigma_{k+1}^P \cap \Pi_{k+1}^P$. ∎

## 9.4 $\#P$ and the Polynomial-Time Hierarchy

In this section, we prove that $PH \subseteq P^{\#P}$. From Theorem 9.26, we only need to show that $BP(\oplus P) \subseteq P^{\#P}$. The main idea is that the number of accepting computations of an NTM can be *amplified* without changing its parity. Recall that $\varphi_M(x)$ denotes the number of accepting computations of an NTM $M$ on input $x$.

**Lemma 9.28** *Let $A \in \oplus P$. Then, for any polynomial $q$, there exists a polynomial-time NTM $M$ such that for any input $x$ of length $n$,*

$$x \in A \Rightarrow \varphi_M(x) \equiv -1 \bmod 2^{q(n)},$$
$$x \notin A \Rightarrow \varphi_M(x) \equiv 0 \bmod 2^{q(n)}.$$

*Proof.* Let $M_1$ be a polynomial-time NTM such that $\varphi_{M_1}(x) \equiv 1 \bmod 2$ if $x \in A$, and $\varphi_{M_1}(x) \equiv 0 \bmod 2$ if $x \notin A$. We define another polynomial-time NTM $M_2$ that repeats $M_1$ on $x$ a number of times to achieve the required value $\varphi_{M_2}(x)$. More precisely, let $M_2$ be the following NTM, defined in a recursive form:

*Nondeterministic Machine $M_2$:*
Input: $\langle x, i \rangle$.

(1) If $i = 0$ then simulate $M_1(x)$;

(2) Otherwise, if $i > 0$ then do the following:

    (2.1) Nondeterministically choose an integer $m \in \{1, 2, \ldots, 7\}$;

    (2.2) If $m \leq 4$ then let $k := 3$ else let $k := 4$;

    (2.3) For $j := 1$ to $k$ do

        begin

            recursively run $M_2(\langle x, i - 1 \rangle)$;

            if $M_2(\langle x, i - 1 \rangle)$ rejects then reject and halt

        end;

    (2.4) Accept and halt.

Let $f(x, i) = \varphi_{M_2}(\langle x, i \rangle)$. It is clear that $f$ satisfies the following recurrence equation:

$$f(x, i + 1) = 3f(x, i)^4 + 4f(x, i)^3, \qquad i \geq 0.$$

It is easy to prove by induction that if $f(x, 0)$ is even, then $f(x, i) \equiv 0 \bmod 2^{2^i}$, and if $f(x, 0)$ is odd, then $f(x, i) \equiv -1 \bmod 2^{2^i}$. Let $M(x) = M_2(x, \lceil \log q(n) \rceil)$. Then, from the above analysis, for any $x$ of length $n$,

$$x \in A \Rightarrow \varphi_M(x) \equiv -1 \bmod 2^{2^{\lceil \log q(n) \rceil}} \equiv -1 \bmod 2^{q(n)},$$
$$x \notin A \Rightarrow \varphi_M(x) \equiv 0 \bmod 2^{2^{\lceil \log q(n) \rceil}} \equiv 0 \bmod 2^{q(n)}.$$

Let $t_{M_1}$ be the time complexity function of $M_1$. Note that the machine $M$ halts in time $4^{\lceil \log q(n) \rceil} \cdot t_{M_1}(n)$ on any input of length $n$. So, $M$ runs in polynomial time. ∎

**Theorem 9.29** $BP(\oplus P) \subseteq P^{\#P}$.

*Proof.* Let $L \in BP(\oplus P)$; that is, there exist a set $A \in \oplus P$ and a polynomial $p$ such that for all $x$,

$$\Pr_y[x \in L \iff \langle x, y \rangle \in A] \geq 3/4,$$

where $y$ ranges over all strings of length $p(|x|)$. By Lemma 9.28, there exists a polynomial-time NTM $M$ such that for all $\langle x, y \rangle$, with $|x| = n$ and $|y| = p(n)$,

$$\varphi_M(\langle x, y \rangle) \equiv \begin{cases} -1 \bmod 2^{p(n)} & \text{if } \langle x, y \rangle \in A, \\ 0 \bmod 2^{p(n)} & \text{if } \langle x, y \rangle \notin A. \end{cases}$$

Let $g(x) = |\{y : |y| = p(|x|), \langle x, y \rangle \in A\}|$ and $h(x) = \sum_{|y| = p(|x|)} \varphi_M(\langle x, y \rangle)$. Then, for any $x$ of length $n$,

$$\begin{aligned} h(x) &= \sum_{\langle x, y \rangle \in A} \varphi_M(\langle x, y \rangle) + \sum_{\langle x, y \rangle \notin A} \varphi_M(\langle x, y \rangle) \\ &\equiv (g(x) \cdot (-1) + (2^{p(n)} - g(x)) \cdot 0) \bmod 2^{p(n)} \\ &\equiv -g(x) \bmod 2^{p(n)}. \end{aligned}$$

So, we can decide whether $x \in L$ from $h(x)$; namely, from $h(x)$ and $p(n)$, we can compute $g(x)$ and decide that $x \in L$ if and only if $g(x) > 2^{p(n)-1}$. Note that the function $h$ is in $\#P$: we can define a new NTM $M_1$ that on input $x$ first nondeterministically guesses a string $y$ of length $p(|x|)$ and then simulates $M$ on $\langle x, y \rangle$. Therefore, $L \in P^{\#P}$.                                   ∎

**Corollary 9.30** $PH \subseteq P^{\#P} = P^{PP}$.

The above proof actually gives a stronger result. Let $MAJ$ be the following operator: $L \in MAJ(A)$ if for all $x$, $\Pr_{|y|=p(|x|)}[\langle x, y \rangle \in A \iff x \in L] > 1/2$ for some polynomial $p$. Also define, for a class $C$ of sets, $MAJ(C) = \bigcup_{A \in C} MAJ(A)$.

**Corollary 9.31** $MAJ(\oplus P) \subseteq P^{\#P}$.

*Proof.* In the proof of Theorem 9.29, $h(x)$ actually tells us the exact number of $y$ such that $\langle x, y \rangle \in A$. Thus, even if $L \in MAJ(A)$, we can decide whether $x \in L$ from the oracle $h$.                                   ∎

## 9.5   Circuit Complexity and Relativized $\oplus P$ and $\#P$

We have seen in the last two sections that the class $\oplus P$ is hard for $PH$ under the polynomial-time randomized reduction, and the class $\#P$ is hard for $PH$ under the polynomial-time Turing reduction. These results imply that if $\oplus P$ or $\#P$ is actually contained in $PH$, then $PH$ collapses into a finite hierarchy. On the other hand, if either $\oplus P$ or $\#P$ is provably not contained in $PH$, then $PSPACE \neq NP$, since $\oplus P \subseteq P^{\#P} \subseteq PSPACE$. Therefore, it appears difficult to determine the precise relations between $\oplus P$, $\#P$ and the polynomial-time hierarchy.

In this section, we investigate the relativized relations between these counting complexity classes and the polynomial-time hierarchy. We prove that in the relativized form, the complexity classes $\oplus P$ and $P^{\#P}$ are not contained in $PH$. The proof technique is particularly interesting, as it demonstrates a simple relation between the relativized separation results on classes in $PH$ and the lower bound results on circuit complexity. Using the lower bound results on constant-depth circuits of Section 6.6, we are able to separate the relativized counting complexity classes from the polynomial-time hierarchy and, moreover, to separate the polynomial-time hierarchy into a properly infinite hierarchy by oracles.

We fist show how to translate a relativized separation problem involving the polynomial-time hierarchy into a lower bound problem about constant-depth circuits. Recall from Chapter 3 a set $B$ in $\Sigma_k^{P,A}$ can be characterized by a $\Sigma_k^{P,A}$-predicate $\sigma(A; x)$: for each $x$ of length $n$,

$$
\begin{aligned}
x \in B &\iff \sigma(A; x) \\
&\iff (\exists y_1, |y_1| \leq q(n)) \, (\forall y_2, |y_2| \leq q(n)) \cdots \\
&\qquad (Q_k y_k, |y_k| \leq q(n)) \, R(A; x, y_1, y_2, \ldots, y_k),
\end{aligned} \tag{9.3}
$$

where $Q_k$ is $\exists$ if $k$ is odd and is $\forall$ if $k$ is even, and $R$ is a polynomial-time computable predicate relative to any oracle $A$. Also recall from Section 6.6 the notion of constant-depth levelable Boolean circuits. Such a circuit has alternating AND and OR gates at each level. Each gate of the circuit has an unlimited fanin, and its input nodes can take an input value as it is or take its negated value. In the following, by a circuit we mean a constant-depth levelable circuit. We label each input node of a circuit $C$ by a string $z$ or its negation $\bar{z}$. These labels are called *variables* of $C$. We say that a variable $z$ in $C$ is assigned with a Boolean value $b$ to mean that each input node with the label $z$ takes the input value $b$ and each input node with the label $\bar{z}$ takes the input value $1 - b$.

**Lemma 9.32** *Let $k \geq 1$ and $q(n)$ and $r(n)$ be two polynomial functions. Let $\sigma(A; x)$ be a $\Sigma_k^{P,A}$-predicate defined as in (9.3), where $R(A; x, y_1, \ldots, y_k)$ is computable in time $r(n)$ by an oracle DTM $M$ using oracle $A$. Then, for each string $x$, there exists a circuit $C$ having the following properties:*

*(a) The depth of $C$ is $k + 1$.*

*(b) The top gate of $C$ is an OR gate.*

*(c) The fanin of each gate in $C$ is $\leq 2^{q(n)+r(n)}$.*

*(d) The bottom fanin of $C$ is $\leq r(n)$.*

*(e) The variables of $C$ are strings (and their negations) which are queried by $M$ on input $(x, y_1, \ldots, y_k)$ over all possible $y_1, \ldots, y_k$ of length $\leq q(n)$ and all possible oracles $A$.*

*(f) For each set $A$, if we assign the value $\chi_A(z)$ to each variable $z$ in $C$ then $C$ outputs 1 if and only if $\sigma(A; x)$ is true.*

*Proof.* We prove the lemma by induction on $k$. For $k = 1$, let $\sigma(A; x) = (\exists y, |y| \leq q(n)) R(A; x, y)$, where $R$ is computable by an oracle Turing machine $M$ in time $r(n)$. Then, for each $y$ of length $\leq q(n)$, consider the computation tree $T$ of $M(x, y)$. Each path of $T$ corresponds to a sequence of answers to the queries made by $M$. For each accepting path in $T$, let $U$ be the set of strings answered positively by the oracle and $V$ the set of strings answered negatively by the oracle. For this accepting path, define an AND gate with $|U \cup V|$ many children, each attached with a string in $U$ or the negation of a string in $V$. Then, the OR of all these AND gates, corresponding to all accepting paths in $T$, is a circuit $G_y$ (depending only on $x$ and $y$), which computes $R(A; x, y)$ when each variable $z$ in $G_y$ is assigned the value $\chi_A(z)$. Note that each AND gate of the circuit $G_y$ has $\leq r(n)$ children. Now consider the circuit $C$, which is the OR of all $G_y$'s for $y$ of length $\leq q(n)$. We know that $C$ computes $\sigma(A; x)$ when each variable $z$ in $C$ is given value $\chi_A(z)$. Furthermore, combining all OR gates of $G_y$'s into one, $C$ can be written as a depth-2 circuit with top fanin $\leq 2^{q(n)+r(n)}$ and bottom fanin $\leq r(n)$.

For the inductive step, assume that $\sigma(A; x)$ is a $\Sigma_k^{P,A}$-predicate as defined in (9.3) with $k > 1$. Then, for each $y_1$ of length $\leq q(n)$ there is a depth-$k$

circuit $C_{y_1}$ satisfying the conditions (a)—(f) with respect to predicate

$$\tau(A; x, y_1) = \quad (\exists y_2, |y_2| \leq q(n)) \cdots (Q_{k-1}y_k, |y_k| \leq q(n))$$
$$[\neg R(A; x, y_1, \ldots, y_k)].$$

Now for each $y_1$ define a new circuit $C'_{y_1}$ (called the *dual circuit* of $C_{y_1}$), which has the same structure as $C_{y_1}$ but each AND gate of $C_{y_1}$ is replaced by an OR gate, each OR gate of $C_{y_1}$ is replaced by an AND gate, and each variable in $C_{y_1}$ is replace by its negation. Then the circuit $C'_{y_1}$ computes the negation of the function computed by $C_{y_1}$. We let $C$ be the OR of all these circuits $C'_{y_1}$. Then, $C$ is the circuit we need.  ∎

In Corollary 6.38, it has been proved that any circuit of depth $k$ that computes the parity of $n$ variables must have size $2^{cn^{1/k}}$ for some constant $c > 0$. We apply this lower bound on parity circuits to show that $\oplus P^A$ is not in $PH^A$ for some oracle $A$.

The relativized counting complexity classes $\oplus P^A$ and $\#P^A$ can be defined naturally. Namely, a set $B$ is in $\oplus P^A$ if there exist a set $C \in P^A$ and a polynomial $q$ such that for each $x$, $x \in B$ if and only if there exists an odd number of $y$ of length $q(|x|)$ satisfying $\langle x, y \rangle \in C$. That is, $\oplus P^A = \oplus(P^A)$. Similarly, a function $f$ is in $\#P^A$ if there exist a set $C \in P^A$ and a polynomial $q$ such that for each $x$, $f(x)$ is the number of strings $y$ of length $q(|x|)$ such that $\langle x, y \rangle \in C$.

**Theorem 9.33** *There exists a set $A$ such that $\oplus P^A \not\subseteq PH^A$ and, hence, $PH^A \subsetneq P^{\#P^A} \subseteq PSPACE^A$.*

*Proof.* For any set $A$, let $parity_A(n)$ denote the parity of the $2^n$ bits $\chi_A(z)$, $|z| = n$. Let $L_A = \{0^n : parity_A(n) = 1\}$. Then it is clear that $L_A \in \oplus P^A$. By enumerating all polynomial-time oracle Turing machines and all polynomial functions, we may assume that there is an effective enumeration $\sigma_i^{(k)}(A; x)$ of all $\Sigma_k^{P,A}$-predicates for all $k \geq 1$. In addition, we assume that $\sigma_i^{(k)}(A; x)$ is defined by equation (9.3) with function $q(n) = q_i(n)$ and predicate $R$ computable by an oracle DTM in time $r_i(n)$, where $\{q_i\}$ and $\{r_i\}$ are two enumerations of polynomial functions. We will construct an oracle $A$ to satisfy the following requirements:

$R_{k,i}$: For the $i$th $\Sigma_k^{P,A}$-predicate $\sigma_i^{(k)}(A; x)$, there exists an $n$ such that
$$0^n \in L_A \iff \neg\sigma_i^{(k)}(A, 0^n).$$

We construct set $A$ in a standard stage construction, satisfying requirement $R_{k,i}$ at stage $e = \langle k, i \rangle$. First, let $A(0) = \overline{A}(0) := \emptyset$ and $n_0 := 1$.

Stage $e = \langle k, i \rangle$.

Define $s_{k,i}(n) = 2^{(k+1)(q_i(n)+r_i(n))}$. Let $n$ be an integer such that (a) $n > n_{e-1}$, and (b) $2^{(0.1)2^{n/(k+1)}} > s_{k,i}(n)$. Consider the circuit $C$ described in Lemma 9.32 with respect to the predicate

$\sigma_i^{(k)}(A; 0^n)$. Assign all variables $x$ of length less than $n$ with the value $\chi_{A(e-1)}(x)$, and all variables $x$ of length greater than $n$ with the value 0. Then, we obtain a circuit $C'$ of depth $k+1$ and size $\leq s_{k,i}(n)$, and variables in $C'$ are those strings of length $n$. By Corollary 6.38, this circuit $C'$ does not compute the parity of $2^n$ variables. So, there must exist an assignment $\rho$ of values of 0 or 1 to the variables $x$ of length $n$ such that the circuit $C'$ outputs a value different from the parity of the input values. Let $Y := \{x : |x| = n, \rho(x) = 1\}$, and $Z := \{z : |z| > n, z \text{ is a variable in } C\}$. We let $A(e) := A(e-1) \cup Y$ and $\overline{A}(e) := \overline{A}(e-1) \cup Z$. Finally, let $n_e := \max(\{|z| : z \text{ is a variable in } C\} \cup \{n+1\})$.

End of Stage $e = \langle k, i \rangle$.

We let $A := \bigcup_{e=1}^{\infty} A(e)$. From our construction in stage $e = \langle k, i \rangle$, we know that, on input values $\chi_{A(e)}(x)$, the output of circuit $C$ is different from $parity_{A(e)}(n)$. Furthermore, by the choice of $n_e$, $A(e)$ and $\overline{A}(e)$, it is clear that $A(e)$ and $A$ agree on strings of length $n$. Thus, requirement $R_{k,i}$ is satisfied in stage $\langle k, i \rangle$.     ∎

The above theorem can actually be improved so that for a random oracle $A$, $\oplus P^A \not\subseteq PH^A$. To prove this result, we recall the result of Theorem 6.47, in which it was proved that the parity function cannot be $(3/4)$-approximated by polynomial-size, constant-depth circuits. In fact, Theorem 6.47 can be improved to the following form (see Exercise 9.15).

**Theorem 9.34** *Let $C$ be a circuit of depth $d$ that $(3/4)$-approximates the parity of $n$ inputs. Then, $C$ must have size $2^{c \cdot n^{1/(d+6)}}$ for some constant $c > 0$.*

**Theorem 9.35** *Let $k \geq 0$. Relative to a random oracle $A$, $\oplus P^A \not\subseteq \Sigma_k^{P,A}$.*

*Proof.* We follow the notation of Theorem 9.33. Let $L_A = \{0^n : parity_A(n) = 1\}$. Let $\sigma_i^{(k)}(A; x)$ be an enumeration of all $\Sigma_k^{P,A}$-predicates, and let $s_{k,i}(n) = 2^{(k+1)(q_i(n)+r_i(n))}$. Let $E_{n,i}$ denote the event of $parity_A(n) = \sigma_i^{(k)}(A; 0^n)$. Since there are only countably many $\Sigma_k^{P,A}$-predicates, it is sufficient to show that $\Pr[(\forall n)\, E_{n,i}] = 0$ for all $i > 0$.

For any fixed $i > 0$, we first show that there exists a constant $n_0$ such that, for all $n > n_0$, $\Pr[E_{n,i}] \leq 3/4$. Assume, by way of contradiction, that $\Pr[E_{n,i}] > 3/4$. Consider the depth-$(k+1)$ circuit $C_n$ associated with predicate $\sigma_i^{(k)}(A; 0^n)$ as given by Lemma 9.32. We divide the variables $V$ of $C_n$ into two groups: $V_1 = \{x \in V : |x| = n\}$ and $V_2 = \{y \in V : |y| \neq n\}$. Our assumption that $\Pr[E_{n,i}] > 3/4$ implies that, over all the assignments to variables in $V_1 \cup V_2$, the probability is greater than $3/4$ that $C_n$ outputs $parity_A(n)$. Therefore, there exists an assignment $\rho$ of variables in $V_2$ such that over all the assignments to variables in $V_1$, the probability is greater than $3/4$ that $C_n|_\rho$ outputs $parity_A(n)$. In other words, the circuit $C_n|_\rho$ $(3/4)$-approximates the parity function of $2^n$ inputs $\chi_A(z)$, $|z| = n$. However, $C_n$ has only depth $k + 1$ and size $s_{k,i}(n)$. This is a contradiction to Theorem 9.34 for $n$ satisfying $2^{c2^{n/(d+6)}} > s_{k,i}(n)$.

Now, consider two integers, $n_1$ and $n_2$, both greater than $n_0$ with $n_2 > r_i(n_1)$. We note that the same argument as above can be used to show that $\Pr[E_{n_2,i} \mid E_{n_1,i}] \leq 3/4$, since the variables in circuit $C_{n_1}$ corresponding to predicate $\sigma_i^{(k)}(A; 0^{n_1})$ are all of length $\leq r_i(n_1) < n_2$ and, hence, the computation of $C_{n_1}$ is independent of $parity_A(n_2)$. Therefore,

$$\Pr[E_{n_1,i} \text{ and } E_{n_2,i}] \leq \left(\frac{3}{4}\right)^2.$$

Applying this argument for $k$ times on $n = n_1, n_2, \ldots, n_k$, with $n_{j+1} > r_i(n_j)$, for $j = 1, 2, \ldots, k - 1$, we get

$$\Pr[(\forall n) E_{n,i}] \leq \left(\frac{3}{4}\right)^k.$$

Thus, $\Pr[(\forall n) E_{n,i}]$ must be equal to 0, and the theorem is proven. ∎

## 9.6   Relativized Polynomial-Time Hierarchy

In this section, we apply the lower bound results on constant-depth circuits to separate the polynomial-time hierarchy by oracles. Let $C_d^n$ be the depth-$d$ levelable circuit that has the following structure:

(a)  $C_d^n$ has a top OR gate.

(b)  The fanin of all gates is $n$.

(c)  $C_d^n$ has exactly $n^d$ variables, each variable occurring exactly once in a leaf in the positive form.

Let the function computed by $C_d^n$ be $f_d^n$. From Exercise 6.20, we know that for any $\ell$ and sufficiently large $n$, a depth-$d$ circuit $D$ with fanin $\leq 2^{n^\ell}$ and bottom fanin $\leq n^\ell$ cannot compute $f_d^{2^n}$ (since such a circuit has size bounded by $2^{(d+1)n^\ell} < 2^{c2^{n/3}}$ for large $n$, where $c$ is the constant given in Exercise 6.20). We apply this lower bound result to separate the relativized polynomial-time hierarchy.

**Theorem 9.36** *There exists a set $A$ such that for all $k > 0$, $\Sigma_k^{P,A} \neq \Sigma_{k+1}^{P,A}$.*

*Proof.* For each $k$, let $L_{A,k}$ be the set of all $0^n$ such that the following is true:

$$(\exists y_1, |y_1| = n) \cdots (Q_{k+1} y_{k+1}, |y_{k+1}| = n) \, y_1 \cdots y_{k+1} \in A, \qquad (9.4)$$

where $Q_i = \exists$ if $i$ is even and $Q_i = \forall$ if $i$ is odd. It is clear that $L_{A,k} \in \Sigma_{k+1}^{P,A}$. We need to construct a set $A$ such that $L_{A,k} \notin \Sigma_k^{P,A}$. To do this by a stage construction, we need to satisfy the following requirements:

$$R_{k,i}: (\exists n = n_{k,i}) [0^n \in L_{A,k} \iff \neg \sigma_i^{(k)}(A; 0^n)].$$

In the above, $\sigma_i^{(k)}$ is the $i$th $\Sigma_k^{P,A}$-predicate as in the proof of Theorem 9.33. The general setting for the stage construction is the same as that of Theorem 9.33.

*Stage $e = \langle k, i \rangle$.*

We choose a sufficiently large $n > n_{e-1}$ such that $2^{c2^{n/3}} > s_{k,i}(n)$, where $c$ is the constant of Exercise 6.20 and $s_{k,i}(n) = 2^{(k+1)(q_i(n)+r_i(n))}$. We define a circuit $C_{k+1}^{2^n}$ with each of its gates labeled by a string as follows:

(1) For $j \geq 1$, each gate at the $j$th level is labeled by a string of length $2^{(j-1)n}$ (and, hence, the input gates are labeled by strings of length $2^{(k+1)n}$).

(2) The top gate of $C_{k+1}^{2^n}$ is labeled with $\lambda$.

(3) If a gate is labeled by a string $u$, then the $\ell$th child of this gate is labeled by $uv_\ell$, where $v_\ell$ is the $\ell$th string of length $n$.

Then, it is clear that if each input gate with label $z$ takes the value $\chi_A(z)$, then the circuit $C_{k+1}^{2^n}$ outputs the value 1 if and only if $0^n \in L_{A,k}$.

Let $D$ be the depth-$(k+1)$ circuit associated with predicate $\sigma_i^{(k)}(A; 0^n)$ as given in Lemma 9.32. Assign all variables $z$ in $D$ of length $< (k+1)n$ with the value $\chi_{A(e-1)}(z)$, and assign all variables $z$ in $D$ of length $> (k+1)n$ with value 0. Then, we obtain a depth-$(k+1)$ circuit $D'$ with size $\leq s_{k,i}(n) < 2^{c2^{n/3}}$, fanin $\leq 2^{q_i(n)+r_i(n)}$ and bottom fanin $\leq q_i(n)+r_i(n)$. By Exercise 6.20, this circuit $D'$ does not compute function $f_{k+1}^{2^n}$. Choose one assignment $\rho$ to input values such that $D'|_\rho$ and $C_{k+1}^{2^n}|_\rho$ compute different output values. Let $Y = \{y : |y| = (k+1)n, \rho(y) = 1\}$ and $Z = \{z : |z| > (k+1)n, z$ is a variable in $D\}$. Let $A(e) := A(e-1) \cup Y$ and $\overline{A}(e) := \overline{A}(e-1) \cup Z$. Finally, let $n_e := \max(\{|z| : z \in Z\} \cup \{n+1\})$.

*End of Stage $e$.*

Let $A := \bigcup_{e=1}^{\infty} A(e)$. It is clear that Stage $e = \langle k, i \rangle$ satisfies the requirement $R_{k,i}$. Thus the final set $A$ satisfies all requirements. ∎

The above proof technique is not only strong enough to separate the polynomial-time hierarchy by oracles, but it also can be applied to collapse the polynomial-time hierarchy to any finite level. These results suggest that the relativized results on the polynomial-time hierarchy do not bear much relation to the nonrelativized questions, though the proof technique of using lower bound results on constant-depth circuits is itself interesting.

**Theorem 9.37** *For each integer* $k > 0$, *there exists a set* $A_k$ *such that* $\Sigma_{k+1}^{P,A_k} = \Pi_{k+1}^{P,A_k} \neq \Sigma_k^{P,A_k}$.

*Proof.* We leave it as an exercise.  ∎

Finally, we remark that it is unknown whether the relativized polynomial-time hierarchy is a properly infinite hierarchy relative to a random oracle. In Theorem 9.35, we applied the stronger lower bound result on parity circuits to show that $\oplus P^A$ is not contained in $PH^A$ relative to a random oracle $A$. Namely, the stronger lower bound implies that no circuit of a constant depth $d$ and size $2^{p(n)}$ for any polynomial $p$ can approximate the parity function correct on more than $(3/4) \cdot 2^n$ inputs. Similarly, if we could show a stronger lower bound result on depth-$(k+1)$ circuits which $(3/4)$-approximate $f_{k+1}^{2^n}$, then it follows that the polynomial-time hierarchy is properly infinite relative to a random oracle. Note, however, that the symmetry property of the parity function is critical to the proof of Theorem 9.34. Without the symmetry property, the lower bound on the approximation circuits for function $f_k^n$ appears more difficult.

# Exercises

**9.1** Let #HC be the problem of counting the number of Hamiltonian circuits of a given graph $G$ and #SS be the problem of counting the number of ways to select a sublist, from a given list of integers, so that the sublist sums to a given integer $K > 0$. Prove that the problems #HC and #SS are #P-complete.

**9.2** A matching in a bipartite graph $G$ is a set of edges which do not share common vertices. The problem GENERAL MATCHING (GM) is to count the number of matchings of a given bipartite graph. Prove that the problem GM is #P-complete.

**9.3** Prove that the problem #SAT remains #P-complete even if we restrict the input formulas to be of the form 2-CNF and be monotone. (A CNF Boolean formula is *monotone* if all literals in each clause appear in the nonnegated form.)

**9.4** Let LE be the problem of counting the number of linear extensions of a given partial ordering on a given set $A$; that is, for a given set $A = \{a_1, a_2, \ldots, a_n\}$ and a set $B = \{(a_{j_1}, a_{k_1}), \ldots, (a_{j_m}, a_{k_m})\}$ of ordered pairs of elements in $A$, find the number of total ordering $<_0$ over $A$ such that for each pair $(a_{j_r}, a_{k_r})$ in $B$, $a_{j_r} <_0 a_{k_r}$. Prove that the problem LE is #P-complete. [*Hint:* Similar to the problem PERM, first show that for each fixed but sufficiently large prime number $p$, each Boolean formula $F$ can be reduced to a partial ordering $P_F$ on a set $A$ such that #SAT$(F)$ mod $p$ is reducible to the number of linear extensions of $P_F$.]

**9.5** For a set $A = \{a_1, a_2, \ldots, a_n\}$ and a subset $B \subseteq A$, a pair $(T, a)$, with $T \subseteq A$ and $a \in \{0, 1\}$, is said to be *consistent* with $B$ if $[a = 0$ and $T \cap B = \emptyset]$ or $[a = 1$ and $T \cap B \neq \emptyset]$.

(a) Prove that the following counting problem is #$P$-complete: for a given set $A = \{a_1, a_2, \ldots, a_n\}$ and a set $X = \{(T_1, a_1), \ldots, (T_m, a_m)\}$, with $T_i \subseteq A$ and each $a_i \in \{0, 1\}$ for each $1 \leq i \leq m$, find the number of subsets $B \subseteq A$ such that each pair $(T_i, a_i) \in X$, $1 \leq i \leq m$, is consistent with $B$.

(b) Prove that the above counting problem remains #$P$-complete if we only count the number of subsets $B$ of $A$ which are of size $d$ and consistent with all pairs in $X$, where $1 \leq d \leq n$ is an additional input parameter.

**9.6** In the proof of Lemma 9.23, we used the second Swapping Lemma to prove (9.1). This seems unnecessary: By Lemma 9.22(a), the error probability of (9.2) can be reduced to $2^{-2p(n)}$, and then the inequality (9.1), with $B = B_1$, follows immediately. Discuss what was wrong with this simpler argument.

**9.7** Let $n \geq 1$. Identify the set $\{0, 1\}^n$ with the vector space $GF(2^n)$ of all $n$-dimensional vectors over the field $\mathbf{Z}_2$. Then, for any two strings $u, v \in \{0, 1\}^n$, their inner product is $u \cdot v = \bigoplus_{i=1}^n u_i \cdot v_i$, where $u_i$ denotes the $i$th bit of $u$ and $\oplus$ denotes the exclusive-or, or the odd parity, operation.

(a) For any random strings $w_1, \ldots, w_n \in \{0, 1\}^n$, define $H_i = \{v \in \{0, 1\}^n : v \cdot w_i = 0\}$, for $1 \leq i \leq n$. Prove that $\Pr[\bigcap_{i=1}^n H_i = \{0^n\}] \geq 1/4$ and $\Pr[\text{rank}(\bigcap_{i=1}^n H_i) = 1] \geq 1/2$.

(b) Let $S$ be a nonempty subset of $\{0, 1\}^n$. For any random strings $w_1, \ldots, w_n \in \{0, 1\}^n$, define $S_0 = S$, $S_i = \{v \in S : v \cdot w_1 = v \cdot w_2 = \cdots = v \cdot w_i = 0\}$, $1 \leq i \leq n$. Prove that $\Pr[(\exists i, 0 \leq i \leq n) \, |S_i| = 1] \geq 1/4$.

(c) Apply part (b) above, instead of the Isolation Lemma, to prove Theorem 9.24.

**9.8** Let PRIMESAT $= \{F \in$ SAT $:$ the number of truth assignments $t$ that satisfies $F$ is a prime number$\}$. Prove that SAT $\in BP(\text{PRIMESAT})$.

**9.9** In this exercise, we prove a straight-line program characterization of the class #$P$. A *straight-line arithmetic program* (or, simply, *program*) is a finite sequence $(p_1, p_2, \ldots, p_t)$ of instructions such that each $p_k$, $1 \leq k \leq t$, is of one of the following forms:

(i) $p_k$ is a constant 0 or 1,

(ii) $p_k = x_i$ or $p_k = 1 - x_i$ for some $i \leq k$,

(iii) $p_k = p_i + p_j$ for some $1 \leq i, j < k$,

(iv) $p_k = p_i p_j$ for some $1 \leq i + j \leq k$, or

(v) $p_k = p_j|_{x_i=0}$ or $p_k = p_j|_{x_i=1}$ for some $1 \leq i, j < k$. ($p_j|_{x_i=0}$ means the polynomial obtained from $p_j$ by substituting 0 for the variable $x_i$.)

Such a program $Q = (p_1, \ldots, p_t)$ is said to compute the polynomial $p_t$; we denote it by $\tilde{Q}$. A family of programs $Q_1, Q_2, \ldots$ is a *uniform* family if each

$\tilde{Q}_k$ has at most $k$ variables $x_1, \ldots, x_k$ and if there is a polynomial-time DTM transducer that prints the instructions of program $Q_k$ from input $1^k$. Prove that a counting function $f : \{0,1\}^* \to \mathbf{N}$ is in $\#P$ if and only if there is a uniform family of programs $Q_k$ such that, for all $w \in \{0,1\}^*$ of length $n$, $f(w) = \tilde{Q}_n(w_1, w_2, \ldots, w_n)$, where $w_j$ denotes the $j$th bit of $w$.

**9.10** Let $f, g : \Sigma^* \to \mathbf{N}$ be two counting functions, and $r : \mathbf{N} \to \mathbf{N}$ an integer function. We say that function $g$ *approximates* $f$ with errors bounded by $r(n)$ if for any $x \in \Sigma^*$,

$$\left(1 - \frac{1}{r(|x|)}\right) \cdot g(x) \le f(x) \le \left(1 + \frac{1}{r(|x|)}\right) \cdot g(x).$$

(a) Let $F\Delta_3^P$ denote the class of functions $f : \Sigma^* \to \mathbf{N}$ that are computable by a polynomial-time oracle DTM with respect to an oracle in $\Sigma_2^P$. Prove that for any $f \in \#P$ and any polynomial function $r$, there exists a function $g \in F\Delta_3^P$ such that $g$ approximates $f$ with errors bounded by $r$. In addition, the oracle DTM for $g$ queries only the oracle for at most $O(\log n)$ times.

(b) Let $FBPP^{NP}$ denote the class of functions $f : \Sigma^* \to \mathbf{N}$ that are computable by a polynomial-time oracle PTM with respect to an oracle in $NP$ such that the error probability is bounded by $1/4$. Prove that for any $f \in \#P$ and any polynomial function $r$, there exists a function $g \in FBPP^{NP}$ such that $g$ approximates $f$ with errors bounded by $r$. In fact, the oracle machine for $g$ queries only the oracle for at most $O(\log n)$ times.

**9.11** A *fully probabilistic polynomial-time approximation scheme* for a counting function $f : \Sigma^* \to \mathbf{N}$ is a PTM that on inputs $x$ and $\epsilon$ computes, in time polynomial in $|x| + 1/\epsilon$ and with probability $3/4$, an approximate value to $f(x)$ with errors bounded by $\epsilon$.

(a) We say a bipartite graph $G = (A, B, E)$ with $|A| = |B| = n$ is *dense* if each node has degree at least $n/2$. A 0-1-matrix is called *dense* if it is the adjacency matrix of a dense bipartite graph. Prove that the problem PERM restricted to dense 0-1-matrices remains $\#P$-complete.

(b) Prove that there is a fully probabilistic polynomial-time approximation scheme for computing the permanents of dense 0-1-matrices.

**9.12** Prove that the class $\oplus P$ is closed under the Boolean operations union, intersection, and complementation.

**9.13** Let *FewP* denote the class of languages $L$ that are accepted by polynomial-time NTMs that have at most a polynomial number of accepting computation paths for any $x \in L$. Prove that the class *FewP* $\subseteq \oplus P$.

**9.14** Prove that there exists an oracle $A$ such that $\oplus P^A$ and $NP^A$ are incomparable.

**9.15** Prove Theorem 9.34. That is, do a more careful analysis of the proofs of Theorems 6.46 and 6.47 to get an exponential lower bound for the size of any depth-$d$ circuit $(3/4)$-approximating the parity function $p_n$.

**9.16** Prove Theorem 9.37. In particular, prove that:

(a) For each integer $k > 0$, there exists a set $A_k$ such that $PSPACE = \Sigma_{k+1}^{P,A_k} = \Pi_{k+1}^{P,A_k} \neq \Sigma_k^{P,A_k}$; and

(b) For each integer $k > 0$, there exists a set $B_k$ such that $PSPACE \neq \Sigma_{k+1}^{P,B_k} = \Pi_{k+1}^{P,B_k} \neq \Sigma_k^{P,B_k}$. [*Hint:* Use Exercise 6.21.]

**9.17** (a) Prove that there exists an oracle $C$ such that $BP(\Sigma_k^{P,C}) \not\subseteq \Sigma_{k+1}^{P,C}$ for all $k \geq 1$.

(b) Prove that there exists an oracle $D$ such that $\Sigma_k^{P,D} = BP(\Sigma_k^{P,D}) \neq \Sigma_{k+1}^{P,D}$ for all $k \geq 1$.

**9.18** In this exercise, we consider the *low* and *high hierarchies* in *NP*. For each $k \geq 0$, let $L_k^P$ be the class of sets $A \in NP$ such that $\Sigma_k^{P,A} = \Sigma_k^P$, and $H_k^P$ be the class of sets $B \in NP$ such that $\Sigma_k^{P,A} = \Sigma_{k+1}^P$.

(a) Show that $L_k^P \subseteq L_{k+1}^P$ and $H_k^P \subseteq H_{k+1}^P$ for all $k \geq 0$.

(b) Show that $L_1^P = NP \cap coNP$ and $H_1^P$ equals the class of sets $A$ that are $\leq_T^{SNP}$-complete for *NP*.

(c) Show that, for any $k \geq 0$, if $\Sigma_k^P \neq \Pi_k^P$ then $L_k^P \cap H_k^P = \emptyset$.

(d) Show that, for any $k \geq 0$, if $\Sigma_k^P = \Pi_k^P$ then $L_k^P = H_k^P = NP$.

**9.19** We consider the relativized low and high hierarchies in *NP*. For any set $X$, we let $L_k^{P,X}$ denote the class of sets $A \in NP^X$ such that $\Sigma_k^{P,X \; join \; A} = \Sigma_k^{P,X}$, and let $H_k^{P,X}$ denote the class of sets $B \in NP^X$ such that $\Sigma_k^{P,X \; join \; B} = \Sigma_{k+1}^{P,X}$. Note that Theorem 4.25(c) showed that there exists an oracle $A$ such that $L_0^{P,A} = L_1^{P,A} \neq L_2^{P,A}$.

(a) Show that there exists an oracle $X$ such that $L_k^{P,X} \neq L_{k+1}^{P,X}$ and $H_k^{P,X} \neq H_{k+1}^{P,X}$ for all $k \geq 0$.

(b) Show that, for any $k \geq 0$, there exists an oracle $Y$ such that

$$L_0^{P,Y} \neq L_1^{P,Y} \neq \cdots \neq L_k^{P,Y} = L_{k+1}^{P,Y} = L_{k+2}^{P,Y} = \cdots$$

and

$$H_0^{P,Y} \neq H_1^{P,Y} \neq \cdots \neq H_k^{P,Y} = H_{k+1}^{P,Y} = H_{k+2}^{P,Y} = \cdots .$$

(c) Recall the class $UP$ defined in Chapter 4. For $k \geq 0$ and any set $Z$, let $\Sigma_k^{P,UP^Z} = \bigcup_{A \in UP^Z} \Sigma_k^{P,A}$. Show that there exists an oracle $Z$ such that $UP^Z$ is not contained in the low and high hierarchies relative to $Z$; that is,

$$\Sigma_k^{P,UP^Z} \not\subseteq \Sigma_k^{P,Z} \text{ and } \Sigma_{k+1}^{P,Z} \not\subseteq \Sigma_k^{P,UP^Z}.$$

## Historical Notes

The class $\#P$ was first defined in Valiant [1979a, 1979b], in which the $\#P$-completeness of PERM and other counting problems associated with *NP*-

complete problems are proven. The natural reductions among many *NP*-complete problems are known to preserve the number of solutions; see, for example, Simon [1975] and Berman and Hartmanis [1977]. The #*P*-completeness of the problem LE (Exercise 9.4) is proved by Brightwell and Winkler [1991]. Exercise 9.5 is from Du and Ko [1987]. The straight-line program characterization of #*P* (Exercise 9.9) is from Babai and Fortnow [1991]. Papadimitirou and Zachos [1983] studied the class ⊕*P* and proved that ⊕*P*(⊕*P*) = ⊕*P*. Exercise 9.13 is from Cai and Hemachandra [1990]. Theorem 9.24 was first proved by Valiant and Vazirani [1986], using a different probabilistic lemma (see Exercise 9.7). The use of the Isolation Lemma simplifies the proof; this was pointed out by Mulmuley et al. [1987]. Lemma 9.23, Theorem 9.26, and the results of Section 9.4 are all from Toda [1991]. Approximation to counting functions were studied by Stockmeyer [1985], Jerrum et al. [1986], Broder [1986], and Jerrum and Sinclair [1989]. Exercises 9.10 and 9.11 are from these works.

The reduction from the relativization problem to the lower bound problem for constant-depth circuits was given in Furst et al. [1984] and Sipser [1983a]. They also proved that the parity circuit complexity is super-polynomial. The separation of *PSPACE* from *PH* by oracles and the separation of *PH* into a properly infinite hierarchy by oracles are given in Yao [1985] and Hastad [1986a, 1986b]. The collapsing of *PH* to any finite levels and the separation of *PSPACE* from a collapsed *PH* by oracles was proved in Ko [1989a]. Theorem 9.35, showing that random oracles separate *PSPACE* from *PH*, was first proven by Cai [1986]. Our simplified version is from Babai [1987]. The low and high hierarchies were first introduced by Schöning [1983]. Exercise 9.18 is from there. The relativized separation and collapsing of the low and high hierarchies (Exercise 9.19) was proved by Ko [1991b] and Sheu and Long [1992]. Ko [1989b] contains a survey of the proof techniques that use the lower bound results for circuits to construct oracles.

# 10

# *Interactive Proof Systems*

*Those who consider a thing proved simply
because it is in print are fools.*
— Maimonides

The complexity classes *RP* and *BPP* are the probabilistic extensions of the deterministic feasible class *P*. In this chapter, we study interactive proof systems as the probabilistic extension of the polynomial-time hierarchy and the class *PSPACE*. Although the original motivation of interactive proof systems is for their applications in the theory of cryptography, they have found many interesting applications in complexity theory too. In particular, we will discuss characterizations of complexity classes between *NP* and *PSPACE* in terms of interactive proof systems. We also demonstrate an interesting application of this study, in which some intractable problems in *NP* are proved to be *not NP*-complete unless the polynomial-time hierarchy collapses.

## 10.1 Examples and Definitions

The notion of interactive proof systems is most easily explained from a game-theoretic view of the complexity classes. In the general setting, each problem $A$ is interpreted as a two-person game in which the first player, the prover, tries to convince the second player, the verifier, that a given instance $x$ is in $A$. On a given instance $x$, each player takes turn sending a string $y_i$ to the other player, where the $i$th string $y_i$ may depend on the input $x$ and the previous strings $y_1, \ldots, y_{i-1}$. After a number $m$ of moves, the prover wins the game if the verifier is able to verify that the strings $x, y_1, \ldots, y_m$ satisfy a predetermined condition; otherwise, the verifier wins. Depending on the computational power of the two players, and on the type of protocols allowed between the two players, this game-theoretic view of the computational problems can be used to characterize some familiar complexity classes.

As a simple example, consider the famous *NP*-complete problem SAT. To prove that a Boolean formula $F$ is satisfiable, the prover simply sends a truth assignment $t$ on the variables that occurred in $F$ to the verifier, then the verifier verifies, in polynomial time, whether $t$ satisfies $F$ or not and accepts $F$ if $t$ indeed satisfies $F$. Thus, we say that SAT has a proof system with the following properties:

(1) The verifier has the power of a polynomial-time DTM, and the prover has the unlimited computational power.

(2) The game lasts for one round only, with the prover making the first move.

(3) A formula $F$ is in SAT if and only if the prover is able to win the game on $F$.

Indeed, it is easy to see that the class *NP* consists of exactly those sets having proof systems of the above properties.

A more interesting example is the proof systems for the polynomial-time hierarchy. Recall the following alternating quantifier characterization of the polynomial-time hierarchy (Theorem 3.8): A set $A$ is in $\Pi_k^P$ if and only if there exist a polynomial-time computable predicate $R$ and a polynomial $q$ such that for all $x$,

$$x \in A \iff (\forall y_1, |y_1| = q(|x|))\, (\exists y_2, |y_2| = q(|x|))$$
$$\cdots (Q_k y_k, |y_k| = q(|x|))\, R(x, y_1, \ldots, y_m),$$

where $Q_k = \forall$ if $k$ is odd and $Q_k = \exists$ if $k$ is even. From the game-theoretic point of view, the above characterization means that a set $A$ in $\Pi_k^P$ has a proof system of the following properties:

(1) The verifier has the power of a polynomial-time DTM, and the prover has the unlimited computational power.

(2) The game lasts for exactly $k$ rounds, with the verifier making the first move.

(3) An instance $x$ is in $A$ if and only if the prover has a winning strategy on $x$.

The above game-theoretic view of the complexity classes in *PH*, though helpful in our understanding of the problems in these classes, does not help us to *prove* them efficiently in practice. For instance, consider the above game for some set $A \in \Pi_2^P$. This game consists of only two moves: the verifier sends to the prover a string $y_1$ and the prover responds with a string $y_2$. Following these two moves, the verifier verifies whether $R(x, y_1, y_2)$ holds and decides who wins the game. In this game, even if the prover is able to win the games on the same input $x$ polynomially many times (with respect to different strings

$y_1$), the verifier is not necessarily convinced that the prover has a winning strategy. The reason is that the verifier has an exponential number of choices for the first string $y_1$, and only polynomially many of them have been defeated by the prover. The prover might simply be lucky in winning in these special cases. An exhaustive proof by the prover would convince the verifier that $x$ indeed belongs to $A$ but it would, unfortunately, take an exponential amount of time.

Following the spirit of probabilistic computing we studied in Chapter 8, it is possible, however, that for a subclass of problems $A$ in $\Pi_2^P$, the prover is able to demonstrate *probabilistically* that a string $x$ is in $A$ in polynomial time. To do this, the prover first shows the verifier that if $x \notin A$ then there exist not just one but many bad strings $y_1$ such that $(\forall y_2, |y_2| = q(|x|)) \, \neg R(x, y_1, y_2)$. (Therefore, only a subclass of problems $A$ in $\Pi_2^P$ qualify.) Assume, for instance, there are $c \cdot 2^{q(|x|)}$ such bad $y_1$'s for some constant $c > 0$. Then, the verifier may randomly choose a string $z_1$ of length $q(|x|)$ and has the probability $c$ to win the game with this string $z_1$. If this game on $x$ is played repeatedly for a number of times and the prover is able to win the game consistently, then the verifier is more convinced that $x$ is indeed in $A$ (or, the verifier is extremely unlucky). This type of probabilistic game is called a *probabilistic interactive proof system* or, simply, an interactive proof system. The main difference between the probabilistic proof systems and the deterministic proof systems described earlier is that the verifier is allowed to use a random number generator to help him/her to compute new strings $y_i$, as well as verifying the final accepting condition. That is, an interactive proof system for a set $A$ is a proof system with the following properties:

(1) The verifier has the power of a polynomial-time PTM, and the prover has the power of a PTM, without a limit on time or space bound.

(2) If an instance $x$ is in $A$, then the prover is able to win the game of $x$ with probability $> 1 - \epsilon$; and if $x \notin A$, then the probability that the prover is able to win the game of $x$ is $< \epsilon$, where $\epsilon$ is a constant strictly less than $1/2$.[1]

We look at some examples before giving the formal definition of interactive proof systems.

**Example 10.1** (GRAPH NONISOMORPHISM) Recall that the graph isomorphism problem is GIso $= \{\langle G_1, G_2 \rangle : G_1 \cong G_2\}$, where $G_1 \cong G_2$ means that $G_1$ are $G_2$ are isomorphic. It is known that GIso $\in NP$ and so $\overline{\text{GIso}} \in coNP$. We show that $\overline{\text{GIso}}$ has a probabilistic interactive proof system. In the following, for any graph $G = (V, E)$ and any permutation $\pi$ on $V$, we write $\pi(G)$ to denote the graph $G' = (V, E')$, where $\{u, v\} \in E'$ if and only if $\{\pi^{-1}(u), \pi^{-1}(v)\} \in E$; that is, the graph $\pi(G)$ is the isomorphic copy of $G$ induced by the permutation $\pi$.

---

[1] Similar to the probabilistic class *BPP*, we will see that the class of sets having interactive proof systems is robust with respect to any constant error probability $\epsilon$.

*Interactive Proof System for* $\overline{\text{GIso}}$.
Input: $\langle G_1, G_2 \rangle$.
*Round* 1. The verifier chooses a random bit $b \in \{1, 2\}$, and chooses a random permutation $\pi$ on $V_b$. Then, the verifier sends the graph $G' = \pi(G)$ to the prover.
*Round* 2. The prover finds $b' \in \{1, 2\}$ such that $G_{b'} \cong G'$ and sends $b'$ to the verifier.
The verifier accepts the input if and only if $b = b'$.

First assume that $G_1$ and $G_2$ are not isomorphic. Since the prover has unlimited computational power, he/she can always win the game by finding the correct bit $b'$ for the verifier. Therefore, the error probability is 0.[2] Conversely, assume that $G_1 \cong G_2$. What is the probability that the prover can cheat the verifier that $G_1 \not\cong G_2$? Since $G_1 \cong G_2 \cong G'$, and since the verifier uses a random bit $b$ to get $G'$, the best the prover can do is to randomly guess a bit $b'$ and has only the probability 1/2 to win the game. To get an arbitrarily small error probability, say $\epsilon$, it is only necessary to repeat the game on the same input $\langle G_1, G_2 \rangle$ for $k = -\lceil \log \epsilon \rceil$ times, using an independent random number generator for the bit $b$. $\qquad\square$

**Example 10.2** (QUADRATIC NONRESIDUES, or QNR) Recall that an integer $x$ is a quadratic residue modulo a positive integer $n$ ($x \in QR_n$) if there exists an integer $y$ such that $y^2 \equiv x \bmod n$. Also recall that $\mathbf{Z}_n^* = \{x \in \mathbf{N} : 0 < x < n, \gcd(x, n) = 1\}$. (See Examples 4.1 and 8.4.) Let $QNR = \{\langle x, n \rangle : x \notin QR_n\}$. It is known that QNR $\in$ *coNP*, but it is not known whether it is in *NP*.

*Interactive Proof System for* QNR.
Input: $\langle x, n \rangle$.
*Round* 1. The verifier chooses a random bit $b \in \{0, 1\}$, and a random $z \in \mathbf{Z}_n^*$. Then, the verifier defines

$$w = \begin{cases} z^2 \bmod n & \text{if } b = 1, \\ x \cdot z^2 \bmod n & \text{if } b = 0, \end{cases}$$

and sends $w$ to the prover.
*Round* 2. The prover decides whether $w \in QR_n$ and sends the answer $b' \in \{0, 1\}$ to the verifier ($b' = 1$ if and only if $w \in QR_n$).
The verifier accepts the input if and only if $b = b'$.

---

[2]It is possible that the graph $G'$ received by the prover is not isomorphic to either graph $G_1$ or $G_2$. In this case, the prover cannot prove that $G_1 \not\cong G_2$. However, the goal of the prover is only to convince an honest verifier who follows a predetermined protocol of communication. In other words, we have just showed that the error probability is 0 *if* the verifier follows the protocol honestly.

Assume that $\langle x, n \rangle \in$ QNR. Then, the number $w$ generated by the verifier is a quadratic residue modulo $n$ if and only if $b = 1$. Thus, the prover, with the unlimited computational power, can always win the game by sending the bit $b' = 1$ if and only if $w \in QR_n$. Conversely, if $\langle x, n \rangle \notin$ QNR, then $w$ is always a quadratic residue modulo $n$ (regardless of $b = 0$ or $b = 1$). Furthermore, $w$ is a random quadratic residue modulo $n$ since $z$ was chosen randomly. Therefore, the prover has only the probability $1/2$ to guess the correct bit $b'$. As in Example 10.1 above, the error probability can be reduced to $2^{-k}$ by repeating the game $k$ times. □

The above two examples demonstrate simple interactive proof systems for some problems in *coNP* that are not known to be complete for *coNP*. The next example shows a more complicated interactive proof system for a *#P*-complete problem, with a hint of the general theorem that all problems in *PSPACE* have interactive proof systems (Theorem 10.26).

**Example 10.3** (PERMANENT, or PERM) Recall that the counting problem PERM is a *#P*-complete problem (Theorem 9.13). We first consider a weak interactive proof system for PERM that allows large error probability.

*Weak Interactive Proof System for PERM.*
Input: An $n \times n$ Boolean matrix $A$ and an integer $q$. (The prover needs to prove that $perm(A) = q$.)

Initially, let $A_1 := A$ and $q_1 := q$.
*Stage i*, $1 \leq i < n$. $A_i$ is an $(n - i + 1) \times (n - i + 1)$ matrix. The prover sends integers $q_{i,j} = perm(B_{i,j})$, $1 \leq j \leq n - i + 1$, to the verifier, where $B_{i,j}$ is the submatrix of $A_i$ obtained by deleting the first row and the $j$th column of $A_i$. Next, the verifier verifies that

$$q_i = \sum_{j=1}^{n-i+1} A_i[1, j] \cdot q_{i,j}. \tag{10.1}$$

If the above equation does not hold, then the verifier rejects the input $(A, q)$. Otherwise, the verifier chooses a random $j \in \{1, \ldots, n - i + 1\}$, and lets $A_{i+1} := B_{i,j}$ and $q_{i+1} := q_{i,j}$. He/she sends $A_{i+1}$ and $q_{i+1}$ to the prover and goes to the next stage.
*Stage n.* The verifier has a $1 \times 1$ matrix $A_n$ and an integer $q_n$. The verifier accepts the input if and only if $q_n = perm(A_n) = A_n[1, 1]$.

Assume that $perm(A) = q$. Then the prover can always compute the correct values $q_{i,j} = perm(B_{i,j})$ at each stage $i$ and, hence, can prove that $perm(A) = q$ with probability one.

On the other hand, when $perm(A) \neq q$, the error probability could be very large. More precisely, assume that at stage $i$, $perm(A_i) \neq q_i$. Then, the prover could give $n - i$ correct values $q_{i,j} = perm(B_{i,j})$ and only one incorrect value

$q_{i,j}$ and could still satisfy (10.1). The probability that the verifier chooses this incorrect value to be $q_{i+1}$ is $1/(n - i + 1)$. As a whole, the probability that all pairs $(A_i, q_i)$ are incorrect in all stages $i$ is only $1/(n!)$. Thus, the above proof system is too weak.

To remedy this problem, we modify the above proof system to reduce the error probability. Let us call a pair $(C, r)$ of a matrix $C$ and an integer $r$ *incorrect* if $perm(C) \neq r$. We have just seen that if at each stage $i$, the verifier only recursively tests the correctness of a random pair $(B_{i,j}, q_{i,j})$, then the error probability is too large. On the other hand, if the verifier recursively tests every pair $(B_{i,j}, q_{i,j})$ at stage $i$, then it would take an exponential amount of time. The idea here is to combine all $n-i+1$ pairs $(B_{i,j}, q_{i,j})$ to create a single pair which, with a large probability, is incorrect as long as one of the original pairs is incorrect. Therefore, the total number of pairs $(B_{i,j}, q_{i,j})$ considered is polynomially bounded, and yet the error probability is kept small.

> *Interactive Proof System for* PERM.
> Input: $(A, q)$.
>
> Choose a large prime number $p$ such that $n! < p < 2^{n^2}$. (It could be done in polynomial time by the verifier; or, the prover could send the prime number, together with a short proof that it is indeed a prime, to the verifier.)
>
> Let $A_1 := A$ and $q_1 := q$.
>
> *Stage $i$, $1 \leq i < n$.* The prover needs to prove that $perm(A_i) \equiv q_i \bmod p$.
>
> *Substage $\langle i, 1 \rangle$.* The prover finds $q_{i,j} = perm(B_{i,j}) \bmod p$, $1 \leq j \leq n - i + 1$, and sends them to the verifier, where $B_{i,j}$ is the submatrix of $A_i$ obtained by deleting the first row and the $j$th column of $A_i$. Next, the verifier verifies that
>
> $$q_i \equiv \left( \sum_{j=1}^{n-i+1} A_i[1,j] \cdot q_{i,j} \right) \bmod p.$$
>
> If the above equation does not hold, then the verifier rejects the input $(A, q)$. Otherwise, let $C := B_{i,1}$ and $r_{i,1} := q_{i,1}$, and go to substage $\langle i, 2 \rangle$.
>
> *Substage $\langle i, j \rangle$, $2 \leq j \leq n - i + 1$.* The verifier forms an $(n - i) \times (n - i)$ linear-polynomial matrix
>
> $$D(x) = C + x(B_{i,j} - C),$$
>
> and sends $D(x)$ to the prover, asking for its permanent $f(x)$ (a polynomial of degree $\leq n - i$). When the verifier receives the permanent polynomial $f(x)$ from the prover, he/she checks that $f(0) \equiv r_{i,j-1} \bmod p$ and $f(1) \equiv q_{i,j} \bmod p$; if either equation fails, he/she rejects the input. Otherwise, the verifier chooses a random integer $a \in \mathbf{Z}_p$ and let $C := D(a)$ (i.e., $C$ is the $(n - i) \times$

$(n - i)$ integer matrix obtained by substituting the integer $a$ for the variable $x$ in $D$) and $r_{i,j} := f(a) \bmod p$.

At the end of substage $\langle i, n - i + 1 \rangle$, let $A_{i+1} := C$ and $q_{i+1} := r_{i, n-i+1}$.

*Stage n.* The verifier has a $1 \times 1$ matrix $A_n$ and an integer $q_n$. The verifier accepts the input if and only if $q_n = perm(A_n) = A_n[1, 1]$.

First notice that an $n \times n$ Boolean matrix $A$ must have $perm(A) \leq n!$, and so, for any $0 \leq q < p$, $perm(A) \equiv q \bmod p$ implies $perm(A) = q$. Thus, we don't lose any generality using arithmetic modulo $p$. (In the following lemma, we say $(C, r)$ is incorrect if and only if $perm(C) \not\equiv r \bmod p$.) Next, we observe again that if $perm(A) \equiv q \bmod p$, then the prover can always compute correct permanent values, including the permanent $f$ of the polynomial matrix $D$ in each substage $\langle i, j \rangle$, and wins the game with probability one. We only need to prove that when $perm(A) \not\equiv q \bmod p$, the probability of the prover winning the game is small.

**Lemma 10.4** *Let* $1 \leq i < n$ *and* $2 \leq j \leq n - i + 1$. *Assume that at the start of stage* $\langle i, j \rangle$ *either* $(C, r_{i,j-1})$ *or* $(B_{i,j}, q_{i,j})$ *is incorrect, and assume that the verifier does not reject the input in substage* $\langle i, j \rangle$. *Then, at the end of substage* $\langle i, j \rangle$, $(C, r_{i,j})$ *is incorrect with probability* $\geq 1 - (n - i + 1)/p$.

*Proof.* Suppose that the prover sends the correct permanent $f$ of the polynomial matrix $D$ to the verifier in substage $\langle i, j \rangle$. Then, it must hold that $f(0) \equiv perm(C) \bmod p$ and $f(1) \equiv perm(B_{i,j}) \bmod p$. Thus, the verifier would find either $f(0) \not\equiv r_{i,j-1} \bmod p$ or $f(1) \not\equiv q_{i,j} \bmod p$, and would reject the input. Thus, the prover must have sent an incorrect polynomial function $f$ to the verifier. Since each entry of the matrix $D$ is a linear polynomial, the permanent $g$ of $D$ has degree at most $n - i + 1$. Thus, the two polynomials $f$ and $g$ can agree at most at $n - i + 1$ points. It follows that the probability of a random integer $a \in \mathbf{Z}_p$ chosen by the verifier satisfying $f(a) \equiv g(a) \bmod p$ is $\leq (n - i + 1)/p$ and, hence, the probability that the final pair $(C, r_{i,j})$ at the end of substage $\langle i, j \rangle$ is correct is $\leq (n - i + 1)/p$. ∎

We observe that if $(A_i, q_i)$ is incorrect, then after substage $\langle i, 1 \rangle$, one of the pairs $(B_{i,j}, q_{i,j})$ is incorrect. Now, following the above lemma, the error in the pair $(A_i, q_i)$ at stage $i$ will be passed to the next pair $(A_{i+1}, q_{i+1})$ at stage $i + 1$ with probability $\geq 1 - (n - i + 1)^2/p$. The total error probability of the above interactive proof system is thus bounded by

$$\sum_{i=1}^{n-1} \frac{(n - i + 1)^2}{p} \leq \frac{n^3}{p},$$

which is negligible when $n$ is reasonably large.

From the above examples, we can see that an interactive proof system consists of an honest verifier who wants to learn whether or not a given instance $x$

is in a set $A$ and an unreliable prover whose task is to answer questions of the verifier to maximize his winning probability. Since the prover is potentially treacherous, it is the verifier's job to detect the cheating to minimize the error probability. Thus, it is natural to define the interactive proof systems from the verifier's point of view; that is, we define the verifier as a probabilistic machine and let the prover take the role of an oracle. Formally, we define an *interactive Turing machine* to be a polynomial time oracle PTM $M_v$ that uses an oracle function $f_p$ with the following special rule of making queries: The machine $M_v$ makes a query $y$ to the oracle $f_p$ by writing on the query tape (1) the input $x$, (2) the whole history of queries and answers exchanged between the machine $M_v$ and the oracle $f_p$, and (3) the new query $y$. (This special rule reflects the assumption that the prover can send a message $y_i$ based on the input $x$ as well as the history of the messages exchanged between the prover and the verifier.) Note that the polynomial time bound on $M_v$ implies that it can only work with polynomial length-bounded oracle functions $f_p$. The notion of interactive proof systems can be defined by interactive Turing machines, working with unreliable oracles. (See Exercise 10.1 for an alternative formulation of the notion of interactive proof systems.)

**Definition 10.5** *We say a set $A$ has an* interactive proof system *with an error probability $\epsilon$ if there exists an interactive Turing machine $M_v$ with the following properties:*

*(1) There exists an oracle function $f_p$ such that for every $x \in A$, the probability of $M_v^{f_p}$ accepting $x$ is at least $1 - \epsilon$; and*

*(2) For every $x \notin A$, and any oracle $f$, the probability of $M_v^f$ accepting $x$ is at most $\epsilon$.*

*If the machine $M_v$ and the function $f_p$ satisfy conditions (1) and (2) above, then we say that $(M_v, f_p)$ is an interactive proof system for set $A$.*

The time complexity of an interactive Turing machine is restricted to be polynomially bounded. In addition to the time complexity, an important complexity measure of an interactive proof system $(M_v, f_p)$ is the number of queries made by $M_v$ to the oracle $f_p$. Recall that a *computation* of a probabilistic oracle machine $M$, relative to an oracle $f$, is a sequence of successive configurations of $M^f$, from a starting configuration to a halting configuration. We define the number of *rounds* of a computation of an interactive proof system $(M_v, f_p)$ to be the number of messages (queries and answers) exchanged between the machine $M_v$ and the oracle $f_p$ in this computation. For an integer function $r$, we say that a set $A$ has an $r(n)$-round interactive proof system if there exists an interactive proof system $(M_v, f_p)$ for $A$ such that the number of rounds of all the computations of the system $(M_v, f_p)$ on all $x$ is bounded by $r(|x|)$.[3]

---

[3] For each query and its answer, we count two rounds of exchanging messages. In the literature, they are sometimes counted as only one round.

**Definition 10.6** *We let IP be the class of sets that have interactive proof systems with error probability* $1/4$, *and for each* $k \geq 0$, *let* $IP_k$ *be the class of sets that have k-round interactive proof systems with error probability* $1/4$.

Similar to the probabilistic complexity classes *RP* and *BPP*, the classes *IP* and $IP_k$ remain the same if we change the error bound $1/4$ to any constant $\epsilon < 1/2$.

**Proposition 10.7** *The following are equivalent:*

*(a) There exists a constant* $\epsilon$, $0 < \epsilon < 1/2$, *such that set A has an interactive proof system with the error probability* $\epsilon$.

*(b) For every constant* $\epsilon$, $0 < \epsilon < 1/2$, *set A has an interactive proof system with the error probability* $\epsilon$.

*(c) For every polynomial q, set A has an interactive proof system with the error probability bounded by* $2^{-q(n)}$.

*Proof.* The proof is similar to that of Theorem 8.19. The idea is to simulate the computation on each input for many times *in parallel* and then accept the input if and only if the majority of simulations accept. We leave the detail to the reader (Exercise 10.2). ∎

Finally, we observe the simple relations that $IP_0 = BPP$ and $NP \subseteq IP_1$, and that the problems $\overline{\text{GIso}}$ and QNR of Examples 10.1 and 10.2 above are both in $IP_2$.

## 10.2 Arthur-Merlin Proof Systems

In order to analyze the computational power of interactive proof systems, we first study a weaker type of probabilistic proof system called the *Arthur-Merlin proof system*. An Arthur-Merlin proof system is similar to an interactive proof system where Arthur is a verifier with the power of a polynomial-time PTM and Merlin is a prover, except that Merlin is even more powerful than an ordinary prover so that he is able to read the whole history of the computation of Arthur on the given input, including the random numbers generated by Arthur. If we examine the interactive proof systems of Examples 10.1 and 10.2, we can see that the secrecy of the random bits used by the verifier is critical to the correctness of the system. Indeed, if in Example 10.1 the prover knows the random bit $b$ chosen by the verifier, then he/she can always win the game by sending back the bit $b$, regardless of whether $G_1 \cong G_2$ or not. Thus, the extra power of Merlin appears to be a strong restriction to the notion of interactive proof systems. Later, however, we will show that this restriction is not really so strong, and that the two notions of proof systems are essentially equivalent. Yet, the simplicity of the Arthur-Merlin proof systems allows us to perform detailed analysis of its computational power.

Since Merlin is so powerful that Arthur is not able to hide his random numbers from Merlin, we may as well require that in an Arthur-Merlin proof system, Arthur always sends the random numbers he used to Merlin. Furthermore, Merlin can always simulate Arthur's computation to compute the next query to be asked by Arthur from the history of the computation and the new random numbers received from Arthur. Thus, Arthur really does not have to send Merlin anything except the new random numbers generated by the random number generator. In other words, Arthur really plays a passive role whose only task is to verify, at the end, whether the computation satisfies a predetermined condition and whether to accept the input. Formally, we define an Arthur machine to be an interactive Turing machine such that:

(1) Each new query of Arthur is simply a new sequence of random bits; and

(2) Every random number generated by Arthur must be sent to Merlin as a query, even if Arthur does not expect an answer from Merlin (so that the last random number generated by Arthur counts as one round of communication between the two players).

To understand how such a simplistic communication protocol could work, it is helpful to review the interactive proof system of Example 10.3 for the problem PERM and observe that it is actually an Arthur-Merlin system. In particular, at each substage $\langle i,j \rangle$, $2 \leq j \leq n-i+1$, the verifier only needs to verify that the answer $f$ satisfies the two equations, $f(0) \equiv r_{i,j-1} \bmod p$ and $f(1) \equiv q_{i,j} \bmod p$, and then sends a random number $a \in \mathbf{Z}_p$ to the prover. At the next substage, the prover can compute the matrix $D$ by itself and sends its permanent $f$ to the verifier.

**Definition 10.8** *We say that a set $L$ has an* Arthur-Merlin *proof system with the error probability $\epsilon$ if there exists an Arthur machine $M_A$ with the following properties:*

(1) *There exists an oracle $f_M$ such that for every $x \in L$, the probability of $M_A^{f_M}$ accepting $x$ is at least $1 - \epsilon$; and*

(2) *For every $x \notin L$ and any oracle $f$, the probability of $M_A^f$ accepting $x$ is at most $\epsilon$.*

*If the machine $M_A$ and the oracle $f_M$ satisfy the above conditions, then we say that $(M_A, f_M)$ is an Arthur-Merlin proof system for set $L$.*

The notion of the number of rounds in an Arthur-Merlin proof system is defined exactly the same as that of an interactive proof system. Here, we distinguish the proof systems in which Merlin sends the messages first from those in which Arthur sends the messages first.

**Definition 10.9** *(a) We let AM be the class of sets $A$ that have Arthur-Merlin proof systems with error probability $\leq 1/4$.*

*(b) For any function $r(n)$, we let $AM_{r(n)}$ (and $MA_{r(n)}$) be the class of sets $A$ that have $r(n)$-round Arthur-Merlin proof systems with error probability $\leq 1/4$ in which Arthur (and, respectively, Merlin) sends the messages first.*

Again, the classes $AM$ and $AM_{r(n)}$ remain unchanged for any constant error probability $\epsilon > 0$. We omit the proof.

**Proposition 10.10** *The following are equivalent:*

*(a) There exists a constant $\epsilon$, $0 < \epsilon < 1/2$, such that set $L$ has an Arthur-Merlin proof system with error probability $\epsilon$.*

*(b) For every constant $\epsilon$, $0 < \epsilon < 1/2$, set $L$ has an Arthur-Merlin proof system with error probability $\epsilon$.*

*(c) For every polynomial $q$, set $L$ has an Arthur-Merlin proof system with error probability bounded by $2^{-q(n)}$.*

In the above, we have defined the Arthur-Merlin proof system from Arthur's point of view; that is, we view such a proof system as Arthur working with an unreliable oracle Merlin trying to prove that an instance $x$ is in $A$. Since Merlin knows every random bit Arthur uses and has the unlimited computational power to simulate Arthur's computation, we may also look at this proof system from Merlin's point of view. Namely, we may think that Merlin tries to convince Arthur, who is a passive verifier, that an instance $x$ is in $A$ by showing the complete *probabilistic proof* to him, where the two players share a common, trustable random number generator. That is, Merlin can demonstrate the proof by playing both roles in the game: he nondeterministically generates his messages for himself, and randomly generates messages for Arthur. To formalize this idea, we consider a new computational model of probabilistic, nondeterministic Turing machines.

A *probabilistic, nondeterministic Turing machine* is an NTM $M$ attached with a random number generator $\phi$. The machine $M$ has two nonstandard states: the nondeterministic state and the random state. When the machine enters the nondeterministic state, it generates a bit nondeterministically, and when it enters a random state, it obtains a random bit from the random number generator. We call a probabilistic, nondeterministic Turing machine a Merlin machine if there exists a polynomial $p$ such that for each input $x$, it always halts in $p(|x|)$ moves.

To simplify the notion of the accepting probability of a Merlin machine, we assume that such a machine $M$ has two extra read-only tapes, called the Merlin tape and the Arthur tape, in addition to the input/output/work tape. The Arthur tape is just the random-bit tape of a PTM, where a sequence of random bits is supplied by the random number generator $\phi$. The Merlin tape contains the messages to be sent from Merlin to Arthur. We assume that when a Merlin machine enters the nondeterministic state, it reads the next bit from the Merlin tape, and when the machine enters the random state, it reads the next bit from the Arthur tape. Thus, a configuration of a Merlin machine consists of the following information: (1) the current state $q$, (2) the current string $s$ in the input/output/work tape to the left of the tape head, (3) the current string $t$ at and on the right of the tape head in the input/output/work tape, (4) the current string $\alpha$ on the Merlin tape (from the beginning cell to the cell scanned by the tapehead), and (5) the current string $\beta$ on the Arthur tape (from the beginning cell to the cell scanned by the tapehead). We define the successor relation $\vdash_M$ on configurations and the concept of accepting configurations similar to that for PTMs (see Section

9.2). In particular, a configuration $c = (q, s, t, \alpha, \beta)$ in the random state (or, in the nondeterministic state) has two successor configurations, $c_0 = (q_0, s, t, \alpha, \beta 0)$ and $c_1 = (q_1, s, t, \alpha, \beta 1)$ (or, respectively, $c_0 = (q_0, s, t, \alpha 0, \beta)$ and $c_1 = (q_1, s, t, \alpha 1, \beta)$) for some states $q_0$ and $q_1$.

To define the notion of the accepting probability, we first define the accepting probability $acc(c)$ of a configuration $c$. For any configuration $c = (q, s, t, \alpha, \beta)$, we let $acc(c) = 1$ if $q$ is an accepting state, $acc(c) = 0$ if $q$ is a rejecting state. If $q$ is a regular, deterministic state, and $c_1$ is the unique successor configuration of $c$, then let $acc(c) = acc(c_1)$. If $q$ is a random state, and $c_0$ and $c_1$ are two successor configurations of $c$, then let $acc(c)$ be the average of $acc(c_0)$ and $acc(c_1)$. If $q$ is a nondeterministic state, and $c_0$ and $c_1$ are two successor configurations of $c$, then let $acc(c)$ be the maximum of $acc(c_0)$ and $acc(c_1)$. For any input $x$, the *accepting probability* of a Merlin machine on $x$ is equal to $acc(c_x)$, where $c_x$ is the starting configuration for $x$. We also write $\Pr[M(x) = 1]$ to denote the accepting probability $acc(c_x)$ of $M$ on $x$. Since a Merlin game always halts in polynomial time, the probability $acc(c_x)$ is well defined.

The concept of rounds can also be defined in a Merlin machine. We define a *random round* of a computation of a Merlin machine $M$ to be a maximum-length sequence of configurations that are either in a deterministic state or in the random state. Similarly, a *nondeterministic round* of a computation of a Merlin machine $M$ is a maximum-length sequence of configurations that are either in a deterministic state or in the nondeterministic state. For $k \geq 1$, a $k$-round Merlin machine $M$ is one all of whose computations have at most $k$ random and nondeterministic rounds.

**Proposition 10.11** *A set $L$ is in AM if and only if there exist a Merlin machine $M$ and a constant $0 < \epsilon < 1/2$ such that:*
*(1) For all $x \in L$, $\Pr[M(x) = 1] > 1 - \epsilon$; and*
*(2) For all $x \notin L$, $\Pr[M(x) = 1] < \epsilon$.*

*Proof.* We first show how to convert an Arthur machine $M_A$ to a Merlin machine $M_M$. We change each query of $M_A$ to the action of generating a sequence of random bits followed by the action of generating a sequence of nondeterministic bits. Let $f$ be the function that answers each query of the machine $M_A$ to maximize the accepting probability. Then, the accepting probability of $M_M$ on any input $x$ is equal to the accepting probability of $M_A^f$ on input $x$, and $\Pr[M_A^f(x) = 1] \geq \Pr[M_A^g(x) = 1]$ for all oracles $g$. It follows that if $(M_A, g)$ is an Arthur-Merlin proof system for $L$ with error probability $\epsilon$, then $M_M$ satisfies conditions (1) and (2).

Conversely, to change a Merlin machine $M_M$ to an Arthur machine $M_A$, we convert each round of random moves of machine $M_M$ to a query move by Arthur in $M_A$, and each round of nondeterministic move to the action of answering the query. Again, the accepting probability of the machine $M_M$ is equal to the accepting probability of the machine $M_A$ relative to the optimum oracle $f$. Thus, the machine $M_A$, together with the optimum oracle $f$, forms an Arthur-Merlin proof system for $L$. ∎

It is interesting to observe that the above conversions between the two types of machines preserve the number of rounds in each machine.

## 10.3 *AM* Hierarchy Versus Polynomial-Time Hierarchy

In this section, we consider the complexity classes $AM_k$ of sets having Arthur-Merlin proof systems of a bounded number of rounds. It is clear that these classes form a hierarchy:

$$AM_0 \subseteq AM_1 \subseteq \cdots \subseteq AM_k \subseteq AM_{k+1} \subseteq \cdots.$$

We are going to see that this hierarchy collapses to the second level. In addition, they are contained in the second level $\Pi_2^P$ of the polynomial-time hierarchy.

To begin with, we first give an alternating quantifier characterization for the $AM$ hierarchy like that for the polynomial-time hierarchy in Theorem 3.8. Intuitively, a move by Merlin demonstrates a new piece of proof to Arthur, and so it corresponds to an existential quantifier $\exists$, while a move by Arthur is simply a sequence of random bits and that corresponds to a probabilistic quantifier $\exists^+$. (Recall that $(\exists_r^+ y, |y| = m)$ means "for at least $r \cdot 2^m$ strings $y$ of length $m$," and $\exists^+$ is the abbreviation for $\exists_{3/4}^+$.)

Recall that two predicates $R_1$ and $R_0$ are *complementary* if $R_1$ implies $\neg R_0$.

**Definition 10.12** *(a) For each $k \geq 0$ and function $r(n)$ such that $0 \leq r(n) \leq 1$, we say predicates $R_1(x)$ and $R_0(x)$ are two complementary $AM_k(r)$-predicates if there exist a polynomial-time predicate $R$ and a polynomial $q$ such that for each $x$ of length $n$,*

$$R_1(x) \equiv (\exists_r^+ y_1)(\exists y_2)(\exists_r^+ y_3)(\exists y_4) \cdots (Q_k y_k)\, R(x, y_1, \ldots, y_k),$$
$$R_0(x) \equiv (\exists_r^+ y_1)(\forall y_2)(\exists_r^+ y_3)(\forall y_4) \cdots (Q_k' y_k)\, \neg R(x, y_1, \ldots, y_k),$$

*where $Q_k = Q_k' = \exists_r^+$ if $k$ is odd, $Q_k = \exists$ and $Q_k' = \forall$ if $k$ is even, $r = r(n)$, and the scope of the quantifiers is the set of all strings of length $q(n)$.*

*When $r(n) = 3/4$ for all $n$, we simply say that $R_1$ and $R_0$ are complementary $AM_k$-predicates.*

*(b) Using the same notation, we define complementary $MA_k(r)$-predicates $R_1$ and $R_0$ by the relation:*

$$R_1(x) \equiv (\exists y_1)(\exists_r^+ y_2)(\exists y_3)(\exists_r^+ y_4) \cdots (Q_{k+1} y_k)\, R(x, y_1, \ldots, y_k),$$
$$R_0(x) \equiv (\forall y_1)(\exists_r^+ y_2)(\forall y_3)(\exists_r^+ y_4) \cdots (Q_{k+1}' y_k)\, \neg R(x, y_1, \ldots, y_k).$$

A simple induction shows that complementary $AM_k(r)$- and $MA_k(r)$-predicates are indeed complementary if $r > 1/2$. We are going to show that if $A \in AM_k$, then the predicates $[x \in A]$ and $[x \notin A]$ are complementary

$AM_k$-predicates. Before we prove that, we need to show that the notion of complementary $AM_k$-predicates remains unchanged with respect to different values $r$. This can be proved by the majority vote technique. Since it is more involved than the proofs of Theorems 10.7 and 10.10, we give a complete proof below. In the following, we write (M $i \leq m$) to mean "for more than $m/2$ numbers $i \in \{1, \ldots, m\}$."

**Lemma 10.13** *Let $m > 0$ and write $\vec{x}$ to denote $\langle x_1, \ldots, x_m \rangle$, with all strings $x_i$ of an equal length $n$. Assume that $R_1'(x)$ and $R_0'(x)$ are two complementary $AM_k(r)$-predicates, with $r^m > 1/2$. Then, the predicates $R_1(\vec{x}) \equiv (M\, i \leq m) R_1'(x_i)$ and $R_0(\vec{x}) \equiv (M\, i \leq m) R_0'(x_i)$ are complementary $AM_k(r^m)$-predicates.*

*Proof.* We prove this lemma by induction on the number of probabilistic quantifiers (or, $\lceil k/2 \rceil$). If $k = 0$, then $R_1'$ and $R_0'$ are complementary polynomial-time predicates and so are $R_1$ and $R_0$. For $k \geq 1$, assume that

$$R_1'(x) \equiv (\exists_r^+ y_1)(\exists y_2) S_1(x, y_1, y_2)$$
$$R_0'(x) \equiv (\exists_r^+ y_1)(\forall y_2) S_0(x, y_1, y_2),$$

for some complementary $AM_{k-2}(r)$-predicates $S_1$ and $S_0$, where the scope of the quantifiers is the set of all strings of length $q(|x|)$ for some polynomial $q$. (For $k = 1$, let $R_1'(x) \equiv (\exists_r^+ y_1) S_1(x, y_1)$ and $R_0'(x) \equiv (\exists_r^+ y_1) S_0(x, y_1)$, with $S_1, S_0 \in P$, and the proof is the same as the general case.) Then, we get

$$R_1(\vec{x}) \equiv (M\, i \leq m)(\exists_r^+ y_1)(\exists y_2) S_1(x_i, y_1, y_2).$$

Fix an $\vec{x}$ with each string $x_i$ of length $n$, Let $Y_i = \{u : |u| = q(n), (\exists y_2) S_1(x_i, u, y_2)\}$. Then, $R_1(\vec{x})$ implies that for more than $m/2$ numbers $1 \leq i \leq m$, the probability of $Y_i$ (in the space of all strings of length $q(n)$) is at least $r$. We fix a set $I$ of size $\lceil (m+1)/2 \rceil$ of integers $i$ such that $\Pr[Y_i] \geq r$, and let

$$Y = \{\langle u_1, \ldots, u_m \rangle : |u_1| = \cdots = |u_m| = q(n), (\forall i \in I)\, u_i \in Y_i\}.$$

Consider the set $Y$ in the probability space of all $\vec{u} = \langle u_1, \ldots, u_m \rangle$. We can see that $\Pr[Y] \geq r^{\lceil (m+1)/2 \rceil} \geq r^m$.

The above observation implies that

$$R_1(\vec{x}) \Rightarrow (\exists_{r^m}^+ \vec{u})(M\, i \leq m)(\exists y_2)\, S_1(x_i, u_i, y_2)$$
$$\Rightarrow (\exists_{r^m}^+ \vec{u})(\exists \vec{v})(M\, i \leq m)\, S_1(x_i, u_i, v_i).$$

Similarly, we can derive that

$$R_0(\vec{x}) \Rightarrow (\exists_{r^m}^+ \vec{u})(M\, i \leq m)(\forall y_2)\, S_0(x_i, u_i, y_2)$$
$$\Rightarrow (\exists_{r^m}^+ \vec{u})(\forall \vec{v})(M\, i \leq m)\, S_0(x_i, u_i, v_i).$$

By the inductive hypothesis, the pair of predicates (M $i \leq m$) $S_1(x_i, u_i, v_i)$ and (M $i \leq m$) $S_0(x_i, u_i, v_i)$ are complementary $AM_{k-2}(r^m)$-predicates and, hence, $R_1$ and $R_0$ are complementary $AM_k(r^m)$-predicates.  ∎

**Theorem 10.14** *Let $k \geq 0$. The following are equivalent:*

*(a) $R_1$ and $R_0$ are complementary $AM_k(c)$-predicates for some $c > 1/2$.*

*(b) $R_1$ and $R_0$ are complementary $AM_k(c)$-predicates for every $c > 1/2$.*

*(c) $R_1$ and $R_0$ are complementary $AM_k(1 - 2^{-p(n)})$-predicates for any polynomial function $p$.*

*Proof.* We show the implication (a) $\Rightarrow$ (c). The proof is by induction on $k$. It is obvious for the case $k = 0$. For the case $k = 1$, it is the same as Theorem 8.19.

Assume that $k \geq 2$. Let $R_1'$ and $R_0'$ be two complementary $AM_{k-2}(c)$-predicates so that

$$R_1(x) \equiv (\exists_c^+ y_1)(\exists y_2) \, R_1'(x, y_1, y_2),$$
$$R_0(x) \equiv (\exists_c^+ y_1)(\forall y_2) \, R_0'(x, y_1, y_2),$$

where the scope of the quantifiers on $y_1$ and $y_2$ is the set of all strings of length $q(|x|)$ for some polynomial $q$. From the majority vote technique of Theorem 8.19, if $R_1(x)$ then there exists a constant $m_0$ such that the probability $r$ of the set

$$Y_1 = \{\langle u_1, \ldots, u_m \rangle : (\mathsf{M} \, i \leq m)(\exists y_2) \, R_1'(x, u_i, y_2)\}$$

(in the probability space of all $\vec{u} = \langle u_1, \ldots, u_m \rangle$, with each $|u_i| = q(|x|)$) is at least $1 - 2^{-p(n)}$ if we let $m = m_0 \cdot p(n)$. Therefore, we can write $\vec{u} = \langle u_1, \ldots, u_m \rangle$ and get

$$R_1(x) \Rightarrow (\exists_r^+ \vec{u})(\mathsf{M} \, i \leq m)(\exists y_2) \, R_1'(x, u_i, y_2)$$
$$\Rightarrow (\exists_r^+ \vec{u})(\exists \vec{v})(\mathsf{M} \, i \leq m) \, R_1'(x, u_i, v_i).$$

Similarly, we get

$$R_0(x) \Rightarrow (\exists_r^+ \vec{u})(\mathsf{M} \, i \leq m)(\forall y_2) \, R_0'(x, u_i, y_2)$$
$$\Rightarrow (\exists_r^+ \vec{u})(\forall \vec{v})(\mathsf{M} \, i \leq m) \, R_0'(x, u_i, v_i).$$

From the inductive hypothesis, we can assume that $R_1'$ and $R_0'$ are actually complementary $AM_{k-2}(s)$-predicates, where $s = 1 - 2^{-(m_0+1)p(n)}$ (note that $m_0$ is a constant). Then, $s^m \geq 1 - 2^{-p(n)}$. It follows from Lemma 10.13 that the predicates $(\mathsf{M} \, i \leq m) \, R_1'(x, u_i, v_i)$ and $(\mathsf{M} \, i \leq m) \, R_0'(x, u_i, v_i)$ are complementary $AM_{k-2}(1 - 2^{-p(n)})$-predicates. This completes the proof of the theorem. ∎

**Theorem 10.15** *Let $k \geq 0$.*

*(a) A set $A$ is in $AM_k$ if and only if the predicates $R_1(x) \equiv [x \in A]$ and $R_0(x) \equiv [x \notin A]$ are complementary $AM_k$-predicates.*

*(b) A set $A$ is in $MA_k$ if and only if the predicates $R_1(x) \equiv [x \in A]$ and $R_0(x) \equiv [x \notin A]$ are complementary $MA_k$-predicates.*

*Proof.* (a) Assume that $A$ is accepted by a Merlin machine $M$ in the sense of Proposition 10.11, with the error probability $\epsilon \leq 2^{-(k+3)}$. Let $(y_1, \ldots, y_k)$ be the sequence of random and nondeterministic strings generated alternatingly by the machine $M$ on input $x$, padded with dummy bits to make them of length exactly $q(n)$. (In terms of the equivalent Arthur machine, $(y_1, \ldots, y_k)$ is the sequence of queries and answers exchanged between Arthur and Merlin.) Let $c(x, y_1, \ldots, y_i)$ be the configuration of the Merlin machine $M$ on input $x$, after the strings $y_1, \ldots, y_i$ have been generated. Assume that $x \in A$. Then, $acc(c(x)) > 1 - \epsilon$. This implies that for at least $(3/4)2^{q(n)}$ choices of $y_1$, $acc(c(x, y_1)) > 1 - 4\epsilon$. For each such string $y_1$, there exists a string $y_2$ such that $acc(c(x, y_1, y_2)) > 1 - 4\epsilon$. Continuing this argument, we get, for each odd $i \leq k$,

$$(\exists^+ y_1)(\exists y_2) \cdots (Q_i y_i)\,[acc(c(x, y_1, \ldots, y_i)) > 1 - 2^{i+1}\epsilon],$$

where $Q_j = \exists^+$ if $j$ is odd and $Q_j = \exists$ if $j$ is even. It follows that $acc(c(x, y_1, \ldots, y_k)) > 1 - 2^{k+1}\epsilon \geq 3/4$. Note that if $c$ is a halting configuration and $acc(c) > 3/4$, then $acc(c) = 1$. Thus, we have proved that

$$x \in A \Rightarrow (\exists^+ y_1)(\exists y_2) \cdots (Q_k y_k)\,[acc(c, y_1, \ldots, y_k) = 1].$$

For the case $x \notin A$, we can prove in a similar way that

$$x \notin A \Rightarrow (\exists^+ y_1)(\forall y_2) \cdots (Q_k' y_k)\,[acc(c, y_1, \ldots, y_k) = 0],$$

where $Q_j' = \exists^+$ if $j$ is odd and $Q_j' = \forall$ if $j$ is even.

For the converse, assume, by Theorem 10.14, that $[x \in A]$ and $[x \notin A]$ are complementary $AM_k(r)$-predicates, where $r = 1 - 1/(4k)$. That is,

$$x \in A \;\Rightarrow\; (\exists_r^+ y_1)(\exists y_2) \cdots (Q_k y_k)\, R(x, y_1, \ldots, y_k),$$
$$x \notin A \;\Rightarrow\; (\exists_r^+ y_1)(\forall y_2) \cdots (Q_k' y_k)\, \neg R(x, y_1, \ldots, y_k),$$

for some polynomial-time predicate $R$, where $Q_k$ and $Q_k'$ and the scope of the quantifiers are defined as in Definition 10.12. From this relation, we define a Merlin machine $M$ that works on each input $x$ as follows:

> At an odd stage $i$, $1 \leq i \leq k$, $M$ enters the random state and generates a random string $y_i$ of length $q(n)$. At an even stage $i$, $1 \leq i \leq k$, $M$ enters the nondeterministic state and generates a string $y_i$ of length $q(n)$ nondeterministically. Then, in the final stage $k + 1$, $M$ verifies whether $R(x, y_1, \ldots, y_k)$ holds.

By a simple induction proof, we can see that for each $x \in A$, the above machine has the accepting probability $\geq r^{(k+1)/2} > 3/4$. Conversely, it is also easy to check that for each $x \notin A$, the accepting probability is at most $1 - r^{(k+1)/2} < 1/4$.

(b) The proof for sets in $MA_k$ is similar. ∎

From the above characterization, we may say, informally, that $AM_k$ is defined by a pair of alternating (polynomial length-bounded) quantifier sequences $(\exists^+\exists\exists^+\exists\cdots, \exists^+\forall\exists^+\forall\cdots)$. More specifically, $AM_2$ is defined by $(\exists^+\exists, \exists^+\forall)$. Using the generalized BPP theorem, we can see that the first pair of quantifiers $(\exists^+, \exists^+)$ may be converted to the pair $(\forall\exists^+, \exists^+\forall)$, and so $AM_2$ can be defined by $(\forall\exists^+\exists, \exists^+\forall\forall)$, or simply $(\forall\exists, \exists^+\forall)$. In other words, the probabilistic quantifier of the first sequence could be replaced by the universal quantifier. In terms of the Arthur-Merlin game, it means that the game is equivalent to the stronger game that has only a one-sided error (the error only happens when $x \notin A$). In the following, we prove this important result formally.

**Theorem 10.16** *(a) Let $R_1$ and $R_0$ be two complementary $AM_2$-predicates. Then, there exist a polynomial-time predicate $R$ and a polynomial $p$ such that for each $x$,*

$$R_1(x) \iff (\forall z_1)(\exists z_2)\, R(x, z_1, z_2),$$
$$R_0(x) \iff (\exists^+ z_1)(\forall z_2)\, \neg R(x, z_1, z_2),$$

*where the scope of the quantifiers is the set of all strings of length $p(|x|)$.*

*(b) Let $R_1$ and $R_0$ be two complementary $MA_2$-predicates. Then, there exist a polynomial-time predicate $R$ and a polynomial $p$ such that for each $x$,*

$$R_1(x) \iff (\exists z_1)(\forall z_2)\, R(x, z_1, z_2),$$
$$R_0(x) \iff (\forall z_1)(\exists^+ z_2)\, \neg R(x, z_1, z_2),$$

*where the scope of the quantifiers is the set of all strings of length $p(|x|)$.*

*Proof.* In the following, we write $\vec{w}$ to denote the list $\vec{w} = \langle w_1, \ldots, w_{q(n)} \rangle$, with each string $w_i$ of length $q(n)$, whenever $q(n)$ is understood. Assume that $|x| = n$.

(a) From Theorem 10.14, assume that $R_1$ and $R_0$ are of the form

$$R_1(x) \equiv (\exists_r^+ y_1)(\exists y_2)\, R(x, y_1, y_2),$$
$$R_0(x) \equiv (\exists_r^+ y_1)(\forall y_2)\, \neg R(x, y_1, y_2),$$

where $r = 1 - 2^{-n}$, $R$ is polynomial-time computable and the scope of the quantifiers is the set of strings of length $q(n)$ for some polynomial $q$. By the generalized $BPP$ theorem (Theorem 8.34), we can derive the desired relations:

$$\begin{aligned}
R_1(x) &\Rightarrow (\forall \vec{w})(\exists^+ y_1)(\forall i \le q(n))(\exists y_2)\, R(x, y_1 \oplus w_i, y_2)] \\
&\Rightarrow (\forall \vec{w})(\exists^+ y_1)(\exists \vec{u})(\forall i \le q(n))\, R(x, y_1 \oplus w_i, u_i) \\
&\Rightarrow (\forall \vec{w})(\exists \langle y_1, \vec{u} \rangle)(\forall i \le q(n))\, R(x, y_1 \oplus w_i, u_i),
\end{aligned}$$

and

$$\begin{aligned}
R_0(x) &\Rightarrow (\exists^+ \vec{w})(\forall y_1)(\exists i \le q(n))(\forall y_2)\, \neg R(x, y_1 \oplus w_i, y_2) \\
&\Rightarrow (\exists^+ \vec{w})(\forall y_1)(\forall \vec{u})(\exists i \le q(n))\, \neg R(x, y_1 \oplus w_i, u_i) \\
&\Rightarrow (\exists^+ \vec{w})(\forall \langle y_1, \vec{u} \rangle)(\exists i \le q(n))\, \neg R(x, y_1 \oplus w_i, u_i).
\end{aligned}$$

(b) Assume that $R_1$ and $R_0$ are of the form

$$R_1(x) \equiv (\exists y_1)(\exists_r^+ y_2)\, R(x, y_1, y_2),$$
$$R_0(x) \equiv (\forall y_1)(\exists_r^+ y_2)\, \neg R(x, y_1, y_2),$$

where $r = 1 - 2^{-n}$, $R$ is polynomial-time computable and the scope of the quantifiers is the set of strings of length $q(n)$ for some polynomial $q$. We apply the generalized BPP theorem on $R_0$ and $R_1$ (reversing the roles of $R_1$ and $R_0$), and we can obtain

$$\begin{aligned}
R_1(x) &\Rightarrow (\exists y_1)(\exists^+ \vec{w})(\forall y_2)(\exists i \leq q(n))\, R(x, y_1, y_2 \oplus w_i) \\
&\Rightarrow (\exists \langle y_1, \vec{w} \rangle)(\forall y_2)(\exists i \leq q(n))\, R(x, y_1, y_2 \oplus w_i),
\end{aligned}$$

and

$$\begin{aligned}
R_0(x) &\Rightarrow (\forall y_1)(\forall \vec{w})(\exists^+ y_2)(\forall i \leq q(n))\, \neg R(x, y_1, y_2 \oplus w_i) \\
&\Rightarrow (\forall \langle y_1, \vec{w} \rangle)(\exists^+ y_2)(\forall i \leq q(n))\, \neg R(x, y_1, y_2 \oplus w_i),
\end{aligned}$$

and the proof is complete.                                                                 ∎

*Remark.* By induction, the above results can be generalized to complementary $AM_k$- or $MA_k$-predicates. In particular, for complementary $MA_3$-predicates, Theorem 10.16(b) holds with respect to a $\Sigma_1^P$-predicate $R$, since $R(x, y) \in \Sigma_1^P$ implies the predicate $(\exists i, 1 \leq i \leq q(n))\, R(x, y_i)$ is also in $\Sigma_1^P$.

**Corollary 10.17** *(a) $AM_2 \subseteq \Pi_2^P$.*
   *(b) $MA_2 \subseteq \Sigma_2^P$.*

Using this new, stronger characterization of $AM_2$, we can prove that the $AM$ hierarchy collapses to $AM_2$.

**Theorem 10.18** *(a) $MA_3 \subseteq AM_2$.*
   *(b) If $k \geq 2$, then $AM_{k+1} = AM_k$. So, $AM_k = MA_k = AM_2$ for all $k > 2$.*

*Proof.* (a) We follow the same notation established in Theorem 10.16. Let $R_1$ and $R_0$ be two complementary $MA_3$-predicates. By the generalized form of Theorem 10.16(b) (see Remark above) and Theorem 10.14, there exist a polynomial-time predicate $R$ and a polynomial $q$ such that for each $x$, with $|x| = n$,

$$R_1(x) \Rightarrow (\exists y_1)(\forall y_2)(\exists y_3)\, R(x, y_1, y_2, y_3)$$
$$R_0(x) \Rightarrow (\forall y_1)(\exists_r^+ y_2)(\forall y_3)\, \neg R(x, y_1, y_2, y_3),$$

where $r = 1 - 2^{-n}$ and the scope of the quantifiers is the set of strings of length $q(n)$ for some polynomial $q$. Applying the Swapping Lemma (Lemma 8.31) to the predicate $R_0(x)$, we get

$$\begin{aligned}
R_0(x) &\Rightarrow (\forall y_1)(\exists_r^+ y_2)(\forall y_3)\, \neg R(x, y_1, y_2, y_3) \\
&\Rightarrow (\exists^+ \vec{w})(\forall y_1)(\exists i \leq q(n))(\forall y_3)\, \neg R(x, y_1, w_i, y_3) \\
&\Rightarrow (\exists^+ \vec{w})(\forall y_1)(\forall \vec{u})(\exists i \leq q(n))\, \neg R(x, y_1, w_i; u_i) \\
&\Rightarrow (\exists^+ \vec{w})(\forall \langle y_1, \vec{u} \rangle)(\exists i \leq q(n))\, \neg R(x, y_1, w_i, u_i).
\end{aligned}$$

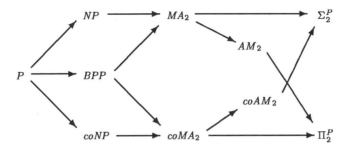

Figure 10.1: Inclusive relations between the $AM$ hierarchy and the polynomial-time hierarchy. $A \to B$ means $A \subseteq B$.

Next, observe that the quantifier sequence $\exists\forall$ can easily be swapped to $\forall\exists$:

$$
\begin{aligned}
R_1(x) &\Rightarrow (\exists y_1)(\forall y_2)(\exists y_3)\, R(x, y_1, y_2, y_3) \\
&\Rightarrow (\forall \vec{w})(\exists y_1)(\forall i \leq q(n))(\exists y_3)\, R(x, y_1, w_i, y_3) \\
&\Rightarrow (\forall \vec{w})(\exists y_1)(\exists \vec{u})(\forall i \leq q(n))\, R(x, y_1, w_i, u_i) \\
&\Rightarrow (\forall \vec{w})(\exists \langle y_1, \vec{u}\rangle)(\forall i \leq q(n))\, R(x, y_1, w_i, u_i).
\end{aligned}
$$

Thus, $R_1$ and $R_0$ are actually complementary $AM_2$-predicates. It implies that $MA_3 \subseteq AM_2$.

(b) Let $A \in AM_{k+1}$, with $k \geq 2$. If $k$ is even, then let $j = 2$; otherwise let $j = 3$. By Theorem 10.15, we may assume that there exist two complementary $MA_j$-predicates, $R_1$ and $R_0$, and a polynomial $q$ such that for all $x$, with $|x| = n$,

$$
\begin{aligned}
x \in A &\Rightarrow (\exists_r^+ y_1)(\exists y_2) \cdots (\exists_r^+ y_{k-j+1})\, R_1(x, y_1, \ldots, y_{k-j+1}) \\
x \notin A &\Rightarrow (\exists_r^+ y_1)(\forall y_2) \cdots (\exists_r^+ y_{k-j+1})\, R_0(x, y_1, \ldots, y_{k-j+1}),
\end{aligned}
\tag{10.2}
$$

where $r = 7/8$, and the scope of the quantifiers is the set of strings of length $q(n)$ for some polynomial $q$. From the proof of part (a), we know that $R_1$ and $R_0$ are actually two complementary $AM_2$-predicates, and can be written in the following form:

$$
\begin{aligned}
R_1(x, y_1, \ldots, y_{k-j+1}) &\Leftrightarrow (\forall z_1)(\exists z_2)\, R_2(x, y_1, \ldots, y_{k-j+1}, z_1, z_2) \\
R_0(x, y_1, \ldots, y_{k-j+1}) &\Leftrightarrow (\exists_r^+ z_1)(\forall z_2)\, \neg R_2(x, y_1, \ldots, y_{k-j+1}, z_1, z_2),
\end{aligned}
\tag{10.3}
$$

where $R_2$ is polynomial-time computable, and the scope of the quantifiers is the set of strings of length $p(n)$ for some polynomial $p$. Now, we can substitute equations (10.3) to relations (10.2) and combine the two quantifiers on $y_{k-j+1}$ and on $z_1$ into a single probabilistic quantifier $\exists^+$. Thus, $A \in AM_{k-j+2} \subseteq AM_k$. ∎

The above results established the relations between the $AM$-hierarchy and the polynomial-time hierarchy as shown in Figure 10.1. In particular, we see that $NP \subseteq MA_2 \subseteq \Sigma_2^P \cap \Pi_2^P$, and $MA_2 \subseteq AM_2 \subseteq \Pi_2^P$. The remaining relations, such as whether $coNP \subseteq AM_2$ and whether $AM_2$ collapses to $MA_2$, are open. In the following, we show that $AM_2$ is not powerful enough to contain $coNP$ unless the polynomial-time hierarchy collapses to $AM_2$.

**Lemma 10.19** *If* $coNP \subseteq AM_2$ *then* $AM_2 = coAM_2$.

*Proof.* Let $A \in coAM_2$. By Theorem 10.16, there exist a polynomial-time predicate $R$ and a polynomial $q$ such that for all $x$,

$$x \in A \Rightarrow (\exists^+ y_1)(\forall y_2) R(x, y_1, y_2)$$
$$x \notin A \Rightarrow (\forall y_1)(\exists y_2) \neg R(x, y_1, y_2),$$

where the scope of the quantifiers on $y_1$ and $y_2$ is the set of strings of length $q(|x|)$. Let $B = \{\langle x, y_1 \rangle : (\forall y_2) R(x, y_1, y_2)\}$. Then $B \in coNP \subseteq AM_2$. So, by Theorem 10.16 again, we have, for some polynomial-time predicate $R'$,

$$x \in A \Rightarrow (\exists^+ y_1)(\forall z_2)(\exists z_3) R'(x, y_1, z_2, z_3)$$
$$\Rightarrow (\exists^+ \langle y_1, z_2 \rangle)(\exists z_3) R'(x, y_1, z_2, z_3),$$

and

$$x \notin A \Rightarrow (\forall y_1)(\exists^+ z_2)(\forall z_3) \neg R'(x, y_1, z_2, z_3)$$
$$\Rightarrow (\exists^+ \langle y_1, z_2 \rangle)(\forall z_3) \neg R'(x, y_1, z_2, z_3),$$

where $y_1$ denotes a string of length $q(|x|)$ and $z_2$ and $z_3$ denote strings of length $p(|x|)$ for some polynomial $p$. This implies that $A \in AM_2$. ∎

**Theorem 10.20** *If* $coNP \subseteq AM_2$ *then* $PH \subseteq AM_2$.

*Proof.* We prove the theorem by induction. Assume that we know that $\Sigma_k^P \subseteq AM_2$, and let $A \in \Sigma_{k+1}^P$. Then, there exists a set $B \in \Pi_k^P$ and a polynomial $q$ such that for all $x$, with $|x| = n$,

$$x \in A \iff (\exists y_1, |y_1| = q(n)) \langle x, y_1 \rangle \in B,$$
$$x \notin A \iff (\forall y_1, |y_1| = q(n)) \langle x, y_1 \rangle \notin B. \tag{10.4}$$

By the above lemma, we have $B \subseteq \Pi_k^P \subseteq coAM_2 \subseteq AM_2$. Substituting two complementary $AM_2$-predicates for the predicates $[\langle x, y_1 \rangle \in B]$ and $[\langle x, y_1 \rangle \notin B]$ in the above relations (10.4), we see that $A$ is in $MA_3$ and, hence, in $AM_2$. ∎

## 10.4   *IP* Versus *AM*

As we discussed in Section 10.2, Arthur-Merlin proof systems appear to be weaker than interactive proof systems since the random numbers generated by Arthur are also known to Merlin, while the random numbers generated by the verifier in an interactive proof system are kept secret from the prover. Nevertheless, we show in this section that $IP = AM$ and, in addition, $IP_k = AM_2$ for all $k \geq 2$. The proof is based on the idea of random hashing functions and is quite involved. We first present a proof showing that all sets $L \in IP_k$, $k \geq 2$, that have a restricted form of interactive proof systems are in $AM_2$.

Recall that the two most important complexity measures of an interactive proof system are the error probability and the number of rounds of communication. Two other important complexity measures are the length $q(n)$ of each query and answer and the total number $r(n)$ of the random bits used by the verifier. To emphasize these two complexity measures, we write $IP_k(q(n), r(n))$ to denote the class of sets $L$ that have $k$-round interactive proof systems in which the length of each query and each answer is bounded by $q(n)$, and the number of random bits used is bounded by $r(n)$, when the input is of length $n$. Let $Q$ and $R$ be two classes of functions. We write

$$IP_k(Q, R) = \bigcup_{q \in Q, r \in R} IP_k(q(n), r(n)).$$

Thus, if we let *poly* denote the class of all polynomial functions, then $IP_k = IP_k(poly, poly)$. In the following, we let *log* denote the class of all functions $f(n) = c \cdot \log n$ for $c > 0$.

**Theorem 10.21** *For each $k \geq 2$, $IP_k(log, log) \subseteq AM_2$.*

*Proof.* We consider a set $L \in IP_{2k}(q(n), r(n))$, where $k \geq 1$, $q(n) = O(\log n)$ and $r(n) = O(\log n)$. By padding the messages and random bits with dummy bits, we may assume that $(M_v, f_p)$ is an interactive proof system for $L$ such that, on inputs of length $n$, the error probability is $\epsilon \leq 1/5$, the queries and answers in this system are all of length *exactly* $q(n)$ and the number of random bits used by $M_v$ is exactly $r(n)$. For any string $\alpha \in \{0,1\}^{r(n)}$, $M_v$ on input $x$ and using $\alpha$ as its random string behaves like a deterministic oracle machine. Thus, for each oracle $f$, the computation of $M_v^f(x, \alpha)$ can be encoded in a unique *history* string $h_{x,\alpha} = y_1 z_1 y_2 z_2 \ldots y_k z_k$, where $y_1, \ldots, y_k$ are the $k$ queries of $M_v$ and $z_1, \ldots, z_k$ are the corresponding answers to these queries from the oracle $f$ (i.e., $f(xy_1z_1 \ldots y_i) = z_i$ for $1 \leq i \leq k$). Each history $h_{x,\alpha}$ is of length exactly $2kq(n)$. We say that $h_{x,\alpha}$ is an *accepting history* if it is the history string of $M_v^f(x, \alpha)$ for some oracle $f$ and $M_v^f$ accepts $x$ with respect to the random string $\alpha$. Note that, given input $x$, a string $\alpha$, and a string $s$ of length $2kq(n)$, it can be determined in polynomial time whether $s$ is an accepting history $h_{x,\alpha}$ with respect to some oracle $f$. Furthermore, we can do this without making queries to the oracle $f$: if $s = y_1z_1 \ldots y_kz_k$, we simply assume that the oracle function $f$ satisfies $f(xy_1z_1 \ldots y_i) = z_i$ for $i = 1, \ldots, j$.

For any strings $x$ and $s$ of lengths $|x| = n$ and $|s| \leq 2kq(n)$, let $A(s, x) = \{\alpha : |\alpha| = r(n), s \text{ is a prefix of an accepting history } h_{x,\alpha} \text{ of } M_v(x, \alpha)\}$. The error bound of the proof system $(M_v, f_p)$ satisfying $\epsilon \leq 1/5$ means that if $x \in L$ then $|A(\lambda, x)| > (4/5)2^{r(n)}$ and if $x \notin L$ then $|A(\lambda, x)| \leq (1/5)2^{r(n)}$.

We observe that if $|s| = 2iq(n)$ for some $0 \leq i \leq k - 1$, then

$$|A(s, x)| = \sum_{|y|=q(n)} |A(sy, x)| = \sum_{|y|=q(n)} \left( \max_{|z|=q(n)} |A(syz, x)| \right). \qquad (10.5)$$

We are going to describe an Arthur-Merlin proof system for $L$. In this proof system, Merlin tries to convince Arthur that if the interactive proof system $(M_v, f_p)$ runs on an input $x$ of length $n$, then $|A(\lambda, x)| > (4/5)2^{r(n)}$. At each stage, Merlin and Arthur simulate one query/answer session of the system $(M_v, f_p)$ and Merlin demonstrates what the corresponding values of $|A(s, x)|$ are at that stage, where $s$ is the prefix of the history of the computation up to this stage.

> *Arthur-Merlin System $A_1$ for $L$.*
> Input: $x$, $|x| = n$.
>     *Stage* 0. Merlin computes $a_0 = |A(\lambda, x)|$ and sends it to Arthur. Arthur checks that $a_0 > (4/5)2^{r(n)}$; Arthur rejects if this is not true. Let $s_0 = \lambda$.
>     *Stage* $i$, $1 \leq i \leq k$. (1) For each $y$ of length $q(n)$, Merlin computes $b_{i,y} = |A(s_{i-1}y, x)|$ and sends them to Arthur.
>     (2) Arthur verifies that $\sum_{|y|=q(n)} b_{i,y} = a_{i-1}$ (he rejects if this does not hold). Then, Arthur selects a random string $y_i$ of length $q(n)$ according to the probability distribution such that $\Pr[y_i = y] = b_{i,y}/a_{i-1}$, and let $a_i = b_{i,y_i}$. Arthur sends $y_i$ to Merlin.
>     (3) Merlin computes $z_i = f_p(xs_{i-1}y_i)$ ($s_{i-1}$ is the history so far and $y_i$ is the new query) and sends $z_i$ to Arthur. Let $s_i = s_{i-1}y_iz_i$.
>     *Stage* $k+1$. (1) Merlin sends all strings $\alpha \in A(s_k, x)$ to Arthur.
>     (2) Arthur verifies that Merlin sends him exactly $a_k$ strings $\alpha$, and then verifies that for each $\alpha$, $s_k$ is an accepting history $h_{x,\alpha}$. Arthur accepts the input if both verifications succeed; he rejects otherwise.

To see that the above protocol works correctly, first assume that $x \in L$. In this case, Merlin can always be honest and send Arthur the correct values $|A(s_{i-1}, x)|$ at each stage $i$. Note that $|A(s, x)| = \sum_{|y|=q(n)} |A(sy, x)|$. Therefore, Arthur never rejects in any stage $i$, $1 \leq i \leq k$. Furthermore, at each stage $i$, Arthur chooses $y_i$ only if $b_{i,y_i} > 0$ and so $a_i$ is positive for all $i = 1, \ldots, k$. In particular, because $a_k > 0$, or $A(s_k, x) \neq \emptyset$, Merlin can send correct strings $\alpha$ to Arthur and Arthur must accept. That is, the probability of this Arthur-Merlin system accepting the input $x$ is 1.

Conversely, assume that $x \notin L$. We need to argue that no matter how Merlin tries to cheat, the accepting probability on $x$ is at most $1/4$. We prove a general property of the Arthur-Merlin system $A_1$. We let $\text{ACC}(x, i, y_1z_1 \ldots y_iz_i)$ be the event of Arthur accepting when he starts from Stage $i + 1$ with respect to the prefix $y_1z_1 \ldots y_iz_i$ of the history. Also let $\text{ACC}(x, i, y_1z_1 \ldots y_i)$ be the event of Arthur accepting when he starts from step (3) of Stage $i$ with respect to the prefix $y_1 z_1 \ldots y_i$ of the history. We claim that for each $i = 0, 1, \ldots, k$, and for each $s_i$ of length $2iq(n)$,

$$\Pr[\text{ACC}(x, i, s_i)] \leq \frac{|A(s_i, x)|}{a_i} . \tag{10.6}$$

We prove this property by the reverse induction on $i$. First, consider the case $i = k$. If $a_k \leq |A(s_k, x)|$, then inequality (10.6) holds trivially. Assume that $a_k > |A(s_k, x)|$. Then, in stage $k + 1$, Merlin must send $a_k$ strings $\alpha$ to Arthur, and at least one of them $\alpha_0$ is wrong (i.e., $\alpha_0 \notin A(s_k, x)$), and Arthur must reject. Therefore, the probability of Arthur accepting is zero, and so (10.6) holds.

Next, assume that the inequality (10.6) holds for $i+1$ for some $0 \leq i \leq k-1$, and let $s_i$ be a string of length $2iq(n)$. At stage $i + 1$, Merlin sends $b_{i+1,y}$ to Arthur for each $y$ of length $q(n)$. Then, Arthur selects a string $y_{i+1}$ of length $q(n)$ and sends it to Merlin and gets a string $z_{i+1}$ from Merlin. For the purpose of calculating the accepting probability, we may assume that Merlin selects $b_{i+1,y}$'s and $z_{i+1}$ to maximize the accepting probability. Thus, we have the relation

$$\Pr[\text{ACC}(x, i, s_i)] \leq \max_{\Sigma\, b_{i+1,y} = a_i} \sum_{|y| = q(n)} \frac{b_{i+1,y}}{a_i} \Pr[\text{ACC}(x, i, s_i y)]$$

$$\leq \max_{\Sigma\, b_{i+1,y} = a_i} \sum_{|y| = q(n)} \frac{b_{i+1,y}}{a_i} \left( \max_{|z| = q(n)} \Pr[\text{ACC}(x, i + 1, s_i y z)] \right) .$$

By the inductive hypothesis, we have

$$\Pr[\text{ACC}(x, i + 1, s_i y z)] \leq \frac{|A(s_i y z, x)|}{b_{i+1,y}} .$$

(Note that, in this case of Arthur choosing $y$ as $y_{i+1}$ in step (2) of stage $i+1$, he also selects $a_{i+1} = b_{i+1,y}$.) Therefore,

$$\Pr[\text{ACC}(x, i, s_i)] \leq \max_{\Sigma\, b_{i+1,y} = a_i} \sum_{|y| = q(n)} \frac{b_{i+1,y}}{a_i} \cdot \frac{\max_{|z| = q(n)} |A(s_i y z, x)|}{b_{i+1,y}}$$

$$= \frac{1}{a_i} \cdot \sum_{|y| = q(n)} \left( \max_{|z| = q(n)} |A(s_i y z, x)| \right) = \frac{|A(s_i, x)|}{a_i} ,$$

where the last equality follows from (10.5). The above completes the induction proof of inequality (10.6).

From the inequality (10.6), we get that the probability of Arthur accepting the input is

$$\Pr[\text{ACC}(x, 0, \lambda)] \leq \frac{|A(\lambda, x)|}{a_0} \leq \frac{(1/5)2^{r(n)}}{(4/5)2^{r(n)}} = \frac{1}{4} .$$

This completes the proof that the Arthur-Merlin proof system $A_1$ for $L$ is correct within error bound $1/4$.

We note that the above proof system $A_1$ runs in polynomial time since both $q(n)$ and $r(n)$ are of size $O(\log n)$ and, hence, each message sent by Merlin to Arthur is of polynomial length. Altogether, there are $2k+1$ messages that have been sent from either party to the other (we combine the consecutive messages from Merlin into one message). Therefore, we conclude that $L \in AM_{2k+2}$. It follows from Theorem 10.18 that $L \in AM_2$. ∎

In the above theorem, we have required that $q(n) = O(\log n)$ and $r(n) = O(\log n)$, and so the Arthur Merlin proof system runs in polynomial time. Suppose we allow both $q(n)$ and $r(n)$ to be any polynomial functions; then the above Arthur-Merlin proof system would take exponential time. However, we observe that any $k$-round interactive proof system with message length $q(n)$ can be regarded as a $(k \cdot q(n))$-round interactive proof system with message length 1. This observation gives us the following theorem.

**Theorem 10.22** $IP = AM$. *In addition, for any $L \in IP$, there is an Arthur-Merlin proof system for $L$ that has the one-sided error only.*

*Proof.* Assume that $L \in IP$ has a $p(n)$-round interactive proof system $(M_v, f_p)$ such that for any input $x$ of length $n$, $M_v^{f_p}(x)$ uses $r(n)$ random bits and each query and answer has length $\leq q(n)$ for some polynomials $p, q, r$. Then, we may modify $(M_v, f_p)$ to a new $(4p(n)q(n) + 2r(n))$-round interactive proof system $(M_v', f_p')$ such that for any input $x$ of length $n$, $M_v^{f_p}(x)$ uses $r(n)$ random bits and each query and each answer is of length 1. Namely, we use $2q(n)$ rounds of queries and (dummy) answers of length 1 to simulate a query made by $M_v$ and another $2q(n)$ rounds of (dummy) queries and answers to simulate an answer from $f_p$. In the last $2r(n)$ rounds, we require $M_v'$ to send all $r(n)$ random bits $\alpha$ it used to $f_p'$ as dummy queries.

Now, we apply the above algorithm $A_1$ to obtain an Arthur-Merlin proof system for $L$ that simulates the computation of $(M_v', f_p')$. It is clear that all but the last stage run in polynomial time. Also, by the end of stage $k = 4p(n)q(n) + 2r(n)$, set $A(s_k, x)$ has at most a single string $\alpha$, since $\alpha$ is part of $s_k$. Therefore, in stage $k + 1$, Merlin only sends one string $\alpha$ back to Arthur, and stage $k + 1$ also runs in polynomial time. This shows that the Arthur-Merlin proof system corresponding to the interactive proof system $(M_v, f_p)$ is a polynomial-time system.

Finally we observe that the Arthur-Merlin system we constructed above always accepts with probability 1 if the input $x$ is in $L$, and the theorem follows. ∎

The above Arthur-Merlin simulation system $A_1$ of the interactive proof systems appears too weak to maintain the same number of rounds of interaction, if the message size is big. In the following, we use an idea from the universal hashing functions to give a stronger simulation, and show that $IP_k = AM_2$ for all $k \geq 2$.

To prepare for this new simulation, we first prove the following properties of the *random hashing functions*: Let $D$ be an $a \times b$ Boolean matrix, and $h_D : \{0,1\}^a \to \{0,1\}^b$ be the function defined by $h_D(x) = x \cdot D$, where $x \in \{0,1\}^a$ is treated as a Boolean vector of size $a$ and $x \cdot D$ is the matrix multiplication modulo 2. A *linear hashing function* from $\{0,1\}^a$ to $\{0,1\}^b$ is just a function $h_D : \{0,1\}^a \to \{0,1\}^b$ for some Boolean matrix $D$. It is a *random* linear hashing function if $D$ is a random $a \times b$ Boolean matrix (with respect to the uniform distribution). Let $H = \{h_1, \ldots, h_m\}$ be a collection of $m$ linear hashing functions from $\{0,1\}^a$ to $\{0,1\}^b$ and let $U \subseteq \{0,1\}^a$ and $V \subseteq \{0,1\}^b$. We write $H(U)$ to denote the set $\bigcup_{i=1}^m h_i(U)$ and $H^{-1}(V)$ to denote $\bigcup_{i=1}^m h_i^{-1}(V)$.

**Lemma 10.23** *Let $a, b > 0$, $m \geq \max\{b, 8\}$, and let $U \subseteq \{0,1\}^a$ with $|U| = n$. Let $H = \{h_1, \ldots, h_m\}$ be a collection of $m$ random linear hashing functions from $\{0,1\}^a$ to $\{0,1\}^b$ and $V = \{v_1, \ldots, v_{m^2}\}$ be a set of $m^2$ random strings from $\{0,1\}^b$. Then, the following hold:*

*(a) If $b = \lceil \log n \rceil + 2$, then $\Pr[|H(U)| \geq n/m] \geq 1 - 2^{-m}$.*

*(b) If $b = \lceil \log n \rceil + 2$, then $\Pr[H(U) \cap V \neq \emptyset] \geq 1 - 2^{-m/8+1}$.*

*(c) If $n \leq 2^b/d$ for some $d > 0$, then $\Pr[H(U) \cap V \neq \emptyset] \leq m^3/d$.*

*Proof.* (a) Let $h_1, \ldots, h_m$ be any linear hashing functions. For each $i$, $1 \leq i \leq m$, let $X_i = \{x \in U : (\forall y \in U - \{x\}) h_i(x) \neq h_i(y)\}$. Then, it is easy to see that if $U = \bigcup_{i=1}^m X_i$ then $|H(U)| \geq n/m$: If $U = \bigcup_{i=1}^m X_i$, then there must be an index $i$, $1 \leq i \leq m$, such that $|X_i| \geq n/m$. For this index $i$, $h_i$ is one-to-one on $X_i$ and, hence, $|H(U)| \geq |h_i(U)| \geq |h_i(X_i)| \geq n/m$.

Next, we observe that for fixed $x, y \in U$, $x \neq y$, if $h$ is a random linear hashing function from $\{0,1\}^a$ to $\{0,1\}^b$, then

$$\Pr[h(x) = h(y)] = \prod_{j=1}^b \Pr[\text{the } j\text{th bit of } h(x) = \text{the } j\text{th bit of } h(y)] = 2^{-b}.$$

Therefore, for any fixed $x \in U$,

$$\Pr[\,(\forall i, 1 \leq i \leq m)\, x \notin X_i]$$
$$= \prod_{i=1}^m \Pr[(\exists y \in U - \{x\})\, h_i(x) = h_i(y)]$$
$$\leq \prod_{i=1}^m |U| \cdot \Pr[h_i(x) = h_i(y)] = \prod_{i=1}^m \frac{n}{2^b} \leq \prod_{i=1}^m \frac{1}{4} = 4^{-m}.$$

It follows that

$$\Pr\left[|H(U)| \geq \frac{n}{m}\right] \geq \Pr[(\forall x \in U)(\exists i \leq m)\, x \in X_i]$$
$$= 1 - \Pr[(\exists x \in U)(\forall i \leq m)\, x \notin X_i]$$
$$\geq 1 - \sum_{x \in U} \Pr[(\forall i \leq m)\, x \notin X_i]$$
$$\geq 1 - n \cdot 4^{-m} \geq 1 - 2^{-m}.$$

(Note that $m \geq b$ implies that $2^m \geq 2^b > n$.)

(b) If $|H(U)| \geq n/m$, then the probability of a random string $v \in \{0,1\}^b$ is in $H(U)$ is $\geq (n/m)/2^b \geq 2^{b-3}/(m2^b) = 1/(8m)$. Therefore,

$$\Pr[H(U) \cap V = \emptyset]$$

$$\leq \Pr\left[|H(U)| < \frac{n}{m}\right] + \Pr\left[|H(U)| \geq \frac{n}{m} \text{ and } H(U) \cap V = \emptyset\right]$$

$$\leq 2^{-m} + \prod_{i=1}^{m^2} \Pr\left[|H(U)| \geq \frac{n}{m} \text{ and } v_i \notin H(U)\right]$$

$$\leq 2^{-m} + \prod_{i=1}^{m^2}\left(1 - \frac{1}{8m}\right) = 2^{-m} + \left(1 - \frac{1}{8m}\right)^{m^2} < 2^{-m/8+1}.$$

(c) Note that $|H(U)| \leq nm$. Therefore, the probability of a random $v$ being in $H(U)$ is at most $nm/2^b \leq m/d$. It follows that $\Pr[H(U) \cap V \neq \emptyset] \leq \sum_{i=1}^{m^2} \Pr[v_i \in H(U)] \leq m^3/d$.                                              ∎

**Theorem 10.24** *For all $k \geq 2$, $IP_k = AM_2$.*

*Proof.* We are going to show that $IP_{2k} = AM_{2k+3}$ for all $k \geq 1$. The theorem then follows from Theorem 10.18. Let $L \in IP_{2k}$, with $k \geq 1$. Let $(M_v, f_p)$ be an interactive proof system for $L$ with error $\epsilon(n) > 0$, such that each query and each answer of $(M_v, f_p)$ is of length exactly $q(n)$ and $(M_v, f_p)$ uses exactly $r(n)$ random bits, where $q(n)$ and $r(n)$ are two polynomial functions. Without loss of generality, we assume that $r(n) \geq 32$ for all $n \geq 0$. From Proposition 10.7(c) (and Exercise 10.2), we may assume that $\epsilon(n) \leq 1/r(n)^9$. We let $A(s, x)$ be defined as in the proof of Theorem 10.21.

We first describe the idea of how the new Arthur-Merlin proof system simulates the interactive proof system $(M_v, f_p)$. In Stage 1 of the previous Arthur-Merlin proof system $A_1$, Merlin tried to convince Arthur that $|A(\lambda, x)| \geq (1 - \epsilon(n))2^{r(n)}$ by sending the values $b_{1,y} = |A(y, x)|$ to Arthur for all $y$ of length $q(n)$ and letting Arthur randomly select a string $y$ to test whether Merlin's values $b_{1,y}$ are correct. Since there are exponentially many such values $b_{1,y}$, we cannot afford to do this. Instead, in our new Arthur-Merlin proof system $A_2$, Merlin makes an assertion that for "many" strings $u$, $A(u, x)$ is "large." That is, Merlin sends parameters $c_1$ and $c_2$ to Arthur, claiming that there are about $2^{c_1-2}$ strings $u$ satisfying $|A(u, x)| \geq 2^{c_2-2}$. How does Arthur test this claim? Note that if Merlin's claim is true, then $U = \{u : |A(u, x)| \geq 2^{c_2-2}\}$ satisfies $c_1 = 2 + \lceil \log |U| \rceil$ and so, by Lemma 10.23(b), with high probability $H(U) \cap V \neq \emptyset$ for the set $H$ of random hashing functions and the random set $V \subseteq \{0,1\}^c$ chosen by Arthur. On the other hand, if Merlin was cheating, then $|U|$ is very small and, by Lemma 10.23(c), with high probability $H(U) \cap V = \emptyset$. Based on the above observations, Arthur tests Merlin's claim as follows: He selects $m = r(n)$ random linear hashing functions $h_1, \ldots, h_m$ and $m^2$ random strings $v_1, \ldots, v_{m^2}$ of length $c_1$ and sends

$H = \{h_1, \ldots, h_m\}$ and $V = \{v_1, \ldots, v_{m^2}\}$ to Merlin, asking Merlin to find a string $u \in U$ such that $H(u) \cap V \neq \emptyset$. If Merlin passes this test, then they enter the next stage, in which Merlin tries to convince Arthur that $|A(u, x)|$ is indeed of size about $2^{c_2-2}$, and Arthur applies the same test to this claim.

In the following, we give a formal proof for the case $k = 1$, that is, when $M_v$ makes only one query to $f_p$. The proof is easily extended to the general case when $k > 1$. We leave it as an exercise (Exercise 10.12).

> *Arthur-Merlin Proof System $A_2$ for $L$ ($k = 1$).*
> Input: $x$, $|x| = n$.
>     *Stage* 1. (1) Merlin sends an integer $c_1$ to Arthur. (Merlin claims that $|U| \geq 2^{c_1-2}$, where $U = \{u \in \{0,1\}^{q(n)} : |A(u, x)| \geq 2^{c_2-2}\}$ for some $c_2$ to be sent later.)
>     (2) Arthur selects a collection of random linear hashing functions $H_1 = \{h_1, \ldots, h_{r(n)}\}$ from $\{0,1\}^{q(n)}$ to $\{0,1\}^{c_1}$, and a set of random strings $V_1 = \{v_1, \ldots, v_{r(n)^2}\} \subseteq \{0,1\}^{c_1}$, and sends them to Merlin.
>     (3) Merlin sends two strings $y, z \in \{0,1\}^{q(n)}$ to Arthur.
>     (4) Arthur verifies that $y \in H_1^{-1}(V_1)$; he rejects if this does not hold.
>     *Stage* 2. (1) Merlin sends an integer $c_2$ to Arthur. (Merlin claims that $|A(yz, x)| \geq 2^{c_2-2}$.)
>     (2) Arthur selects random linear hashing functions $H_2 = \{g_1, \ldots, g_{r(n)}\}$ from $\{0,1\}^{r(n)}$ to $\{0,1\}^{c_2}$, and $r(n)^2$ random strings $V_2 = \{w_1, \ldots, w_{r(n)^2}\} \subseteq \{0,1\}^{c_2}$.
>     (3) Merlin sends a string $\alpha$ of length $r(n)$ to Arthur.
>     (4) Arthur checks that $\alpha \in H_2^{-1}(V_2)$. Arthur also checks whether $s = yz$ is an accepting history $h_{x,\alpha}$ with respect to string $\alpha$.
>     (5) Arthur accepts if $c_1 + c_2 \geq r(n) - \log(r(n))$; he rejects otherwise.

To see that the above Arthur-Merlin proof system $A_2$ works correctly, we first show that for all $x \in L$, the system $A_2$ accepts $x$ with probability at least $1 - 2^{-r(n)/8+2}$, which is $\geq 3/4$ by the assumption of $r(n) \geq 32$. To do this, we describe Merlin's algorithm for calculating $c_1$ and $c_2$:

> *Honest Merlin of System $A_2$.*
>     Stage 1, step (1). For any $j$, $1 \leq j \leq r(n)$, let
>
> $$B_j = \{u \in \{0,1\}^{q(n)} : 2^{j-1} < |A(u, x)| \leq 2^j\}.$$
>
> Let $j_1$ be the integer between 1 and $r(n)$ which maximizes $|\bigcup_{u \in B_j} A(u, x)|$. Merlin lets $c_1 = 2 + \lceil \log |B_{j_1}| \rceil$.
>     Stage 1, step (3). After he receives $H_1$ and $V_1$, Merlin finds a string $y \in H_1^{-1}(V_1) \cap B_{j_1}$, if such a string exists; otherwise, Merlin

sends "failure." Next, Merlin finds $z$ that maximizes $|A(yz, x)|$ and sends $y, z$ to Arthur.

Stage 2, step (1). Merlin lets $c_2 = 2 + \lceil \log |A(yz, x)| \rceil$.

Stage 2, step (3). Merlin finds a string $\alpha \in H_2^{-1}(V_2) \cap A(yz, x)$ if such a string exists; otherwise he sends "failure."

We note that since $c_1 = 2 + \lceil \log |B_{j_1}| \rceil$, the probability of $H_1(B_{j_1}) \cap V_1 \neq \emptyset$ is at least $1 - 2^{-r(n)/8+1}$. This means that the probability of Merlin failing at step (3) of Stage 1 is at most $2^{-r(n)/8+1}$. Similarly, the probability of Merlin failing at step (3) of Stage 2 is also bounded by $2^{-r(n)/8+1}$. Finally, we claim that it is always true that $c_1 + c_2 \geq r(n) - \log(r(n))$. It then follows that the probability of Arthur accepting the input is at least $1 - 2^{-r(n)/8+2}$.

To prove that $c_1 + c_2 \geq r(n) - \log(r(n))$, we first show that for any $u \in B_{j_1}$,

$$|A(u, x)| \geq \frac{|A(\lambda, x)|}{r(n)2^{c_1}} . \tag{10.7}$$

Notice that $|A(\lambda, x)| = \sum_{|u|=q(n)} |A(u, x)|$, and so

$$\left| \bigcup_{u \in B_{j_1}} A(u, x) \right| = \sum_{u \in B_{j_1}} |A(u, x)| \geq \frac{|A(\lambda, x)|}{r(n)} ,$$

since $j_1$ is chosen to maximize $|\bigcup_{u \in B_{j_1}} A(u, x)|$. Note that among all $u \in B_{j_1}$, $|A(u, x)|$ differ from each other by at most a factor of 2. Therefore, for any $u \in B_{j_1}$,

$$|A(u, x)| \geq \frac{|\bigcup_{u \in B_{j_1}} A(u, x)|}{2|B_{j_1}|} \geq \frac{|A(\lambda, x)|}{2|B_{j_1}|r(n)} .$$

Since $c_1 = 2 + \lceil \log |B_{j_1}| \rceil$, we have $2^{c_1} \geq 2|B_{j_1}|$ and (10.7) follows.

Now, we let $y = u$ and take logarithm of both sides of (10.7). We get

$$c_1 + \log(r(n)) + \log |A(y, x)| \geq \log |A(\lambda, x)|$$
$$\geq \log((1 - \epsilon)2^{r(n)}) \geq r(n) - 1.$$

Or, $c_1 \geq r(n) - 1 - \log(r(n)) - \log |A(y, x)|$. Since Merlin chose $z$ to maximize $|A(yz, x)|$, we know that $|A(yz, x)| = |A(y, x)|$ and $c_2 = 2 + \lceil \log |A(y, x)| \rceil$. So, we get $c_1 + c_2 \geq r(n) - \log(r(n))$. This completes the proof that system $A_2$ works correctly when $x \in L$.

Next, we show that if $x \notin L$, then, with probability at least 3/4, Arthur rejects $x$. Assume that Arthur does not reject by the end of step (4) of Stage 2. Let $y$ and $z$ be the strings sent by Merlin to Arthur at step (3) of Stage 1, and let $H_1, V_1, H_2, V_2$ be the sets sent by Arthur to Merlin at step (2) of Stages 1 and 2. Let $d$ be any positive integer. We claim

(a) $\Pr\left[ |A(yz, x)| \geq \frac{|A(\lambda, x)|}{2^{c_1}/d} \right] \leq \frac{r(n)^3}{d}$, and

(b) $\Pr\left[ |A(yz, x)| \leq \frac{2^{c_2}}{d} \text{ and } H_2^{-1}(V_2) \cap A(yz, x) \neq \emptyset \right] \leq \frac{r(n)^3}{d}$.

To see that (a) holds, we let $U_d$ be the set of all strings $u$ of length $q(n)$ such that $|A(uv_u, x)| \geq |A(\lambda, x)|/(2^{c_1}/d)$, where $v_u$ is the string of length $q(n)$ that maximizes $|A(uv_u, x)|$ (when $u$ and $x$ are fixed). We observe that $|U_d| \leq 2^{c_1}/d$ since $\sum_{|u|=q(n)} |A(uv_u, x)| = |A(\lambda, x)|$. Since Arthur does not reject at step (4) of Stage 1, we know that $y \in H_1^{-1}(V_1)$ and, hence, by Lemma 10.23(c),

$$
\begin{aligned}
\Pr\left[|A(yz, x)| \geq \frac{|A(\lambda, x)|}{2^{c_1}/d}\right] &= \Pr[y \in U_d \cap H_1^{-1}(V_1)] \\
&= \Pr[U_d \cap H_1^{-1}(V_1) \neq \emptyset] \leq \frac{r(n)^3}{d} .
\end{aligned}
$$

Claim (b) follows similarly from Lemma 10.23(c), noticing that if Arthur does not reject at step (4) of Stage 2, it must be true that $A(yz, x) \cap H_2^{-1}(V_2) \neq \emptyset$.

Now, let $d = 8(r(n))^3$. Then, we see that the probability is at least $3/4$ such that

$$
|A(yz, x)| < \frac{|A(\lambda, x)|}{2^{c_1}/d} \text{ and } |A(yz, x)| > \frac{2^{c_2}}{d} .
$$

It implies that $2^{c_1+c_2}/d^2 < |A(\lambda, x)| \leq \epsilon(n)2^{r(n)}$. Recall that $\epsilon(n) \leq r(n)^{-9}$, and so $2^{c_1+c_2} < 64\epsilon(n)r(n)^6 2^{r(n)} \leq 64 \cdot 2^{r(n)}/r(n)^3$. Taking logarithm on both sides, we get $c_1 + c_2 < r(n) + 6 - 3\log(r(n)) < r(n) - \log(r(n))$ (note that $r(n) \geq 32$ and, hence, $2\log(r(n)) > 6$).

We have just proved that if Arthur does not reject before step (5) of Stage 2, he would reject at step (5) of Stage 2 with probability at least $3/4$. This completes the proof of the correctness of system $A_2$.

Finally, we observe that the proof system $A_2$ lasts totally five rounds on any input $x$. (In general, for $k \geq 2$, the system $A_2$ needs $2k + 3$ rounds; see Exercise 10.12.) This shows that $L \in MA_5$ and, hence, in $AM_2$.   ∎

In Chapter 4, we saw some examples of problems that are known to be in *NP* but are not known to be in *P* or to be *NP*-complete. In Examples 10.1 and 10.2, we showed that the complements of two of these problems, GIso and QR, have two-round interactive proof systems. From the above result, we know that the complements of GIso and QR are actually in $AM_2$ and, hence, by Theorem 10.20, GIso and QR cannot be *NP*-complete unless the polynomial-time hierarchy collapses to the second level. Thus, the study of interactive proof systems yields some surprising results regarding the *NP*-completeness of some natural problems.

**Corollary 10.25** *The graph isomorphism problem* GIso *and the quadratic residuosity problem* QR *are not NP-complete, unless the polynomial-time hierarchy collapses to* $AM_2$.

*Proof.* It is clear that $AM_2$ is closed under the $\leq_m^P$-reducibility. Therefore, if GIso or QR is *NP*-complete, then $NP \subseteq coAM_2$.   ∎

## 10.5   *IP* Versus *PSPACE*

In the past few sections, we have established that the computational power of the interactive proof systems of a constant number of rounds is quite limited and is just a level beyond *NP*. A critical step of this collapsing result is Theorem 10.14, where two complementary $AM_k(3/4)$-predicates are converted to two complementary $AM_k(1 - 2^{-n})$-predicates. A careful analysis shows that this result cannot be extended to complementary $AM_{k(n)}(3/4)$-predicates if $k(n)$ is not bounded by a constant (cf. Exercise 10.6). Therefore, the collapsing result stops at the constant-round *AM* hierarchy. Indeed, we show in the following that *IP* is actually equal to *PSPACE*, and so any result collapsing *IP* to the polynomial-time hierarchy would be a major breakthrough.

**Theorem 10.26** *IP = PSPACE*.

*Proof.* First we show that $IP \subseteq PSPACE$. By the Merlin machine characterization of the class *AM* (Proposition 10.11), we only need to show a polynomial space-bounded simulation of polynomial time-bound, probabilistic, nondeterministic machine $M$.

On an input $x$ of length $n$, the computation of $M$ on $x$ is a tree of depth $p(n)$ for some polynomial $p$. Each node of the tree is a configuration $c$ and has a value $acc(c)$. We can simply perform a depth-first traversal of this tree, computing $acc(c)$ of each node $c$ from the values of their child nodes. (It is the maximum of the values of the child nodes if it is a nondeterministic node, and it is the average of the values of the child nodes if it is a random node.) It is clear that the simulation uses only polynomial space (cf. Theorem 3.17).

Next, we show that each $L \in PSPACE$ has an Arthur-Merlin proof system. Assume that $L$ is accepted by a deterministic TM $M = (Q, q_0, F, \Sigma, \Gamma, \delta)$ in space $q(n)$ for some polynomial $q$. As proved in Theorem 1.28, we may assume that $M$ always halts on an input of length $n$ in $2^{q(n)}$ moves. That is, for any input $x$ of length $n$, the computation of $M$ on $x$ is a sequence of $2^{q(n)}$ configurations, each of length $q(n)$. The following setup is similar to that in Savitch's Theorem (Theorem 1.30). We let $reach(\alpha, \beta, k)$ be the predicate stating that the configuration $\beta$ of $M$ can be reached from the configuration $\alpha$ in exactly $k$ moves. We assume that there is a unique halting configuration $\alpha_f$ of $M$. Thus, the question of whether $x \in L$ is reduced to the predicate $reach(\alpha_0, \alpha_f, 2^{q(n)})$, where $\alpha_0$ is the initial configuration of the computation of $M$ on $x$.

The main idea of the Arthur-Merlin proof system for $L$ is to arithmetize the predicate $reach(\alpha, \beta, k)$ into an integer polynomial over $O(q(n))$ variables. Then, Merlin tries to convince Arthur that the polynomial $reach(\alpha, \beta, 2^{q(n)})$ evaluates to 1 when $\alpha$ is set to $\alpha_0$ and $\beta$ is set to $\alpha_f$. At each stage, Arthur and Merlin work together to reduce the problem of whether a polynomial $reach(\alpha, \beta, k)$ evaluates to 1 to the problem of whether some simpler polynomials $reach(\alpha', \beta', k/2)$ evaluate to 1. The reduction is similar to the reduction in the interactive proof system for PERM of Example 10.3.

We first describe the arithmetization of the predicate $reach(\alpha, \beta, k)$. We assume that all configurations $\alpha$ are strings over alphabet $\Gamma \cup (Q \times \Gamma)$ of length exactly $q(n)$. We let $\alpha_i$ denote the $i$th symbol of $\alpha$. In the proof of Cook's Theorem, we have constructed a Boolean formula $\phi_4$ that encodes the predicate $reach(\alpha, \beta, 1)$. Let us examine this formula more carefully. Suppose $\alpha \vdash_M \beta$ via an instruction $\delta(s, a) = (r, b, R)$, where $s, r \in Q$ and $a, b \in \Gamma$. Then, there must be a position $i$, $1 \leq i \leq q(n) - 1$, such that $\alpha_i = \langle s, a \rangle$, $\beta_i = b$, $\beta_{i+1} = \langle r, \alpha_{i+1} \rangle$, and $\alpha_j = \beta_j$ for all $j \leq i - 1$ and all $j \geq i + 2$. Suppose $\alpha \vdash_M \beta$ via an instruction $\delta(s, a) = (r, b, L)$. Then, there must be a position $i$, $1 < i \leq q(n)$, such that $\alpha_i = \langle s, a \rangle$, $\beta_i = b$, $\beta_{i-1} = \langle r, \alpha_{i-1} \rangle$, and $\alpha_j = \beta_j$ for all $j \leq i - 2$ and all $j \geq i + 1$. Therefore, we may encode $reach(\alpha, \beta, 1)$ as a formula of the following form:

$$\phi_1(\alpha, \beta) = \sum_{(s,a) \in Q \times \Gamma} \sum_{i=1}^{q(n)} \sum_{d \in \Sigma} (\alpha_i = \langle s, a \rangle)(\alpha_{i'} = d)$$
$$(\beta_i = b)(\beta_{i'} = \langle r, d \rangle) \prod_{j \neq i, i'} (\alpha_j = \beta_j), \tag{10.8}$$

where $i'$ (equal to either $i + 1$ or $i - 1$), $b$, and $r$ are to be determined from the instruction $\delta(s, a)$. Notice that in the right-hand side of (10.8), there is at most one nonzero term, since $M$ is deterministic.

Now we use Boolean variables to encode tape symbols. Namely, let $\alpha_{i,a}$ be the Boolean variable indicating $\alpha_i = a$. Then, each equation of the form $\alpha_j = \beta_j$ in formula (10.8) becomes the product of $k$ equations $\alpha_{j,a} = \beta_{j,a}$ for some constants $k$, and each equation of the form $\alpha_i = a$ becomes equations $\alpha_{i,a} = 1$ and $\alpha_{i,b} = 0$ for all $b \in \Gamma \cup (Q \times \Gamma) - \{a\}$. Thus, we obtain a Boolean formula $\psi_1(\alpha, \beta)$ over Boolean variables of the form $\alpha_{i,a}$ and $\beta_{i,a}$ such that it evaluates to 1 with respect to the intended values of $\alpha_{i,a}$ and $\beta_{i,a}$ if and only if $reach(\alpha, \beta, 1)$. In addition, $\psi_1(\alpha, \beta)$ is a sum of the products of the form $(x = c)$ or $(x = y)$, where $x$, $y$ are Boolean variables and $c$ is a Boolean constant. Replacing $u = v$ by $2uv - u - v + 1$, we obtain an arithmetic formula $f_1(\alpha, \beta)$ with the following properties:

(i) $reach(\alpha, \beta, 1)$ if and only if $f_1(\alpha, \beta) = 1$, when variables $\alpha_{i,a}$ take the value 1 if $\alpha_i = a$, and 0 if $\alpha_i \neq a$, and variables $\beta_{i,a}$ take the value 1 if $\beta_i = a$, and 0 if $\beta_i \neq a$.

(ii) $f_1(\alpha, \beta)$ is multi-linear; that is, $f_1$ has degree one in each variable.

The property (i) above follows from the observation that formula $\phi_1$ (and, hence, $\psi_1$) contains at most one nonzero term. It implies that even if we use integer arithmetic instead of Boolean arithmetic, the formula $f_1$ always evaluates to either 0 or 1 on inputs of values 0 or 1. (It may assume other values if inputs to variables are not limited to 0 or 1.)

Next, for each $k > 1$, consider the predicate $reach(\alpha, \beta, k)$. It is clear that

$$reach(\alpha, \beta, k) = \sum_{\gamma} reach(\alpha, \gamma, \lceil k/2 \rceil) \cdot reach(\gamma, \beta, \lfloor k/2 \rfloor).$$

Therefore, we just inductively define, for $2 \leq k \leq 2^{q(n)}$,

$$f_k(\alpha, \beta) = \sum_{\gamma_{q(n),a}=0}^{1} \cdots \sum_{\gamma_{1,a}=0}^{1} f_{\lceil k/2 \rceil}(\alpha, \gamma) f_{\lfloor k/2 \rfloor}(\gamma, \beta), \qquad (10.9)$$

where the summations are taken over all possible variables $\gamma_{i,a}$, for all $1 \leq i \leq q(n)$ and all $a \in \Gamma \cup (Q \times \Gamma)$. Then, we claim that $f_k(\alpha, \beta)$ is equivalent to $reach(\alpha, \beta, k)$ in the sense of property (i) above.

Let us call the assignments of values to variables $\alpha_{i,a}$ *illegal* if they do not correspond to a configuration $\alpha$ (e.g., if there exist $a, b \in \Gamma$, $a \neq b$, such that $\alpha_{i,a} = \alpha_{i,b} = 1$ for some $1 \leq i \leq q(n)$). First, we can prove, by a simple induction, that if the assignments to $\alpha_{i,a}$'s are legal and the assignments to $\beta_{i,a}$'s are illegal, then $f_k(\alpha, \beta) = 0$. Indeed, for $k = 1$, this follows from the formula $\psi_1(\alpha, \beta)$. For $k > 1$, $f_k(\alpha, \beta) \neq 0$ implies for some $\gamma$, $f_{\lceil k/2 \rceil}(\alpha, \gamma) \neq 0$ and $f_{\lfloor k/2 \rfloor}(\gamma, \beta) \neq 0$. So, by the inductive hypothesis, if assignments to $\alpha_{i,a}$'s are legal, then the assignments to $\gamma_{i,a}$'s and to $\beta_{i,a}$'s must also be legal.

Next, we check that for any configurations $\alpha$, $\beta$, there is at most one configuration $\gamma$ such that $reach(\alpha, \gamma, \lceil k/2 \rceil)$ and $reach(\gamma, \beta, \lfloor k/2 \rfloor)$ both hold. Therefore, if all variables $\alpha_{i,a}$ and $\beta_{i,a}$ have legal values, there is at most one nonzero term on the right-hand side of (10.9), corresponding to the legal values of the unique configuration $\gamma$ and, hence, $f_k(\alpha, \beta)$ has value either 0 or 1. By a simple induction, it is easy to see that $f_k(\alpha, \beta) = 1$ if and only if $reach(\alpha, \beta, k)$. This completes the proof of the claim.

We also observe that for each $k \geq 1$, $f_k$ remains a multi-linear polynomial. (It may appear that some variables $\gamma_{i,a}$ may occur twice in the product $f_{\lceil k/2 \rceil}(\alpha, \gamma) f_{\lfloor k/2 \rfloor}(\gamma, \beta)$ and, hence, the degree of such variables may double. However, we note that $\gamma_{i,a}$ is not really a free variable and, hence, has nothing to do with the degree of the function $f_k$.)

The above completes the arithmetization of the predicate $reach(\alpha, \beta, k)$. Following this arithmetization, we see that Merlin needs to prove that the formula $f_{2^{q(n)}}(\alpha_0, \alpha_f)$ evaluates to 1. Before we present such a proof system, we first introduce some more notations.

Let $q'(n) = q(n) \cdot |\Gamma \cup (Q \times \Gamma)|$. We note that each function $f_k$ has $2q'(n)$ variables. In general, we write $\alpha$ or $\beta$ to denote a vector of $q'(n)$ integers (not necessarily 0 or 1), and write $f_k(\alpha, \beta)$ to denote the function value of $f_k$ with $q'(n)$ integers from vector $\alpha$ assigned to variables $\alpha_{i,a}$ and $q'(n)$ integers from vector $\beta$ assigned to variables $\beta_{i,a}$. If $d_1, \ldots, d_i$, $1 \leq i \leq q'(n)$, are integer constants and if $k > 1$, then we write $f_{k,d_1,\ldots,d_i}(\alpha, \beta)$ to denote the function value

$$\sum_{e_{i+1}=0}^{1} \cdots \sum_{e_{q'(n)}=0}^{1} f_k(\alpha, d_1, \ldots, d_i, e_{i+1}, \ldots, e_{q'(n)})$$
$$\cdot f_k(d_1, \ldots, d_i, e_{i+1}, \ldots, e_{q'(n)}, \beta).$$

(In the above, $\alpha$ and $\beta$ are two vectors of $q'(n)$ integers each and each $d_j$ is a single integer and each $e_j$ is a single variable.) In addition, for constants

$d_1, \ldots, d_{i-1}$ and variable $x$, $f_{k,d_1,\ldots,d_{i-1},x}(\alpha, \beta)$ denotes the degree-2, univariate polynomial

$$\sum_{e_{i+1}=0}^{1} \cdots \sum_{e_{q'(n)}=0}^{1} f_k(\alpha, d_1, \ldots, d_{i-1}, x, e_{i+1}, \ldots, e_{q'(n)})$$
$$\cdot f_k(d_1, \ldots, d_{i-1}, x, e_{i+1}, \ldots, e_{q'(n)}, \beta).$$

Our Arthur-Merlin proof system uses the following relations to reduce the evaluation of the polynomial $f_{k,d_1,\ldots,d_{i-1}}$ to the evaluation of polynomials $f_{k,d_1,\ldots,d_i}$: For $1 \leq s \leq q(n)$ and $1 \leq i \leq q'(n)$,

$$f_{2^s}(\alpha, \beta) = \sum_{x=0}^{1} f_{2^{s-1},x}(\alpha, \beta),$$

and

$$f_{2^s,d_1,\ldots,d_{i-1}}(\alpha, \beta) = \sum_{x=0}^{1} f_{2^s,d_1,\ldots,d_{i-1},x}(\alpha, \beta).$$

*Arthur-Merlin Sum Check Proof System for L.*
Input: $w$, $|w| = n$.

*Stage* 0. Merlin finds a prime number $p$ in the range $n \cdot q(n)^2 \leq p \leq 2^n \cdot q(n)^2$. Merlin sends to Arthur the prime $p$ together with a proof that $p$ is indeed a prime (see Theorem 4.5). In the rest of the proof system, the arithmetic is done in the field $\mathbf{Z}_p$. Let $\alpha = \alpha_0$, $\beta = \alpha_f$, and $c_{1,0} = 1$. (Merlin needs to convince Arthur that $f_{2^{q(n)}}(\alpha, \beta) = c_{1,0}$.)

*Stage* $k$, $1 \leq k \leq q(n)$. Let $s = q(n) + 1 - k$. Merlin tries to convince Arthur that $f_{2^s}(\alpha, \beta) = c_{k,0}$.

*Substage* $\langle k, i \rangle$, $1 \leq i \leq q'(n)$. Before this substage, Arthur has already chosen random numbers $d_1, \ldots, d_{i-1}$ from $\mathbf{Z}_p$, and a constant $c_{k,i-1}$ has been defined. At this substage, Merlin tries to convince Arthur that $c_{k,i-1} = f_{2^{s-1},d_1,\ldots,d_{i-1}}(\alpha, \beta)$ (except when $i = 1$, in which case Merlin needs to prove $c_{k,0} = f_{2^s}(\alpha, \beta)$). Merlin sends a degree-2, univariate polynomial $g_{k,i}(x)$ to Arthur, claiming that $g_{k,i}(x) = f_{2^{s-1},d_1,\ldots,d_{i-1},x}(\alpha, \beta)$. Arthur checks that $g_{k,i}(0) + g_{k,i}(1) = c_{k,i-1}$ (he rejects if this does not hold). Then, Arthur picks a random $d_i \in \mathbf{Z}_p$ and sends it to Merlin. Let $c_{k,i} = g_{k,i}(d_i)$.

*Substage* $\langle k, q'(n) + 1 \rangle$. Merlin needs to convince Arthur that

$$c_{k,q'(n)} = f_{2^{s-1}}(\alpha, d_1, \ldots, d_{q'(n)}) \cdot f_{2^{s-1}}(d_1, \ldots, d_{q'(n)}, \beta).$$

First, Merlin sends to Arthur a univariate polynomial $h_k(x)$ of degree at most $2q'(n)$, claiming that

$$h_k(x) = f_{2^{s-1}}((\delta - \alpha)x + \alpha, (\beta - \delta)x + \delta), \qquad (10.10)$$

where $\delta = (d_1, d_2, \ldots, d_{q'(n)})$. (We treat $\alpha, \beta$ and $\delta$ as size-$q'(n)$ vectors over $\mathbf{Z}_p$, and $x$ as a scaler.) Arthur verifies that $h_k(0) \cdot h_k(1) = c_{k,q'(n)}$. Then, Arthur picks a random $e \in \mathbf{Z}_p$ and sends it to Merlin. Let $c_{k+1,0} = h_k(e)$ and reset $\alpha := (\delta - \alpha)e + \alpha$ and $\beta := (\beta - \delta)e + \delta$. Go to Stage $k + 1$.

*Stage $q(n)+1$.* Arthur finds the polynomial $f_1(\alpha, \beta)$ and verifies that $f_1(\alpha, \beta) = c_{q(n)+1,0}$. He accepts if and only if this equation holds.

To see that this proof system works correctly, we first assume that $w \in L$. Then, Merlin can simply be honest. That is, at each substage $\langle k, i \rangle$, $1 \le k \le q(n)$ and $1 \le i \le q'(n)$, Merlin sends the correct polynomial $g_{k,i}(x) = f_{2^s-1, d_1, \ldots, d_{i-1}, x}(\alpha, \beta)$ to Arthur, and it will pass Arthur's test. At substage $\langle k, q'(n) + 1 \rangle$, Merlin also sends the correct polynomial $h_k(x)$ to Arthur. Note that if $h_k$ satisfies (10.10), then $h_k(0) = f_{2^s-1}(\alpha, \delta)$, and $h_k(1) = f_{2^s-1}(\delta, \beta)$. Thus, it also passes Arthur's test at this substage. Furthermore, at the end of each stage $k$, it remains true that $c_{k+1,0} = f_{2^s}(\alpha, \beta)$. In particular, at stage $q(n) + 1$, Arthur accepts. In other words, if $w \in L$, then Arthur accepts $w$ with probability one.

Conversely, if $w \notin L$, we show that the probability of Arthur accepting $w$ is at most $1/4$. We first make two critical observations.

*Claim 1.* If at the start of substage $\langle k, i \rangle$, for any $1 \le k \le q(n)$ and $1 \le i \le q'(n)$, $c_{k,i-1} \ne f_{2^s-1, d_1, \ldots, d_{i-1}}(\alpha, \beta)$, and if Arthur does not reject in substage $\langle k, i \rangle$, then by the end of substage $\langle k, i \rangle$, we have, with probability $1 - 2/p$, that $c_{k,i} \ne f_{2^s-1, d_1, \ldots, d_i}(\alpha, \beta)$.

*Proof of Claim 1.* If Merlin gives Arthur the correct polynomial $f_{2^s-1, d_1, \ldots, d_{i-1}, x}(\alpha, \beta)$ as $g_{k,i}(x)$, then it cannot pass Arthur's test, because

$$\sum_{x=0}^{1} f_{2^s-1, d_1, \ldots, d_{i-1}, x}(\alpha, \beta) = f_{2^s-1, d_1, \ldots, d_{i-1}}(\alpha, \beta) \ne c_{k,i-1}.$$

Therefore, Merlin must have given Arthur a polynomial $g_{k,i}(x)$ that is different from the polynomial $f_{2^s-1, d_1, \ldots, d_{i-1}, x}(\alpha, \beta)$. Since both of these polynomials are polynomials of degree two, they agree at most at two points. That is, the probability that they agree at a random point $d_i \in \mathbf{Z}_p$ is $\le 2/p$. So, with probability $\ge 1 - 2/p$, $c_{k,i} = g_{k,i}(d_i)$ is not equal to $f_{2^s-1, d_1, \ldots, d_i}(\alpha, \beta)$.

*Claim 2.* If at the start of substage $\langle k, q'(n) + 1 \rangle$, for any $1 \le k \le q(n)$, $c_{k,q'(n)} \ne f_{2^s-1}(\alpha, \delta) f_{2^s-1}(\delta, \beta)$, and if Arthur does not reject in substage $\langle k, q'(n) + 1 \rangle$, then by the end of substage $\langle k, q'(n) + 1 \rangle$, we have, with probability $\ge 1 - 2q'(n)/p$, that $c_{k+1,0} \ne f_{2^s-1}(\alpha, \beta)$.

*Proof of Claim 2.* It is similar to Claim 1, except here the polynomial $h_k(x)$ has degree at most $2q'(n)$ and so the probability of $h_k(e) = f_{2^s-1}((\delta - \alpha)e + \alpha, (\beta - \delta)e + \delta)$ is at most $2q'(n)/p$.

Now, if this proof system is given an input $w \notin L$, then it begins with $c_{1,0} \ne f_{2^q(n)}(\alpha_0, \alpha_f)$. Suppose Arthur does not reject in the first $q(n)$ stages;

Then, by the above claims, we have, by Stage $q(n) + 1$, probability at least $1 - 4q(n)q'(n)/p \geq 3/4$ (for sufficiently large $n$) that $c_{q(n)+1,0} \neq f_1(\alpha, \beta)$, and Arthur will reject it at Stage $q(n) + 1$. This completes the proof of the correctness of the Arthur-Merlin proof system for $L$.

Finally, we note that in each substage, Arthur's task is only to evaluate a polynomial of degree at most $2q'(n)$ in the field $\mathbf{Z}_p$, with $|p| = O(n \log n)$. This can be done in polynomial time. Furthermore, there are only $(q'(n) + 1)q(n)$ substages. Therefore, the whole proof system operates in polynomial time. ∎

We remark that the above proof also establishes Theorem 10.22. In particular, the above sum check proof system has only a one-sided error.

In Chapter 4, we mentioned that most results that separate complexity classes by the diagonalization techniques or collapsing complexity classes by the simulation techniques are relativizable; that is, these results hold relative to all oracles (if the relativized complexity classes can be defined in a natural way). It is therefore interesting to point out here that the above result of $IP = PSPACE$ is not relativizable. (Exercise 10.13 constructs an oracle $A$ such that $coNP^A \not\subseteq IP^A$.) Thus, this algebraic proof technique may potentially lead to more separating or collapsing results that are not provable by the standard diagonalization or the simulation techniques. We will see more results using the algebraic proof techniques in Chapter 11.

## Exercises

**10.1** In this exercise, we present a balanced view of the interactive proof systems (in contrast to the verifier's view of the interactive proof systems defined in Definition 10.5). We define an *interactive protocol* as a pair of machines $(M_v, M_p)$, where $M_v$ is a polynomial-time PTM, $M_p$ is a DTM, and they share a common read-only input tape and a common communication tape. The two machines take turns in being active. During its active stage, the machine (except for the first stage) first reads the string in the communication tape, performs some local computation on its local tapes, and then writes a string on the communication tape (or, if the machine is $M_v$, it may choose to halt the computation and accept or reject). We say a set $L \subseteq \Sigma^*$ has an *interactive proof system* with error bound $\epsilon$ if there exists a polynomial-time PTM $M_v$ such that the following hold:

(i) There exists a DTM $M_p$ such that $(M_v, M_p)$ is an interactive protocol and for each $x \in L$, the probability of $M_v$ and $M_p$ together accepting $x$ is at least $1 - \epsilon$; and

(ii) For any DTM $M$ such that $(M_v, M)$ forms an interactive protocol and for any $x \notin L$. the probability of $M_v$ and $M$ together accepting $x$ is at most $\epsilon$.

Show that the above definition of interactive proof systems is equivalent to Definition 10.5.

**10.2** Assume that $A$ has a $k(n)$-round interactive proof system $(M_v, f_p)$ with the error probability $\epsilon \leq 1/4$. Further assume that in the system $(M_v, f_p)$, each message exchanged between the prover and the verifier is of length at most $q(n)$ and the machine $M_v$ uses at most $r(n)$ random bits, if the input is of length $n$. Prove that for any polynomial $p$, there is a $k(n)$-round interactive proof system for $A$ with the error probability $\epsilon' \leq 2^{-p(n)}$, in which the message length is bounded by $O(p(n)q(n))$ and the number of random bits used is bounded by $O(p(n)r(n))$. In particular, for any $m > 0$, there is a $k(n)$-round interactive proof system $(M_v', f_p')$ for $A$ with the error probability $\leq 1/r'(n)^m$, where $r'(n)$ is the number of random bits used by $M_v'$.

**10.3** It is shown through Theorem 10.24 that the problem $\overline{\text{GIso}}$ is in $AM_2$. Give a direct proof for this fact by demonstrating an Arthur-Merlin proof system for $\overline{\text{GIso}}$. [*Hint*: Let $G = (V, E)$ be a graph, and $\pi : V \to V$ be a permutation of the vertices of $G$. We write $G|_\pi$ to denote the graph $H = (V, E_\pi)$, where $\{u, v\} \in E_\pi$ if and only if $\{\pi^{-1}(u), \pi^{-1}(v)\} \in E$. Let $\text{Aut}(G)$ be the set of all automorphisms of $G$, that is, the set of all permutations $\pi$ over $V$ such that $G|_\pi$ is identical to $G$. We observe that in the interactive proof system of Example 10.1, if the inputs $G_1$ and $G_2$ are isomorphic, then there are only $n!/|\text{Aut}(G_1)|$ different messages that could be sent by the verifier to the prover in round 1 (called *legal messages*). Otherwise, if $G_1$ and $G_2$ are not isomorphic, then there are $n!/|\text{Aut}(G_1)| + n!/|\text{Aut}(G_2)|$ different legal messages. Design an Arthur-Merlin proof system for $\overline{\text{GIso}}$ in which Merlin tries to convince Arthur that there are "many" legal messages for the given inputs $G_1$ and $G_2$.]

**10.4** Design an interactive proof system for #SAT.

**10.5** There is a natural extension of Merlin machines to oracle Merlin machines in which Merlin can make queries to an oracle $X$. Define the notion of relativized $AM_k^X$ based on oracle Merlin machines.

(a) Show that the class $AM_k^X$ has an alternating quantifier characterization like that of Theorem 10.15.

(b) Prove that $AM_2^{AM_2 \cap coAM_2} = AM_2$.

**10.6** It is straightforward to extend the notion of complementary $AM_k(c)$-predicates to complementary $AM_{k(n)}(c)$-predicates, where $k(n)$ is a function of the input $n$.

(a) Can you prove an alternating quantifier characterization for $AM_{k(n)}$ by the notion of complementary $AM_{k(n)}(c)$-predicates? If so, what is the best bound for the probability $c$ you can prove?

(b) Assume that $R_1$ and $R_0$ are two complementary $AM_n(c_1)$-predicates, with $c_1 = 1 - 2^{-n}$ for some polynomial $p$. Show that $R_1$ and $R_0$ are two complementary $AM_{n-1}(3/4)$-predicates. Can you generalize this result to show that $R_1$ and $R_0$ are two complementary $AM_{n/2}(3/4)$-predicates (if necessary, assuming a smaller $c_1$ to begin with)?

**10.7** Show that $AM_2 \subseteq NP/poly$.

**10.8** Recall that *almost-P* denotes the class of languages $A$ such that $\mathrm{Pr}_X[A \in P^X] = 1$ (see Exercise 8.16). Extend this notion to *almost-NP* $= \{A : \mathrm{Pr}_X[A \in NP^X] = 1\}$. Show that *almost-NP* $= AM_2$.

**10.9** Prove that $NP^{BPP} \subseteq MA_2 \subseteq ZPP^{NP}$. Are these three classes equal?

**10.10** Prove that if $NP = R$ then $AM_2 = BPP$.

**10.11** Prove that for each of the following statements, there is an oracle $X$ relative to which the statement is false.
(a) $MA_2 \subseteq coAM_2$.
(b) $AM_2 \subseteq ZPP^{NP}$.
(c) $coMA_2 \subseteq AM_2$.

**10.12** Extend the Arthur-Merlin proof system $A_2$ of Theorem 10.24 to work for every $k \geq 1$. In particular, for any $L \in IP_k$, find a $(2k+3)$-round Arthur-Merlin proof system for $L$.

**10.13** Prove that there exists an oracle $A$ such that $coNP^A \not\subseteq AM^A$.

**10.14** Prove that if $f(n)$ and $g(n)$ are two functions bounded by some polynomial functions such that $g(n) = o(f(n))$, then there exists an oracle $A$ such that $AM^A_{f(n)} \not\subseteq \Sigma^{P,A}_{g(n)}$, where $\Sigma^P_{g(n)}$ is the class of sets in *PSPACE* which have the form (3.7) in Corollary 3.20, with $g(n)$ levels of quantifiers, beginning with $\exists$, and $\Sigma^{P,A}_{g(n)}$ is this class relative to set $A$.

The following three problems are about *zero-knowledge proof systems*. Intuitively, a proof session of an interactive proof system provides *zero knowledge* if the prover is able to convince the verifier that an input $x$ is in the set $L$ and yet the verifier learns no extra knowledge other than the fact that $x \in L$. For instance, consider the graph isomorphism problem GIso. For two given graphs $G_1 = (V_1, E_1)$ and $G_2 = (V_2, E_2)$, a simple proof system is to let the prover send a mapping $\pi : V_1 \rightarrow V_2$ to the verifier and let the verifier check that $\pi$ is an isomorphism between the two graphs. This proof system, however, reveals exactly what the isomorphism is between $G_1$ and $G_2$. Alternatively, the following proof system for GIso does not reveal this information and yet can convince the verifier that $G_1 \cong G_2$: The prover first sends a graph $G_3 = (V_3, E_3)$ that is isomorphic to $G_1$ (and $G_2$) to the verifier. Then, the verifier picks a random number $i \in \{1, 2\}$ and sends it to the prover. The prover sends a function $\pi : V_i \rightarrow V_3$ that is an isomorphism between $G_i$ and $G_3$. Note that although the verifier learns the isomorphism function between $G_i$ and $G_3$, he/she is not able to find the isomorphism between $G_1$ and $G_2$ from this *side* knowledge.

We are going to define the notion of zero knowledge proof systems based on the balanced view of interactive proof systems defined in Exercise 10.1.

Recall that on an input $x$ and for a sequence $\alpha$ of random bits, a $k$-round interactive proof system $(M_v, M_p)$ has a unique *history* $h_{x,\alpha} = y_1 z_1 \ldots y_k z_k$ (see the proof of Theorem 10.21). Here, we modify it and attach $\alpha$ itself to the history and define $h_{x,\alpha} = y_1 z_1 \ldots y_k z_k \alpha$. We notice that for a fixed prover machine $M_p$, an interactive protocol $(M_v, M_p)$ defines, for each input $x$, a probability distribution $D_{M_v,x}$ on the histories $h_{x,\alpha}$. We say a PTM $M$ on input $x$ generates a probability distribution $D'_x$ if for each string $w$, the probability of $M(x)$ prints $w$ is $D'_x(w)$. Formally, we define that an interactive proof system $(M_v, M_p)$ for set $L$ is a *perfect zero-knowledge proof system for* $M_v$, if there exists a polynomial-time PTM $M$ that on each input $x$ generates a probabilistic distribution $D'_x$ that is identical to $D_{M_v,x}$. We say $(M_v, M_p)$ is a *perfect zero-knowledge proof system*, if for every verifier machine $M_{v'}$ that forms an interactive protocol with $M_p$, there exists a polynomial-time PTM $M$ that on each input $x$ generates a probabilistic distribution $D'_x$ that is identical to $D_{M_{v'},x}$.

**10.15** Prove that the above second interactive protocol for GIso is a perfect zero-knowledge proof system.

**10.16** The interactive protocol for $\overline{\text{GIso}}$ of Example 10.1 is *not* perfect zero-knowledge, because a verifier $M_{v'}$ may send a graph $G'$ to the prover that is not randomly generated from $G_1$ or $G_2$ and gain information of whether $G'$ is isomorphic to $G_1$ or $G_2$. Modify this interactive protocol to a perfect zero-knowledge proof system for $\overline{\text{GIso}}$.

**10.17** Prove that if a set $L$ has a perfect zero-knowledge proof system, then $L \in AM \cap coAM$.

> *If you are going to do something wrong,*
> *at least enjoy it.*
> — Leo Rosten

## Historical Notes

Interactive proof systems were first introduced by Goldwasser, Micali, and Rackoff [1989]. Their formulation is close to the one presented in Exercise 10.1. Our equivalent formulation is based on the work of Fortnow et al. [1994]. Arthur-Merlin proof systems were introduced by Babai [1985] and Babai and Moran [1988] from a different motivation. A number of survey papers have been written about the properties and applications of these proof systems, including Johnson [1988, 1992]. Example 10.1 is from Goldreich et al. [1991], and Example 10.3 is from Lund et al. [1992]. The robustness of the notion of complementary $AM_k$-predicates and the collapsing of the $AM$ hierarchy are contained in Babai and Moran [1988]. Theorem 10.20 was first proved by Boppana et al. [1987]. The equivalence between $IP_k$ and $AM_k$ was first proved by Goldwasser and Sipser [1989]. The simpler proof of the weaker form of the equivalence presented in Theorem 10.21 is due to Killian (see Goldwasser [1988]). The notion of universal hashing functions as that used in Lemma 10.23 was first defined by Carter and Wegman [1979]. The interactive proof system for the problem $\overline{\text{GIso}}$ was first presented by Goldreich et al. [1991]. Schöning [1987] contains a direct construction of an Arthur-Merlin proof system for the graph nonisomorphism problem (Exercise 10.3). The equivalence of *IP* and *PSPACE* was first proved by Shamir [1990], based on the idea of Lund et al. [1992]. Both proofs worked on the *PSPACE*-complete problem QBF. Our proof of Theorem 10.26 was based on the unpublished idea of Hartmanis [1991]. The characterization of *AM* as *almost-NP* (Exercise 10.8) is from Nisan and Wigderson [1994]. Exercise 10.14 is from Fortnow and Sipser [1988] and Aiello et al. [1990]. Exercise 10.9 is from Goldreich and Zuckerman [1997] and Arvind and Köbler [1998]. Zero-knowledge proof systems were first introduced in Goldwasser, Micali, and Rackoff [1989]. Exercises 10.15 and 10.16 are from Goldreich et al. [1991]. Exercise 10.17 is from Fortnow [1989] and Aiello and Hastad [1991]. Goldreich [1997] contains a recent survey on zero-knowledge proof systems.

# 11

# *Probabilistically Checkable Proofs and NP-Hard Optimization Problems*

The notion of interactive proof systems may be extended to a more general notion of presenting short proofs whose correctness can be checked in probabilistic polynomial time. It is shown that the class of sets that have these types of probabilistically checkable proofs is exactly the class of nondeterministic exponential-time computable sets. Based on this characterization, we derive a new characterization for the class *NP* in terms of the notion of probabilistically checkable proofs. That is, the membership of an instance in a set in *NP* has a proof of a polynomially bounded length that can be verified by only checking randomly a constant number of bits. The proofs of these characterization results use algebraic proof techniques and the ideas from coding theory. The new characterization of *NP* has an important application to the study of the approximation to *NP*-hard combinatorial optimization problems. We present a number of combinatorial optimization problems which are not approximable within a constant factor of the optimum solutions if $P \neq NP$.

## 11.1  Probabilistically Checkable Proofs

We have studied in Chapter 10 the notion of interactive proof systems for problems in *PSPACE*. In this chapter, we make further detailed analysis of the power of interactive proof systems. In particular, we are interested in two additional complexity measures of interactive proof systems, namely, the number of random bits used by the verifier and the number of bits of the proofs the verifier actually reads. The number of random bits is a natural complexity measure of any probabilistic computation. To understand the meaning of

393

the second measure, it is best to convert an interactive proof system into a noninteractive (or, nonadaptive) proof system in which the prover sends only a single message (not necessarily of polynomial length) to the verifier. This message is stored in a random access storage so that the verifier is able to read any bit randomly. Thus, although the total number of bits sent by the prover may be very long, the verifier may be able to decide the membership of the input instance by reading only a small portion of those bits sent by the prover. Let us first look at a simple example. Assume that $F$ is a 3-CNF formula of $m$ clauses, and $r$ is a constant satisfying $r > 1$. We say that $F$ is *r-unsatisfiable* if for any Boolean assignment $\tau$ of variables in $F$, at least $m/r$ clauses are not satisfied by $\tau$.

**Example 11.1** Let (SAT, $r$-UNSAT) be the following decision problem: Given a 3-CNF Boolean formula $F$ which is guaranteed to be either satisfiable or $r$-unsatisfiable, determine whether it is satisfiable. This problem is in *NP* and, hence, there is a simple deterministic proof system for it. Namely, the prover can send a Boolean assignment $\tau$ of all variables in $F$ to the verifier and the verifier checks that $\tau$ satisfies all clauses. In this proof system, the number of random bits used by the verifier is 0, but the number of bits read by the verifier is, in general, linear in the input size.

Alternatively, we may use the following probabilistic proof system: the prover puts an assignment $\tau$ in a random access storage, and the verifier chooses randomly $K = \lceil 2r \rceil$ clauses and reads only the Boolean values of $\tau$ on variables that appear in those $K$ clauses. The verifier accepts if $\tau$ satisfies these $K$ clauses and rejects otherwise. Note that if $F$ is $r$-unsatisfiable, then for any $\tau$ the probability that $\tau$ satisfies a randomly chosen clause is at most $1 - 1/r$, and so the probability that $\tau$ satisfies $K$ random clauses is at most $(1 - 1/r)^K < 1/4$. Therefore, this is a proof system for (SAT, $r$-UNSAT) with the error probability $< 1/4$. The number of random bits used by the verifier is $\leq \log(m^K) = O(\log m)$. The total number of bits sent by the prover is, in general, proportional to $m$, but the number of bits actually read by the verifier is bounded by a constant $3K$.                                   □

Note that in the second system above it is necessary for the prover to send all Boolean assignments $\tau(x)$ to the verifier *before* the verifier reads any values, otherwise the prover may be able to cheat by adaptively creating different Boolean functions $\tau$ depending on the particular clauses chosen by the verifier. In other words, the *proof* of the instance must be independent of the random bits. We may imagine that the proof is stored by the prover in a random access storage and so the verifier can read only what he/she needs. Formally, we will define this random access storage as an oracle and require the system to work for a fixed oracle function.

We now proceed to show that, in general, every interactive proof system can be converted to such a nonadaptive proof system. Assume that $(M, f_p)$ is an interactive proof system for set $A$. That is, $M$ is a probabilistic oracle TM. Since we are interested in the number of random bits used by $M$, we assume that $M$ is a probabilistic oracle TM of the type defined in Exercise 8.4; that is, the random bit generator $\phi$ only generates a new random bit when $M$ enters a special *random* state. Thus, with the optimum prover $f_p$, the computation of $M^{f_p}$ on an input $x \in A$ is a computation tree in which a node branches into two child nodes when and only when its configuration is in the random state. Let $q_x$ be the maximum number of answer bits sent by the prover in any accepting computation path of this computation tree, and let $r_x$ be the maximum number of random bits used in any accepting path of the tree. Then, the computation tree of $M^{f_p}(x)$ has at most $2^{r_x}$ accepting paths, each containing at most $q_x$ bits of answers from $f_p$. Altogether, there are at most $q_x 2^{r_x}$ answer bits in the tree. Therefore, the prover could instead send to the verifier in one round all these answers. The verifier first tosses coins to get $r_x$ random bits, and then simulates the computation path defined by these random bits, reading only those bits used in this path. Suppose this computation path halts with the verifier reading no more than $q_x$ bits; then the verifier accepts or rejects as the simulation path does. Otherwise, if this computation path needs to read more than $q_x$ bits, then the verifier simply rejects.

We call this resulting nonadaptive proof system a *probabilistically checkable proof system* (*PCP*). It is obvious that this *PCP* system has the same error probability and the same random-bit complexity as the original interactive proof system. In the following sections, we will actually show that the notion of *PCP* systems is more general than the notion of interactive proof systems.

Our formal definition of *PCP* systems is based on the notion of probabilistic oracle TMs, with the oracle representing the proof stored in the random access storage. We recall that a probabilistic oracle TM $M$ consists of a random-bit generator $\phi$ and an oracle DTM $\tilde{M}$ such that on each input $x$, $\phi$ first generates a string $y$ of random bits and stores it on a random tape, and then $\tilde{M}$ works on $(x, y)$ deterministically to decide to accept or reject $x$. We say $M$ is a *nonadaptive oracle PTM* if $M$ contains a random-bit generator $\phi$ and three deterministic machines $\tilde{M}_j$, $1 \le j \le 3$. The first machine $\tilde{M}_1$ is a DTM that on input $(x, y)$ generates $m$ strings $z_1, \ldots, z_m$. The second machine $\tilde{M}_2$ is a deterministic oracle machine that asks the oracle $m$ queries: "Is $z_j$ in the oracle?" for $j = 1, \ldots, m$. The third machine $\tilde{M}_3$ is a DTM that outputs 1 or 0 from input $(x, y, b_1, \ldots, b_m)$, where $b_j$ is the answer to the query $z_j$ from the oracle. We say a nonadaptive oracle PTM $M$ *uses* $r(n)$ *random bits* and *asks* $q(n)$ *queries* to the oracle, if the random-bit generator $\phi$ generates at most $r(n)$ random bits on any input $x$ of length $n$, and on each $(x, y)$, with $|x| = n$ and $|y| \le r(n)$, $\tilde{M}_1$ generates at most $q(n)$ queries to the oracle.

**Definition 11.2** *(a) Let $r(n)$ and $q(n)$ be two integer functions. A set $A \subseteq \Sigma^*$ is in $PCP(r(n), q(n))$ if there exists a polynomial-time, nonadaptive oracle PTM $M$ that uses $r(n)$ random bits and asks $q(n)$ queries such that:*

*(i) If $x \in A$, then there is an oracle $Q_x$ such that $\Pr[M^{Q_x} \text{ accepts } x] = 1$; and*

*(ii) If $x \notin A$, then for any oracle $Q$, $\Pr[M^Q \text{ accepts } x] \leq 1/4$.*

*(b) Let $F$ and $G$ be two classes of functions. We let $PCP(F,G) = \bigcup_{r \in F, q \in G} PCP(r(n), q(n))$.*

We are interested in the power of the classes $PCP(F,G)$ with respect to the classes $F$ and $G$ of polynomial functions, logarithm functions, and constant functions. For convenience, we write *poly* to denote the class of all polynomial functions of nonnegative integer coefficients,[1] $O(log)$ or simply *log* to denote the class of functions $c \log n$, and $O(1)$ or simply *const* to denote the class of constant functions $f(n) = c$. The following relations between *PCP* classes and other complexity classes are straightforward. Let $NEXPPOLY = \bigcup_{p \in poly} NTIME(2^{p(n)})$.

**Proposition 11.3** *(a) If $r_1(n) \leq r_2(n)$ and $q_1(n) \leq q_2(n)$ for all $n \geq 1$, then $PCP(r_1, q_1) \subseteq PCP(r_2, q_2)$.*

*(b) $PCP(poly, poly) \subseteq NEXPPOLY$.*

*(c) $PCP(const, poly) = PCP(log, poly) = NP$.*

*Proof.* Part (a) is trivial. To prove parts (b) and (c), assume that a set $S$ is in $PCP(r(n), q(n))$. Then, for each input $x$ of length $n$, the verifier $V$ reads totally (over all random strings) at most $2^{r(n)}q(n)$ query bits, and so we may assume that the proof $\Pi$ is of length $\leq 2^{r(n)}q(n)$. Thus, if $x \in S$, then there is a proof $\Pi$ of length $\leq 2^{r(n)}q(n)$ such that the verifier $V$ accepts for all strings $y$ of length $r(n)$; and if $x \notin S$ then for all proofs $\Pi$ of length $\leq 2^{r(n)}q(n)$, the verifier $V$ does not accept for most strings $y$ of length $r(n)$. We can perform a nondeterministic simulation of $V$ on $x$ as follows: We first guess a proof $\Pi$ and then deterministically verify that for all strings $y$ of length $r(n)$ the proof works correctly. This simulation takes nondeterministic time $2^{r(n)}p(n)$ for some polynomial $p$, $p(n) \geq q(n)$. Therefore, $S$ is accepted by a nondeterministic TM in time $2^{r(n)}p(n)$ for some polynomial $p$. The above argument shows part (b). It also shows that $PCP(log, poly) \subseteq NP$. It is obvious from the existential quantifier characterization of $NP$ that $NP = PCP(0, poly)$, and so part (c) follows. ∎

## 11.2  *PCP* Characterization of Nondeterministic Exponential Time

In this section, we give a precise characterization of $NEXPPOLY$ in terms of probabilistically checkable proof systems; that is, we prove the following theorem:

**Theorem 11.4** $PCP(poly, poly) = NEXPPOLY$.

---

[1]Note that here *poly* represents a different class of functions than the one defined in Chapter 6.

The proof of the theorem is long. We present it in three parts. In Section 11.2.1, we present the proof with two critical *PCP* systems, the *multilinearity test* system and the *sum check* system, omitted. These two *PCP* systems are presented in Sections 11.2.2 and 11.2.3. In this and subsequent sections, we will use multivariate polynomial functions to encode strings. We will use the term "polynomial" to mean a multivariate polynomial function, and use the term "degree of a polynomial" to mean the total degree of the polynomial over all variables.

## 11.2.1 Proof

In this and the following sections, we will use both of the terms *proof* and *oracle* to denote the oracle $Q_x$ or $Q$ of Definition 11.2. The use of *proof* emphasizes the idea that the whole proof is sent by the prover in one around, before any queries are made. The use of *oracle* emphasizes that the verifier does not read the whole proof but only queries for a few bits.

Let $A \in$ *NEXPPOLY* be accepted by an NTM $M$ in time $2^{p(n)}$ for some polynomial $p$. Then, for each input $w \in A$ of length $n$, there is an accepting computation of $M$ that consists of at most $2^{p(n)}$ configurations and each of length at most $2^{p(n)}$. Now, applying Cook's Theorem to this computation, we obtain a 3-CNF formula $\Phi_w$ such that $w \in A$ if and only if $\Phi_w$ is satisfiable. By adding dummy variables and dummy clauses, we may assume that $\Phi_w$ has exactly $2^m$ variables and has exactly $2^m$ clauses for some $m$ bounded by $q(n)$ for some polynomial $q$.

We now arithmetize the formula $\Phi_w$. We identify the names of variables and clauses with the strings in $\{0,1\}^m$; that is, we assume the variables in $\Phi_w$ are $v_s$, $s \in \{0,1\}^m$, and clauses of $\Phi_w$ are $C_s$, $s \in \{0,1\}^m$. For each $j = 1, 2, 3$, define Boolean functions $\phi_j : \{0,1\}^m \times \{0,1\}^m \to \{0,1\}$ and $\psi_j : \{0,1\}^m \to \{0,1\}$ by

$$\phi_j(s,t) = \begin{cases} 1 & \text{if } v_s \text{ is the } j\text{th variable in } C_t, \\ 0 & \text{otherwise,} \end{cases}$$

and

$$\psi_j(t) = \begin{cases} 1 & \text{if the } j\text{th variable in } C_t \text{ occurs positively,} \\ 0 & \text{otherwise.} \end{cases}$$

Let $G$ be an *oracle* Boolean formula of $4m$ variables that, on inputs $s_1, s_2, s_3, t \in \{0,1\}^m$ (each $s_j$ and $t$ representing $m$ Boolean values) and oracle function $f : \{0,1\}^m \to \{0,1\}$, is defined as

$$G^f(s_1, s_2, s_3, t) = \prod_{j=1}^{3} \phi_j(s_j, t)(\psi_j(t) - f(s_j)). \tag{11.1}$$

(The sign "$-$" denotes the Boolean subtraction.) It is easy to see that for any function $f$, $G^f(s_1, s_2, s_3, t) = 0$ for all $s_1, s_2, s_3, t \in \{0,1\}^m$ if and only if function $f$ (or, more precisely, the assignment $\tau(v_s) = f(s)$) satisfies $\Phi_w$. Notice that, from the proof of Cook's Theorem, formulas $\phi_j$

and $\psi_j$ can be constructed in time polynomial in $m$, and so can the oracle Boolean formula $G$ (in the sense that the formula $G_1(s_1, s_2, s_3, t, z_1, z_2, z_3) = \prod_{j=1}^3 \phi_j(t, s_j)(\psi_j(t) - z_j)$, is constructible in polynomial time).

We can summarize the above construction into a simple-minded deterministic proof system for $A$: [2]

> *Deterministic Proof System D.* For the input $w$, the proof consists of a function $f : \{0,1\}^m \to \{0,1\}$ (given in the form of an oracle function). The verifier constructs formula $G$ as above, and then verifies that
> $$G^f(x) = 0 \text{ for all } x \in \{0,1\}^{4m}. \tag{11.2}$$

Unfortunately, the above proof system did not use the idea of probabilistic checking and takes time exponential in $|w|$ to verify the proof $f$. In the following, we describe how to convert the above proof system into a probabilistic one. In short, the prover needs to present a second part of the proof (in addition to function $f$) that helps the verifier to check that $f$ indeed satisfies (11.2). The main idea here is to use the arithmetization technique similar to that used in Theorem 10.26 to convert the problem of checking the condition (11.2) to the problem of checking a polynomial equation over a finite field.

Let $f$ be a Boolean function of $m$ variables, that is, $f : \{0,1\}^m \to \{0,1\}$, and let $\mathcal{F}$ be a finite field of size $\alpha$. A function $\hat{f} : \mathcal{F}^m \to \mathcal{F}$ is a *multilinear extension* of $f$ over $\mathcal{F}$ if $\hat{f}$ is a polynomial function over $\mathcal{F}$ of degree 1 on each individual variable such that $\hat{f}(y) = f(y)$ for all $y \in \{0,1\}^m \subseteq \mathcal{F}^m$.

**Lemma 11.5** *For any Boolean function $f$ of $m$ variables and any finite field $\mathcal{F}$, there is a unique multilinear extension $\hat{f}$ of $f$ over $\mathcal{F}$.*

*Proof.* We denote an element $(a_1, \ldots, a_m) \in \mathcal{F}^m$ by $\vec{a}$. Let $\ell_0(x) = 1 - x$ and $\ell_1(x) = x$. Define

$$\hat{f}(z_1, \ldots, z_m) = \sum_{\vec{a} \in \{0,1\}^m} f(\vec{a}) \prod_{j=1}^m \ell_{a_j}(z_j).$$

It is clear that for any $\vec{a} \in \{0,1\}^m$, $\hat{f}(\vec{a}) = f(\vec{a})$. Furthermore, for any $j$, $1 \leq j \leq m$, $\hat{f}$ has degree 1 on the variable $z_j$.

For the uniqueness, we claim that if $g : \mathcal{F}^m \to \mathcal{F}$ is multilinear and if $g(\vec{a}) = 0$ for all $\vec{a} \in \{0,1\}^m$, then $g(\vec{b}) = 0$ for all $\vec{b} \in \mathcal{F}^m$. For any $\vec{b} = (b_1, \ldots, b_m) \in \mathcal{F}^m$, let $k(\vec{b})$ denote the number of coordinates $b_j$ that are not equal to 0 or 1. We prove the claim by induction on $k(\vec{b})$. First, if $k(\vec{b}) = 0$, then $\vec{b} \in \{0,1\}^m$, and so, by the assumption, $g(\vec{b}) = 0$. If $k(\vec{b}) > 0$, let $j = \min\{i : b_i \notin \{0,1\}\}$, and consider the function $g_j(x) = g(b_1, \ldots, b_{j-1}, x, b_{j+1}, \ldots, b_m)$. By the

---

[2]The above construction actually proved that the problem ORACLE-SAT of Exercise 3.20 is complete for *NEXPPOLY*.

inductive hypothesis, $g_j(0) = g_j(1) = 0$. Therefore, $g_j(x) = 0$ for all $x \in \mathcal{F}$ since $g_j$ is a linear function. In particular, $g_j(b_j) = g(\vec{b}) = 0$. We have completed the proof of the claim.

Now, for any $\tilde{f}$ that is a multilinear extension of $f$, define $g = \hat{f} - \tilde{f}$. By the above claim, $g$ is identical to zero on $\mathcal{F}^m$, and so $\hat{f} = \tilde{f}$. ∎

We will use the notation $\hat{f}$ for the unique multilinear extension of $f$ when the field $\mathcal{F}$ is understood.

We now can describe the general idea of the approach. First, we choose a field $\mathcal{F}$ of an appropriate size $\alpha$. The prover then presents a polynomial function $f$ together with a proof $\Pi$ for (11.2), from which:

(a) The verifier can verify that $f$ is indeed a multilinear function, or that $f$ is *very close* to a multilinear function over $\mathcal{F}$; and

(b) Assuming that $f$ is a multilinear function or is very close to a multilinear function, the verifier can use $\Pi$ to verify that $f$ indeed satisfies (11.2).

Part (a) is to be accomplished by the multilinearity test system and part (b) is to be done by the sum check system. Part (a) is necessary because of the technical reason that our sum check system only works for polynomials or functions that are close to polynomials.

We will leave the details of parts (a) and (b) to later subsections. Here, we summarize the properties of these two proof systems. Let us denote an element in $\mathcal{F}^m$ by a vector $\vec{z} = (z_1, \ldots, z_m)$. Define the *distance* $\Delta(f, g)$ of two functions $f$ and $g$ on $\mathcal{F}^m$ to be the fraction of vectors $\vec{z} \in \mathcal{F}^m$ such that $f(\vec{z}) \neq g(\vec{z})$, that is, the number of such vectors divided by $\alpha^m$. We say that a function $f : \mathcal{F}^m \to \mathcal{F}$ is *$\delta$-close to a multilinear function* (or, simply *$\delta$-close*) if there exists a multilinear function $g : \mathcal{F}^m \to \mathcal{F}$ such that $\Delta(f, g) \leq \delta$. The multilinearity test system has the following properties:

*Multilinearity Test System.* Given constants $\delta, \epsilon > 0$, and a function $f : \mathcal{F}^m \to \mathcal{F}$ (in the form of an oracle), with $12m/(\alpha - 2) < \delta$, there is a verifier such that:

(i) If $f$ is multilinear, then the verifier accepts with probability 1;

(ii) If $f$ is not $\delta$-close, then the verifier rejects with probability $\geq 1 - \epsilon$; and

(iii) The verifier uses $O(m^2 \log \alpha)$ random bits, reads $O(m \log \alpha)$ bits from $f$ and runs in time polynomial in $m + \log \alpha$.

This proof system resolves the problem of part (a).

For part (b), we first transfer the *universal* condition (11.2) to a *probabilistic* condition.

**Lemma 11.6** *For any field $\mathcal{F}$ of size $\alpha$, and any integer $k$ such that $2k < \alpha$, there exists a family $\{R_i\}_{i=1}^{\alpha^k}$ of polynomial functions of degree $k$ such that for* *any nonzero function* $f : \{0,1\}^k \to \mathcal{F}$,

$$\Pr\left[ \sum_{\vec{z} \in \{0,1\}^k} f(\vec{z})R_i(\vec{z}) = 0 \right] < \frac{2k}{\alpha},$$

*when the polynomial $R_i$ is chosen randomly from the family (over the uniform distribution). Furthermore, each $R_i$ can be constructed from $i$ in time polynomial in $k$.*

*Proof.* For each $u \in \mathcal{F}$, let $\ell_u(x) = (2u-1)x + (1-u)$ so that $\ell_u(0) = 1-u$ and $\ell_u(1) = u$. For each $\vec{u} = (u_1, \ldots, u_k) \in \mathcal{F}^k$, define

$$R_{\vec{u}}(z_1, \ldots, z_k) = \prod_{j=1}^{k} \ell_{u_j}(z_j).$$

Then, $R_{\vec{u}}$ is a multilinear function such that for any $\vec{u}, \vec{z} \in \{0,1\}^k$, $R_{\vec{u}}(\vec{z}) = 1$ if $\vec{u} = \vec{z}$ and $R_{\vec{u}}(\vec{z}) = 0$ if $\vec{u} \neq \vec{z}$. Define $g : \mathcal{F}^k \to \mathcal{F}$ by

$$g(\vec{u}) = \sum_{\vec{z} \in \{0,1\}^k} f(\vec{z})R_{\vec{u}}(\vec{z}).$$

Then, it is easy to see that $g$ is identical to zero if and only if $f$ is identical to zero. In particular, if $f(\vec{z}_0) \neq 0$ then, from the above property of $R_{\vec{u}}$, $g(\vec{z}_0) \neq 0$.

Furthermore, we claim that the roots of $g$ constitute at most a fraction of $2k/\alpha$ of all $\vec{u} \in \mathcal{F}^k$, if $g$ is not identical to zero. This claim can be proved by induction on $k$ for all multilinear functions on $k$ variables. First, if $k = 1$, then $g$ is a univariate linear function and so if $g$ is not identical to zero then $g$ has at most $1 < 2k = \alpha(2k/\alpha)$ root. Assume that $k > 1$ and that $g : \mathcal{F}^k \to \mathcal{F}$ has more than $2k\alpha^{k-1}$ roots. For any $\vec{b} = (b_2, \ldots, b_k) \in \mathcal{F}^{k-1}$, the function $g_{\vec{b}}(x) = g(x, b_2, \ldots, b_k)$ is a univariate linear function, and so it has either $\leq 1$ root or it is identical to zero. By a simple counting argument, we can see that for more than $2(k-1)\alpha^{k-2}$ vectors $\vec{b} \in \mathcal{F}^{k-1}$, $g_{\vec{b}}$ is identical to zero on $\mathcal{F}$. In other words, for each $a \in \mathcal{F}$, the function $h_a(b_2, \ldots, b_k) = g(a, b_2, \ldots, b_k)$ is an $(k-1)$-variate multilinear function with more than $2(k-1)/\alpha$ fractions of roots. By the inductive hypothesis, $h_a$ is identical to zero on $\mathcal{F}^{k-1}$. Since this is true for all $a \in \mathcal{F}$, it follows that $g$ is identical to zero on $\mathcal{F}^k$.   ∎

Based on the above lemma, the problem of checking (11.2) is reduced to the problem of checking that

$$\sum_{x \in \{0,1\}^{4m}} \hat{G}^f(x)R_i(x) = 0 \tag{11.3}$$

for a random $R_i$ from the family $\{R_i\}_{i=1}^{\alpha^{4m}}$. In the above, $\hat{G}^f$ is the function defined from the multilinear extensions of $\hat{\phi}_j$ and $\hat{\psi}_j$:

$$\hat{G}^f(s_1, s_2, s_3, t) = \prod_{j=1}^{3} \hat{\phi}_j(s_j, t)(\hat{\psi}_j(t) - f(s_j)).$$

We say a function $f : \mathcal{F}^m \to \mathcal{F}$ is $\delta$-*close to a polynomial function of degree $d$* if there is a polynomial function $g : \mathcal{F}^m \to \mathcal{F}$ of degree $\leq d$ such that $\Delta(f, g) \leq \delta$. By Lemma 8.3, we know that two different $m$-variate, degree-$d$ polynomials can agree at most at a fraction of $d/\alpha$ points in $\mathcal{F}^m$. This implies that if $f$ is $\delta$-close to a degree-$d$ polynomial $g$ with $\delta < (1 - d/\alpha)/2$, then $g$ is unique.

In Section 11.2.3, we will develop the sum check system, which has the following properties:

> *Sum Check System.* Given a function $g : \mathcal{F}^m \to \mathcal{F}$ (in the form of an oracle), a constant $c \in \mathcal{F}$, a constant $d > 0$, and a constant $\epsilon$ such that $2md/\alpha < \epsilon < 1$ (all in the form of inputs), there is a verifier having the following properties:
>
> (i) If $g$ is a polynomial function of degree $d$, and if $\sum_{\vec{z} \in \{0,1\}^m} g(\vec{z}) = c$, then there is an additional oracle $\Pi$ with respect to which the verifier accepts with probability 1; if $g$ is a polynomial function of degree $d$ but $\sum_{\vec{z} \in \{0,1\}^m} g(\vec{z}) \neq c$, then the verifier rejects with probability $\geq 1 - \epsilon$ for all $\Pi$.
>
> (ii) If $g$ is $(\epsilon/2)$-close to a polynomial function $\tilde{g}$ of degree $d$, and if $\sum_{\vec{z} \in \{0,1\}^m} \tilde{g}(\vec{z}) \neq c$, then the verifier rejects with the probability $\geq 1 - \epsilon$ for all $\Pi$.
>
> (iii) The verifier uses $O(m \log \alpha)$ random bits, reads a single value of $g$ at a *random* point and totally $O(md \log \alpha)$ bits from $g$ and $\Pi$, and runs in time polynomial in $m + d + \log \alpha$.

The above proof system can be applied to solve the problem in part (b).

In summary, our proof system for the set $A$ can be described as follows:

> *Probabilistic Proof System $S_1$.* The prover and the verifier agree at a finite field $\mathcal{F}$ of size $\alpha = \Theta(m^2)$ and a bound $\epsilon_0$ for the error probability. (For instance, let $\alpha$ be the least prime greater than $104m^2/\epsilon_0$, and let $\mathcal{F} = \mathbf{Z}_\alpha$.) The proof consists of a function $f : \mathcal{F}^m \to \mathcal{F}$ and additional $\alpha^{4m}$ strings $\Pi_i$, $1 \leq i \leq \alpha^{4m}$, in the form of oracles. The algorithm for the verifier is the following:
>
> (1) Construct function $G^f$ as described above.
>
> (2) Construct the polynomial function $\hat{G}^f$.
>
> (3) Randomly pick an integer $i$, $1 \leq i \leq \alpha^{4m}$, and construct the polynomial $R_i$ as in Lemma 11.6 (with $k = 4m$). Let $Q_i^f(\vec{z}) = \hat{G}^f(\vec{z}) R_i(\vec{z})$.
>
> (4) Run the multilinearity test system to check whether $f$ is $\delta$-close, with the parameter $\epsilon = \epsilon_0$ and $\delta = \epsilon_0/8$; reject if this test fails.
>
> (5) Use $\Pi_i$ as the additional proof to run the sum check system to see whether (11.3) holds, with the parameters $d = 13m$ and $\epsilon = \epsilon_0/4$; reject if the test fails, else accept.

Step (5) above needs some explanation. In the sum check system, the verifier checks that the values of $g(\vec{z})$ on $\vec{z} \in \{0,1\}^m$ sum to a constant $c$, where $g$ is the oracle function given by the prover. In step (5) above, the verifier needs to check instead that the values $Q_i^f(\vec{z})$ on $\vec{z} \in \{0,1\}^{4m}$ sum to 0, where $f$ is the oracle given by the prover. (If $f$ is a polynomial of degree $m$, then $Q_i^f$ is a polynomial of degree $13m$.) This difference involves a minor change in the procedure of the sum check system. Namely, the verifier expects the prover to provide the additional proof $\Pi_i$ for the condition that $\sum_{\vec{z} \in \{0,1\}^{4m}} Q_i^f(\vec{z}) = 0$, and behaves as if the oracle is actually the function $Q_i^f$. Then, when he/she needs a value of $Q_i^f(\vec{z})$, the verifier reads some values of $f$ and uses them to compute the desired value of $Q_i^f(\vec{z})$. Since both $\hat{G}$ and $R_i$ are computable in time polynomial in $m$, the values of $Q_i^f(\vec{z})$ can be computed in polynomial time, as long as the corresponding values of $f$ are known. Notice that in the original sum check system, the verifier only reads a single value of $g$. That is, to check whether $Q_i^f(\vec{z})$'s sum to 0, the verifier only needs one value of $Q_i^f(\vec{z})$ at a random point $\vec{z}$, and that translates to three values of the real oracle $f$ at three random points.

**Lemma 11.7** *(a) If $w \in A$, then there exist two functions $f$ and $\Pi$ with respect to which the verifier accepts with probability 1.*

*(b) If $w \notin A$, then the verifier rejects with probability $\geq 1 - \epsilon_0$ for all oracles $f$ and $\Pi$.*

*Proof.* We first check that in order to apply the multilinearity test system with $\epsilon = \epsilon_0$ and $\delta = \epsilon_0/8$, we need $\delta > 12m/(\alpha - 2)$, or $\alpha - 2 > 96m/\epsilon_0$. In order to apply the sum check system with $\epsilon = \epsilon_0/4$, we need $\epsilon_0/4 > 2md/\alpha$, or $\alpha > 104m^2/\epsilon_0$. Our choice of $\alpha > 104m^2/\epsilon_0$ satisfies both conditions, and so the properties of these two proof systems listed above hold for the proof system $S_1$ too.

(a) If $w \in A$, then there exists a function $f_1 : \{0,1\}^m \to \{0,1\}$ satisfying (11.2). Thus, the prover can provide its multilinear extension $f : \mathcal{F}^m \to \mathcal{F}$ as the oracle, together with $\Pi = (\Pi_1, \ldots, \Pi_{\alpha^{4m}})$, where each $\Pi_i$ is a proof aimed for the sum check system for $\sum_{\vec{z} \in \{0,1\}^m} Q_i^f(\vec{z}) = 0$. They will pass all tests since $f$ is multilinear and $\hat{G}^f(\vec{z})$ is identical to zero, and the verifier will accept with probability 1.

(b) Assume that $w \notin A$, and so there is no function $f_1 : \{0,1\}^m \to \{0,1\}$ satisfying (11.2). We consider three cases.

*Case* 1. The oracle function $f$ is multilinear. Then, $\hat{G}^f(\vec{z})$ is not identical to zero on $\{0,1\}^{4m}$. The verifier accepts only when one of the following holds:

(i) The random polynomial $R_i$ is chosen so that $\sum_{\vec{z} \in \{0,1\}^{4m}} Q_i^f(\vec{z}) = 0$; or

(ii) $\sum_{\vec{z} \in \{0,1\}^{4m}} Q_i^f(\vec{z}) \neq 0$ but the sum check system does not catch it.

The probability of (i) is, by Lemma 11.6, at most $8m/\alpha < \epsilon_0/4$. In the second case, since $f$ is multilinear and, hence, a polynomial of degree $m$, the function $Q_i^f$ is a polynomial function of degree $13m$. Therefore, by property

(i) of the sum check system, the probability of (ii) is at most $\epsilon_0/4$. The total error probability is thus less than $\epsilon_0/2$.

*Case* 2. The oracle function $f$ is not $(\epsilon_0/8)$-close to any multilinear function. Then, the verifier rejects in the multilinearity test system with probability $\geq 1 - \epsilon_0$.

*Case* 3. The oracle function $f$ is $(\epsilon_0/8)$-close to a multilinear function $\tilde{f}$. Then, $\hat{G}^{\tilde{f}}(\vec{z})$ is not identical to zero on $\vec{z} \in \{0,1\}^{4m}$. From Case 1 above, we know that if the prover gives $\tilde{f}$ as the oracle function instead of $f$, then the probability of accepting is at most $\epsilon_0/2$. In other words, extra errors for Case 3 may occur only if at least one queried value $f(\vec{z})$ of step (5) is different from the value $\tilde{f}(\vec{z})$. Notice that the verifier asks in step (5) for only three values of $f$ at three random points. Since $\Delta(f, \tilde{f}) \leq \epsilon_0/8$, the probability that $f$ and $\tilde{f}$ do not agree at any of these three queried points is at most $3\epsilon_0/8$. We conclude that the error probability is $< \epsilon_0/2 + 3\epsilon_0/8 < \epsilon_0$. ∎

**Lemma 11.8** *The proof system $S_1$ for $A$ uses $O(m^2 \log m)$ random bits, reads $O(m^2 \log m)$ bits from the oracles, and runs in time polynomial in $m$.*

*Proof.* In step (3), the verifier chooses a random $i$ between 1 and $\alpha^{4m}$. It takes $\lceil \log(\alpha^{4m}) \rceil = O(m \log \alpha) = O(m \log m)$ random bits. In step (4), the verifier uses $O(m^2 \log \alpha) = O(m^2 \log m)$ random bits. In step (5), the verifier uses $O(m \log \alpha) = O(m \log m)$ random bits. The total number of random bits used is $O(m^2 \log m)$.

The number of queried bits of step (4) is $O(m \log \alpha) = O(m \log m)$ and that of step (5) is $O(md \log \alpha) = O(m^2 \log m)$. Therefore, the verifier queries only $O(m^2 \log m)$ bits to the oracles.

We have seen that it only takes time polynomial in $m$ to construct $\hat{G}$ in step (2) and $R_i$ in step (3). From the properties of the two sub-proof systems, we know that steps (4) and (5) also take only time polynomial in $m$. So the total runtime of the proof system $S_1$ is only polynomial in $m$ (although the proof $(f, \Pi)$ is of length exponential in $m$). ∎

The proof of Theorem 11.4 is now complete by noticing that $m$ is bounded by a polynomial in the input length $n$.

### 11.2.2 Multilinearity Test System

Let $\mathcal{F}$ be a field of size $\alpha$, and $m$ an integer such that $96m < \alpha - 2$. Let $1 \leq i \leq m$ and $a_1, a_2, \ldots, a_m \in \mathcal{F}$. The set $L_i = \{(a_1, \ldots, a_{i-1}, x, a_{i+1}, \ldots, a_m) : x \in \mathcal{F}\}$ is called an *axis-parallel line*, or simply a *line*. Any three points $\vec{a}, \vec{b}, \vec{c}$ in $\mathcal{F}^m$ are called an *aligned triple*, or simply a *triple* if they are in a same line $L_i$. We say a function $f : \mathcal{F}^m \rightarrow \mathcal{F}$ is *linear* on an aligned triple $\{\vec{a}, \vec{b}, \vec{c}\}$, and $\{\vec{a}, \vec{b}, \vec{c}\}$ is *f-linear*, if $f$ is a linear function when restricted to these three points. It is easy to see that a function $f : \mathcal{F}^m \rightarrow \mathcal{F}$ is multilinear if and only if $f$ is linear on all aligned triples (see Exercise 11.2). The algorithm of the multilinearity test system is based on the stronger fact that $f$ is $\delta$-close to a multilinear function with a small $\delta$ if and only if $f$ is linear on *most* aligned triples. To be more precise, define $\Delta_{ml}(f)$ to be the minimum distance $\Delta(f, g)$ between $f$ and a multilinear function $g$. Let $\delta_{nl}(f)$ be the fraction of aligned triples on which $f$ is not linear; that is, $\delta_{nl}(f)$ is equal to the number of non-$f$-linear aligned triples divided by $M$, where $M = m\binom{\alpha}{3}\alpha^{m-1}$ is the total number of aligned triples in the space $\mathcal{F}^m$. We will prove that $\delta_{nl}(f) \geq \Delta_{ml}(f)/32m$ if $\Delta_{ml}(f)$ is reasonably large. We first consider the case when $\Delta_{ml}(f) \leq 1/2$.

**Lemma 11.9** *Let $f$ be a function from $\mathcal{F}^m$ to $\mathcal{F}$. If $12m/(\alpha-2) \leq \Delta_{ml}(f) \leq 1/2$, then $\delta_{nl}(f) \geq \Delta_{ml}(f)/2m$.*

*Proof.* Let $\delta = \Delta_{ml}(f)$, and let $g$ be a multilinear function on $\mathcal{F}^m$ such that $\Delta(f, g) = \delta$. For any $\vec{a} \in \mathcal{F}^m$, we say $\vec{a}$ has *color* 1 if $f(\vec{a}) = g(\vec{a})$, and has *color* 0 otherwise. A subset $H \subseteq \mathcal{F}^m$ is called *two-colored* if some of its elements has color 1 and some has color 0; it is called *one-colored* otherwise. In the following, we show that the fraction of two-colored aligned triples is large but the fraction of two-colored, $f$-linear triples is small, and so the fraction of $f$-linear aligned triples is small. Let $tc(f)$ be the number of two-colored aligned triples divided by $M$.

*Claim* 1. $tc(f) \leq \delta_{nl}(f) + 3/(\alpha - 2)$.

*Proof.* Assume that $\{\vec{a}, \vec{b}, \vec{c}\}$ is a two-colored, $f$-linear aligned triple. Then, there is exactly one point in the triple having color 1, for otherwise $f$ would agree with $g$ at two points and so, by the linearity of $f$ on the triple, also agree at the third point. Assume that $\vec{c}$ has color 1 and $\vec{a}$ and $\vec{b}$ have color 0. We observe that no other point $\vec{d}$ has the property that $\{\vec{a}, \vec{b}, \vec{d}\}$ is two-colored and is $f$-linear. To see this, assume that $f$ and $g$ agree at two points $\vec{c} \neq \vec{d}$ that are in the same line of $\vec{a}$ and $\vec{b}$ and that both $\{\vec{a}, \vec{b}, \vec{c}\}$ and $\{\vec{a}, \vec{b}, \vec{d}\}$ are $f$-linear. Then, $f$ is linear on $\{\vec{a}, \vec{c}, \vec{d}\}$. It follows that $f(\vec{a})$ is linearly dependent on $f(\vec{c}) = g(\vec{c})$ and $f(\vec{d}) = g(\vec{d})$ and, hence, is equal to $g(\vec{a})$. This is a contradiction.

From the above observation, we see that the number of two-colored, $f$-linear aligned triples is bounded by $m\binom{\alpha}{2}\alpha^{m-1}$, the number of pairs $\{\vec{a}, \vec{b}\}$ that are in the same axis-parallel line. It follows that

$$tc(f) \leq \delta_{nl}(f) + \frac{m\binom{\alpha}{2}\alpha^{m-1}}{M} = \delta_{nl}(f) + \frac{3}{\alpha - 2} \ .$$

This completes the proof of Claim 1.

Next, we show that $tc(f)$ is large. It is easy to see that the number of two-colored triples is exactly $(\alpha-2)/2$ times the number of two-colored pairs $\{\vec{a}, \vec{b}\}$ that lie on the same axis-parallel line (called *two-colored aligned pairs*): from each two-colored aligned pair, we can form $\alpha - 2$ two-colored aligned triples (by adding a third point in the same line), and each two-colored aligned triple can be formed in this way from two two-colored subpairs. So, we only need to count the number of two-colored aligned pairs.

*Claim 2.* There are at least $\delta(1-\delta)\alpha^{m+1}/2$ two-colored aligned pairs.

*Proof.* For $i = 0, 1$, let $S_i$ be the set of points $\vec{a}$ in $\mathcal{F}^m$ that have color $i$. By the assumption that $\Delta(f, g) = \delta$, $|S_0| = \delta\alpha^m$ and $|S_1| = (1-\delta)\alpha^m$. A two-colored pair must have one element in $S_1$ and the other in $S_0$. Let $\vec{a} = (a_1, \ldots, a_m) \in S_1$ and $\vec{b} = (b_1, \ldots, b_m) \in S_0$. Define, for $1 \leq j \leq m$, $\vec{c}_j = (b_1, \ldots, b_j, a_{j+1}, \ldots, a_m)$. Note that $\vec{c}_0 = \vec{a}$ and $\vec{c}_m = \vec{b}$, and that each pair $\{\vec{c}_{j-1}, \vec{c}_j\}$, $1 \leq j \leq m$, is an aligned pair. The sequence $\{\vec{c}_j\}_{j=0}^m$ is called the *path* from $\vec{a}$ to $\vec{b}$. Since $\vec{a}$ and $\vec{b}$ have different colors, there must be at least one two-colored aligned pair $\{\vec{c}_{j-1}, \vec{c}_j\}$, for some $1 \leq j \leq m$, in the path from $\vec{a}$ to $\vec{b}$. Furthermore, if $\vec{c} = (c_1, \ldots, c_m)$ and $\vec{d} = (c_1, \ldots, c_{j-1}, d_j, c_{j+1}, \ldots, c_m)$ are a two-colored aligned pair, and if it belongs to a path from $\vec{x} = (x_1, \ldots, x_m)$ to $\vec{y} = (y_1, \ldots, y_m)$, then we must have that $y_k = c_k$, for all $1 \leq k \leq j-1$ and $x_k = c_k$ for all $j+1 \leq k \leq m$, and that $\{x_j, y_j\} = \{c_j, d_j\}$. It follows that each two-colored aligned pair belongs to at most $2\alpha^{m-1}$ paths. Therefore, there are at least

$$\frac{|S_1| \cdot |S_0|}{2\alpha^{m-1}} = \frac{\delta(1-\delta)\alpha^{m+1}}{2}$$

two-colored aligned pairs. This completes the proof of Claim 2.

From Claim 2, we know that there are at least $\delta(1-\delta)(\alpha-2)\alpha^{m+1}/4$ two-colored aligned triples and so

$$tc(f) \geq \frac{\delta(1-\delta)(\alpha-2)\alpha^{m+1}}{4m\binom{\alpha}{3}\alpha^{m-1}} \geq \frac{3\delta}{4m},$$

since $1 - \delta \geq 1/2$. From Claim 1 and the assumption that $\delta > 12m/(\alpha-2)$, we get $\delta_{nl}(f) \geq 3\delta/4m - 3/(\alpha-2) \geq \delta/2m$. ∎

**Lemma 11.10** *Let $f$ be a function from $\mathcal{F}^m$ to $\mathcal{F}$. If $\Delta_{ml}(f) \geq 12m/(\alpha-2)$ then $\delta_{nl}(f) \geq \Delta_{ml}(f)/32m$.*

*Proof.* Let $\delta = \Delta_{ml}(f)$. First, Lemma 11.9 has already established this bound for $\delta_{nl}(f)$ if $\delta \leq 1/2$. So, we may assume that $\delta > 1/2$. In the following, we prove by induction on $m$ a stronger statement: If $\delta \geq 1/8$, then there are at least $\binom{\alpha}{3}(\alpha-1)^{m-1}/16$ non-$f$-linear aligned triples.

First, consider the case of $m = 1$. For any pair $\{a, b\}$, with $a \neq b \in \mathcal{F}$, there are $\alpha - 2$ choices of a third point $c \in \mathcal{F}$ to make an aligned triple $\{a, b, c\}$. Among them, for at least $\delta\alpha$ choices of $c$, the triple $\{a, b, c\}$ is non-$f$-linear, since $\Delta_{ml}(f) = \delta$. Since each non-$f$-linear triple can be formed in three ways from three subpairs, there are at least $\delta\alpha\binom{\alpha}{2}/3 > \binom{\alpha}{3}/16$ non-$f$-linear aligned triples in $\mathcal{F}$.

Next, assume that $m > 1$. For any $a \in \mathcal{F}$, let $S_a$ be the subspace $\{(a, z_2, \ldots, z_m) : z_2, \ldots, z_m \in \mathcal{F}\}$ of $\mathcal{F}^m$. Define $T_a$ to be the set of all aligned triples in the direction of the first coordinate that passes through $S_a$; that is,

$\vec{b} = (b_1, \ldots, b_m)$, $\vec{c} = (c_1, \ldots, c_m)$, and $\vec{d} = (d_1, \ldots, d_m)$ are an aligned triple in $T_a$ if and only if $b_j = c_j = d_j$ for all $2 \le j \le m$, and one of $b_1, c_1$ or $d_1$ is equal to $a$. For any $a \in \mathcal{F}$, the size of $T_a$ is $|T_a| = \binom{\alpha-1}{2}\alpha^{m-1}$. Let $T_a'$ be the set of aligned triples in $T_a$ that are not $f$-linear. Let $A = \{a \in \mathcal{F} : |T_a'|/|T_a| \ge 1/8\}$.

For each $a \in \mathcal{F}$, let $f_a : \mathcal{F}^{m-1} \to \mathcal{F}$ be $f_a(z_2, \ldots, z_m) = f(a, z_2, \ldots, z_m)$. Let $B = \{a \in \mathcal{F} : \Delta_{ml}(f_a) \ge 1/8\}$. We consider two cases.

*Case 1.* $|A \cup B| \le \alpha - 2$. Then, there are two points $a, b \in \mathcal{F}$ that are neither in $A$ nor in $B$. Thus, both $\Delta_{ml}(f_a)$ and $\Delta_{ml}(f_b)$ are less than $1/8$. Let $g_a$ and $g_b$ be the multilinear functions such that $\Delta(f_a, g_a) = \Delta_{ml}(f_a)$ and $\Delta(f_b, g_b) = \Delta_{ml}(f_b)$. Let $g : \mathcal{F}^m \to \mathcal{F}$ be the unique multilinear extension of $g_a$ and $g_b$ such that $g(a, \vec{z}) = g_a(\vec{z})$ and $g(b, \vec{z}) = g_b(\vec{z})$ for all $\vec{z} \in \mathcal{F}^{m-1}$. That is, $g(x, \vec{z}) = g_a(\vec{z})(x - b)/(a - b) + g_b(\vec{z})(x - a)/(b - a)$. We claim that $\Delta(f, g) \le 1/2$.

*Proof of Claim.* Let $\vec{c} = (c_1, c_2, \ldots, c_m)$ be a random point of $\mathcal{F}^m$. We consider two cases. In the first case, if $c_1 \in \{a, b\}$, then, by the property that both $\Delta(f_a, g_a)$ and $\Delta(f_b, g_b)$ are less than $1/8$, we know that

$$\Pr[f(\vec{c}) = g(\vec{c}) \mid c_1 \in \{a, b\}] \ge \frac{7}{8}.$$

Next, assume that $c_1 \notin \{a, b\}$. Define $\vec{u} = (a, c_2, \ldots, c_m)$ and $\vec{v} = (b, c_2, \ldots, c_m)$. As shown above, we have

$$\Pr[f(\vec{u}) = g(\vec{u}), f(\vec{v}) = g(\vec{v})] \ge \frac{3}{4}.$$

Let us randomly pick a point $d_1 \in \mathcal{F} - \{a, b, c_1\}$, and let $\vec{d} = (d_1, c_2, \ldots, c_m)$. Then, the triple $\{\vec{u}, \vec{c}, \vec{d}\}$ is a randomly chosen triple in $T_a$, and so, by the assumption that $a \notin A$,

$$\Pr[\{\vec{u}, \vec{c}, \vec{d}\} \text{ is } f\text{-linear}] \ge \frac{7}{8}.$$

Similarly,

$$\Pr[\{\vec{v}, \vec{c}, \vec{d}\} \text{ is } f\text{-linear}] \ge \frac{7}{8}.$$

Together, the probability is greater than or equal to $1/2$ that $f$ and $g$ agree at $\vec{u}$ and $\vec{v}$, and $\{\vec{u}, \vec{v}, \vec{c}, \vec{d}\}$ is $f$-linear. This implies that

$$\Pr[f(\vec{c}) = g(\vec{c}) \text{ and } f(\vec{d}) = g(\vec{d})] \ge \frac{1}{2}.$$

The above shows that

$$\Pr[f(\vec{c}) = g(\vec{c}) \mid c_1 \notin \{a, b\}] \ge \frac{1}{2}.$$

Since this probability is at least $1/2$ for both cases, the claim follows.

From the above claim and Lemma 11.9, we know that $\delta_{nl}(f) \geq \delta/2m \geq 1/16m$. That is, the number of non-$f$-linear triples is $\geq M/16m > \binom{\alpha}{3}(\alpha - 1)^{m-1}/16$, and the induction statement is proved for Case 1.

*Case 2.* $|A \cup B| \geq \alpha - 1$. For each $a \in A$, there are $\binom{\alpha-1}{2}\alpha^{m-1}$ aligned triples in $T_a$ and more than $1/8$ of them are in $T_a'$. Therefore, $|T_a'| \geq \binom{\alpha-1}{2}\alpha^{m-1}/8$. Since each non-$f$-linear triple may appear in at most three $T_a'$'s, there are totally at least $|A|\binom{\alpha-1}{2}\alpha^{m-1}/(3 \cdot 8) = |A|\binom{\alpha}{3}\alpha^{m-2}/8$ non-$f$-linear triples in $\bigcup_{a \in A} T_a'$.

For each $a \in B$, $\Delta_{ml}(f_a) \geq 1/8$. By the inductive hypothesis, we know that there are at least $\binom{\alpha}{3}(\alpha-1)^{m-2}/16$ non-$f_a$-linear aligned triples in $\mathcal{F}^{m-1}$. Each of these triples $\{\vec{b}, \vec{c}, \vec{d}\}$ in $\mathcal{F}^{m-1}$ can be mapped uniquely to a non-$f$-linear triple $\{(a, \vec{b}), (a, \vec{c}), (a, \vec{d})\}$. Notice that none of these triples belongs to $\bigcup_{a \in A} T_a'$, since they have the same first coordinate value, while those in $\bigcup_{a \in A} T_a'$ have different first coordinate values.

Altogether, we have at least

$$\frac{1}{8}|A|\binom{\alpha}{3}\alpha^{m-2} + \frac{1}{16}|B|\binom{\alpha}{3}(\alpha - 1)^{m-2}$$

$$\geq \frac{1}{16}|A \cup B|\binom{\alpha}{3}(\alpha - 1)^{m-2} \geq \frac{1}{16}\binom{\alpha}{3}(\alpha - 1)^{m-1}$$

non-$f$-linear aligned triples. This completes the proof of Case 2 as well as the induction statement.

Now, from the induction statement, we know that

$$\delta_{nl}(f) \geq \frac{\binom{\alpha}{3}(\alpha - 1)^{m-1}}{16M} = \left(1 - \frac{1}{\alpha}\right)^{m-1}\frac{1}{16m} \geq \left(1 - \frac{m-1}{\alpha}\right)\frac{1}{16m} \geq \frac{1}{32m},$$

since $\alpha > 12m$. This completes the proof of the lemma. ∎

Now we are ready to describe the proof system for the multilinearity test.

*Multilinearity Test System.* The verifier and the prover agree in advance at the parameters $\alpha = |\mathcal{F}|$ and $m$, with $96m < \alpha - 2$. The inputs are two rational numbers $0 < \epsilon < 1$ and $12m/(\alpha - 2) < \delta < 1$. The prover presents a function $f : \mathcal{F}^m \to \mathcal{F}$ in the form of an oracle. The verifier does the following:

(1) Let $d_1 = \lceil -\ln\epsilon \rceil$ and $d_2 = \lceil 1/\delta \rceil$. Select $N = 32md_1d_2$ random aligned triples $\{\vec{a}_i, \vec{b}_i, \vec{c}_i\}$, $1 \leq i \leq N$, from $\mathcal{F}^m$.

(2) For each $i$, $1 \leq i \leq N$, ask the oracle $f$ for its values at $\vec{a}_i, \vec{b}_i$ and $\vec{c}_i$. Verify whether they are $f$-linear. Reject if any triple $\{\vec{a}_i, \vec{b}_i, \vec{c}_i\}$ is not $f$-linear; accept otherwise.

*Correctness.* We show that the above proof system works correctly. That is, if $f$ is multilinear, then the verifier accepts with probability 1, and if $f$ is not $\delta$-close to a multilinear function, then the verifier rejects with probability $\geq 1 - \epsilon$.

The first half is obvious, since all aligned triples are $f$-linear if $f$ is multi-linear. For the second half, assume that $f$ is not $\delta$-close. By Lemma 11.10, the probability that a single random aligned triple passes the $f$-linearity test is at most $1 - \delta/32m$. So the probability that the verifier accepts is $\leq (1 - \delta/32m)^{d_1 32m/\delta} < e^{-d_1} \leq \epsilon$.

*Complexity.* The verifier needs to pick $O(m)$ random points from $\mathcal{F}^m$ (assuming that $\delta$ and $\epsilon$ are constants independent of $m$), or, $O(m^2)$ points from $\mathcal{F}$. Each point in $\mathcal{F}$ needs $\lceil \log \alpha \rceil$ bits. Therefore, the total number of random bits needed by the verifier is $O(m^2 \log \alpha)$. The verifier queries $f$ for $O(m)$ values each of $O(\log \alpha)$ bits. So the total number of query bits is $O(m \log \alpha)$. It is easy to see that the total runtime of the verifier is $O(m)$ field operations, and so is polynomial in $m + \log \alpha$.

### 11.2.3    Sum Check System

In the following, let $\mathcal{F}$ be a field of size $\alpha$ and let $m$, $d$ be two integers such that $2md < \alpha$.

Assume that $f : \mathcal{F}^m \to \mathcal{F}$ is an $m$-variate polynomial function of degree $d$ in $\mathcal{F}$, $H$ is a subset of $\mathcal{F}$, and $c$ is a constant in $\mathcal{F}$. The sum check system decides whether $\sum_{\vec{z} \in H^m} f(\vec{z}) = c$. For any $i$, $1 \leq i \leq m$, and any $a_1, \ldots, a_{i-1} \in \mathcal{F}$, define

$$f_{a_1, \ldots, a_{i-1}}(x) = \sum_{z_{i+1}, \ldots, z_m \in H} f(a_1, \ldots, a_{i-1}, x, z_{i+1}, \ldots, z_m).$$

Thus, $f_{a_1, \ldots, a_{i-1}}$ is a univariate polynomial of degree at most $d$. (When $i = 1$, we let $f_\lambda(x) = \sum_{z_2, \ldots, z_m \in H} f(x, z_2, \ldots, z_m)$.)

> *Sum Check System.* The prover and the verifier agree at the parameters $\alpha$ and $m$. The inputs are an integer $d > 0$ such that $2md < \alpha$, a constant $c \in \mathcal{F}$, a constant $\epsilon$ such that $2md/\alpha < \epsilon < 1$, and a subset $H \subseteq \mathcal{F}$. The proof consists of an oracle function $f : \mathcal{F}^m \to \mathcal{F}$, and a table $\Pi$ of $\sum_{i=0}^{m-1} \alpha^i$ items, each item consisting of $d + 1$ constants in $\mathcal{F}$. That is, for each $i$, $1 \leq i \leq m$, and each $a_1, \ldots, a_{i-1} \in \mathcal{F}$, the item $\Pi(a_1, \ldots, a_{i-1})$ consists of $d + 1$ constants in $\mathcal{F}$, representing a degree-$d$ univariate polynomial function $g_{a_1, \ldots, a_{i-1}}$. The verifier does the following to check the condition $\sum_{\vec{z} \in H^m} f(\vec{z}) = c$:
>
> (1) Select random elements $a_1, \ldots, a_m$ from $\mathcal{F}$; let $c_0 := c$.
> (2) For each $i$ from 1 to $m$, perform the following:
>>     (2.$i$) If $\sum_{x \in H} g_{a_1, \ldots, a_{i-1}}(x) \neq c_{i-1}$ then reject, else let $c_i := g_{a_1, \ldots, a_{i-1}}(a_i)$.
> (3) If $c_m \neq f(a_1, \ldots, a_m)$ then reject, else accept.

The following two lemmas prove the correctness of the sum check system.

**Lemma 11.11** *Assume that the function $f$ is a polynomial function of degree $\leq d$.*

*(a) If $\sum_{\vec{z} \in H^m} f(\vec{z}) = c$, then there is an additional proof $\Pi$ on which the verifier accepts with probability 1.*

*(b) If $\sum_{\vec{z} \in H^m} f(\vec{z}) \neq c$, then for any $\Pi$, the verifier rejects with probability $\geq 1 - \epsilon/2$.*

*Proof.* (a) In this case, the prover can let $g_{a_1,\ldots,a_{i-1}} = f_{a_1,\ldots,a_{i-1}}$ for all $1 \leq i \leq m$ and all $a_1, \ldots, a_m \in \mathcal{F}$. Then, $c_0 = \sum_{\vec{z} \in H^m} f(\vec{z}) = \sum_{x \in H} f_\lambda(x)$ and so it passes step (2.1). Next, for each $1 \leq i < m$, $c_i$ is defined in step (2.$i$) as

$$f_{a_1,\ldots,a_{i-1}}(a_i) = \sum_{z_{i+1},\ldots,z_m \in H} f(a_1, \ldots, a_i, z_{i+1}, \ldots, z_m),$$

and so it passes the test of step (2.($i+1$)). Similarly, $c_m$ passes the test of step (3). Therefore, the verifier accepts with probability 1.

(b) We make a simple observation:

*Claim.* Let $1 \leq i \leq m$. Assume that at the start of step (2.$i$),

$$c_{i-1} \neq \sum_{z_i,\ldots,z_m \in H} f(a_1, \ldots, a_{i-1}, z_i, \ldots, z_m),$$

and assume that the verifier does not reject at step (2.$i$). Then, at the end of (2.$i$), with probability $\geq 1 - d/\alpha$.

$$c_i \neq \sum_{z_{i+1},\ldots,z_m \in H} f(a_1, \ldots, a_i, z_{i+1}, \ldots, z_m).$$

*Proof of Claim.* If $g_{a_1,\ldots,a_{i-1}} = f_{a_1,\ldots,a_{i-1}}$, then the verifier would reject at step (2.$i$). Therefore, it must be true that $g_{a_1,\ldots,a_{i-1}} \neq f_{a_1,\ldots,a_{i-1}}$. Since both $g_{a_1,\ldots,a_{i-1}}$ and $f_{a_1,\ldots,a_{i-1}}$ are univariate polynomials of degree $\leq d$, they agree at most at $d$ points in $\mathcal{F}$. Thus, $c_i = g_{a_1,\ldots,a_{i-1}}(a_i)$ is equal to $f_{a_1,\ldots,a_{i-1}}(a_i)$ with probability $\leq d/\alpha$.

Now, assume that the verifier does not reject at step (2). Then, by the end of step (2), from the above claim, the probability of $c_m \neq f(a_1, \ldots, a_m)$ is at least $1 - dm/\alpha \geq 1 - \epsilon/2$, and so the verifier rejects at step (3) with probability $\geq 1 - \epsilon/2$, if not sooner. ∎

**Lemma 11.12** *Assume that $f$ is $(\epsilon/2)$-close to a polynomial $\tilde{f}$ of degree $d$.*

*(a) If $\sum_{\vec{z} \in H^m} \tilde{f}(\vec{z}) = c$, then there is a proof $\Pi$ such that the verifier accepts with probability $\geq 1 - \epsilon/2$.*

*(b) If $\sum_{\vec{z} \in H^m} \tilde{f}(\vec{z}) \neq c$, then for any $\Pi$, the verifier rejects with probability $\geq 1 - \epsilon$.*

*Proof.* (a) From Lemma 11.11(a), the prover could design $g$ according to $\tilde{f}$ instead of $f$ to pass step (2) with probability 1. Then, by the beginning of step (3), $c_m = \tilde{f}(a_1, \ldots, a_m)$. Thus, it also passes the test at step (3) with probability $1 - \epsilon/2$ since $\Delta(f, \tilde{f}) \leq \epsilon/2$.

(b) From Lemma 11.11(b), if the verifier does not reject in step (2), then $c_m \neq \tilde{f}(a_1, \ldots, a_m)$ with probability $\geq 1 - \epsilon/2$. Since $\Delta(f, \tilde{f}) \leq \epsilon/2$, $c_m \neq f(a_1, \ldots, a_m)$ with probability $\geq 1 - \epsilon$. So, the verifier rejects at step (3) with probability at least $1 - \epsilon$.                                                        ∎

*Complexity.* We first remark that the algorithm for the verifier was written in an adaptive form. It is easy, though, to see how to implement it in a non-adaptive way: the verifier can first read $f(a_1, \ldots, a_m)$ and the polynomials $g_{a_1, \ldots, a_{i-1}}$, $1 \leq i \leq m$, before executing steps (2) and (3).

The only random bits used in the algorithm are those to produce $a_1, \ldots, a_m$. Each of these elements are from $\mathcal{F}$ of size $\alpha$ and so needs $\lceil \log \alpha \rceil$ bits, and the total number of random bits used is $O(m \log \alpha)$.

The verifier queries function $f$ only at one random point that needs $\lceil \log \alpha \rceil$ bits. For each $i$, $1 \leq i \leq m$, the verifier reads $d + 1$ coefficients for $g_{a_1, \ldots, a_{i-1}}$. So the total number of bits read from the oracle is $O(md \log \alpha)$.

The total runtime of the verifier is $O(mdh)$ operations in $\mathcal{F}$, where $h = |H|$.

## 11.3   *PCP* Characterization of *NP*

In the last section, we have proved that a set $A$ in *NEXPPOLY* has a polynomial-time probabilistically checkable proof. Let us review this proof for the special case when $A$ is in *NP*. First, we reduce the question of whether $w \in A$ to the question of whether a Boolean formula $\Phi_w$ is satisfiable. This formula $\Phi_w$ has $2^m$ variables and $2^m$ clauses, where $2^m = p(n)$ for some polynomial $p$ since we are dealing with a set $A \in NP$. That is, $m = c \log n$ for some constant $c$. Then, the probabilistic proof system $S_1$ checks whether $\Phi_w$ has a satisfying assignment $f$ in time polynomial in $m$. It uses $O(m^2 \log m)$ random bits and reads $O(m^2 \log m)$ bits from the oracles. Therefore, we have the following corollary, in which *polylog* denotes the collection of functions $c(\log n)^k$ for all $c, k > 0$.

**Corollary 11.13** $NP \subseteq PCP(polylog, polylog)$.

We remark that, in the multilinearity test system, we may use the de-randomization techniques to reduce the number of random bits used from $O(m^2 \log m)$ to $O(m \log m)$ (see Exercise 11.4). This improvement, however, is not sufficient for our need. In this section, we use a new approach to give a much stronger improvement and show the following optimal characterization for *NP*.

**Theorem 11.14** $NP = PCP(log, const)$.

The proof of this result is a modification of that of Theorem 11.4, but is much more involved. We divide the proof into three steps, each step improving the complexity of the proof system by a big factor. Each step also uses a new proof technique. We give a sketch first.

(1) Our first goal is to reduce the number of random bits used in Corollary 11.13 from $(\log n)^{O(1)}$ to $O(\log n)$ (but the verifier still reads $(\log n)^{O(1)}$ oracle bits).[3] If we temporarily ignore step (4) of the proof system $S_1$ and consider only steps (3) and (5), then we see that the number of random bits used is $O(m \log \alpha)$, where $m$ is the number of variables of the multilinear function $\hat{f}$ and $\alpha$ is the size of the field $\mathcal{F}$. In the proof of Theorem 11.4, we defined $\hat{f}$ to be the unique multilinear extension of the truth assignment $f$ for $\Phi_w$, and so the number of Boolean variables in $\Phi_w$ is equal to $2^m$. Here, we use the idea of *error-detecting/correcting* encoding to design a different function $\bar{f}$ to represent the truth assignment $f$ for $\Phi_w$. This new function $\bar{f}$ is a polynomial function of $m = O(\log n / \log \log n)$ variables and of degree $d = (\log n)^{O(1)}$, where $n$ is the size of $\Phi_w$. Notice that the number of random bits $O(m \log \alpha)$ used by the sum check system is independent of the degree $d$, and the error probability is as small as $O(md/\alpha)$. Therefore, by choosing a little larger $\alpha$, say $\alpha = m^{O(1)}$, the sum check system works also for the function $\bar{f}$ and only uses $O(m \log \alpha) = O(\log n)$ random bits. Similarly, step (3) works for any function $\bar{f}$ and so it also uses only $O(m \log \alpha) = O(\log n)$ random bits.

The main obstacle in this approach is thus step (4). Since our new function $\bar{f}$ is no longer a multilinear function, the multilinearity test system is no longer applicable. We need to design a new *low-degree test system* that tests whether a given oracle function is $\delta$-close to a polynomial function of a given degree. The idea of this low-degree test system uses the error-detecting property of our encoding scheme and it uses only $O(m \log \alpha)$ random bits. With this new low-degree test system, we obtain a new proof system $S_2$ for $A \in NP$ that uses $O(\log n)$ random bits but still queries $(\log n)^{O(1)}$ oracle bits.

(2) The second step is to reduce the number of query bits in the above system $S_2$ from $(\log n)^{O(1)}$ to $O(\log n)$. In the proof system $S_2$, the verifier $V$ works as follows: $V$ first tosses coins to get a sequence $r$ of $O(\log n)$ random bits, then $V$ reads $(\log n)^{O(1)}$ bits from the oracle, and finally $V$ decides whether to accept or to reject. We notice that once the sequence $r$ of random bits is fixed, the verifier actually behaves like a deterministic verifier. That is:

(i) The locations $\ell_r$ of the proof bits to be read (or, the queries to be made to the oracle) depend *deterministically* on the random bits $r$; and

(ii) There is a *deterministic* algorithm $V_d$ for the verifier that computes a function $V_d(r, x_r) \in \{0, 1\}$ in time $(\log n)^{O(1)}$, where $x_r$ is the sequence of bits obtained from the oracle at locations $\ell_r$.

By Cook's Theorem, for each $r$ of length $O(\log n)$, there is a Boolean formula $\phi_r$ of $(\log n)^{O(1)}$ Boolean variables such that $V_d(r, x_r) = 1$ if and only if

---

[3]In this section, we use the notation $f(n)^{O(1)}$ to denote a function $p(f(n))$ where $p$ is a polynomial.

$\phi_r$ is satisfiable (by $x_r$ and some other Boolean values $y_r$ for auxiliary variables). Now, we may imagine that the verifier does his/her jobs in a different order: $V$ first gets $r$, then computes the formula $\phi_r$, then reads $x_r$, and finally checks whether $x_r$ and some other values $y_r$ satisfy $\phi_r$. Notice that the last step of checking whether $x_r$ and $y_r$ satisfy $\phi_r$ is just another instance for SAT. The key idea here is to apply the probabilistic proof system $S_2$ *recursively* to this formula $\phi_r$ to reduce the number of queries by the verifier. That is, we ask the prover to present the encodings $\bar{x}_r$ and $\bar{y}_r$ of the truth assignments $x_r$ and $y_r$ for $\phi_r$ and some additional proofs showing that $x_r$ and $y_r$ indeed satisfy $\phi_r$. Since $\phi_r$ has only $k = (\log n)^{O(1)}$ Boolean variables, this proof can be checked by the verifier using additional $O(\log k) = O(\log\log n)$ random bits, and reading only $(\log k)^{O(1)} = (\log\log n)^{O(1)}$ bits from the oracle. Notice that the original verifier need not read any bits from the oracle. Thus, the total number of query bits is reduced to $(\log\log n)^{O(1)}$. In summary, this recursive proof system $S_3$ for $A$ uses only $O(\log n)$ random bits and reads only $(\log\log n)^{O(1)}$ oracle bits.

(3) The number of oracle bits read by the verifier of the proof system $S_3$ is $(\log\log n)^{O(1)}$. We may apply the idea of recursive proofs to system $S_3$ again. It will not, however, reduce the number of query bits to a constant. So, we design yet another proof system $S_4$ for set $A \in NP$ that queries only a constant number of bits to the oracle but uses $n^{O(1)}$ random bits. This proof system does not use the sum check system and avoids the reading of the big table $\Pi$ in that system. Now, we apply system $S_4$ to the subformulas $\phi_r$ of the system $S_3$ as described in step (2) above. This reduces the number of query bits to a constant but increases the number of random bits by only $(\log\log n)^{O(1)} = O(\log n)$ because each $\phi_r$ in system $S_3$ has only $(\log\log n)^{O(1)}$ Boolean variables. Thus this final system $S_5$, the composition of systems $S_3$ and $S_4$, accomplishes the task of verifying the proof by using $O(\log n)$ random bits and reading $O(1)$ bits of the oracle.

A number of technical problems arise in the above approach. In addition to the need of a stronger low-degree test system, we also need to require that the verifier of the proof systems $S_3$ and $S_4$ to be able to read encodings (of Boolean formulas $\phi_r$ and assignments $x_r$ and $y_r$) of a special form in steps (3) and (4). We describe the details in the following subsections.

### 11.3.1   Proof System for *NP* Using $O(\log n)$ Random Bits

In this section, we show that for each set $A \in NP$, there is a *PCP* system for $A$ in which the verifier uses $O(\log n)$ random bits and reads $(\log n)^{O(1)}$ oracle bits on any input of length $n$. We are going to work in a finite field $\mathcal{F}$ of size $\alpha$. Without loss of generality, we may just assume that $\alpha$ is a prime and $\mathcal{F} = \mathbf{Z}_\alpha$. Therefore, if $h < \alpha$, we assume that $\{0, 1, \ldots, h\} \subseteq \mathcal{F}$.

First we discuss the encoding scheme to be used in the proof. Our encoding scheme will encode a sequence $s$ of $l$ elements in $\mathcal{F}$ by a polynomial function $\bar{s}$ in $\mathcal{F}$. To describe the encoding, we first extend Lemma 11.5 to polynomial functions. Let $H = \{0, 1 \ldots, h\}$, in which $h < \alpha$. We say a function $g : \mathcal{F}^m \to \mathcal{F}$ is a *degree-d polynomial extension* of a function $f : H^m \to \mathcal{F}$ if $g$ is a polynomial function of degree $d$ and $g(\vec{a}) = f(\vec{a})$ for all $\vec{a} \in H^m$.

**Lemma 11.15** *For every function $f : H^m \to \mathcal{F}$, there is a polynomial extension $\bar{f}$ of $f$ that is of degree $d = mh$.*

*Proof.* For any $u \in H$, let $\ell_u$ be the unique univariate polynomial of degree $h$ such that $\ell_u(u) = 1$ and $\ell_u(v) = 0$ for all $v \in H - \{u\}$. Let

$$\bar{f}(z_1, \ldots, z_m) = \sum_{a_1, \ldots, a_m \in H} f(a_1, \ldots, a_m) \prod_{j=1}^{m} \ell_{a_j}(z_j). \qquad \blacksquare$$

*Remark.* The function $\bar{f}$ is not necessarily unique. It is unique if we require that each individual variable of $\bar{f}$ has degree $\leq h$. (See Exercise 11.3.)

Assume that $(h + 1)^m \geq l$. Then, there is a one-to-one mapping from set $\{1, \ldots, l\}$ to set $H^m$. (For instance, write each integer $i$, $1 \leq i \leq l$, in its $m$-digit, base-$(h+1)$ representation.) Therefore, each sequence $s$ of $l$ elements in $\mathcal{F}$ may be viewed as a function $f_s : H^m \to \mathcal{F}$, and the code $\bar{s}$ of $s$ is defined to be any degree-$(mh)$ polynomial extension $\bar{f}_s$ of $f_s$. We let $\sigma_1$ be the encoding function mapping $s$ to $\bar{f}_s$.

From the point of view of coding theory, the encoding scheme $\sigma_1$ has the error-detecting property in the following sense. Let $\mathcal{F}$ be a field of size $|\mathcal{F}| = \alpha$. For any two polynomials $f, g : \mathcal{F}^m \to \mathcal{F}$ of degree $d$, if $f \neq g$ then, by Lemma 8.3, $\Delta(f, g) \geq 1 - d/\alpha$. Thus, if $s_1 \neq s_2$, then $\sigma_1(s_1)$ and $\sigma_1(s_2)$ have a big distance. It implies that using a randomized algorithm, we can efficiently detect the difference between two polynomial codes $\sigma_1(s_1)$ and $\sigma_1(s_2)$. Furthermore, if $g$ is not a *polynomial code* (i.e., $g$ is not an image of a binary sequence under $\sigma_1$) but $g$ is $\delta$-close to a polynomial code $f$ with a small $\delta$, then $f$ is unique, and $f$ can be efficiently recovered from $g$. This is the error-correcting property of $\sigma_1$. We will formalize these properties in Section 11.3.2.

Now we begin the proof. From the proof of Theorem 11.4, it is clear that it suffices to show that SAT $\in PCP(log, polylog)$. Let $\Phi$ be an instance of SAT of $n$ variables and $n$ clauses. Let $h = \lceil \log n \rceil$, and $m$ be the least integer such that $(h+1)^m \geq n$. Then, $m = O(\log n / \log \log n)$ and $O(m \log m) = O(\log n)$.

We identify each $i$, $1 \leq i \leq n$, with a string in $H^m$, where $H = \{0, 1 \ldots, h\}$. Then, following the proof of Theorem 11.4, we define, for each $j = 1, 2, 3$, Boolean formulas $\phi_j : H^{2m} \to \{0, 1\}$ and $\psi_j : H^m \to \{0, 1\}$, and an oracle Boolean function $G^f : H^{4m} \to \{0, 1\}$ by

$$G^f(s_1, s_2, s_3, t) = \prod_{j=1}^{3} \phi_j(s_j, t)(\psi_j(t) - f(s_j)).$$

This oracle Boolean function satisfies the property that there is a Boolean function $f : \{1, \ldots, n\} \to \{0, 1\}$ satisfying $\Phi$ if and only if $G^f(z) = 0$ for all

$z \in H^{4m}$. (Note that in the first half of the above statement, we treat $f$ as a Boolean function from $\{1, \ldots, n\}$ to $\{0, 1\}$, and in the second half we treat $f$ as a function from $H^m$ to $\{0, 1\}$.)

For any $c > 0$ and any function $g : H^{cm} \to \mathcal{F}$, we let $\bar{g} : \mathcal{F}^{cm} \to \mathcal{F}$ denote a degree-$(cmh)$ polynomial extension of $g$. For any function $f : \mathcal{F}^m \to \mathcal{F}$, we let $\bar{G}^f : \mathcal{F}^{4m} \to \mathcal{F}$ be the following function:

$$\bar{G}^f(\vec{u}_1, \vec{u}_2, \vec{u}_3, \vec{v}) = \prod_{j=1}^{3} \bar{\phi}_j(\vec{u}_j, \vec{v})(\bar{\psi}_j(\vec{v}) - f(\vec{u}_j)).$$

If $f$ is a polynomial function of degree $mh$, then $\bar{G}^f$ is a polynomial function of degree $9mh$.

Next, we extend Lemma 11.6 to the following form:

**Lemma 11.16** *There exists a family $\{R_i\}_{i=1}^{\alpha^k}$ of polynomials of $k$ variables and of degree $kh$ such that for any nonzero function $f : H^k \to \mathcal{F}$,*

$$\Pr\left[ \sum_{\vec{z} \in H^k} f(\vec{z}) R_i(\vec{z}) = 0 \right] < \frac{kh}{\alpha},$$

*when $R_i$ is chosen randomly from the family. Furthermore, the polynomial $R_i$ may be computed from $i$ in time polynomial in $k + h$.*

*Proof.* The proof is a simple modification of that of Lemma 11.6. See Exercise 11.7. ∎

For each $\vec{z} \in \mathcal{F}^{4m}$, let $Q_i^f(\vec{z}) = \bar{G}^f(\vec{z}) R_i(\vec{z})$. If $f$ is a polynomial function of degree $mh$, then $Q_i$ is a polynomial function of degree $13mh$. In the proof system $S_1$, step (3) needs a random number $i$, $1 \leq i \leq \alpha^{4m}$. In our setting here, if we have $\alpha = m^{O(1)}$, then the integer $i$ only needs $O(m \log \alpha) = O(m \log m) = O(\log n)$ random bits. Similarly, in step (5) of applying the sum check system, the number of random bits used is also $O(m \log \alpha) = O(\log n)$.

However, the multilinearity test system is no longer applicable to step (4), since $\bar{f}$ is not necessarily a multilinear function. We need a stronger proof system, called the *low-degree test system*, which will be developed in Section 11.3.2. We summarize its properties here. In the following, $\delta_0$ is an absolute constant, $0 < \delta_0 < 1$, which is to be defined in the next section.

> *Low-Degree Test System.* The prover and the verifier agree at the field $\mathcal{F}$ of size $\alpha$ and integers $m, d > 0$ such that $100md^3 < \alpha$. The inputs are constants $\epsilon > 0$ and $0 < \delta \leq \delta_0$. The proof consists of a function $f : \mathcal{F}^m \to \mathcal{F}$ and an additional string $\Pi$ (both in the form of oracles). There is a verifier having the following properties:
>
> (i) If $f$ is a polynomial function of degree $d$, then there exists an additional proof $\Pi$ such that the verifier accepts with probability 1.
>
> (ii) If $f$ is not $\delta$-close to any polynomial of degree $d$, then for any $\Pi$, the verifier rejects with probability $\geq 1 - \epsilon$.

(iii) The verifier uses $O(m \log \alpha)$ random bits, reads $O(1)$ values of $f$ at random points and totally $O(d \log \alpha)$ oracle bits, and runs in time polynomial in $m + d + \alpha$.

Using our encoding scheme and using the low-degree test system to replace the multilinearity test system, we obtain a new proof system $S_2$ for $A \in NP$.

*Proof System $S_2$.* The input is a Boolean formula $\Phi$ of $n$ variables and $n$ clauses. Let $m$ be the least integer such that $(\lceil \log n \rceil + 1)^m \geq n$. The prover and the verifier agree at a finite field $\mathcal{F}$ of size $\alpha = m^{O(1)}$ and the error bound $\epsilon_0 > 0$ with $\alpha > 104m^4h^3/\epsilon_0$. The proof consists of a function $f : \mathcal{F}^m \to \mathcal{F}$ and additional $\alpha^{4m}$ strings $\Pi_i$, $1 \leq i \leq \alpha^{4m}$, in the form of oracles. The algorithm for the verifier is the following:

(1) Construct function $G^f$ as described above.

(2) Construct the polynomial function $\bar{G}^f$.

(3) Randomly pick an integer $i$, $1 \leq i \leq \alpha^{4m}$, and construct the polynomial $R_i$ as in Lemma 11.16 (with $k = 4m$). Let $Q_i^f(\vec{z}) = \bar{G}^f(\vec{z})R_i(\vec{z})$.

(4) Run the low-degree test system to check whether $f$ is $\delta$-close to a degree-$(mh)$ polynomial, with the parameter $\epsilon = \epsilon_0$ and $\delta = \min\{\delta_0, \epsilon_0/8\}$; reject if this test fails.

(5) Use $\Pi_i$ as the additional proof to run the sum check system to see whether $\sum_{\vec{z} \in H^m} Q_i^f(\vec{z}) = 0$, with the parameters $d = 13mh$ and $\epsilon = \epsilon_0/4$. Reject if the test fails; accept otherwise.

**Lemma 11.17** *(a) If $\Phi \in$ SAT, then there exist two functions $f$ and $\Pi$ with respect to which the verifier accepts with probability 1.*

*(b) If $\Phi \notin$ SAT, then the verifier rejects with probability $\geq 1 - \epsilon_0$ for all oracles $f$ and $\Pi$.*

*(c) On input $\Phi$ of $n$ variables and $n$ clauses, the verifier uses $O(\log n)$ random bits, queries $O(1)$ values of $f$ all at random points, reads totally $p(\log n)$ bits from the oracles for some polynomial $p$, and runs in time polynomial in $n$.*

*Proof.* We note that by our choice of $\alpha$, we have $\alpha > 100m^4h^3$ and $\alpha > 104m^2h/\epsilon_0$. Thus, the conditions of both the low-degree test system and the sum check system are satisfied. The rest of the proofs for parts (a) and (b) are essentially the same as Lemma 11.7. For part (c), we note that the total number of random bits used is $O(m \log \alpha) = O(\log n)$. Also, both the low-degree test system and the sum check system read $O(1)$ values of $f$ all at random points, and the total number of query bits is $O(d \log \alpha) + O(md \log \alpha) = (\log n)^{O(1)}$ (where $d = O(mh)$). ∎

From Lemma 11.17, we have

**Theorem 11.18** $NP \subseteq PCP(log, polylog)$.

### 11.3.2   Low-Degree Test System

In this section, we present the low-degree test system that tests whether a given oracle function is $\delta$-close to a degree-$d$ polynomial function. In addition, we also present a companion system, called the *polynomial recovering system*, which, for a given function $f$ that is $\delta$-close to a degree-$d$ polynomial function $\tilde{f}$, recovers the function values of $\tilde{f}$ at some given points $\vec{a}_1, \ldots, \vec{a}_k$. The polynomial recovering system is not needed for the proof system $S_2$. Instead, we observed that if $\vec{a}$ is a random point in $\mathcal{F}^m$, then the probability of $f(\vec{a})$ equal to $\tilde{f}(\vec{a})$ is at least $1 - \delta$ as long as $f$ passes the low-degree test. In the following sections, however, we will encounter the situation in which we need to get $\tilde{f}(\vec{a})$ at some points $\vec{a}$ which are not totally random. In such cases, the polynomial recovering system allows us to randomize the input and makes the output $\tilde{f}(\vec{a})$ more reliable. From the coding-theoretic point of view, the low-degree test system shows that our encoding scheme has the *error-detecting* property and the polynomial recovering system shows that our encoding scheme has the *error-correcting* property.

Let $\mathcal{F}$ be a finite field of size $\alpha$, and $m$ be an integer such that $\alpha = m^{O(1)}$. In Section 11.2.2, we defined the notion of *axis-parallel lines*. We extend this notion to arbitrary lines. For any two points $\vec{a} = (a_1, \ldots, a_m)$, $\vec{b} = (b_1, \ldots, b_m) \in \mathcal{F}^m$, and any $c \in \mathcal{F}$, we write $\vec{a} + \vec{b}$ to denote the point $(a_1 + b_1, \ldots, a_m + b_m)$ and write $c \cdot \vec{a}$ to denote the point $(ca_1, \ldots, ca_m)$. For any $\vec{a}, \vec{b} \in \mathcal{F}^m$, let $L_{\vec{a},\vec{b}}$ denote the set $\{\vec{a} + t \cdot \vec{b} : t \in \mathcal{F}\}$. We call $L_{\vec{a},\vec{b}}$ a *line*. Each line has more than one representations; for instance, for any $\vec{a}, \vec{b} \in \mathcal{F}^m$ and any $c \in \mathcal{F}$, $L_{\vec{a},\vec{b}} = L_{\vec{a}, c \cdot \vec{b}}$. For any line $L$, however, we fix a canonical representation $L_{\vec{a},\vec{b}}$ for it. Let $f : \mathcal{F}^m \to \mathcal{F}$ be a function and $L = L_{\vec{a},\vec{b}}$ a line. We let $f_L$ or $f_{\vec{a},\vec{b}}$ denote the restriction of $f$ on line $L$; that is, $f_{\vec{a},\vec{b}}(t) = f(\vec{a} + t \cdot \vec{b})$. It can be proven that $f$ is a polynomial function of degree $d$ if and only if $f_L$ is a univariate polynomial of degree $d$ for all lines $L$ in $\mathcal{F}^m$ (Exercise 11.6). The algorithm of the low-degree test system is based on the following stronger property: if, for most lines $L$ in $\mathcal{F}^m$, $f_L$ is close to a univariate polynomial of degree $d$, then $f$ itself is close to a polynomial function of degree $d$. To make this statement more precise, we introduce some notation. Let $\Delta_d^L(f)$ be the minimum distance between $f_L$ and a univariate degree-$d$ polynomial. We define $\Delta_d(f)$ to be the average of $\Delta_d^L(f)$ over all lines $L$ in $\mathcal{F}^m$. The proof of the following lemma is quite involved but it is purely algebraic. We omit it here.

**Lemma 11.19** *Let $d, m, \alpha$ satisfy condition $100d^3m < \alpha$. There exist two constants $0 < \delta_0 < 1$ and $c_0 \geq 1$ such that for any function $f : \mathcal{F}^m \to \mathcal{F}$, if $\Delta_d(f) \leq \delta$ for some $\delta < \delta_0$, then $f$ is $(c_0\delta)$-close to a degree-$d$ polynomial.*

Now we are ready to present the low-degree test system.

> *Low-Degree Test System.* The prover and the verifier agree at the parameters $\alpha$, $m$, $d$ such that $100d^3m < \alpha$. The inputs to

the system are two constant rationals $\delta$, $\epsilon$ satisfying $0 < \epsilon < 1$ and $0 < \delta \leq \delta_0$. The prover presents a function $f : \mathcal{F}^m \to \mathcal{F}$, and an additional proof $\Gamma$, that, for each line $L$, consists of $d + 1$ constants in $\mathcal{F}$ representing a univariate polynomial $g_L$ of degree $d$. The algorithm for the verifier is as follows:

(1) Select $k = \lceil -c_0 \ln \epsilon / \delta \rceil$ random lines $L_1, \ldots, L_k$, and for each $i$, $1 \leq i \leq k$, a random element $t_i \in \mathcal{F}$. Assume that $L_i = L_{\vec{a}_i, \vec{b}_i}$, we let $\vec{c}_i = \vec{a}_i + t_i \cdot \vec{b}_i$.

(2) Ask the oracle for the polynomials $g_{L_1}, \ldots, g_{L_k}$, and the values of $f(\vec{c}_1), \ldots, f(\vec{c}_k)$.

(3) For each $i$, $1 \leq i \leq k$, check whether $g_{L_i}(t_i) = f(\vec{c}_i)$. If this is not true for any $i$, then reject, else accept.

*Correctness.* We show that (a) if $f$ is a polynomial function of degree $d$, then there exists a proof $\Gamma$ such that the verifier accepts with probability 1; and (b) if $f$ is not $\delta$-close to any degree-$d$ polynomial, then, for any $\Gamma$, the verifier rejects with probability $\geq 1 - \epsilon$.

Part (a) is easy: if $f$ is a degree-$d$ polynomial, then the function $f$ restricted to each line $L$ is a univariate degree-$d$ polynomial. The prover can simply let each $g_L$ be the polynomial $f_L$.

For part (b), assume that $f$ is not $\delta$-close to any degree-$d$ polynomial. Then, by Lemma 11.19, $\Delta_d(f) > \delta/c_0$. Notice that, by the definition of $\Delta_d^L(f)$, for any fixed line $L$, $\Delta_d^L(f) \leq \Pr_t[f_L(t) \neq g_L(t)]$, when the point $t$ is chosen randomly. This means that the average of $\Pr_t[f_L(t) = g_L(t)]$ over all lines $L$ is at most $1 - \Delta_d(f) < 1 - \delta/c_0$, or, equivalently,

$$\Pr_{L,t}[f_L(t) = g_L(t)] < 1 - \frac{\delta}{c_0},$$

when $L$ and $t$ are chosen randomly and independently. The probability of $\Gamma$ passing all $k$ tests is at most

$$\left(1 - \frac{\delta}{c_0}\right)^k \leq \left(1 - \frac{\delta}{c_0}\right)^{(c_0/\delta)(-\ln \epsilon)} < e^{(-1)(-\ln \epsilon)} = \epsilon.$$

*Complexity.* Each line $L$ is specified by two points in $\mathcal{F}^m$. The verifier selects a constant $k$ number of random lines, and needs $2km\lceil \log \alpha \rceil = O(m \log \alpha)$ random bits. Each polynomial is specified by $d + 1$ constants of $\mathcal{F}$. The verifier needs $k$ values of $f$ and $k$ polynomials. The total number of query bits is thus $O(d \log \alpha)$. The total time used by the verifier is obviously bounded by a polynomial in the number of query bits and so is $p(md \log \alpha)$ for some polynomial $p$.

Next we present the polynomial recovering system. To do this, we further extend the notion of lines to the notion of *curves*. A *degree-$k$ curve* in $\mathcal{F}^m$ is a set of points in $\mathcal{F}^m$ of the form $C_{\vec{a}_0, \ldots, \vec{a}_k} = \{\vec{a}_0 + t \cdot \vec{a}_1 + \cdots + t^k \cdot \vec{a}_k : t \in \mathcal{F}\}$, in which $\vec{a}_0, \ldots, \vec{a}_k \in \mathcal{F}^m$. Similar to lines, a curve does not have a unique representation; we fix for each curve $C$ a canonical representation $C_{\vec{a}_0, \ldots, \vec{a}_k}$.

For each degree-$k$ curve $C$ of the canonical form $C_{\vec{a}_0, \ldots, \vec{a}_k}$, and for each $t \in \mathcal{F}$, we write $C(t)$ to denote the point $\vec{a}_0 + t \cdot \vec{a}_1 + \cdots + t^k \cdot \vec{a}_k$. For every $k < \alpha$ and for every sequence $\vec{z}_0, \ldots, \vec{z}_k$ of $k+1$ points in $\mathcal{F}^m$, there is a unique degree-$k$ curve $C$ such that $C(j) = \vec{z}_j$ for each $0 \le j \le k$.[4] We write $C^{\vec{z}_0, \ldots, \vec{z}_k}$ to denote this curve.

Let $f : \mathcal{F}^m \to \mathcal{F}$ be a function of $m$ variables. For any curve $C$ in $\mathcal{F}^m$, we write $f_C : \mathcal{F} \to \mathcal{F}$ to denote the function $f$ restricted to the curve $C$, that is, $f_C(t) = f(C(t))$. Assume that $f$ is a degree-$d$ polynomial function. Then, for every degree-$k$ curve $C$, $f_C$ is a univariate polynomial of degree $dk$. The algorithm for the polynomial recovering system is based on the stronger property that if $f$ is very close to a degree-$d$ polynomial $\tilde{f}$, then for any $\vec{z}_1, \ldots, \vec{z}_k \in \mathcal{F}^m$, $f_C$ is close to $\tilde{f}_C$ on most degree-$k$ curves $C$ that pass through the points $\vec{z}_1, \ldots, \vec{z}_k$.

**Lemma 11.20** Let $\vec{z}_1, \ldots, \vec{z}_k$ be $k$ distinct points in $\mathcal{F}^m$, with $k < \alpha^m/2$. Let $\vec{x}$ be a random point in $\mathcal{F}^m - \{\vec{z}_1, \ldots, \vec{z}_k\}$. For each $\vec{x}$, we write $C^{\vec{x}}$ for $C^{\vec{z}_1, \ldots, \vec{z}_k, \vec{x}}$. If $f$ is $\delta$-close to a degree-$d$ polynomial $\tilde{f}$, then $\mathrm{Pr}_{\vec{x}}[\Delta(f_{C^{\vec{x}}}, \tilde{f}_{C^{\vec{x}}}) > \sqrt{2\delta}] < \sqrt{2\delta}$.

*Proof.* See Exercise 11.8.                                                    ∎

> *Polynomial Recovering System.* The inputs to the system are an integer $d > 0$, $k$ points $\vec{z}_1, \ldots, \vec{z}_k \in \mathcal{F}^m$ (at which the function values $\tilde{f}(\vec{z}_i)$ are to be computed), with $2dk < \alpha$, and a constant $\delta$, $dk/\alpha < \delta < 1$. The proof consists of a function $f : \mathcal{F}^m \to \mathcal{F}$ and an additional proof $\Pi$. the proof $\Pi$ contains, for each $\vec{x} \in \mathcal{F}^m$, $dk + 1$ elements of $\mathcal{F}$ representing a univariate degree-$dk$ polynomial $g_{\vec{x}}$. The verifier does the following:
>
> (1) Pick a random $\vec{x} \in \mathcal{F}^m$ and a random $t \in \mathcal{F}$. Let $C = C^{\vec{x}}$ be the unique curve such that $C(j) = \vec{z}_{j+1}$ for $0 \le j \le k - 1$, and $C(k) = \vec{x}$.
>
> (2) Read $f_C(t)$ and the polynomial $g_{\vec{x}}$.
>
> (3) Check that $g_{\vec{x}}(t) = f_C(t)$. If this fails, then reject; else output $g_{\vec{x}}(0), \ldots, g_{\vec{x}}(k - 1)$.

**Lemma 11.21** *Assume that $f$ is $\delta$-close to a degree-$d$ polynomial $\tilde{f}$. For any $\Pi$, if the verifier does not reject, then the probability that the verifier outputs correct values of $\tilde{f}(\vec{z}_1), \ldots, \tilde{f}(\vec{z}_k)$ is greater than $1 - 4\sqrt{\delta}$.*

*Proof.* A wrong output can occur only if $g_{\vec{x}}$ is not identical to $\tilde{f}_C$ but $g_{\vec{x}}(t) = f_C(t)$. This can happen in three cases:

*Case 1.* $g_{\vec{x}} \ne \tilde{f}_C$ but $g_{\vec{x}}(t) = \tilde{f}_C(t) = f_C(t)$. Since both $g_{\vec{x}}$ and $\tilde{f}_C$ are univariate polynomials of degree $dk$, this case only happens with probability $\le dk/\alpha$.

---

[4] We assume that $\mathcal{F} = \mathbf{Z}_\alpha$ and so $\{0, 1 \ldots, k\} \subseteq \mathcal{F}$.

*Case 2.* $g_{\vec{x}} \neq \tilde{f}_C$, $g_{\vec{x}}(t) = f_C(t) \neq \tilde{f}_C(t)$, but $\Delta(f_C, \tilde{f}_C) \leq \sqrt{2\delta}$. The probability of this happening is $\leq \sqrt{2\delta}$.

*Case 3.* $g_{\vec{x}} \neq \tilde{f}_C$, $g_{\vec{x}}(t) = f_C(t) \neq \tilde{f}_C(t)$, and $\Delta(f_C, \tilde{f}_C) > \sqrt{2\delta}$. By Lemma 11.20, this case happens with probability $\leq \sqrt{2\delta}$.

Altogether, the error probability is bounded by $2\sqrt{2\delta} + dk/\alpha < 4\sqrt{\delta}$, since $dk/\alpha < \delta < \sqrt{\delta}$. ∎

*Complexity.* It is easy to check that the verifier of the polynomial recovering system uses $O(m \log \alpha)$ random bits and reads one value of $f$ and a polynomial function of $dk+1$ coefficients, with totally $O(dk \log \alpha)$ query bits. The runtime of the verifier is polynomial in $dk \log \alpha$.

### 11.3.3 Composition of Two *PCP* Systems

In this section, we compose the *PCP* system $S_2$ with itself to reduce the number of query bits to $(\log \log n)^{O(1)}$. We have discussed informally how to compose two *PCP* systems $X_1$ and $X_2$ into a new system $X_3$. Now we formalize the ideas here. First, let us assume, without loss of generality, that all the proof systems considered here are proof systems for the problem SAT, since SAT is *NP*-complete. Suppose, for $i = 1, 2$, the system $X_i$ has the complexity bounds $r_i(n)$ on random bits and $q_i(n)$ on query bits. The actions of the verifier $V_1$ of system $X_1$ on an input Boolean formula $\Phi$ of $n$ variables and $n$ clauses can be summarized as follows:

(1) $V_1$ selects a string $r$ of $r_1(n)$ random bits. From $r$, $V_1$ then computes the queries $i_1, \ldots, i_q$, where $q = q_1(n)$, to be made to the oracle $\Pi_1$. (If we consider the oracle as a binary string, then making a query $i$ is the same as reading the $i$th bit of that string.)

(2) $V_1$ reads these $q_1(n)$ bits from the oracle $\Pi_1$. Let $x_r$ be the string of the query bits obtained from $\Pi_1$.

(3) $V_1$ runs a deterministic algorithm $A_1$ on input $(r, x_r)$ to decide whether to accept or to reject. Assume that the algorithm $A_1$ runs in $t_1(n)$ moves. We say that $V_1$ has the *decision time* $t_1(n)$.

We can modify the verifier $V_1$ to get a new verifier $V_1'$ that works in a different order:

(1') Same as (1).

(2') $V_1'$ applies Cook's Theorem to reduce the instance $(r, x_r)$ to $A_1$ to a Boolean formula $\phi_r$. Since $V_1$ already knows what $r$ is but does not know what $x_r$ is, the reduction produces a formula $\phi_r$ with values of $x_r$ left as Boolean variables. From the proof of Cook's Theorem, this formula $\phi_r$ has $p(t_1(n))$ Boolean variables and $p(t_1(n))$ clauses for some polynomial $p$. Let $y_r$ be the additional Boolean variables in $\phi_r$.

(3') $V_1'$ reads $x_r$. $V_1'$ expects the prover to provide an additional proof $\Gamma_1$, that for each string $r$ of length $r_1(n)$ contains a string $y_r$ of length $\leq p(t_1(n))$; $V_1'$ reads $y_r$.

(4') $V_1'$ decides whether $x_r$ and $y_r$ together satisfy $\phi_r$.

It is easy to check that this modified version of $X_1$ has the same error probability as the original version, with the number of query bits increased by $(t_1(n))^{O(1)}$. In particular, we observe that the system $S_2$ has the decision time $t_1(n) = (\log n)^{O(1)}$ and so its modified version still uses $O(\log n)$ random bits and reads $(\log n)^{O(1)}$ query bits.

The idea of the composition is to replace the steps $(3')$ and $(4')$ by the verifier $V_2$ of system $X_2$ on input $\phi_r$. Since $\phi_r$ has size $p(t_1(n))$ for some polynomial $p$, the verifier $V_2$ needs only an additional $r_2(p(t_1(n)))$ random bits and reads $q_2(p(t_1(n)))$ query bits. In the case of $X_1 = X_2 = S_2$, the total number of random bits is $O(\log n) + O(\log(\log n)^{O(1)}) = O(\log n)$, and the number of query bits is $(\log(\log n)^{O(1)}))^{O(1)} = (\log \log n)^{O(1)}$. (Notice that $V_1'$ no longer reads any query bit.) Therefore, this composition reduces the number of query bits by a logarithmic factor but increases the number of random bits only by a constant factor.

The main technical problem with this idea is that, in system $X_2$, the verifier $V_2$ is not supposed to read $x_r$ and $y_r$ directly. Instead, the proofs $x_r$ and $y_r$ for $V_2$ need to be encoded into a polynomial code. Since $x_r$ is a collection of $q_1(n)$ bits from different locations of the proof $\Pi_1$, this double encoding could become very complicated. To solve this problem, we add two requirements to the systems $X_1$ and $X_2$ allowing a simple implementation of the double encoding. First, we require that the verifier $V_1$ only read query bits in a constant number of *entries* from the oracle, each entry containing a number of bits given by the oracle in one query. Second, the verifier $V_2$ must be able to read his/her proof (encoded in $V_2$'s encoding scheme) in the *split form*, in the sense that the proof is split into a constant number of parts and encoded separately. Together, these two requirements mean that for each $r$, $V_1$ only reads $x_r$ and $y_r$ from a constant number of entries of $\Pi_1$, and the prover $P_2$ of system $X_2$ encodes those entries separately to let $V_2$ read them in the split form. We formalize these requirements as follows.

**Definition 11.22** *We say the verifier $V$ of a PCP system $X$ for* SAT *reads a constant number of entries (or, simply, $V$ has property A) if in step (1) above $V$ computes from $r$ only a constant number of queries, that is, $q$ is a constant but the size $e(n)$ of the answer to each query (the entry size) is not limited to a constant.*

If a *PCP* system $X$ has property A, then the two main complexity measures are the number of random bits and the entry size. The number of query bits is a constant number times the entry size. In addition, the decision time is also a useful measure. The entry size is always assumed to be bounded above by the decision time.

**Definition 11.23** *Assume that the proof system $X$ uses a fixed encoding scheme $\sigma$ to encode a binary string to a polynomial code. Let $\delta_\sigma$ be the minimum distance between two polynomial codes. We assume that $\delta_\sigma > 0$. We say that the verifier $V$ of system $X$ can check assignments in split form (or, $V$ has property B), if for any input $\phi$ and any integer $k$ (the extra input), $V$ expects* *the proof to consist of $k + 1$ parts $\Pi_1, \ldots, \Pi_{k+1}$, where $\Pi_1, \ldots, \Pi_k$ are supposed to be split encodings of a truth assignment to $\phi$ and each $\Pi_j$, $1 \leq j \leq k$, encodes $\lceil n/k \rceil$ Boolean values (with the last part $\Pi_k$ possibly containing some dummy values), and $V$ behaves as follows:*

*(i) If for each $j$, $1 \leq j \leq k$, $\Pi_j$ is a polynomial code (with respect to $\sigma$) and if $\sigma^{-1}(\Pi_1)\sigma^{-1}(\Pi_2)\cdots\sigma^{-1}(\Pi_k)$ is a satisfying assignment for $\phi$, then there exists a $\Pi_{k+1}$ such that $V$ accepts with probability 1.*

*(ii) If for any $j$, $1 \leq j \leq k$, $\Pi_j$ is not $\delta_\sigma/4$-close to a polynomial code, then, for all $\Pi_{k+1}$, $V$ rejects with probability $\geq 1 - \epsilon_V$, where $\epsilon_V$ is a fixed error bound.*

*(iii) If for each $j$, $1 \leq j \leq k$, $\Pi_j$ is $\delta_\sigma/4$-close to a polynomial code $g_j$, and if $\sigma^{-1}(g_1)\sigma^{-1}(g_2)\cdots\sigma^{-1}(g_k)$ is not a satisfying assignment for $\phi$, then, for all $\Pi_{k+1}$, $V$ rejects with probability $\geq 1 - \epsilon_V$.*

Notice that the above property B is very restrictive, since it is conceivable that for some sophisticated proof system, the prover may be able to convince a verifier that $\phi$ is satisfiable without even providing the complete satisfying assignments (in or not in the encoded form). Nevertheless, this is a special property which, we will see, system $S_2$ may be modified to have. Indeed, we show in the following that system $S_2$ may be modified to have both properties A and B.

**Theorem 11.24** *There is a PCP system $S_2'$ for* SAT *in which the verifier has the properties A and B, uses $O(\log n)$ random bits, and has the decision time $p(\log n)$ for some polynomial $p$.*

*Proof.* Let $V$ be the verifier of the system $S_2$. We first observe that $V$ is easily modified to be able to check the assignment split into $k$ parts. For each $k > 0$, we describe a subroutine $V_k$ of $V$ that deals with inputs $(\Phi, k)$. We assume that the proof has $k + 1$ parts: $f_1, \ldots, f_k, \Pi$, where $\Pi$ contains $k + 1$ subparts $\Pi_1, \ldots, \Pi_k, \Pi_{k+1}$. The verifier $V_k$ first, for each $i = 1, \ldots, k$, applies the low-degree test system to test whether $f_i$ is $(\delta/k)$-close to a polynomial function, using $\Pi_i$ as the additional proof (assuming that $\delta \leq \delta_\sigma/4$). Here, $\delta$ is the parameter used in system $S_2$. Then, $V_k$ simulates $V$ using $\Pi_{k+1}$ as the additional proof. When $V$ needs a value of $f(\vec{a})$, the verifier $V_k$ simply finds out in which $f_i$ this value is encoded and uses $f_i(\vec{a})$ for it. We can see that $V_k$ works correctly with the same error probability and an increase in complexity by a constant factor $k$. Namely, if each $f_i$ is $(\delta/k)$-close to a polynomial code $\tilde{f}_i$, then the error probability of getting the wrong $f(\vec{a})$ at some random point $\vec{a}$ remains to be $\delta$. Since $V$ needs to perform $k$ low-degree tests, it uses $k$ times many random bits and reads $k$ times many bits from the proofs.

Next we consider the property A. The idea here is to encode the proof $\Pi$ of system $S_2$ one more time, and to get the $q(n) = (\log n)^{O(1)}$ bits of $x_r$ in different positions of $\Pi$ by the polynomial recovering system, which reads only a single *entry* to recover all $q(n)$ bits. We describe in more detail the algorithm $V_k'$ that checks split assignments in $k$ split parts and reads only a constant number of entries. For each input $(\Phi, k)$, where the size of $\Phi$ is $n$, the proof is supposed to be of the form $(f_1, \ldots, f_k, \Pi')$, where each $f_i$, $1 \le i \le k$, encodes $\lceil n/k \rceil$ Boolean variables. The proof $\Pi'$ consists of three parts: $\Pi_1, \Pi_2, \Pi_3$, where $\Pi_1$ is supposed to be the encoding of the additional proof $\Pi$ for $V_k$. Notice that $V_k$ reads totally (over all possible random strings $r$) at most $l = 2^{r(n)} q(n) = n^{O(1)}$ bits, where $r(n) = O(\log n)$ and $q(n) = (\log n)^{O(1)}$. So, we can use our encoding scheme of Section 11.3.1 to encode the proof $\Pi$ for $V_k$ into a polynomial function $f : \mathcal{F}^m \to \mathcal{F}$ of degree $mh$, where the parameters $m = O(\log n / \log \log n)$, $h = O(\log n)$ and $\alpha = h^{O(1)}$ are just a constant factor larger than those used in System $S_2$. The algorithm for $V_k'$ is as follows:

(1) Toss coins to get a random string $r$ of length $r(n)$. Simulate $V_k$ to calculate the queries $i_1, \ldots, i_{q(n)}$ to ask to the oracle $\Pi$.

(2) Use $\Pi_2$ as the additional proof to run the low-degree test system on $\Pi_1$ to check whether it is $\delta$-close to a polynomial of degree $mh$ with error bound $\epsilon = \epsilon_0/2$; reject if this fails. Here, $\delta = \epsilon_0^2/64$, and $\epsilon_0$ is the error bound of $V_k$.

(3) Use $\Pi_3$ as the additional proof to run the polynomial recovering system to get the values $g(i_1), \ldots, g(i_{q(n)})$, where $g$ is the closest polynomial to $\Pi_1$.

(4) Simulate $V_k$ using $g(i_1), \ldots, g(i_{q(n)})$ as the values read from $\Pi$ and using $f_1, \ldots, f_k$ as the first $k$ proofs.

*Correctness.* Suppose input $\Phi$ is satisfiable. Then the prover can provide the correct proofs to make the error probability equal to zero. Suppose $\Phi$ is not satisfiable. We claim that the error probability is at most twice of that of $V_k$. An extra error can occur only if

(i) $\Pi_1$ is not $\delta$-close but $V_k'$ does not catch it, or

(ii) $\Pi_1$ is $\delta$-close but the recovered values are not equal to $g(i_1), \ldots, g(i_{q(n)})$.

The error probability for (i) is $\le \epsilon = \epsilon_0/2$, and that for (ii) is, by Lemma 11.21, at most $4\sqrt{\delta} = \epsilon_0/2$. So, the total error probability increases by, at most, $\epsilon_0$.

To check that $V_k'$ has the property A, we assume that each function value $f_i(\vec{a})$, $1 \le i \le k$, can be obtained in one query. In addition, in the low-degree test system and the polynomial recovering system, we assume that each polynomial $g_L$ of degree $d$ represented by $d+1$ constants in $\mathcal{F}$ can be obtained as a single entry in one query. In the above, the only queries made in step (4) are those queries of $V_k$ to get the function values $f_i(\vec{a})$ for $1 \le i \le k$. All other queries by $V_k$ to $\Pi$ are replaced by steps (2) and (3). We have seen that in each call of the low-degree test, the verifier only asks for a constant number of function values, and in each call of the sum check system, the verifier only asks for three function values. Thus, the total number of queries of step (4) is a constant. The low-degree test of step (2) asks for a constant number of function values of $\Pi_1$ and a constant number of polynomials from $\Pi_2$. By our definition of "entries," they can be obtained in a constant number of queries. Finally, the polynomial recovering of step (3) only needs one function value of $\Pi_1$ and one polynomial in $\Pi_3$. Thus, the total number of queries is a constant. The largest entry in the above queries is a polynomial of $\Pi_3$, which is of degree

$mhq(n)$ and, hence, of $O(mhq(n)\log\alpha) = (\log n)^{O(1)}$ bits.

*Complexity.* We observe that $V_k'$ uses $O(m\log\alpha) = O(\log n)$ extra random bits in steps (2) and (3), and so the total number of random bits remains to be $O(\log n)$. The decision time of $V_k'$ is the decision time of $V_k$ plus the time for steps (2) and (3). The runtime for the low-degree test system and the polynomial recovering system is bounded by $p(m^2 hq(n)\log\alpha) = (\log n)^{O(1)}$ for some polynomial $p$, and so the decision time of $V_k'$ is still $(\log n)^{O(1)}$. $\blacksquare$

Next we show that if a *PCP* system $X_1$ has the property A and if a *PCP* system $X_2$ has the property B, then we can compose them into a new system $X_3$.

**Lemma 11.25** *Assume that $X_1$ is a PCP system for SAT whose verifier $V_1$ has the property A, uses $r_1(n)$ random bits, and has the decision time $t_1(n)$, and that $X_2$ is a PCP system for SAT whose verifier $V_2$ has the property B, uses $r_2(n)$ random bits, and has the decision time $t_2(n)$. Then, there is a composed system $X_3$ for SAT whose verifier $V_3$ uses $r_3(n) = O(r_1(n) + r_2(p(t_1(n))))$ random bits and has the decision time $t_3(n) = t_2(p(t_1(n)))$ for some polynomial $p$. Furthermore, if $V_2$ also has the property A then $V_3$ also has the property A.*

*Proof.* As discussed at the beginning of this section, we may assume that $V_1$ is of the modified form (step (1') to step (4')). In addition, we may assume, because of property A of $V_1$, that $x_r$ can be obtained in a constant number of queries. We further assume that $y_r$ can also be obtained in one query from the proof $\Gamma_1$ (by asking the prover to put them together as a single entry). The size of each entry is at most $t_1(n)$.

The proof for $V_3$ contains three parts: $\Pi_3$, $\Gamma_3$, and $\Lambda_3$. For each entry $\Pi_1(i)$ of $\Pi_1$, $\Pi_3(i)$ is the polynomial code for $\Pi_1(i)$ under the encoding scheme of $V_2$, and for each entry $\Gamma_1(r)$, $\Gamma_3(r)$ is the polynomial code for $\Gamma_1(r)$ under the encoding scheme of $V_2$ (i.e., encoded $y_r$'s), and for each string $r$ of length $r_1(n)$, $\Lambda(r)$ contains some extra proofs to be used by $V_2$.

The algorithm for $V_3$ is simple:

(1) Simulate steps (1') and (2') of $V_1$ to obtain a random string $r$ of $r_1(n)$ bits and the queries $i_1,\ldots,i_q$, (here, $q$ is a constant).

(2) Simulate $V_2$ on input $(\phi_r, q+1)$, using $\Pi_3(i_1),\ldots,\Pi_3(i_q)$ and $\Gamma_3(r)$ as the first $q+1$ parts and $\Lambda(r)$ as the $(q+2)$nd part of the proof for $V_2$.

*Correctness.* The case of the input $\Phi$ belonging to SAT is easy to check. We consider the case $\Phi \notin$ SAT. Let $\epsilon_i$ be the error bound for system $X_i$, $i = 1, 2$.

*Case 1.* Each part $\Pi_3(i)$ of the proof $\Pi_3$ and each $\Gamma_3(r)$ of the proof $\Gamma_3$ is $\delta_\sigma$-close to a polynomial code, where $\sigma$ is the encoding scheme of $V_2$. Let $\tilde{\Pi}_3(i)$ be the closest polynomial code to $\Pi_3(i)$, and let $\tilde{\Pi}_1(i)$ be the string whose code is $\tilde{\Pi}_3(i)$. Also let $\tilde{\Gamma}_3(r)$ be the closest polynomial code to $\Gamma_3(r)$ and let $\tilde{\Gamma}_1(r)$ be the string whose code is $\tilde{\Gamma}_3(r)$. With respect to the proofs $\Pi_1$ and $\Gamma_1$, the probability of $V_1$ accepting $\Phi$ is at most $\epsilon_1$. That is, the probability that $\phi_r$ is satisfied by the strings $x_r$ and $y_r$ is at most $\epsilon_1$, where

$x_r$ is the string from $\Pi_1(i_1), \ldots, \Pi_1(i_q)$, and $y_r = \Gamma_1(r)$. For any fixed $r$ such that $\phi_r$ is not satisfied by the strings $x_r$ and $y_r$, the probability of $V_2$ accepting $\phi_r$ with respect to the proofs $\Pi_3(i_1), \ldots, \Pi_3(i_q)$ and $\Gamma_3(r)$ and any $\Lambda_3$ is at most $\epsilon_2$. This follows from condition (iii) of the property B of $V_2$, as each of $\Pi_3(i_1), \ldots, \Pi_3(i_q), \Gamma_3(r)$ is $\delta_\sigma$-close to a polynomial code and they together encode $x_r$ and $y_r$. Therefore, the total probability of $V_3$ accepting $\Phi$ is bounded by $\epsilon_1 + \epsilon_2$.

*Case* 2. Some of $\Pi_3(i)$ or $\Gamma_3(r)$ is not $\delta_\sigma$-close to a polynomial code under the encoding scheme of $V_2$. Let us replace each part of the proof that is not $\delta_\sigma$-close to a polynomial code by a fixed polynomial code $\Pi_0$. Using this new proof, the accepting probability is exactly the same as that of Case 1; that is, at most $\epsilon_1 + \epsilon_2$. It is not hard to see that this change of *bad* proofs by $\Pi_0$ decreases the accepting probability by at most $\epsilon_2$, since, by condition (ii) of property B, the verifier $V_2$ and, hence, step (2) of $V_3$, would reject with probability $\geq 1 - \epsilon_2$, if any of $\Pi_3(i_1), \ldots, \Pi_3(i_q), \Gamma_3(r)$ is not $\delta$-close to a polynomial code. It follows that the total accepting probability of this case is at most $\epsilon_1 + 2\epsilon_2$.

*Complexity.* Step (1) uses exactly $r_1(n)$ bits. Step (2) is a simulation of $V_2$ applied to $\phi_r$ whose length is bounded by $p(t_1(n))$ for some polynomial in $p$, and so it uses $r_2(p(t_1(n)))$ random bits. Together, the total number of random bits used by $V_3$ is $r_1(n) + r_2(p(t_1(n)))$.

Next, we consider the decision time of $V_3$. We observe that step (1) did not read any query bits. Thus, the decision time of $V_3$ is exactly the decision time of $V_2$ on input $\phi_r$ of length $p(t_1(n))$, or $t_2(p(t_1(n)))$.

Finally, we check that the only queries made to the oracle are those in step (2) simulating $V_2$. Thus, if $V_2$ has property A then $V_3$ has property A.    ∎

**Corollary 11.26** *There is a PCP system $S_3$ for* SAT *whose verifier has the property A, uses $O(\log n)$ random bits, and has the decision time $p(\log \log n)$ for some polynomial $p$.*

*Proof.* Compose $S_2'$ with itself.    ∎

### 11.3.4   Proof System Reading a Constant Number of Oracle Bits

In this section, we construct a *PCP* system $S_4$ for SAT that uses a polynomial number of random bits but only reads a constant number of query bits. In addition, the system $S_4$ also has the property B. The proof of Theorem 11.14 is then completed by composing system $S_3$ with system $S_4$.

The proof system $S_4$ uses a different encoding scheme from $S_2$. For system $S_4$, all computations are to be done in the field $\mathbf{Z}_2$. For each binary string $s \in \{0,1\}^m$, we let $s_i, 1 \leq i \leq m$, be the $i$th bit of $s$; that is, $s = s_1 s_2 \cdots s_m$. Then, $s$ may be treated as a vector $\vec{s} = (s_1, s_2, \ldots, s_m)$ in $\mathbf{Z}_2^m$. Each vector $\vec{s} \in \mathbf{Z}_2^m$ is encoded as a function $\check{s}$ from $\mathbf{Z}_2^m$ to $\mathbf{Z}_2$: $\check{s}(\vec{x}) = \vec{s} \cdot \vec{x}^T$, that is, $\check{s}(\vec{x}) = \sum_{i=1}^m s_i x_i$, where $\vec{s} = (s_1, \ldots, s_m)$, $\vec{x} = (x_1, \ldots, x_m)$, and the addition and multiplication are those in $\mathbf{Z}_2$. Such a function $\breve{s}$ is called a *linear* function. Note that when we represent $\breve{s}$ by its function values over $\mathbf{Z}_2^m$, we need $2^m$ bits.

The proof system $S_4$ needs some subsystems to test the algebraic properties of the proofs. The first is the *linearity test* system, which tests whether a given function is $\delta$-close to a linear function. We observe that a function $f : \mathbf{Z}_2^m \to \mathbf{Z}_2$ is linear if and only if for any two points $\vec{y}, \vec{z} \in \mathbf{Z}_2^m$, $f(\vec{y}+\vec{z}) = f(\vec{y})+f(\vec{z})$, where $\vec{y} + \vec{z}$ denotes the vector $(y_1 + z_1, \ldots, y_m + z_m)$. The linearity test system is based on the following stronger property.

**Lemma 11.27** *Let $f$ be a function from $\mathbf{Z}_2^m$ to $\mathbf{Z}_2$, and $0 < \delta < 1/18$. If, for random $\vec{y}, \vec{z} \in \mathbf{Z}_2^m$, $\Pr[f(\vec{y}+\vec{z}) = f(\vec{y})+f(\vec{z})] \geq 1 - \delta/3$, then $f$ is $\delta$-close to a linear function.*

*Proof.* This is essentially a special case of Lemma 11.10, but the proof is considerably easier. We leave it as an exercise (Exercise 11.12). ∎

> *Linearity Test System.* The input to the system are two constants $0 < \delta < 1/18$ and $\epsilon > 0$. The proof consists of a function $f : \mathbf{Z}_2^m \to \mathbf{Z}_2$, represented by $2^m$ bits. The verifier does the following:
>
> (1) Let $l = \lceil -3 \ln \epsilon/\delta \rceil$. Pick $2l$ random points $\vec{y}_1, \vec{z}_1, \ldots, \vec{y}_l, \vec{z}_l$ from $\mathbf{Z}_2^m$.
>
> (2) Check, for each $1 \leq i \leq l$, whether $f(\vec{y}_i+\vec{z}_i) = f(\vec{y}_i)+f(\vec{z}_i)$. Reject if it fails for any $i$, $1 \leq i \leq l$; accept otherwise.

*Correctness.* If $f$ is not $\delta$-close to a linear function then, by Lemma 11.27, the probability of $f$ passing all $l$ tests in step (2) is at most $(1 - \delta/3)^l < e^{(-1)(-\ln \epsilon)} = \epsilon$. Thus, if $f$ is a linear function, then the verifier accepts with probability 1; if $f$ is not $\delta$-close to a linear function, then the verifier rejects with probability at least $1 - \epsilon$.

*Complexity.* Assume that $\delta$ and $\epsilon$ are constants. It is clear that the linearity test system uses $O(m)$ random bits. In addition, it reads only a constant number of query bits. Indeed, it reads $3l$ bits from the oracle $f$ and makes $l$ additions and $l$ comparisons. Thus, its *decision time* is also bounded by a constant.

As in Section 11.3.2, the above test of $\delta$-closeness to a linear function has a companion system that recovers the values of the closest linear function at some given points.

> *Linear Function Recovering System.* The input to the system is a constant $\delta > 0$ and a point $\vec{z}_0 \in \mathbf{Z}_2^m$. The proof is the same as that of the linearity test system. The verifier picks a random $\vec{y} \in \mathbf{Z}_2^m$ and outputs $f(\vec{y} + \vec{z}_0) - f(\vec{y})$.

*Correctness.* If $f$ is $\delta$-close to a linear function $\tilde{f}$, then $\Pr[f(\vec{y}+\vec{z}_0)-f(\vec{y}) \neq \tilde{f}(\vec{z}_0)] \leq \Pr[f(\vec{y} + \vec{z}_0) \neq \tilde{f}(\vec{y} + \vec{z}_0)] + \Pr[f(\vec{y}) \neq \tilde{f}(\vec{y})] \leq 2\delta$. Therefore, the error probability is bounded by $2\delta$.

*Complexity.* It uses $O(m)$ random bits and $O(1)$ query bits.

Next we need a subsystem that checks whether a given function $h$ is the *tensor-product* of two other given functions $f$ and $g$. Let linear functions $f : \mathbf{Z}_2^m \to \mathbf{Z}_2$ and $g : \mathbf{Z}_2^k \to \mathbf{Z}_2$ be defined by $f(\vec{x}) = \sum_{i=1}^m a_i x_i$, where $\vec{x} = (x_1, \ldots, x_m)$ and $g(\vec{y}) = \sum_{j=1}^k b_j y_j$, where $\vec{y} = (y_1, \ldots, y_k)$. Then, the *tensor-product* $f \otimes g$ of $f$ and $g$ is the linear function from $\mathbf{Z}_2^{mk}$ to $\mathbf{Z}_2$ defined by

$$(f \otimes g)(\vec{z}) = \sum_{1 \le i \le m} \sum_{1 \le j \le k} a_i b_j z_{i,j},$$

where $\vec{z} = (z_{1,1}, \ldots, z_{1,k}, z_{2,1}, \ldots, z_{2,k}, \ldots, z_{m,k})$.

*Tensor-Product Test System.* The input to the system is the error bound $\epsilon > 0$. The proof consists of three functions $f : \mathbf{Z}_2^m \to \mathbf{Z}_2$, $g : \mathbf{Z}_2^k \to \mathbf{Z}_2$, and $h : \mathbf{Z}_2^{mk} \to \mathbf{Z}_2$, represented by their function values. The verifier does the following:

(1) Let $q = \lceil \log \epsilon / \log(3/4) \rceil$. Repeat steps (1.1) and (1.2) for $q$ times.

    (1.1) Choose a random vector $\vec{u} = (u_1, \ldots, u_m)$ from $\mathbf{Z}_2^m$, and a random vector $\vec{v} = (v_1, \ldots, v_k)$ from $\mathbf{Z}_2^k$. Let $\vec{z}_{u,v} = (u_1 v_1, \ldots, u_1 v_k, u_2 v_1, \ldots, u_2 v_k, \ldots, u_m v_k)$.

    (1.2) Test whether $h(\vec{z}_{u,v}) = f(\vec{u})g(\vec{v})$.

(2) Reject if any test of step (1.2) fails; accept otherwise.

The following two lemmas establish the correctness of the above proof system.

**Lemma 11.28** *Let $\vec{a} = (a_1, \ldots, a_m)$ be a nonzero vector in $\mathbf{Z}_2^m$. Then, for a randomly chosen $\vec{u} = (u_1, \ldots, u_m) \in \mathbf{Z}_2^m$, $\Pr[\sum_{i=1}^m a_i u_i = 0] = 1/2$.*

*Proof.* Assume that $a_j \ne 0$ for some $1 \le j \le m$. Then, for any fixed $u_1, \ldots, u_{j-1}, u_{j+1}, \ldots, u_m$ and random $u_j$, $\sum_{i=1}^m a_i u_i$ is equally likely to be 0 or 1. ∎

**Lemma 11.29** *Assume that functions $f$, $g$, and $h$ in the tensor-product test system are all linear functions.*

*(a) If $h = f \otimes g$, then the verifier accepts with probability 1.*

*(b) If $h \ne f \otimes g$, then the verifier rejects with probability $\ge 1 - \epsilon$.*

*Proof.* Assume that $f(\vec{x}) = \vec{a} \cdot \vec{x}^T = \sum_{i=1}^m a_i x_i$, $g(\vec{y}) = \vec{b} \cdot \vec{y}^T = \sum_{j=1}^k b_j y_j$, and $h(\vec{w}) = \sum_{i=1}^m \sum_{j=1}^k c_{i,j} w_{i,j}$ where $\vec{w} = (w_{1,1}, \ldots, w_{1,k}, \ldots, w_{m,1}, \ldots, w_{m,k})$. Define two $m \times k$ matrices $A$ and $B$ by $A[i,j] = a_i b_j$ and $B[i,j] = c_{i,j}$. Then, we have $f(\vec{u})g(\vec{v}) = \sum_{i=1}^m \sum_{j=1}^k a_i b_j u_i v_j = \vec{u} A \vec{v}^T$, and $h(\vec{z}_{u,v}) = \sum_{i=1}^m \sum_{j=1}^k c_{i,j} u_i v_j = \vec{u} B \vec{v}^T$.

(a) If $h = f \otimes g$, then $A = B$ and so they pass all tests.

(b) If $h \neq f \otimes g$, then $A \neq B$, then by Lemma 11.28, $\mathrm{Pr}_{\vec{u}}[\vec{u}A \neq \vec{u}B] = 1/2$ and $\mathrm{Pr}_{\vec{u},\vec{v}}[\vec{u}A\vec{v}^T \neq \vec{u}B\vec{v}^T] = 1/4$. Therefore, the probability of $f, g$, and $h$ passing all $q$ tests is at most $(3/4)^q \leq \epsilon$. ∎

*Complexity.* The verifier of the tensor-product test system uses $O(m + k)$ random bits and reads only $O(1)$ values of $f$, $g$, and $h$.

Now we are ready to develop the proof system $S_4$ for SAT. Recall that $\Phi$ is a Boolean function of $n$ variables $\{v_1, \ldots, v_n\}$ and $n$ clauses $\{C_1, \ldots, C_n\}$, and that a linear function $f : \mathbf{Z}_2^n \to \mathbf{Z}_2$ encodes a satisfying assignment for $\Phi$ (we simply say $f$ *satisfies* $\Phi$) if $f(\vec{x}) = \sum_{i=1}^n a_i x_i$, and $(a_1, \ldots, a_n)$ satisfies $\Phi$. Using the notation of Section 11.2, $f(\vec{x}) = \sum_{i=1}^n a_i x_i$ satisfies $\Phi$ if and only if

$$G_t(s_1, s_2, s_3, a_{s_1}, a_{s_2}, a_{s_3}) = \prod_{j=1}^3 \phi_j(s_j, t)(\psi_j(t) - a_{s_j}) = 0,$$

for all $s_1, s_2, s_3, t \in \{1, \ldots, n\}$, where $\phi_j$ and $\psi_j$ are those functions defined in Section 11.2. We notice that for each fixed $t$, there is a unique triple $(t_1, t_2, t_3)$ such that $\phi_j(t_j, t)$ are nonzero for all $j = 1, 2, 3$, and, in that case, $G_t$ is a degree-3 polynomial $G_t(a_{t_1}, a_{t_2}, a_{t_3})$ over the variables $a_{t_1}, a_{t_2}, a_{t_3}$ (with $t_1, t_2$, and $t_3$ being constants). If $f$ does not satisfy $\Phi$, then the values of these cubic functions $G_t$ evaluated at the point $(a_1, \ldots, a_n)$ are not all zeros. Thus, by Lemma 11.28, for a random vector $\vec{u} = (u_1, \ldots, u_n) \in \mathbf{Z}_2^n$, the probability of

$$\sum_{t=1}^n u_t \cdot G_t(a_{t_1}, a_{t_2}, a_{t_3}) \neq 0 \tag{11.4}$$

is equal to $1/2$.

The main idea of the system $S_4$ is to use the tensor-products $g = f \otimes f$ and $h = f \otimes f \otimes f$ to test whether (11.4) holds. We note that each $G_t$ is a degree-3 polynomial and so the left-hand side of (11.4) is a degree-3 polynomial over variables $a_1, \ldots, a_n$. By combining the similar terms, the left-hand side can be expressed as the sum of the function values of $f$, $g$, and $h$ at some appropriate points. To be more precise, for each $\vec{u} = (u_1, \ldots, u_n) \in \mathbf{Z}_2^n$, there exist vectors $\vec{x} = (x_1, \ldots, x_n)$, $\vec{y} = (y_{1,1}, \ldots, y_{n,n})$ and $\vec{z} = (z_{1,1,1}, \ldots, z_{n,n,n})$ such that

$$\sum_{t=1}^n u_t \cdot G_t(a_{t_1}, a_{t_2}, a_{t_3}) = \sum_{i=1}^n x_i a_i + \sum_{i=1}^n \sum_{j=1}^n y_{i,j} a_i a_j$$
$$+ \sum_{i=1}^n \sum_{j=1}^n \sum_{k=1}^n z_{i,j,k} a_i a_j a_k. \tag{11.5}$$

In addition, $\vec{x}$, $\vec{y}$ and $\vec{z}$ are computable from $\vec{u}$ (and $\Phi$) in time polynomial in $n$. Therefore, the testing of (11.4) can be done by evaluating the functions $f$, $g$, and $h$ at the points $\vec{x}$, $\vec{y}$ and $\vec{z}$, respectively.

*Proof System $S_4$.* The input is a Boolean formula $\Phi$ of $n$ variables and $n$ clauses, and an error bound $\epsilon_0 > 0$. The proof consists of three functions $f : \mathbf{Z}_2^n \to \mathbf{Z}_2$, $g : \mathbf{Z}_2^{n^2} \to \mathbf{Z}_2$, and $h : \mathbf{Z}_2^{n^3} \to \mathbf{Z}_2$, represented by their function values. The verifier does the following:

(1) Apply the linearity test system to check whether each of $f$, $g$, and $h$ is $\delta$-close to a linear function $\tilde{f}$, $\tilde{g}$, and $\tilde{h}$, respectively, with the parameters $\delta = \epsilon_0/(96q)$ and $\epsilon = \epsilon_0/24$, where $q = \lceil \log \epsilon / \log(3/4) \rceil$. Reject if any function fails the test.

(2) Apply the tensor-product test system to check whether $\tilde{g} = \tilde{f} \otimes \tilde{f}$ and $\tilde{h} = \tilde{f} \otimes \tilde{g}$, with the parameter $\epsilon = \epsilon_0/8$. Reject if either test fails. (In the tests, the verifier uses the linear function recovering system to obtain the function values $\tilde{f}(\vec{x})$, $\tilde{g}(\vec{y})$ and $\tilde{h}(\vec{z})$, when needed.)

(3) Construct functions $G_t$ as discussed above, and repeat the following for $N = \lceil -\log(\epsilon_0/4) \rceil$ times: select a random vector $\vec{u} \in \mathbf{Z}_2^n$; compute the vectors $\vec{x}$, $\vec{y}$, and $\vec{z}$ satisfying (11.5); and test whether $\tilde{f}(\vec{x}) + \tilde{g}(\vec{y}) + \tilde{h}(\vec{z}) = 0$. Reject if any of the $N$ tests fails.

*Correctness.* If $\Phi \in$ SAT, then the prover can simply present the satisfying function $f$ together with $g = f \otimes f$ and $h = f \otimes g$. They will pass all tests and so the verifier accepts with probability 1.

If $\Phi \notin$ SAT, then an error can occur in one of the following forms:

(i) $f$ or $g$ or $h$ is not $\delta$-close to any linear function, but they still pass the tests of step (1).

(ii) $f$, $g$, and $h$ are $\delta$-close to linear functions $\tilde{f}$, $\tilde{g}$ and $\tilde{h}$, respectively, but $\tilde{g} \neq \tilde{f} \otimes \tilde{f}$ or $\tilde{h} \neq \tilde{f} \otimes \tilde{g}$, and they pass the tests of step (2).

(iii) $f$, $g$, and $h$ are $\delta$-close to linear functions $\tilde{f}$, $\tilde{g}$ and $\tilde{h}$, respectively, and $\tilde{g} = \tilde{f} \otimes \tilde{f}$, $\tilde{h} = \tilde{f} \otimes \tilde{g}$, and $\tilde{f}(\vec{x}) + \tilde{g}(\vec{y}) + \tilde{h}(\vec{z}) = 0$ for all $N$ tests of step (3).

We observe that the probability of (i) to occur is bounded by $\epsilon_0/8$.

For the case (ii), there are two subcases that can occur. First, it can occur if the recovered values of $\tilde{f}$, $\tilde{g}$, or $\tilde{h}$ are not correct. We observe that to test whether $\tilde{g} = \tilde{f} \otimes \tilde{f}$ and $\tilde{h} = \tilde{f} \otimes \tilde{g}$, we need to recover $3q$ values of $\tilde{f}$, $2q$ values of $\tilde{g}$ and $q$ values of $\tilde{h}$, where $q = \lceil \log \epsilon / \log(3/4) \rceil$. The probability of any of the recovered value is incorrect is $\leq 2\delta$, and so the total error probability for this case is at most $12q\delta$. Next, if all recovered values of $\tilde{f}$, $\tilde{g}$, and $\tilde{h}$ are correct, then the probability of these values passing the tensor-product test system is $\leq \epsilon = \epsilon_0/8$. Therefore, total error probability of case (ii) is at most $12q\delta + \epsilon_0/8 \leq \epsilon_0/4$.

For the case (iii), we observe that we need, for each test, to recover a value from each of $\tilde{f}$, $\tilde{g}$, and $\tilde{h}$. Thus, the probability of recovering some incorrect values in $N$ tests is bounded by $6N\delta \leq 6q\delta \leq \epsilon_0/4$. Next, if all recovered values are correct, then the probability of these values passing the tests of step (3) is $\leq (1/2)^N = \epsilon_0/4$. The error probability of the case (iii) is thus at most $\epsilon_0/2$.

The above analysis shows that the error probability of system $S_4$ is bounded by $\epsilon_0$.

*Complexity.* The number of random bits used in each step is $O(n^3)$, and the number of query bits in each step is $O(1)$. Note that the computation of $\vec{x}$, $\vec{y}$, and $\vec{z}$ from random $\vec{u}$ in step (3) can be done before making queries to the oracles $f$, $g$, and $h$. Therefore, the decision time of system $S_4$ is $O(1)$.

Finally, we modify system $S_4$ to satisfy the property B. Suppose the input is $(\Phi, k)$, where $\Phi$ is a 3-CNF formula having $n$ variables and $n$ clauses. Then, the proof should contain $k + 2$ functions: $f_1, \ldots, f_k : \mathbf{Z}_2^{\lceil n/k \rceil} \rightarrow \mathbf{Z}_2$, and $g : \mathbf{Z}_2^{n^2} \rightarrow \mathbf{Z}_2$ and $h : \mathbf{Z}_2^{n^3} \rightarrow \mathbf{Z}_2$. The verifier $V_k$ for system $S_4$ does the following:

> *Verifier $V_k$ of System $S_4$.* First test for each $i$, $1 \leq i \leq k$, whether $f_i$ is $(\delta/k)$-close to a linear function $\tilde{f}_i$. Reject if any test fails. Let $m = \lceil n/k \rceil$. For any $\vec{x} = (x_1, \ldots, x_n) \in \mathbf{Z}_2^n$, extend it to have size $km$ with $x_{n+1} = \cdots = x_{km} = 0$. Define $f : \mathbf{Z}_2^n \rightarrow \mathbf{Z}_2$ by
>
> $$f(x_1, \ldots, x_n) = \sum_{i=1}^{k} \tilde{f}_i(x_{(i-1)m+1}, \ldots, x_{im}).$$
>
> Then, simulate the verifier of system $S_4$ with the oracles $f$, $g$, and $h$. (To obtain a value of $f(\vec{x})$, the verifier $V_k$ recovers a value from each $\tilde{f}_i$, $1 \leq i \leq k$, and takes the sum as the value of $f$.)

We check that the verifier $V_k$ satisfies the conditions for property B:

(i) If $f_1, \ldots, f_k$ together encode a satisfying assignment to $\Phi$, then, with the right $g$ and $h$, the verifier $V_k$ accepts with probability 1.

(ii) If any function $f_i$, $1 \leq i \leq k$, is not $(\delta/k)$-close to a linear function, then the verifier $V_k$ rejects with probability $\geq 1 - \epsilon_0$.

(iii) If each $f_i$, $1 \leq i \leq k$, is $(\delta/k)$-close to a linear function $\tilde{f}_i$ and if the function $f$ defined above is not a satisfying assignment to $\Phi$, then the verifier $V$ of $S_4$ rejects with probability $\geq 1 - \epsilon_0$, with respect to $f$ and any $g$ and $h$. Since the function values of $f$ can be recovered from $\tilde{f}_1, \ldots, \tilde{f}_k$, with the error probability $2\delta/k$ for each value of $\tilde{f}_i$, the error of recovering each value of $f$ remains to be $2\delta$, and so the verifier $V_k$ also rejects with the same probability $1 - \epsilon_0$.

(iv) The number of query bits used by $V_k$ increases by a constant factor $k$, and remains a constant.

From the above analysis, we obtain the following theorem.

**Theorem 11.30** *There is a PCP system $S_4$ for* SAT *that uses $O(n^3)$ random bits, reads $O(1)$ query bits, and satisfies property B.*

Finally, we apply Lemma 11.25 to compose systems $S_3$ and $S_4$ into a new system $S_5$. System $S_3$ uses $O(\log n)$ random bits and has the decision time $p(\log \log n)$ for some polynomial $p$. System $S_4$ uses $O(n^3)$ random bits and has the decision time $O(1)$. Furthermore, system $S_3$ has the property A and system $S_4$ has the property B. Therefore, by Lemma 11.25, system $S_5$ uses only $O(\log n + (p(\log \log n))^3) = O(\log n)$ random bits and has the decision time $O(1)$. This completes the proof of our main theorem 11.14.

## 11.4    Probabilistic Checking and Nonapproximability

We studied in Section 2.5 the approximation versions of *NP*-hard optimization problems. Recall that for an optimization problem $\Pi$ and a real number $r > 1$, the problem $r$-Approx-$\Pi$ is its approximation version with the approximation ratio $r$. We have seen that some *NP*-hard optimization problems, such as the traveling salesman problem (TSP), are so hard to approximate that, for all approximation ratios $r > 1$, their approximation versions remain to be *NP*-hard. We also showed that the for some problems, there is a threshold $r_0 > 1$ such that its approximation version is polynomial-time solvable with respect to the approximation ratio $r_0$, and it is *NP*-hard with respect to any approximation ratio $r$, $1 < r < r_0$. The traveling salesman problem with the cost function satisfying the triangle inequality is a candidate for such a problem. In this and the next sections, we apply the *PCP* characterization of *NP* to show more nonapproximability results of these two types. We first establish the *NP*-hardness of two prototype problems: the approximation to the maximum satisfiability problem and the approximation to the maximum clique problem. In the next section, other *NP*-hard approximation problems will be proved by reductions from these two problems.

For any 3-CNF Boolean formula $F$, let $sat^*(F)$ be the maximum number of clauses that can be satisfied by a Boolean assignment on the variables of $F$. The maximum satisfiability problem is to find the value $sat^*(F)$ for a given 3-CNF Boolean formula $F$. Its approximation version is formulated as follows:

> $r$-Approx-3Sat: Given a 3-CNF formula $F$, find an assignment to the Boolean variables of $F$ that satisfies at least $sat^*(F)/r$ clauses.

In order to prove the *NP*-hardness of approximation problems like $r$-Approx-3Sat, we first generalize the notion of *NP*-complete decision problems. Let $A, B \subseteq \Sigma^*$ for some fixed alphabet $\Sigma$ such that $A \cap B = \emptyset$. We write $(A, B)$ to denote the *partial* decision problem which asks, for a given input $x \in A \cup B$, whether $x \in A$ or $x \in B$. (It is called a partial decision problem because for inputs $x \notin A \cup B$, we do not care what the decision is.) The notions of complexity and completeness of decision problems can be extended to partial decision problems in a natural way. In particular, we say the partial problem $(A, B)$ is in the complexity class $\mathcal{C}$ if there exists a decision problem $C \in \mathcal{C}$ such that $A \subseteq C$ and $B \subseteq \overline{C}$. Assume that $A \cap B = C \cap D = \emptyset$. Then, we say the problem $(A, B)$ is *polynomial-time (many-one) reducible* to the problem $(C, D)$, and write $(A, B) \leq_m^P (C, D)$, if there exists a polynomial-time computable function $f$ such that for any $x \in \Sigma^*$,

$$x \in A \;\Rightarrow\; f(x) \in C,$$
$$x \in B \;\Rightarrow\; f(x) \in D.$$

For any 3-CNF formula $F$, let $c(F)$ be the number of clauses in $F$. Let $r$-UNSAT be the set of $r$-unsatisfiable Boolean formulas in 3-CNF, that is, the set of formulas $F$ with $sat^*(F) \le (1 - 1/r)c(F)$. Recall the problem (SAT, $r$-UNSAT) defined in Example 11.1. Note that SAT $\cap$ $r$-UNSAT $= \emptyset$; thus, it is a partial decision problem. Furthermore, it is clear that (SAT, $r$-UNSAT) is in *NP*. In the following, we show that (SAT, $r$-UNSAT) is *NP*-complete for some $r > 1$.

**Theorem 11.31** *There exists a real number $r > 1$ such that* (SAT, $r$-UNSAT) *is NP-complete.*

*Proof.* The proof of Theorem 11.14 may be viewed as a reduction from SAT to (SAT, $r$-UNSAT) for some $r > 1$. Indeed, Theorem 11.14 showed that SAT has a proof system with the following properties (cf. the verifier $V_1'$ at the beginning of Section 11.3.3):

(1) For any 3-CNF formula $\Phi$, with $|\Phi| = n$, the prover commits a proof $p_\Phi$ of length $q(n)$, where $q$ is a polynomial.

(2) The verifier generates a random string $r'$ of length $\ell = O(\log n)$.

(3) From $\Phi$ and $r'$, the verifier generates a Boolean formula $\Psi_{r'}$ that contains $c$ variables for some constant $c$.

(4) The proof $p_\Phi$ may be viewed as an assignment to variables in $\Psi_{r'}$, for all $r'$ of length $\ell$. If $\Phi \in$ SAT then $p_\Phi$ satisfies all formulas $\Psi_{r'}$, and if $\Phi \notin$ SAT then $p_\Phi$ satisfies at most $1 - 1/r$ portion of formulas $\Psi_{r'}$.

Thus, the verifier is actually a reduction from SAT to (SAT, $r$-UNSAT). The input to this reduction is a 3-CNF formula which may be viewed as the AND of $n$ Boolean circuits $C_1, \ldots, C_n$, each reading three variables, and the output of the reduction is a Boolean formula $\Psi$ which may be viewed as the AND of $m$ Boolean circuits $D_1, \ldots, D_m$, each reading $c$ variables, where $m = 2^\ell = p_1(n)$ for some polynomial $p_1$ (see Figure 11.1). In addition, property (4) above shows that if $\Phi \in$ SAT then $\Psi \in$ SAT, and if $\Phi \notin$ SAT then $\Psi \in r$-UNSAT for some $r > 1$.

For the sake of completeness, we present a formal proof of this reduction as an application of Theorem 11.14. More precisely, for any $L \in PCP(log, const)$, we will show that $(L, \overline{L}) \le_m^P$ (SAT, $r$-UNSAT).

Let $M$ be a polynomial-time nonadaptive oracle PTM such that for each $x \in L$, $M^{Q_x}$ accepts $x$ with probability 1 for some oracle $Q_x$, and for each $x \notin L$, $M^Q$ rejects $x$ with probability at least $3/4$ for any oracle $Q$. In addition, we assume that for each $x$ of length $n$, $M$ uses exactly $c \log n$ random bits and reads from the oracle exactly $d$ bits, where $c$ and $d$ are constants. For each string $y \in \{0, 1\}^*$ of length $|y| = c \log n$, the computation of $M$ on the

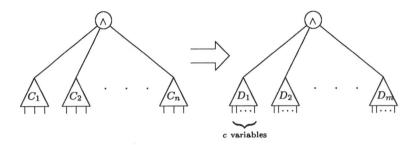

Figure 11.1: The reduction $f$.

input $x$ and with the random string $y$ can be divided into three steps (see the definition of nonadaptive oracle machines in Section 11.1):

(i) $\tilde{M}_1$ computes, from $(x, y)$, $d$ strings $z_1, \ldots, z_d$;

(ii) $\tilde{M}_2$ asks the oracle $Q$ for the $d$ bits $b_{z_1}, \ldots, b_{z_d}$ where $b_{z_j} = \chi_Q(z_j)$ for $1 \leq j \leq d$; and

(iii) $\tilde{M}_3$ computes 1 or 0 deterministically from $(x, y, b_{z_1}, \ldots, b_{z_d})$.

Note that the total number of queries to the oracle is bounded by $d \cdot n^c$.

We observe that, without knowing the actual Boolean values $b_{z_1}, \ldots, b_{z_d}$, we can simulate the computation of $\tilde{M}_1$ and $\tilde{M}_3$ on input $(x, y)$ to obtain a Boolean formula $F_{x,y}$ over $d$ variables $b_{z_1}, \ldots, b_{z_d}$ such that $F_{x,y}$ evaluates to 1 with respect to the Boolean assignment $\tau_Q(b_{z_j}) = \chi_Q(z_j)$ if and only if the above computation of $\tilde{M}_1, \tilde{M}_2, \tilde{M}_3$ on $(x, y)$ accepts relative to the oracle $Q$. In addition, since $d$ is a constant, this Boolean formula $F_{x,y}$ may be written in 3-CNF with $d_1$ clauses, where $d_1$ is a constant depending on $d$. Notice that each variable $b_z$ is identified with the query string $z$. In other words, two different formulas $F_{x,y_1}$ and $F_{x,y_2}$ may have a same Boolean variable $b_z$ if in step (i), $\tilde{M}_1(x, y_1)$ and $\tilde{M}_1(x, y_2)$ both query the same string $z$ to the oracle $Q$. As a consequence, each oracle $Q$ corresponds to a Boolean assignment $\tau_Q$ that assigns 1 to $b_z$ if and only if $z \in Q$.

Now, if $x \in L$, then there exists an oracle $Q_x$ such that $M^{Q_x}(x, y)$ accepts for all $y$, and so the Boolean assignment $\tau_{Q_x}$ satisfies all formulas $F_{x,y}$. On the other hand, if $x \notin L$, then for all oracles $Q$, $M^Q(x, y)$ accepts for at most $n^c/4$ many $y$. It means that no Boolean assignment $\tau$ on variables $b_z$ can satisfy more than $n^c/4$ formulas $F_{x,y}$. Let $F_x = \prod_{|y| = c \log n} F_{x,y}$. Then, $F_x$ is a 3-CNF formula of $d_1 n^c$ clauses such that $F_x \in \text{SAT}$ if $x \in L$ and $F$ has at least $(3/4)n^c$ unsatisfiable clauses (out of a total of $d_1 n^c$ clauses). Thus, the mapping from $x$ to $F_x$ is a reduction from $L$ to $(\text{SAT}, r\text{-UNSAT})$, where $r = 4d_1/3$. ∎

**Corollary 11.32** *There exists a real number $s > 1$ such that the problem $s$-APPROX-3SAT is $\leq_T^P$-hard for NP.*

**Example 11.36** *For any* $b > 1$, VC-$b \leq_L^P$ VC-3.

*Proof.* For $b \leq 3$, the theorem is trivial. In the following, we construct an $L$-reduction from VC-$b$, with $b \geq 4$, to VC-3. Given a graph $G = (V, E)$ in which every vertex has degree at most $b$, we construct a graph $G'$ as follows: For each vertex $x$ of degree $d$ in $G$, construct a path $p_x$ of $2d - 1$ vertices to replace it as shown in Figure 11.3. Note that this path has a unique minimum vertex cover $c_x$ with size $d - 1$. This vertex cover covers only edges in path $p_x$. The set $\bar{c}_x$ of vertices in $p_x$ but not in $c_x$ is also a vertex cover of $p_x$. This $p_x$. The set $\bar{c}_x$ of vertices in $p_x$ but not in $c_x$ is also a vertex cover of $p_x$. This vertex cover has size $d$ but it covers all edges in $p_x$ plus all edges that are originally incident on $x$ and now incident on path $p_x$.

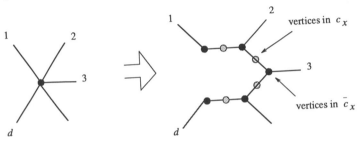

Figure 11.3: Path $p_x$.

Assume that $|V| = n$ and $|E| = m$. For any vertex cover $S$ of graph $G$, the set

$$S' = \left( \bigcup_{x \in S} \bar{c}_x \right) \cup \left( \bigcup_{x \notin S} c_x \right).$$

is a vertex cover of $G'$. In addition, if $S$ is of size $s$, then the size of $S'$ is $s + 2m - n$. (Note that the set $\bigcup_{x \in V} \bar{c}_x$ has size $2m$.)

Conversely, for each vertex cover $S'$ of $G'$, the set

$$S = \{x : \bar{c}_x \cap S' \neq \emptyset\}$$

is a vertex cover of $G$. Furthermore, if $|S| = s$ and $|S'| = s'$ then $s \leq s' - (2m - n)$.

One immediate consequence from the above relation is that

$$vc^*(G') = vc^*(G) + 2m - n,$$

where $vc^*(G)$ denotes the minimum size of vertex covers for $G$. Note that $m \leq b \cdot vc^*(G)$. Thus, we have

$$vc^*(G') \leq (2b + 1) \cdot vc^*(G),$$

that is, condition (L1) holds for this reduction. Also note that $s \leq s' - (2m - n)$ is equivalent to

$$s - vc^*(G) \leq s' - vc^*(G').$$

*Proof.* Let $s, r > 1$ be any real numbers satisfying $s \leq r/(r - 1)$. It is easy to see how to solve the problem (SAT, $r$-UNSAT) in polynomial time, with the help of $s$-APPROX-3SAT as an oracle: On input $F$, ask the oracle for a Boolean assignment $\tau$ that satisfies at least $sat^*(F)/s$ clauses. If $\tau$ satisfies at least $c(F)(1 - 1/r)$ clauses, then we know that $F$ cannot be in $r$-UNSAT and so we decide $F \in$ SAT. If $\tau$ satisfies less than $c(F)(1 - 1/r) \leq c(F)/s$ clauses, then $sat^*(F) < c(F)$ and so $F \notin$ SAT, and we decide that $F \in r$-UNSAT. In other words, if (SAT, $r$-UNSAT) is $NP$-complete for $r > 1$ then $s$-APPROX-3SAT is $NP$-hard for all $s \leq r/(r - 1)$. ∎

It is not hard to see that there exists a polynomial-time algorithm that solves the problem 2-APPROX-3SAT (see Exercise 11.14). Thus the above result establishes that there exists a threshold $r_0 > 1$ such that $r_0$-APPROX-3SAT is in $P$ and $r$-APPROX-3SAT is $NP$-hard for all $1 < r < r_0$. It has recently been proved that $r_0 = 8/7$ [Hastad, 1997; Karloff and Zwick, 1997].

Next we consider the maximum clique problem. For any graph $G$, we let $\omega(G)$ denote the size of its maximum clique.

> $r$-APPROX-CLIQUE: Given a graph $G = (V, E)$, find a clique $Q$ of $G$ whose size is at least $\omega(G)/r$.

We are going to show that for all $r > 1$, the problem $r$-APPROX-CLIQUE is $NP$-hard. Similar to the problem $r$-APPROX-3SAT, we will establish this result through the $NP$-completeness of a partial decision problem. Recall that CLIQUE $= \{(G, k) : \omega(G) \geq k\}$. Let $r$-NOCLIQUE $= \{(G, k) : \omega(G) \leq k/r\}$.

**Theorem 11.33** *For any* $r > 1$, *the problem* (CLIQUE, $r$-NOCLIQUE) *is NP-complete.*

*Proof.* Let $r > 1$ be any real number. We first observe that if $L \in PCP(\log, const)$, then there exists a polynomial-time nonadaptive oracle PTM $M$ that accepts $L$ in the sense of Definition 11.2 with the error bounded by $1/r$. This can be achieved by simulating a PCP machine for $L$ with the error bound $1/4$ for $\lceil \log r \rceil$ times.

Assume $M$ is such a PCP machine for $L$, and let $\tilde{M}_1, \tilde{M}_2, \tilde{M}_3$, constants $c$ and $d$ be defined as in the proof of Theorem 11.31. Let $x$ be an input of length $n$. We say

$$v = (x, y, z_1, b_1, z_2, b_2, \ldots, z_d, b_d),$$

where $|y| = c \log n$ and $b_j \in \{0, 1\}$ for $1 \leq j \leq d$, is an *accepting computation* of $M(x)$ if (i) $\tilde{M}_1(x, y)$ makes $d$ queries $z_1, \ldots, z_d$ to the oracle; (ii) the answers to these queries from the oracle are $b_1, \ldots, b_d$, respectively; and (iii) $\tilde{M}_3$ accepts based on these answers. We say two accepting computations $(x, y_1, z_1, b_1, \ldots, z_d, b_d)$ and $(x, y_2, w_1, a_1, \ldots, w_d, a_d)$ are *consistent* if the answers from the oracle are the same for the same query string, that is, $z_j = w_k$ implies $b_j = a_k$ for any $1 \leq j \leq d$ and $1 \leq k \leq d$.

For each input $x$ to the problem $L$, we define a graph $G_x = (V, E)$ as follows: The set $V$ consists of all the accepting computations of $M(x)$. Assume that $v_1 = (x, y_1, z_1, b_1, \ldots, z_d, b_d)$ and $v_2 = (x, y_2, w_1, a_1, \ldots, w_d, a_d)$ are both accepting computations of $M(x)$. Then, $\{v_1, v_2\} \in E$ if and only if $y_1 \neq y_2$, and $v_1$ and $v_2$ are consistent.

We observe that if $C$ is a clique of the graph $G_x$, then $C$ contains, for each $y$ of length $c \log n$, at most one accepting computation. Furthermore, there is an oracle $Q$ such that for each query $z_j$ of any accepting computation $v$ in $C$, the corresponding answer $b_j$ of $v$ is consistent with oracle $Q$ (i.e., $b_j = \chi_Q(z_j)$). Now, if $x \in L$, then for each $y$ of length $c \log n$, $M(x, y)$ accepts relative to a fixed oracle $Q_x$. Thus, for each $y$ of length $c \log n$, there is a vertex $v$ in $V$ that has the form $v = (x, y, z_1, b_1, \ldots, z_d, b_d)$ such that $b_j = \chi_{Q_x}(z_j)$ for all $1 \leq j \leq d$. These vertices form a clique of size $n^c$ since they are all consistent to the oracle $Q_x$. Conversely, if $x \notin L$, then for any oracle $Q$, $M^Q(x, y)$ accepts for at most $n^c/r$ strings $y$, and so the maximum clique of $G$ has at most $n^c/r$ vertices. In other words, the mapping from $x$ to $(G_x, n^c)$ is a reduction from $L$ to (CLIQUE, $r$-NOCLIQUE). ∎

**Corollary 11.34** *For any $s > 1$, the problem $s$-APPROX-CLIQUE is NP-hard.*

The above result shows that the maximum clique $\omega(G)$ of a graph $G$ cannot be approximated within any constant factor of $\omega(G)$, if $P \neq NP$. Can we approximate $\omega(G)$ within a factor of a small function of the size of $G$, for instance, $\log |G|$ (i.e., can we find a clique of size guaranteed to be at least $\omega(G)/\log n$? The answer is no. Indeed, if we explore the construction of $PCP$ system for $NP$ further, we could actually show that the maximum clique problem is not approximable in polynomial time with ration $n^{1-\epsilon}$ for any $\epsilon > 0$, unless $P = NP$ [Hastad, 1996]. On the positive side, the current best heuristic algorithm for approximating $\omega(G)$ can only achieve the approximation ratio $O(n/(\log n)^2)$ [Boppana and Halldórsson, 1992].

## 11.5 More *NP*-Hard Approximation Problems

Starting from the problems $s$-APPROX-3SAT and $s$-APPROX-CLIQUE, we can use reductions to prove new approximation problems to be $NP$-hard. These reductions are often based on strong forms of reductions between the corresponding decision problems. There are several types of strong reductions in literature. In this section, we introduce a special type of strong reduction, called the *L-reduction*, among optimization problems which can be used to establish the $NP$-hardness of the corresponding approximation problems.

Recall from Section 2.5 that for a given *instance* $x$ of an optimization problem $\Pi$, there are a number of *solutions* $y$ to $x$, each with a *value* $v_\Pi(y)$ (or, $v(y)$, if the problem $\Pi$ is clear from the context). The problem $\Pi$ is to find, for a given input $x$, a solution $y$ with the maximum (or, the minimum) $v(y)$. We let $v_\Pi^*(x)$ (or, simply, $v^*(x)$) denote the value of the optimum solutions to $x$ for problem $\Pi$.

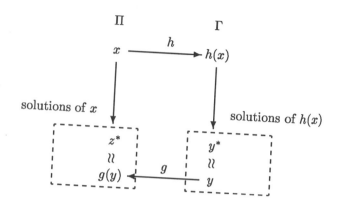

Figure 11.2: The $L$-reduction.

An *L-reduction* from an optimization problem $\Pi$ to another optimization problem $\Gamma$, denoted by $\Pi \leq_L^P \Gamma$, consists of two polynomial-time computable functions $h$ and $g$, and two constants $a, b > 0$, which satisfy the following properties:

(L1) For any instance $x$ of $\Pi$, $h(x)$ is an instance of $\Gamma$ such that $v_\Gamma^*(h(x)) \leq$ $a \cdot v_\Pi^*(x)$.

(L2) For any solution $y$ of instance $h(x)$ with value $v_\Gamma(y)$, $g(y)$ is a soluti of $x$ with value $v_\Pi(g(y))$ such that

$$|v_\Pi(g(y)) - v_\Pi^*(x)| \leq b \cdot |v_\Gamma(y) - v_\Gamma^*(h(x))|.$$

We show the $L$-reduction in Figure 11.2, where $y^*$ denotes the opti solution to $h(x)$ and $z^*$ denotes the optimum solution to $x$.

It is easy to see that the $L$-reduction is transitive and, hence, is re reducibility relation.

**Proposition 11.35** *If $\Pi \leq_L^P \Gamma$ and $\Gamma \leq_L^P \Lambda$ then $\Pi \leq_L^P \Lambda$.*

*Proof.* Suppose $\Pi \leq_L^P \Gamma$ via mappings $h_1$ and $g_1$ and $\Gamma \leq_L^P \Lambda$ via $h_2$ and $g_2$. It is easy to verify that $\Pi \leq_L^P \Lambda$ via $h_2 \circ h_1$ and $g_1 \circ g_2$ (11.2).

For any integer $b > 0$, consider the following two optimization

    VC-$b$: Given a graph $G$ in which every vertex has degree a $b$, find the minimum vertex cover of $G$.

    IS-$b$: Given a graph $G$ in which every vertex has degree $b$, find the maximum independent set of vertices of $G$.

So, condition (L2) also holds. Therefore, this reduction is an $L$-reduction. ∎

**Example 11.37** *For every $b \geq 1$, VC-$b \leq_L^P$ IS-$b$ and IS-$b \leq_L^P$ VC-$b$.*

*Proof.* Consider the identity function as a reduction from VC-$b$ to IS-$b$. For any graph $G = (V, E)$ with the maximum degree at most $b$, its minimum vertex cover must contain at least $|V|/b$ vertices and its maximum independent set contains at most $|V|$ vertices. Therefore, $is^*(G) \leq b \cdot vc^*(G)$, where $is^*(G)$ is the size of the maximum independent set of $G$ and $vc^*(G)$ is the size of the minimum vertex cover of $G$. It follows that condition (L1) holds. Moreover, a set $U$ is an independent set for $G$ if and only if $g(U) = V - U$ is a vertex cover for $G$, and $U^*$ is a maximum independent set for $G$ if and only if $g(U)$ is a minimum vertex cover for $G$. Hence,

$$|U^*| - |U| = |g(U)| - |g(U^*)|,$$

and condition (L2) holds. Therefore, this is an $L$-reduction from VC-$b$ to IS-$b$.

Similarly, we can show that the identity function is also an $L$-reduction from IS-$b$ to VC-$b$. ∎

The following property of $L$-reduction shows the precise relation between the corresponding approximation problems. (Recall the definition of $r$-Approx-$\Pi$ from Section 2.5.)

**Proposition 11.38** *If $\Pi \leq_L^P \Gamma$, then for any $r > 1$ there exists some $t > 1$ such that $r$-Approx-$\Pi \leq_T^P t$-Approx-$\Gamma$.*

*Proof.* Suppose $\Pi \leq_L^P \Gamma$ via $h$ and $g$. Then there exist two positive constants $a$ and $b$ such that conditions (L1) and (L2) hold with respect to $h$ and $g$. Now, given an instance $x$ of $\Pi$, we show how to compute a solution of $r$-Approx-$\Pi$, using an algorithm for $t$-Approx-$\Gamma$ as an oracle, where $t = 1 + (r-1)/(ab)$ if $\Pi$ is a minimization problem and $t = 1 + (r-1)/(abr)$ if $\Pi$ is a maximization problem.

First, we compute the instance $h(x)$ of $\Gamma$. By condition (L1), we have

$$v_\Gamma^*(h(x)) \leq a \cdot v_\Pi^*(x).$$

Next, compute a solution $y$ of $t$-Approx-$\Gamma$ on instance $h(x)$ and we have

$$\frac{1}{t} \leq \frac{v(y)}{v_\Gamma^*(h(x))} \leq t,$$

which implies

$$|v(y) - v_\Gamma^*(h(x))| \leq (t-1) \cdot v_\Gamma^*(h(x)).$$

Now, from the solution $y$ to the instance $h(x)$, we compute the solution $g(y)$ to instance $x$. By condition (L2), we have

$$|v(g(y)) - v_\Pi^*(x)| \leq b \cdot |v(y) - v_\Gamma^*(h(x))|.$$

If $\Pi$ is a minimization problem and $v(g(y)) \geq v_\Pi^*(x)$, then

$$\frac{v(g(y))}{v_\Pi^*(x)} = \frac{v(g(y)) - v_\Pi^*(x)}{v_\Pi^*(x)} + 1$$

$$\leq \frac{a \cdot b \cdot |v(y) - v_\Gamma^*(h(x))|}{v_\Gamma^*(h(x))} + 1 \leq ab(t-1) + 1 = r,$$

and if $\Pi$ is a maximization problem and $v(g(y)) \leq v_\Pi^*(x)$, then

$$\frac{v_\Pi^*(x)}{v(g(y))} = \frac{v_\Pi^*(x)}{v_\Pi^*(x) - (v_\Pi^*(x) - v(g(y)))}$$

$$\leq \frac{v_\Pi^*(x)}{v_\Pi^*(x) - b \cdot |v(y) - v_\Gamma^*(h(x))|}$$

$$\leq \frac{v_\Gamma^*(h(x))}{v_\Gamma^*(h(x)) - a \cdot b \cdot |v(y) - v_\Gamma^*(h(x))|} \leq \frac{1}{1 - ab(t-1)} = r.$$

Therefore, in either case $g(y)$ is a solution of $r$-APPROX-$\Pi$.    ∎

From Proposition 11.38, we have the following relations:

(a) Assume that $\Pi$ is a minimization problem. If $\Pi \leq_L^P \Gamma$ and $\Gamma$ has a polynomial-time linear approximation algorithm (meaning that the approximation ratio $r$ is a constant), then $\Pi$ also has a polynomial-time linear approximation algorithm.

(b) If $\Pi \leq_L^P \Gamma$ and $\Gamma$ has a (fully) polynomial-time approximation scheme, then $\Pi$ has a (fully, respectively) polynomial-time approximation scheme.

We know from Corollary 11.31 that there exists a number $s > 1$ such that the problem $s$-APPROX-3SAT is $\leq_T^P$-hard for *NP*. The $L$-reduction can pass this nonapproximability result from 3SAT to other problems. We first consider the following restricted version of 3SAT, which is a fundamental problem for proving nonapproximability results.

3SAT-3: Given a 3CNF formula[5] $F$ in which each variable appears in at most three clauses, find a Boolean assignment to the variables of $F$ to maximize the number of true clauses.

We are going to show that 3SAT $\leq_L^P$ 3SAT-3. To prove this, we need a combinatorial lemma involving expanders. Let $G = (V, E)$ be a graph. For each set $S \subseteq V$, we let $N(S)$ denote the set of vertices that are not in $S$ but are adjacent to some vertex in $S$; that is, $N(S) = \{v \in V - S : (\exists u \in S)\, \{u, v\} \in E\}$. Also let $A(S)$ be the edges between $S$ and $N(S)$; that is, $A(S) = \{\{u, v\} \in E : |\{u, v\} \cap S| = 1\}$. Note that $|A(S)| \geq |N(S)|$.

---

[5] Here, we allow a 3CNF formula to have clauses with less than three literals.

An $(n, d, \delta)$-*expander* is a $d$-regular bipartite graph $G = (X, Y, E)$ with $|X| = |Y| = n$ such that for any $S \subseteq X$,

$$|N(S)| \geq \left(1 + \delta\left(1 - \frac{|S|}{n}\right)\right)|S|.$$

The first explicit construction of expanders was given by Gabber and Galil [1981] as follows: Consider $n = m^2$. Let $X = Y = \mathbf{Z}_m \times \mathbf{Z}_m$. For each vertex $(x, y) \in X$, connect it to vertices $(x, y)$, $(x, x + y)$, $(x, x + y + 1)$, $(x + y, y)$, and $(x + y + 1, y)$ in $Y$. It has been shown that this is an $(n, 5, \delta)$-expander, where $\delta = (2 - \sqrt{3})/4 \approx 0.067$.

Now, we define a graph with vertex set $X = \mathbf{Z}_m \times \mathbf{Z}_m$ and edges connecting each vertex $(x, y)$ to $(x, x + y)$, $(x, x + y + 1)$, $(x + y, y)$, and $(x + y + 1, y)$. According to the above result about the expanders, this graph must have the property that for any $S \subseteq X$ with $|S| \leq |X|/2$, $|A(S)| \geq |N(S)| \geq (\delta/2)|S|$. We denote this graph by $H(m^2, 4, \delta/2)$. We now prove a lemma using this graph as a tool.

**Lemma 11.39** *There is an absolute constant $c > 0$ such that for any finite set $X$ with $|X| \geq 4$, there exists a strongly connected directed graph $G = (V, E)$ with $X \subseteq V$ having the following properties:*

*(a) $|V| \leq c|X|$.*

*(b) Every vertex in $X$ has in-degree one and out-degree one. Every vertex not in $X$ has degree (in-degree plus out-degree) at most three.*

*(c) For every subset $S \subseteq V$ containing at most $|X|/2$ vertices in $X$, $|E \cap (S \times (V - S))| \geq |S \cap X|$ and $|E \cap ((V - S) \times S)| \geq |S \cap X|$.*

*Proof.* First assume that $|X| = n$ is a perfect square. Let $m = \sqrt{n}$. Assume that $X = \{x_1, x_2, \ldots, x_n\}$. Construct $t = \lceil 10/\delta \rceil$ copies of $H(n, 4, \delta/2)$. Call them $H_1, H_2, \ldots, H_t$, with the vertices in $H_j$ being $x_{1,j}, x_{2,j}, \ldots, x_{n,j}$, for $j = 1, \ldots, t$. Next, for each $i = 1, \ldots, n$, define a binary tree $T_i$ such that its leaves are exactly the set $\{x_{i,1}, x_{i,2}, \ldots, x_{i,t}\}$ (and the internal nodes of $T_i$ are new vertices, occurring only in $T_i$). Connect the root of $T_i$ with $x_i$ for each $i = 1, \ldots, n$. Call the resulting graph $W = (Y, F)$. It is obvious that $W$ is connected. Note that each node in $Y$ has degree at most five, and each $x_i \in X$ has degree one.

Next, we change $W$ into a digraph $W' = (Y, F')$ by replacing each edge $\{u, v\}$ in $F$ by two edges $(u, v)$ and $(v, u)$. Finally, for each node $y$ in $W'$, we perform the following degree-reducing operation on it: If $y$ has out-degree $d \geq 2$, then replace the $d$ out-edges by a tree as shown in Figure 11.4(a); and if $y$ has in-degree $d \geq 2$, then replace the $d$ in-edges by a tree as shown in Figure 11.4(b). Let the resulting digraph be $G = (V, E)$, and let $H'_j$, $1 \leq j \leq t$, be the subgraph resulting from the above degree-reducing operation performed on vertices in $H_j$. Note that $|H'_j| \leq 9n$ for all $j = 1, \ldots, t$.

We now verify that $G$ satisfies our requirements.

(a) The set $Y$ has $n$ vertices from $X$, $n$ trees $T_i$ each of size $2t - 1$. (All vertices in $H_j$, $1 \leq j \leq t$, are the leaves of these trees.) Therefore, $|Y| = 2tn$.

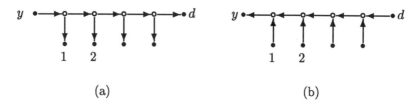

Figure 11.4: Trees for fan out, where the open circles denote new vertices.

Each vertex in $W$ has degree at most five, and so, each vertex in $W'$ has in-degree at most five and out-degree at most five. Therefore, from $W'$ to $G$, we added, for each vertex in $Y$, at most eight more vertices. So, the total number of vertices in $V$ is $18tn$. Note that $t$ is an absolute constant.

(b) This part is obvious from the construction.

(c) First, for any $1 \leq j \leq t$, we claim that for any $S_j \subseteq H'_j$, with $|S_j \cap H_j| = r \leq n/2$, we have

$$|E \cap (S_j \times (H'_j - S_j))| \geq \frac{\delta}{10} \cdot r.$$

From the property of the expander $H(n, 4, \delta/2)$, we know that, in $W'$, it is true that

$$|F' \cap ((S_j \cap H_j) \times (H_j - S_j))| \geq \frac{\delta}{2} \cdot r.$$

Now, when we perform the degree-reducing operation on any $y \in H_j$, we reduce at most five edges from $H_j \cap S_j$ to $H_j - S_j$ to one edge from $S_j$ to $H'_j - S_j$. To see this, we first consider the operation of Figure 11.4(a) on a vertex $y \in H_j$. Assume that $y \in S_j$ and that at least one of its out-neighbors is not in $S_j$. Then, there are two cases:

(1) All newly created vertices (the open circle vertices) are in $S_j$. Then, the number of edges from $S_j$ to $H'_j - S_j$ is unchanged.

(2) At least one of the newly created vertices is not in $S_j$. Then, there is at least one edge from $S_j$ to $H'_j - S_j$ which ends in a vertex in $H'_j - H_j$. In this case, the number of edges decreases at most by a factor of $1/5$.

Next, we perform the operation of Figure 11.4(b) on all vertices in $H_j$. We note that only those edges created in Case (1) above are affected and, again, the number of these edges decreases at most by a factor of $1/5$. In other words, each original edge only participates in one of the two operations. So, we have

$$|E \cap (S_j \times (H'_j - S_j))| \geq \frac{|F' \cap ((S_j \cap H_j) \times (H_j - S_j))|}{5},$$

and the claim is proven.

Since the digraph $W'$ is symmetric, the assumption of the claim also implies that

$$|E \cap ((H'_j - S_j) \times S_j)| \geq \frac{\delta}{10} \cdot r.$$

Now, let $S \subseteq V$ have $|S \cap X| = k \leq n/2$. For each $j = 1, \ldots, t$, let $S_j = S \cap H'_j$, and

$$r = \min_{1 \leq j \leq t} \{|S_j \cap H_j|, |H_j - S_j|\}.$$

*Case 1.* $r \geq k$. Then, for each $j = 1, \ldots, t$, the above claim shows that $|E \cap (S_j \times (H'_j - S_j))| \geq (\delta/10) \cdot r$, and so

$$|E \cap (S \times (V - S))| \geq \sum_{j=1}^{t} |E \cap (S_j \times (H'_j - S_j))|$$

$$\geq t \cdot \left(\frac{\delta}{10}\right) \cdot r \geq k.$$

*Case 2.* $r < k$. For any $x_{i,j}$ in $H_j$, $1 \leq i \leq n$ and $1 \leq j \leq t$, if $|\{x_i, x_{i,j}\} \cap S| = 1$, then we say $x_{i,j}$ is *inconsistent* (with $x_i$). Note that if $x_{i,j}$ is inconsistent, then there must exist an edge in the tree $T_i$ that is in $S \times (V - S)$. Assume that for some fixed $j$, $|S_j \cap H_j| = r$ (or, $|H_j - S_j| = r$). Then, there must be $k - r$ vertices in $H_j$ that are inconsistent. So, there are at least $k - r$ edges between $S$ and $V - S$ in the trees $T_1, \ldots, T_n$. Together, we get

$$|E \cap (S \times (V - S))| \geq (k - r) + \sum_{j=1}^{t} |E \cap (S_j \times (H'_j - S_j))|$$

$$\geq (k - r) + t \cdot \left(\frac{\delta}{10}\right) \cdot r \geq k.$$

From the above two cases, we get $|E \cap (S \times (V - S))| \geq |S \cap X|$. Since the digraph $G$ is symmetric, it follows that $|E \cap ((V - S) \times S)| \geq |S \cap X|$ too. This completes the proof of condition (c).

Finally, consider the case in which $|X|$ is not a perfect square. In this case, let $m = \lceil \sqrt{|X|} \rceil$ and we simply pad $X$ with $m^2 - |X|$ elements into a set $X'$ of size $m^2$ and construct graph $G$ from $X'$ as above. Since $m^2 \leq 2|X|$, the condition (a) holds with the constant $c = 36t$. For condition (c), we assume that $S \subseteq V$ and $|S \cap X| \leq |X|/2$. Then, we consider two cases:

*Case 1.* $|S \cap X'| \leq |X'|/2$. In this case, we have, $|E \cap (S \times (V - S))| \geq |S \cap X'| \geq |S \cap X|$.

*Case 2.* $|S \cap X'| > |X'|/2$. This implies that $|(V - S) \cap X'| \leq |X'|/2$ and so $|E \cap (S \times (V - S))| \geq |(V - S) \cap X'|$. Note that $|(V - S) \cap X'| \geq |(V - S) \cap X| \geq |S \cap X|$, and so condition (c) also holds. This completes the proof of the lemma. ∎

**Theorem 11.40** *There exists a real number $r > 1$ such that the problem $r$-APPROX-3SAT-3 is $\leq_T^P$-hard for NP.*

*Proof.* We will construct an $L$-reduction from 3SAT to 3SAT-3. Let $F$ be a given 3-CNF formula. For each variable $x$ with $k \geq 4$ occurrences in $F$, we replace the occurrences of $x$ by $k$ new variables $x_1, x_2, \ldots, x_k$, The idea, then, is to add a *linear* number of new clauses of the form $(x_i \rightarrow x_j)$ such that, in order to maximize the number of true clauses, it does not pay to assign different values to $x_1, x_2, \ldots, x_k$. (Note also that each $x_i$ can occur in at most two new clauses.) To do this, we construct the directed graph $G_x = (V_x, E_x)$ for $X = \{x_1, x_2, \ldots, x_k\}$ as in Lemma 11.39. Treat all vertices in $V_x - X$ as new Boolean variables and, for each edge $(u, v) \in E_x$, add a new clause $\bar{u} \vee v$ (i.e., $u \rightarrow v$) to the formula. We do this for each variable $x$ and let $h(F)$ be the resulting formula. By condition (b) of Lemma 11.39, each variable occurs at most three times in $h(F)$ and, hence, $h(F)$ is an instance of 3SAT-3.

We now show that $h(F)$ has an interesting property that for any Boolean assignment $\tau$, there exists an assignment $\tau'$ such that:

(1) The number of clauses satisfied by $\tau'$ is no smaller than the number of clauses satisfied by $\tau$; and

(2) All those variables $x_1, x_2, \ldots, x_k$ introduced from the same variable $x$ take the same value in $\tau'$.

In fact, if $\tau$ does not satisfy property (2) with respect to variables in $X = \{x_1, x_2, \ldots, x_k\}$, then we can assign all variables $y$ in $V_x$ the majority value of $\{\tau(x_i) : x_i \in X\}$ as $\tau'(y)$. Let $S = \{y \in V_x : \tau(y) \neq \tau'(y)\}$. We observe that the change from $\tau$ to $\tau'$ may turn at most $|X \cap S|$ satisfying clauses in $F$ to unsatisfying. However, by Lemma 11.39, it also increases at least $|S \cap X|$ satisfying clauses corresponding to edges in $G_x$.

Let $g(\tau)$ be the Boolean assignment for $F$ which assigns each variable $x$ of $F$ with the value taken by the corresponding variables in $G_x$ under assignment $\tau'$. Note that if $F$ has $m$ clauses, then $h(F)$ has at most $3cm$ variables, where $c$ is the constant of Lemma 11.39. Also, each variable occurs in at most three clauses. It follows that $h(F)$ has at most $\lceil 9cm/2 \rceil$ clauses (note that each clause has at least two literals). From the fact that 2-APPROX-3SAT is polynomial-time solvable (see Exercise 11.14), we have

$$sat^*(h(F)) \leq (9c + 1)sat^*(F).$$

That is, condition (L1) holds. Also, from the properties (1) and (2) of $\tau'$, we have

$$|v_F(g(\tau)) - sat^*(F)| \leq |v_{h(F)}(\tau) - sat^*(h(F))|,$$

where $v_F(g(\tau))$ (and $v_{h(F)}(\tau)$) is the number of satisfying clauses in $F$ ($h(F)$) under the assignment $g(\tau)$ ($\tau$, respectively). So, condition (L2) also holds. ∎

**Theorem 11.41** *There exists a real number $r > 1$ such that the problems $r$-APPROX-VC-3 and $r$-APPROX-IS-3 are $\leq_T^P$-hard for NP.*

*Proof.* The reduction from 3SAT to VC in Theorem 2.14 is actually an $L$-reduction from 3SAT-3 to VC-4. The theorem follows from Examples 11.36, 11.37, and Propositions 11.35 and 11.38. ∎

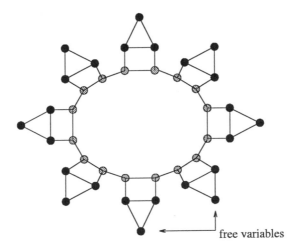

Figure 11.5: Graph $H$.

There is a big class of approximation problems which can be proved to be NP-hard using the $L$-reduction. The following is another example.

VC-in-Cubic-Graphs (VC-CG): Given a graph $G$ in which every vertex has degree exactly three (called a *cubic graph*), find the minimum vertex cover of $G$.

**Theorem 11.42** *There exists a real number $r > 1$ such that $r$-APPROX-VC-CG is NP-hard.*

*Proof.* We construct an $L$-reduction from VC-3 to VC-CG. Let $G$ be an instance of VC-3, that is, a graph in which every vertex has degree at most three. Suppose that $G$ has $i$ vertices of degree one and $j$ vertices of degree two. We construct $2i + j$ triangles and a cycle of size $2(2i + j)$, and connect two vertices of each triangle to two adjacent vertices of the cycle (see Figure 11.5). Denote this graph by $H$. Call, in each triangle, the vertex that is not connected to the cycle a *free vertex* in $H$. Note that $H$ has a minimum vertex cover of size $3(2i+j)$ (the cycle needs $2i+j$ vertices to cover and each triangle needs two vertices to cover) and there exists a minimum vertex cover of $H$ containing all free vertices.

Now, we construct graph $G'$ from $G$ and $H$ as follows: For each vertex $x$ of degree one in $G$, connect $x$ to two free vertices; and for each vertex $x$ of degree two in $G$, connect $x$ to one free vertex. Thus, each free vertex is adjacent to exactly one vertex in $G$. Clearly, $G$ has a vertex cover of size $s$ if and only if $G'$ has a vertex cover of size $s' = s + 3(2i + j)$. Therefore,

$$vc^*(G') = vc^*(G) + 3(2i + j).$$

Note that $G$ has at least $(i + 2j)/2$ edges and each vertex in $G$ can cover at most three edges. Therefore, $i + 2j \le 6 \cdot vc^*(G)$ and, hence, $2i + j \le 12 \cdot vc^*(G)$. It follows that

$$vc^*(G') \le 37 \cdot vc^*(G),$$

and condition (L1) holds. To prove (L2), we note that for each vertex cover $S'$ of size $s'$ in $G'$, we can obtain a vertex cover $S$ of size $s \le s' - 3(2i + j)$ in $G$ by simply removing all vertices in $S'$ which are not in $G$. It follows that

$$s - vc^*(G) \le s' - vc^*(G').$$

Therefore, (L2) also holds.                                                                     ∎

In Section 2.5, we divided the optimization problems into three classes, according to their approximability. The third class contains the problems with polynomial-time approximation schemes. The second class contains the problems for each of which there is a constant $r_0 > 1$ such that it is polynomial-time approximable with respect to ratio $r_0$ but is nonapproximable in polynomial time for any ratio $r < r_0$ unless $P = NP$. We just showed that the maximum satisfiability belongs to this class. Since $L$-reductions preserve linear-ratio approximation, all problems that are $L$-equivalent to 3SAT belong to this class, including the problems VC-3, IS-3, 3SAT-3, and VC-CG studied above (cf. Exercise 11.14).

Next, we turn our attention to problems in the first class, the class of problems for which the corresponding approximation problems are $NP$-hard for all ratios $r > 1$. In the last section, we saw that the problem CLIQUE belongs to this class. We can actually further divide this class into subclasses according to the best approximation ratio $r(n)$, where $r(n)$ is a function of input size $n$. For instance, consider the following problem:

> SET COVER (SC): Given a finite collection $\mathcal{S}$ of subsets of a set $U$ of $m$ elements, find a minimum cardinality subcollection of $\mathcal{S}$ such that the union of all subsets in the subcollection covers $U$.

It is known that SC has a polynomial-time approximation algorithm with approximation ratio $\log n$. In fact, this is the best known result unless $NP$ collapses to a deterministic subexponential time class.

**Theorem 11.43** *For any $0 < \rho < 1$, there is no polynomial-time approximation algorithm with the approximation ratio $\rho \log n$ for SC unless $NP \subseteq \bigcup_{k>0} DTIME(n^{(\log n)^k})$.*

*Proof.* The proof of the theorem, like the problem CLIQUE, comes directly from the proof of the *PCP* characterization of *NP*. We omit it here.           ∎

Note that both the problems SC and CLIQUE belong to the first class of nonapproximable optimization problems, but they still have different approximation ratios: CLIQUE is so hard to approximate that it is not approximable in polynomial time within ratio $n^{1-\epsilon}$ for any $\epsilon > 0$ (if $P \neq NP$), but SC has a polynomial-time approximation algorithm to achieve the approximation ratio $\log n$.

The problem SC is a fundamental problem for classifying optimization problems to the first class. We show an application below.

SUBSET INTERCONNECTION DESIGNS (SID): Given a complete weighted graph $H$ on vertex set $X$ and given subsets $X_1, \ldots, X_m$ of $X$, find a minimum-weight subgraph $G$ (called a *feasible subgraph*) of $H$ such that for every $i = 1, \ldots, m$, $G$ contains a spanning tree for $X_i$.

**Theorem 11.44** *For $0 < \rho < 1$, there is no polynomial-time approximation algorithm for SID with the approximation ratio $\rho \log n$ unless $NP \subseteq \bigcup_{k>0} DTIME(n^{(\log n)^k})$.*

*Proof.* We reduce the problem SC to SID as follows: For each instance $\mathcal{S}$ of SC, assume that $U = \{1, 2, \ldots, m\}$ and $\mathcal{S} = \{S_1, \ldots, S_n\}$. We construct an instance $(H; X_1, \ldots, X_m)$ of SID by setting

$$X = \{0, 1, \ldots, n\},$$
$$X_i = \{j : i \in S_j\} \cup \{0\}, \quad 1 \leq i \leq m,$$

and assigning weights on edges as follows:

$$c(j, k) = \varepsilon, \quad 1 \leq j < k \leq n,$$
$$c(0, j) = 1, \quad 1 \leq j \leq n,$$

where $\varepsilon = 1/(n(n-1))$.

For each set cover $\mathcal{S}' = \{S_{j_1}, \ldots, S_{j_s}\}$ of $\mathcal{S}$, we can construct a feasible subgraph $G$ of $H$ by connecting edges $(0, j_1), \ldots, (0, j_s)$ to the complete graph on vertices $1, \ldots, n$. This subgraph $G$ has total weight $c(G) = s + \varepsilon \cdot n(n-1)/2 = s + 1/2$. If the optimum solution for the instance $\mathcal{S}$ of SC is a subcollection of $s^*$ sets, then the minimum feasible subgraph for $H$ has total weight between $s^*$ and $s^* + 1/2$.

Conversely, if $G$ is a feasible subgraph of $(H; X_1, \ldots, X_m)$ with total weight $c(G)$, then the set

$$\mathcal{S}' = \{S_j : \text{edge } (0, j) \text{ is in } G\}$$

is a set cover of $U$ and

$$c(G) - \frac{n(n-1)}{2} \cdot \varepsilon = c(G) - \frac{1}{2} \leq |\mathcal{S}'| \leq c(G).$$

Now, suppose that there exists a polynomial-time approximation algorithm for SID with approximation ratio $\rho \log m$ for some $0 < \rho < 1$. We compute an approximate solution for an instance $\mathcal{S}$ of the problem SC as follows:

*Step* 1. Check all subcollections of $\mathcal{S}$ of size at most $1/(2(1-\rho))$. If there exists a set cover of $U$ among them, then choose the one with the minimum cardinality as solution; else, go to step 2.

*Step* 2. Construct an instance $(H; X_1, \ldots, X_m)$ of SID from $\mathcal{S}$. Use the polynomial-time approximation algorithm for SID to find a feasible subgraph $G$ of $(H; X_1, \ldots, X_m)$. Define $\mathcal{S}' = \{S_j : \text{edge } (0, j) \text{ is in } G\}$ as the approximate solution to the instance $\mathcal{S}$.

Clearly, if the solution comes from Step 1, then it is optimal; if the solution comes from Step 2, then the optimum solution of $\mathcal{S}$ must have size $s^* > 1/(2(1-\rho))$, or $\rho + 1/(2s^*) < 1$. Therefore, we have

$$|\mathcal{S}'| \le c(G) \le \rho(\log m) \left(s^* + \frac{1}{2}\right) \le \left(\rho + \frac{1}{2s^*}\right)(\log m)\, s^*.$$

That is, this approximate solution is within a factor of $\rho' \log m$ from the optimal, where $0 < \rho' = \rho + 1/(2s^*) < 1$. By Theorem 11.43, $NP \subseteq \bigcup_{k>0} DTIME(n^{(\log n)^k})$. ∎

## Exercises

**11.1** *(Multiple Prover Interactive Proof Systems).* We extend the formulation of the interactive proof systems of Exercise 10.1 to multi-prover interactive proof systems in which more than one provers work together to convince the verifier that the input $x$ is in the language $L$. To make it more powerful than the one-prover systems, the provers in a multi-prover system are not allowed to communicate with each other, once the proof system is activated. Formally, let $M_{p_1}, \ldots, M_{p_k}$ be $k$ DTMs and $M_v$ be a polynomial-time PTM. We say $(M_{p_1}, \ldots, M_{p_k}, M_v)$ is a *k-prover interactive protocol* if they share a common read-only tape and for each $i$, $1 \le i \le k$, $M_{p_i}$ and $M_v$ share a communication tape $T_i$, and they compute in the same manner as the one-prover interactive protocol. That is, $M_v$ and one of $M_{p_i}$, $1 \le i \le k$, take turns in being active. Suppose $M_v$ is activated by $M_{p_i}$; then it reads the string on tape $T_i$, computes a string $y$ and an integer $j$, $1 \le j \le k$, and then writes the string $y$ on tape $T_j$ and activates machine $M_{p_j}$ (or, it may choose to halt and accept or reject). The activity of a prover $M_{p_i}$ is similar, except that it can only read from and write to tape $T_i$; that is, $M_i$ cannot access tape $T_j$ if $j \ne i$. We say that a set $L \subseteq \Sigma^*$ has a *k-prover interactive proof system* with error bound $\epsilon$ if there exists a PTM $M_v$ such that the following hold:

(i) There exist $k$ DTMs $M_{p_1}, \ldots, M_{p_k}$ such that $(M_{p_1}, \ldots, M_{p_k}, M_v)$ forms a $k$-prover interactive protocol and for each $x \in L$, the probability of these $k + 1$ machines together accepting $x$ is at least $1 - \epsilon$; and

(ii) For any $k$ DTMs $M_1, \ldots, M_k$ such that $(M_1, \ldots, M_k, M_v)$ forms a $k$-prover interactive protocol and for any $x \notin L$, the probability of these $k + 1$ machines together accepting $x$ is at most $\epsilon$.

(a) Extend the above definition to allow the number $k$ of provers being a function depending on the input size $|x|$. Show that if a set $L$ is in $PCP(poly, poly)$, then $L$ has a $k(n)$-prover interactive proof system with error bound $1/4$ for some polynomial function $k$.

(b) Prove that if $L$ has a $k(n)$-prover interactive proof system with error bound $1/4$ for some polynomial function $k$, then $L$ has a two-prover interactive proof system with error bound $1/4$.

(c) Show that if $L$ has a two-prover interactive proof system with error bound $1/4$, then $L \in PCP(poly, poly)$.

**11.2** Let $\mathcal{F}$ be a field and $f$ be a function from $\mathcal{F}^m$ to $\mathcal{F}$. Prove that $f$ is multilinear if and only if $f$ is linear on all aligned triples.

**11.3** Let $\mathcal{F}$ be a field and $H \subseteq \mathcal{F}$. Prove that a function $f : H^m \to \mathcal{F}$ may have more than one polynomial extension of degree $d = mh$. However, it has a unique polynomial extension that has degree $h$ on each individual variable.

**11.4** Prove that the multilinearity test system of Section 11.2.2 can be improved so that the verifier only uses $O(m \log \alpha)$ random bits.

**11.5** Prove that if $P \neq NP$ then $NP \not\subseteq PCP(r(n), q(n))$ for all $r(n), q(n) \in o(\log n)$.

**11.6** Let $\mathcal{F}$ be a field of size $\alpha$ and $d$ is an integer less than $\alpha/2$.

(a) Let $k < d$. Assume that $g : \mathcal{F}^m \to \mathcal{F}$ is a polynomial function of degree $\leq d$. If for at least $d/\alpha$ fractions of lines $L$, $g_L$ is $(1/2 + d\alpha)$-close to a univariate polynomial function of degree $k$, then $g$ is a polynomial function of degree $\leq k$.

(b) Prove that $f : \mathcal{F}^m \to \mathcal{F}$ is a polynomial function of degree $d < \alpha/2$ if and only if $f_L$ is a univariate polynomial of degree $d$ for all lines $L$ in $\mathcal{F}^m$.

**11.7** Prove Lemma 11.16

**11.8** Prove Lemma 11.20.

**11.9** At the beginning of Section 11.3.3, we argued that each verifier $V_1$ can be converted to an equivalent verifier $V_1'$ which constructs $\phi_r$ first and then reads $x_r$ and $y_r$ to see whether $x_r$ and $y_r$ satisfy $\phi_r$. Suppose now the verifier does not read any proof bits, and instead performs an exhaustive search for all possible $x_r$ and $y_r$ that satisfy $\phi_r$. Then, this new verifier $V_1''$ uses the same number of random bits as $V_1$, reads no proof bit, and works within randomized time $2^{|\phi_r|}$. If we apply this conversion to system $S_2$, then $|r| = O(\log n)$ and $|\phi_r| = (\log n)^{O(1)}$, and so we have obtained a randomized algorithm for SAT in subexponential time. What is wrong with the above argument?

**11.10** Describe explicitly what $\phi_r$ and $x_r$ of system $S_2$ are, and what $\phi_r$ and $x_r$ of system $S_2'$ are.

**11.11** In this exercise, we show a different approach to the problem of double encoding of Section 11.3.3. The idea is to not require property A on system $X_1$ but require a stronger property B' than the property B on system $X_2$. The requirement B' asks the verifier $V_2$ to read the assignments in the split form $\Pi = (\Pi_1, \ldots, \Pi_m)$, where $m$ is not necessarily a constant. Assume that for each $i$, $1 \le i \le m$, $\Pi_i$ is the polynomial code of string $s_i$ (i.e., $\Pi_i = \sigma(s_i)$, where $\sigma$ is the encoding scheme of $X_2$). Then, we require that $V_2$ be able to reconstruct each value of $\sigma(s_1 s_2 \cdots s_m)$ from at most a constant number of values of $\Pi_i$ for each $1 \le i \le m$. The following shows that the verifier of the system $S_2$ satisfies this requirement.

(a) Let $s_1, \ldots, s_m$ be binary strings of length $(d+1)^l$, and $s = s_1 \cdots s_m$. Let $\mathcal{F}$ be a field of size $\alpha > d+1$. For each $i$, $1 \le i \le m$, let $f_{s_i} : \mathcal{F}^l \to \mathcal{F}$ be the degree-$d$ polynomial code of $s_i$ under the encoding scheme $\sigma_1$ of Section 11.3.1. Also, if $m = (d+1)^k$, let $f_s : \mathcal{F}^{l+k} \to \mathcal{F}$ be the degree-$d$ polynomial code of $s$ under $\sigma_1$. Show that for each $\vec{a} \in \mathcal{F}^l$ and each $\vec{b} \in \mathcal{F}^k$, the value $f_s(\vec{a}, \vec{b})$ can be expressed as the interpolation of $f_{s_i}(\vec{b})$, $1 \le i \le m$.

(b) Apply part (a) above to compose $S_2$ with itself to get a *PCP* system for SAT that has the same complexity as system $S_3$.

**11.12** Prove Lemma 11.27.

**11.13** Extend the *PCP* characterization of *NP* to *PSPACE* and $\Pi_k^P$.

**11.14** Show that the problems 2-APPROX-3SAT, 2-APPROX-VC-3, and 4-APPROX-IS-3 are solvable in polynomial time.

**11.15** Show that if $P \ne NP$ then the problem $r$-APPROX-CLIQUE is not polynomial-time solvable with ratio $r = n^{1-\epsilon}$ for all $\epsilon > 0$.

**11.16** Consider the problem 2SAT-4: Given a 2-CNF formula in which each variable occurs at most four times, find a Boolean assignment to maximize the number of satisfying clauses. Show that there exists a real number $r > 1$ such that $r$-APPROX-2SAT-4 is $\le_T^P$-hard for *NP*.

**11.17** Consider the problem STEINER MINIMUM TREES-IN-GRAPH (SMT-G): Given an edge-weighted graph $G$ and a vertex subset $S$, find a minimum-weight subgraph interconnecting vertices in $S$. Show that there exists a real number $r > 1$ such that $r$-APPROX-SMT-G is $\le_T^P$-hard for *NP*.

**11.18** Consider the problem SMT-IN-DIGRAPH (SMT-DG): Given an edge-weighted directed graph $G$, a vertex $x$, and a vertex subset $S$, find a minimum-weight directed tree rooted at $x$ and connected to every vertex in $S$. Show that:

(a) For some $\rho > 0$, there is no polynomial-time approximation algorithm with the approximation ratio $\rho \log n$ for SMT-DG unless $NP \subseteq \bigcup_{k>0} DTIME(n^{(\log n)^k})$; and

(b) For any $\epsilon > 0$, there exists a polynomial-time approximation algorithm for SMT-DG with the approximation ratio $n^\epsilon$.

**11.19** Consider the problem LONGEST PATH (LP): Given a graph $G$ and two vertices $x$ and $y$ in $G$, find the longest path between $x$ and $y$ in $G$. For any graph $G = (V, E)$, let $G^2$ be the graph with vertex set $V \times V$ and edge set $\{((x, y), (x, y')) : (y, y') \in E\} \cup \{((x, y), (x', y)) : (x, x') \in E\}$. Prove that there exists a path of length $k$ between two vertices $x$ and $y$ in $G$ if and only if there exists a path of length $k^2$ between $(x, x)$ and $(y, y)$ in $G^2$. Use this fact to find the best approximation ratio of any polynomial-time approximation algorithm for LP.

**11.20** Show that 3SAT is $L$-reducible to TSP with the triangle inequality.

**11.21** Recall the problem SHORTEST COMMON SUPERSTRING (SCS) of Exercise 2.20(c). Show that 3SAT is $L$-reducible to SCS.

**11.22** Consider the problem MIN-BISECTION: Given a graph $G$, divide $G$ into two equal halves to minimize the number of edges between the two parts. Study the approximability of this problem.

**11.23** Show that the following problems have no polynomial-time approximation with approximation ratio $\rho \log m$ for $0 < \rho < 1/4$ unless $NP \subseteq \bigcup_{k>0} DTIME(n^{(\log n)^k})$:

(a) Given a finite collection $S$ of subsets of a set $U$ of $m$ elements, find a minimum subset intersecting every subset in $S$.

(b) Given a finite collection $S$ of subsets of a set $U$ of $m$ elements, find a minimum cardinality subcollection of $S$ such that all subsets in the subcollection are disjoint and the union of all subsets in the subcollection covers $U$.

**11.24** Show that for each of the following problems, there exists a real number $0 < \rho < 1$ such that the problem has no polynomial-time approximation with approximation ratio $2^{(\log n)^{1-\rho}}$ unless $NP \subseteq \bigcup_{k>0} DTIME(n^{(\log n)^k})$:

(a) LP (defined above in Exercise 11.19).

(b) Given a system of $n$ linear equations in $m$ variables $x_1, x_2, ..., x_m$ with integer coefficients and $n > m$, find the minimum number of equations such that removal of them makes the remaining system feasible (in the sense that the system has at least one solution).

**11.25** An optimization problem $\Pi$ is *A-reducible* to an optimization problem $\Gamma$ if there exist three polynomial-time computable functions $h$, $g$, and $c$ such that (i) for any instance $x$ of $\Pi$, $h(x)$ is an instance of $\Gamma$; (ii) for any solution $y$ to $h(x)$, $g(h(x), y)$ is a solution to $x$; and (iii) if the error ratio of $y$ to $h(x)$ is bounded by $e$, then the error ratio of $g(h(x), y)$ to $x$ is bounded by $c(e)$, where the error ratio of a solution $z$ to an instance $x$ is $|z - v^*(x)|/v^*(x)$.

Prove that if $\Pi$ is $A$-reducible to $\Gamma$ and $\Gamma$ is approximable in polynomial time with ratio $r(n)$, then $\Pi$ is approximable in polynomial time with ratio $O(r(n))$.

## Historical Notes

The celebrated result of the *PCP* characterization of *NP* started with the extension of the notion of interactive proof systems to the multi-prover proof systems (see Exercise 11.1) by Ben-Or et al. [1988]. They defined *MIP* as the class of languages accepted by a multi-prover proof system in polynomial time. Fortnow et al. [1994] presented an equivalent formulation of the class *MIP*, based on the notion of randomized oracle TMs. Babai, Fortnow, and Lund [1991] showed that *MIP* = *NEXPPOLY*. Babai, Fortnow, Levin, and Szegedy [1991] and Feige et al. [1996] further improved this result on problems in *NP*. Particularly, Feige at al. [1996] introduced a new notion of nonadaptive proof systems with restrictions on the number of random bits used and the number of queries asked by the verifier. Arora and Safra [1992, 1998] generalized it and called it the probabilistically checkable proof system (*PCP*). They also introduced the idea of composing two proof systems into one and showed that *NP* = *PCP*(*log*, *log*). The final characterization of *NP* = *PCP*(*log*, *const*) (Theorem 11.14) was proved by Arora et al. [1992, 1998]. Part of our proof of Theorem 11.14 is based on the presentation of Arora's Ph.D. thesis [1994].

An important idea in the above development is to encode the proofs by some error-correcting coding scheme. For a complete treatment of the coding theory, see, for instance, Berlekamp [1968] and Roman [1992]. The application of error-correcting coding to program checking has been studied by Blum and Kannan [1989], Blum et al. [1993], Gemmell et al. [1991], Lipton [1991], and Rubinfeld and Sudan [1996]. The sum-check system is based on the Arthur-Merlin sum-check system of Chapter 10, which was first proved by Lund et al. [1992]. There are a number of proof systems about low-degree polynomials in the literature. For linearity and multilinearity tests, see Blum et al. [1993], Babai, Fortnow, and Lund [1991], and Feige et al. [1996]. For low-degree tests, see Babai, Fortnow, Levin, and Szegedy [1991], Rubinfeld and Sudan [1992], Gemmell and Sudan [1992], Arora and Safra [1998], and Arora et al. [1998].

The study of approximation algorithms for optimization problems has been around for more than thirty years (see, e.g., Graham [1966] and Garey et al. [1972]). While a number of approximation algorithms have been found, the progress on negative results was slow before the 1990s. The first important work on nonapproximability was given by Papadimitriou and Yannakakis [1991]. They defined a class *MAX-SNP* that contains many NP-hard optimization problems which have polynomial-time linear approximation algorithms. They also studied complete problems for *MAX-SNP* under the *L*-reduction and showed that if any one of these problems has a polynomial-time approximation scheme, then every problem in *MAX-SNP* has a polynomial-time approximation scheme. The *MAX-SNP*-hardness was then widely used to establish the nonapproximability results (see, e.g., Bern and Plassmann [1989]). In addition to the *L*-reduction, similar reductions, such as the *A*-reduction [Crescenzi and Panconesi, 1991] and *AP*-reduction [Crescenzi et al., 1995], have been proposed to prove nonapproximability results. However, all these results relied upon the unproved assumption that $r$-APPROX-3SAT is not in $P$ for some $r > 1$.

The breakthrough came in 1991, when Feige discovered the connection between the multi-prover proof systems of $NP$ and approximability of the problem CLIQUE (see Feige et al. [1991, 1996]; also see Condon [1993]). Following this piece of work and the $PCP$ characterization of $NP$, the nonapproximability results mushroomed. Arora et al. [1992, 1998] established that 3SAT is nonapproximable in polynomial time for some constant ratio $r > 1$ and, hence, all $MAX$-$SNP$-hard problems are nonapproximable for some constant ratio. The best ratio 8/7 for the maximum 3SAT problem was found by Hastad [1997] and Karloff and Zwick [1997]. The problem VC-4 was shown to be one of the $MAX$-$SNP$-complete problems in Papadimitriou and Yannakakis [1991]. Our result on VC-3 (Theorem 11.41) is from Du et al. [1998]. Lund and Yannakakis [1994] built some new $PCP$ systems to prove the nonapproximability of problems beyond the class $MAX$-$SNP$, including the problem SET COVER. The optimum ratio $\log n$ for SET COVER in Theorem 11.43 was obtained by Feige [1996]. Other results in this direction include Arora and Lund [1996], Bellare et al. [1993], Raz and Safra [1997], and Du et al. [1998]. The hardest approximation problems belong to the maximum clique problem and the minimum chromatic number problem, which are known to be nonapproximable with respect to ratio $n^{1-\epsilon}$ for any $\epsilon > 0$ [Feige and Killian, 1996; Hastad, 1996]. These nonapproximability results also motivated some positive results. Among them, Arora [1996] proved that TSP on the Euclidean plane has a polynomial approximation scheme (see also Mitchell [1999]). Arora [1998] contains a survey of nonapproximable results. Condon et al. [1993] and Kiwi et al. [1994] extended the $PCP$ characterizations to classes $PSPACE$ and $PH$. These papers and that by Ko and Lin [1995a, 1995b] also contain nonapproximability results on optimization problems in $PSPACE$ and $PH$.

# Bibliography

Adleman, L. [1978], Two theorems on random polynomial time, *Proceedings of the 19th IEEE Symposium on Foundations of Computer Science*, IEEE Computer Society Press, Los Angeles, pp. 75–83.

Adleman, L. and Huang, M. A. [1987], Recognizing primes in random polynomial time, *Proceedings of the 19th ACM Symposium on Theory of Computing*, Association for Computing Machinery, New York, pp. 461–469.

Adleman, L. and Manders, K. [1977], Reducibility, randomness and intractibility, *Proceedings of the 9th ACM Symposium on Theory of Computing*, Association for Computing Machinery, New York, pp. 151–163.

Adleman, L., Manders, K. and Miller, G. [1977], On taking roots in finite fields, *Proceedings of the 20th IEEE Symposium on Foundations of Computer Science*, IEEE Computer Society Press, Los Angeles, pp. 175–178.

Aho, A., Hopcroft, J. and Ullman, J. [1974], *The Design and Analysis of Computer Algorithms*, Addison-Wesley, Reading, MA.

Aiello, W., Goldwasser, S. and Hastad, J. [1990], On the power of interaction, *Combinatorica* **10**, 3–25.

Aiello, W. and Håstad, J. [1991], Statistical zero-knowledge languages can be recognized in two rounds, *J. Comput. Systems Sci.* **42**, 327–345.

Ajtai, M. [1983], $\Sigma_1^1$-formula on finite structures, *Ann. Pure Appl. Logic* **24**, 1–48.

Ajtai, M. and Ben-Or, M. [1984], A theorem on probabilistic constant depth circuits, *Proceedings of the 16th ACM Symposium on Theory of Computing*, Association for Computing Machinery, New York, pp. 471–474.

Alon, N. and Boppana, R. B. [1987], The monotone circuit complexity of Boolean functions, *Combinatorica* **7**, 1–22.

Arora, S. [1994], *Probabilistic Checking of Proofs and Hardness of Approximation Problems*, Ph.D. dissertation, University of California, Berkeley.

Arora, S. [1996], Polynomial-time approximation schemes for Euclidean TSP and other geometric problems, *Proceedings of the 37th IEEE Symposium on Foundations of Computer Science*, IEEE Computer Society Press, Los Angeles, pp. 2–13.

Arora, S. [1998], The approximability of NP-hard problems, *Proceedings of the 30th ACM Symposium on Theory of Computing*, Association for Computing Machinery, New York.

Arora, S. and Lund, C. [1996], Hardness of approximations, in *Approximation Algorithms for NP-hard Problems*, D. Hochbaum, ed. PWS Publishing, Boston.

Arora, S., Lund, C., Motwani, R., Sudan, M. and Szegedy, M. [1992], Proof verification and intractability of approximation problems, *Proceedings of the 33rd IEEE Symposium on Foundations of Computer Science*, IEEE Computer Society Press, Los Angeles, pp. 13–22.

Arora, S., Lund, C., Motwani, R., Sudan, M. and Szegedy, M. [1998], Proof verification and the hardness of approximation problems, *J. Assoc. Comput. Mach.* **45**, 505–555.

Arora, S. and Safra, S. [1992], Probabilistic checking of proofs: A new characterization of NP, *Proceedings of the 33rd IEEE Symposium on Foundations of Computer Science*, IEEE Computer Society Press, Los Angeles, pp. 2–13.

Arora, S. and Safra, S. [1998], Probabilistic checking of proofs: A new characterization of NP, *J. Assoc. Comput. Mach.* **45**, 70–122.

Arvind, V. and Köbler, J. [1998], On pseudorandomness and resource-bounded measure, preprint.

Babai, L. [1985], Trading group theory for randomness, *Proceedings of the 17th ACM Symposium on Theory of Computing*, Association for Computing Machinery, New York, pp. 496–505.

Babai, L. [1987], Random oracles separate *PSPACE* from the polynomial-time hierarchy, *Info. Proc. Lett.* **26**, 51–53.

Babai, L. and Fortnow, L. [1991], Arithmetization: A new method in structural complexity theory, *comput. complex.* **1**, 41–66.

Babai, L., Fortnow, L., Levin, L. and Szegedy, M. [1991], Checking computations in polylogarithmic time, *Proceedings of the 23rd ACM Symposium on Theory of Computing*, Association for Computing Machinery, New York, pp. 21–31.

Babai, L., Fortnow, L. and Lund, C. [1991], Nondeterministic exponential time has two-prover interactive protocols, *comput. complex.* **1**, 3–40.

Babai, L. and Moran, S. [1988], Arthur-Merlin games: A randomized proof-system and a hierarchy of complexity classes, *J. Comput. Systems Sci.* **36**, 254–276.

Bach, E. and Shallit, J. [1996], *Algorithmic Number Theory; Vol. 1, Efficient Algorithms*, MIT Press, Cambridge, MA.

Baker, T., Gill, J. and Solovay, R. [1975], Relativizations of the $P =? NP$ question, *SIAM J. Comput.* 4, 431–442.

Balcázar, J., Diaz, J. and Gabarró, J. [1988], *Structural Complexity I*, Springer-Verlag, Berlin.

Balcázar, J., Diaz, J. and Gabarró, J. [1990], *Structural Complexity II*, Springer-Verlag, Berlin.

Balcázar, J., Gavaldà, R. and Watanabe, O. [1997], Coding complexity: The computational complexity of succinct descriptions, in *Advances in Algorithms, Languages, and Complexity*, Du, D. and Ko, K., eds., Kluwer Academic Publishers, Boston, pp. 73–91.

Balcázar, J. and Schöning, U. [1985], Bi-immune sets for complexity classes, *Math. Systems Theory* 18, 1–10.

Barrington, D. [1990], Bounded-width polynomial-size branching programs recognizes exactly those languages in $NC^1$, *J. Comput. Systems Sci.* 38, 150–164.

Barrington, D. and Straubing, H. [1991], Superlinear lower bounds for bounded-width branching programs, *Proceedings of the 6th Structure in Complexity Theory Conference*, IEEE Computer Society Press, Los Angeles, pp. 305–313.

Beals, R., Nishino, T. and Tanaka, K. [1995], More on the complexity of negation-limited circuits, *Proceedings of the 27th ACM Symposium on Theory of Computing*, Association for Computing Machinery, New York, pp. 585–595.

Beame, P., Cook, S. and Hoover, H. J. [1984], Log depth circuits for division and related problems, *Proceedings of the 25th IEEE Symposium on Foundations of Computer Science*, IEEE Computer Society Press, Los Angeles, pp. 1–6.

Beigel, R., Hemachandra, L. and Wechsung, G. [1991], Probabilistic polynomial time is closed under parity reductions, *Info. Proc. Lett.* 37, 91–94.

Beigel, R., Reingold, N. and Spielman, D. [1991], *PP* is closed under intersection, *Proceedings of the 23rd ACM Symposium on Theory of Computing*, Association for Computing Machinery, New York, pp. 1–9.

Bellare, M., Goldwasser, S., Lund, C. and Russell, A. [1993], Efficient probabilistically checkable proofs and applications to approximation, *Proceedings of the 25th ACM Symposium on Theory of Computing*, Association for Computing Machinery, New York, pp. 294–304.

Bennett, C. and Gill, J. [1981], Relative to a random oracle A, $P^A \neq NP^A \neq coNP^A$ with probability 1, *SIAM J. Comput.* 10, 96–113.

Ben-Or, M., Goldwasser, S., Killian, J. and Wigderson, A. [1988], Multi-prover interactive proofs: How to remove intractability assumptions, *Proceedings of the 20th ACM Symposium on Theory of Computing*, Association for Computing Machinery, New York, pp. 113–131.

Berkowitz, S. J. [1982], On some relationships between monotone and non-monotone circuit complexity, Technical Report, Computer Science Department, University of Toronto, Toronto, Ontario, Canada.

Berlekamp, E. [1968], *Algebraic Coding Theory*, McGraw-Hill, New York.

Berman, L. [1976], On the structure of complete sets: Almost everywhere complexity and infinitely often speed-up, *Proceedings of the 17th IEEE Symposium on Foundations of Computer Science*, IEEE Computer Society Press, Los Angeles, pp. 76–80.

Berman, L. [1977], *Polynomial Reducibilities and Complete Sets*, Ph.D. thesis, Cornell University, Ithaca, NY.

Berman, L. and Hartmanis, J. [1977], On isomorphisms and density of *NP* and other complete sets, *SIAM J. Comput.* **6**, 305–322.

Berman, P. [1978], Relationships between density and deterministic complexity of *NP*-complete languages, *Proceedings of the 5th International Colloquium on Automata, Languages and Programming*, Lecture Notes in Computer Science **62**, Springer-Verlag, Berlin, pp. 63–71.

Bern, M. and Plassmann, P. [1989], The Steiner problem with edge length 1 and 2, *Info. Proc. Lett.* **32**, 171–176.

Bernstein, E. and Vazirani, U. [1997], Quantum complexity theory, *SIAM J. Comput.* **26**, 1411–1473.

Best, M. R., van Emde Boas, P. and Lenstra, H. W. [1974], A sharpened version of the Aanderaa-Rosenberg Conjecture, Report ZW 30/74, Mathematisch Centrum Amsterdam, Amsterdam, The Netherlands.

Blum, A., Jiang, T., Li, M., Tromp, J. and Yannakakis, M. [1991], Linear approximation of shortest superstrings, *Proceedings of the 23rd ACM Symposium on Theory of Computing*, Association for Computing Machinery, New York, pp. 328–336.

Blum, L., Blum, M. and Shub, M. [1986], A simple secure unpredictable pseudorandom number generator, *SIAM J. Comput.* **15**, 364–383.

Blum, M. [1967], A machine-independent theory of the complexity of recursive functions, *J. Assoc. Comput. Mach.* **14**, 143–172.

Blum, M. and Kannan, S. [1989], Designing programs that check their work, *Proceedings of the 21st ACM Symposium on Theory of Computing*, Association for Computing Machinery, New York, pp. 86–97.

Blum, M., Luby, M. and Rubinfeld, R. [1993], Self-testing/corresting with applications to numerical problems, *J. Comput. Systems Sci.* **47**, 549–595.

Blum, N. [1984], A Boolean function requiring $3n$ network size, *Theoret. Comput. Sci.* **28**, 337–345.

Bollobas, B. [1976], Complete subgraphs are elusive, *J. Combinatorial Theory (Series B)* **21**, 1–7.

Bollobas, B. [1978], *Extremal Graph Theory*, Academic Press, San Diego, CA.

Book, R. [1974a], Comparing complexity classes, *J. Comput. Systems Sci.*, **9**, 213–229.

Book, R. [1974b], Tally languages and complexity classes, *Info. Contr.* **26**, 186–193.

Book, R., Long, T. and Selman, A. [1984], Quantitative relativizations of complexity classes, *SIAM J. Comput..* **13**, 461–487.

Book, R., Long, T. and Selman, A. [1985], Quanlitative relativizations of complexity classes, *J. Comput. Systems Sci.* **30**, 395–413.

Book, R. and Wrathall, C. [1982], A note on complete sets and transitive closure, *Math. Systems Theory* **15**, 311–313.

Boppana, R. and Halldórsson, M. [1992], Approximating maximum independent sets by excluding subgraphs, *BIT* **32**, 180–196.

Boppana, R., Håstad, J. and Zachos, S. [1987], Does co-NP have short interactive proofs?, *Info. Proc. Lett.* **25**, 127–132.

Borodin, A. [1977], On relating time and space to size and depth, *SIAM J. Comput.* **6**, 733–744.

Borodin, A. Cook, S. and Pippenger, N. [1983], Parallel computation for well-endowed rings and space-bounded probabilistic machins, *Info. Contr.* **58**, 113–136.

Borodin, A., Dolev, D., Fich, F. and Paul, W. [1983], Bounds on width two branching programs, *Proceedings of the 15th ACM Symposium on Theory of Computing*, Association for Computing Machinery, New York, pp. 87–93.

Borodin, A., von zur Gathen, J. and Hopcroft, J. [1982], Fast parallel matrix and GCD computations, *Info. Contr.* **52**, 241–256.

Brightwell, G. and Winkler, P. [1991], Counting linear extensions is #*P*-complete, *Proceedings of the 23rd ACM Symposium on Theory of Computing*, Association for Computing Machinery, New York, pp. 175–181.

Broder, A. Z. [1986], How hard is to marry at random? (on the approximation of the permanent), *Proceedings of the 18th ACM Symposium on Theory of Computing*, Association for Computing Machinery, New York, pp. 50–58.

Broder, A. Z. [1988], Errata of Broder [1986], *Proceedings of the 20th ACM Symposium on Theory of Computing*, Association for Computing Machinery, New York, p. 551.

Buss, J. [1986], Relativized alternation, *Proceedings of the Structure in Complexity Theory Conference*, Lecture Notes in Computer Science **223**, Springer-Verlag, Berlin, pp. 66–76.

Cai, J.-Y. [1986], With probability one, a random oracle separates PSPACE from the polynomial-time hierarchy, *Proceedings of the 18th ACM Symposium on Theory of Computing*, Association for Computing Machinery, New York, pp. 21–29.

Cai, J.-Y. [1991], Private communication.

Cai, J.-Y., Gundermann, T., Hartmanis, J., Hemachandra, L., Sewelson, V., Wagner, K. and Wechsung, G. [1988], The Boolean hierarchy I: Structural properties, *SIAM J. Comput.* **17**, 1232–1252.

Cai, J.-Y., Gundermann, T., Hartmanis, J., Hemachandra, L., Sewelson, V., Wagner, K. and Wechsung, G. [1989], The Boolean hierarchy II: Applications, *SIAM J. Comput.* **18**, 95–111.

Cai, J.-Y. and Hemachandra, L. A. [1990], On the power of parity polynomial time, *Math. Systems Theory* **23**, 95–106.

Cai, J.-Y. and Lipton, R. J. [1994], Subquadratic simulations of balanced formulae by branching programs, *SIAM J. Comput.* **23**, 563–572.

Cai, J.-Y. and Ogihara, M. [1997], Sparse sets versus complexity classes, in *Complexity Theory Retrospective II*, Hemaspaandra, L.A. and Selman, A.L., eds., Springer-Verlag, Berlin, pp. 53–80.

Cai, J.-Y. and Sivakumar, D. [1995], The resolution of a Hartmanis conjecture, *Proceedings of the 36th IEEE Symposium on Foundations of Computer Science*, IEEE Computer Society Press, Los Angeles, pp. 362–373.

Carter, J.L. and Wegman, M. N. [1979], Universal classes of hash functions, *J. Comput. Systems Sci.* **18**, 143–154.

Chaitin, G. [1987], *Algorithmic Information Theory*, Cambridge University Press, Cambridge, UK.

Chandra, A., Kozen, D. and Stockmeyer, L. [1981], Alternation, *J. Assoc. Comput. Mach.* **28**, 114–133.

Christofides, N. [1976], Worst-case analysis of a new heuristic for the travelling salesman problem, *Technical Report*, Graduate School of Industrial Administration, Carnegie-Mellon University, Pittsgurgh, PA.

Cobham, A. [1964], The intrinsic computational difficulty of functions, *Proceedings of International Congress for Logic Methodology and Philosophy of Science*, North-Holland, Amsterdam, pp. 24–30.

Coffman, E., Garey, M. and Johnson, D. [1985], Approximation algorithms for bin-packing–An updated survey, in *Algorithm Design for Computer System Design*, Ausiello, G. et al., eds., Springer-Verlag, Berlin, pp. 49–106.

Condon, A. [1993], The complexity of the max-word problem and the power of one-way interactive proof systems, *comput. complex.* **3**, 292–305.

Condon, A., Feigenbaum, J., Lund, C. and Shor, P. [1993], Probabilistically checkable debate systems and approximation algorithms for PSPACE-hard functions, *Proceedings of the 25th ACM Symposium on Theory of Computing*, Association for Computing Machinery, New York.

Cook, S. [1971], The complexity of theorem-proving procedures, *Proceedings of the 3rd ACM Symposium on Theory of Computing*, Association for Computing Machinery, New York, pp. 151–158.

Cook, S. [1973a], A hierarchy for nondeterministic time complexity, *J. Comput. Systems Sci.* **7**, 343–353.

Cook, S. [1973b], An observation on time-storage trade-offs, *Proceedings of the 5th ACM Symposium on Theory of Computing*, Association for Computing Machinery, New York, pp. 29–33.

Cook, S. [1985], A taxonomy of problems with fast parallel algorithms, *Info. Contr.* **64**, 2–22.

Cook, S. and McKenzie, P. [1987], Problems complete for deterministic logarithmic space, *J. Algorithms* **8**, 385–394.

Cormen, T. H., Leiserson, C. E. and Rivest, R. L. [1990], *Introduction ot Algorithms*, MIT Press, Cambridge, MA.

Crescenzi, P., Kann, V., Silvestri, R. and Trevisan, L. [1995], Structure in approximation classes, *Proceedings of the 1st Conference on COCOON*, Lecture Notes on Computer Science **959**, Springer-Verlag, Berlin, pp. 539–548.

Crescenzi, P. and Panconesi, A. [1991], Completeness in approximation classes, *Info. Comput.* **93**, 241–262.

Csansky, L. [1976], Fast parallel matrix inversion algorithms, *SIAM J. Comput.* **5**, 618–623.

Daley, R. [1980], The busy beaver method, *Proceedings of the Kleene Symposium*, North Holland, Amsterdam, pp. 333–345.

Deutsch, D. [1985], Quantum theory, the Church-Turing principle and the universal quantum computer, *Proc. Royal Soc. London*, Ser. A **400**, 97–117.

Diffie, W. and Hellman, M. [1976], New directions in cryptography, *IEEE Trans. Info. Theory* **IT 22**, 644–654.

Du, D.-Z. and Book, R. [1989], On inefficient special cases of $NP$-complete problems, *Theoret. Comput. Sci.* **63**, 239–252.

Du, D.-Z., Gao, B. and Wu, W. [1997], A special case for subset interconnection designs, *Discr. Appl. Math.* **78**, 51–60.

Du, D.-Z., Isakowitz, T. and Russo, D. [1984], Structural properties of complexity cores, manuscript, Department of Mathematics, University of California, Santa Barbara.

Du, D.-Z. and Ko, K. [1987], Some completeness results on decision trees and group testing, *SIAM J. Alge. Disc. Methods* **8**, 762–777.

Du, X., Wu, W. and Kelly, D. [1998], Approximations for subset interconnection designs, *Theoret. Comput. Sci.* **207**, 171–180.

Dunne, P. E. [1988], *The Complexity of Boolean Networks*, Academic Press, San Diego.

Edmonds, J. [1965], Paths, trees and flowers, *Canada J. Math.* **17**, 449–467.

Elgot, C. C. and Robinson, A. [1964], Random access stored program machines, *J. Assoc. Comput. Mach.* **11**, 365–399.

Even, S. and Tarjan, R. E. [1976], A combinatorial problem which is complete in polynomial space, *J. Assoc. Comput. Mach.* **23**, 710–719.

Feather, T. [1984], The parallel complexity of some flow and matching problem, Technical Report, 1984-174, Computer Science Department, University of Toronto, Toronto, Ontario, Canada.

Feige, U. [1996], A threshold of ln $n$ for approximating set cover, *Proceedings of the 28th ACM Symposium on Theory of Computing*, Association for Computing Machinery, New York, pp. 314–318.

Feige, U., Goldwasser, S., Lovasz, L., Safra, S. and Szegedy, M. [1991], Approximating clique is almost NP-complete, *Proceedings of the 32nd IEEE Symposium on Foundations of Computer Science*, IEEE Computer Society Press, Los Angeles, pp. 2–12.

Feige, U., Goldwasser, S., Lovasz, L., Safra, S. and Szegedy, M. [1996], Interactive proofs and the hardness of approximating cliques, *J. Assoc. Comput. Mach.* **43**, 268–292.

Feige, U. and Kilian, J. [1996], Zero knowledge and the chromatic number, *Proceedings of the 11th IEEE Conference on Computational Complexity*, IEEE Computer Society Press, Los Angeles.

Fenner, S., Fortnow, L. and Kurtz, S. A. [1992], An oracle relative to which the isomorphism conjecture holds, *Proceedings of the 33rd IEEE Symposium on Foundations of Computer Science*, IEEE Computer Society Press, Los Angeles, pp. 29–37.

Fischer, M. [1974], *Lectures on network complexity*, preprint, University of Frankfurt, Germany.

Fortnow, L. [1989], The complexity of perfect zero-knowledge, in *Advances in Computing Research 5: Randomness and Computation*, Micali, S., ed., JAI Press, Greenwich, CT, pp. 327–343.

Fortnow, L. and Laplante, S. [1995], Circuit complexity a la Kolmogorov complexity, *Info. Comput.* **123**, 121-126.

Fortnow, L., Rompel, J. and Sipser, M. [1994], On the power of multi-prover interactive proof protocols, *Theoret. Comput. Sci.* **134**, 545–557.

Fortnow, L. and Sipser, M. [1988], Are there interactive protocols for *co-NP* languages?, *Info. Proc. Lett.* **28**, 249–252.

Fortune, S. [1979], A note on sparse complete sets, *SIAM J. Comput.* **8**, 431–433.

Fortune, S and Wyllie, J. [1978], Parallelism in random access machines, *Proceedings of the 10th ACM Symposium on Theory of Computing*, Association for Computing Machinery, New York, pp. 114–118.

Frieze, A. M., Kannan, R. and Lagarias, J. C. [1984], Linear congruencial generators do not produce random sequences, *Proceedings of the 25th IEEE Symposium on Foundations of Computer Science*, IEEE Computer Society Press, Los Angeles, pp. 480–484.

Fürer, M. [1984], Data structures for distributed counting, *J. Comput. Systems Sci.* **28**, 231–243.

Furst, M., Saxe, J. and Sipser, M. [1984], Parity, circuits and the polynomial-time hierarchy, *Math. Systems Theory* **17**, 13–27.

Gabber, O. and Galil, Z. [1981], Explicit construction of linear-sized superconcentrators, *J. Comput. Systems Sci.* **22**, 407–420.

Gao, S.-X., Hu, X.-D. and Wu, W. [1999], Nontrivial monotone weakly symmetric Boolean functions with six variables are elusive, *Theoret. Comput. Sci.* **223**, 193–197.

Gao, S.-X., Wu, W., Du, D.-Z. and Hu, X.-D. [1999], Rivest-Vuillemin conjecture on monotone Boolean functions is true for ten variables, *J. Complex.* (to appear).

Garey, M. R., Graham, R. L. and Ullman, J. D. [1972], Worst case analysis of memory allocation algorithm, *Proceedings of the 4th ACM Symposium on Theory of Computing*, Association for Computing Machinery, New York, pp. 143–150.

Garey, M. R. and Johnson, D. [1976], The complexity of near-optimal graph coloring, *J. Assoc. Comput. Mach.* **23**, 43–49.

Garey, M. R. and Johnson, D. [1979], *Computers and Intractability, a Guide to the Theory of NP-Completeness*, W.H. Freeman, San Francisco.

Gemmell, P., Lipton, R., Rubinfeld, R., Sudan, M. and Wigderson, A. [1991], Self-testing/correcting for polynomials and for approximate functions, *Proceedings of the 23rd ACM Symposium on Theory of Computing*, Association for Computing Machinery, New York, pp. 32–42.

Gemmell, P. and Sudan, M. [1992], Highly resilient correctors for polynomials, *Info. Comput.* **43**, 169–174.

Gill, J. [1977], Computational complexity of probabilistic Turing machines, *SIAM J. Comput.* **6**, 675–695.

Glaser, L. C. [1970], *Geometrical Combinatorial Topology*, Vol. 1, Van Nostrand, New York.

Goldman, M. and Håstad, J. [1995], Monotone circuits for connectivity have depth $(log n)^{2-o(1)}$, *Proceedings of the 27th ACM Symposium on Theory of Computing*, Association for Computing Machinery, New York, pp. 569–574.

Goldreich, O. [1997], On the foundation of modern cryptography, in *Advances in Cryptology—Crypto 97*, Lecture Notes in Computer Science **1294**, Springer-Verlag, Berlin, pp. 46–74.

Goldreich, O., Micali, S. and Wigderson, A. [1991], Proofs that yield nothing but their validity, or all languages in NP have zero-knowledge proof systems, *J. Assoc. Comput. Mach.* **38**, 691–729.

Goldreich, O. and Zuckerman, D. [1997], Another proof that $BPP \subseteq PH$ (and more), ECC Research Report TR97-045.

Goldschlager, L. M. [1977], The monotone and planar circuit value problems are log space complete for P, *ACM SIGACT News* **9**, 25–29.

Goldschlager, L. M. [1978], A unified approach to models of synchronous parallel machines, *Proceedings of the 10th ACM Symposium on Theory of Computing*, Association for Computing Machinery, New York, pp. 89–94.

Goldschlager, L. M., Shaw, R. A. and Staples, J. [1982], The maximum flow problem is log space complete for P, *Theoret. Comput. Sci.* **21**, 105–111.

Goldsmith, J. and Joseph, D. [1986], Three results on the polynomial isomorphism of complete sets, *Proceedings of the 27th IEEE Symposium on Foundations of Computer Science*, IEEE Computer Society Press, Los Angeles, pp. 390–397.

Goldwasser, S. [1988], *Lecture Notes on Interactive Proof Systems*, Laboratory for Computer Science, MIT, Cambridge, MA.

Goldwasser, S., Micali, S. and Rackoff, C. [1989], The knowledge complexity of interactive proof systems, *SIAM J. Comput.* **18**, 186–208.

Goldwasser, S. and Sipser, M. [1989], Private coins versus public coins in interactive proof systems, in *Advances in Computing Research 5: Randomness and Computation*, Micali, S., ed., JAI Press, Greenwich, CT.

Graham, R. L. [1966], Bounds for certain multiprocessing anomalies, *Bell System Tech. J.* **45**, 1563–1581.

Greenlaw, R., Hoover, H. J. and Ruzzo, W. L. [1995], *Limits to Parallel Computation: P-Completeness Theory*, Oxford University Press, Oxford, UK.

Grollman, J. and Selman, A. [1988], Complexity measures for public-key cryptosystems, *SIAM J. Comput.* **17**, 309–335.

Hartmanis, J. [1978], On log-tape isomorphisms of complete sets, *Theoret. Comput. Sci.* **7**, 273–286.

Hartmanis, J. [1991], Notes on $IP = PSPACE$, preprint.

Hartmanis, J. and Hemachandra, L. [1991], One-way functions and the nonisomorphism of NP-complete sets, *Theoret. Comput. Sci.* **81**, 155–163.

Hartmanis, J. and Hopcroft, J. [1976], Independence results in computer science, *ACM SIGACT News* **8**(4), 13–23.

Hartmanis, J., Sewelson, V. and Immerman, N. [1983], Sparse sets in $NP - P$: *EXP-TIME* versus *NEXPTIME*, *Proceedings of the 15th ACM Symposium on Theory of Computing*, Association for Computing Machinery, New York, pp. 382–391.

Hartmanis, J. and Stearns, R. E. [1965], On the computational complexity of algorithms, *Trans. Amer. Math. Soc.* **117**, 285–306.

Håstad, J. [1986a], *Computational Limitations for Small-Depth Circuits*, Ph.D. dissertation, MIT, MIT Press, Cambridge, MA.

Håstad, J. [1986b], Almost optimal lower bounds for small depth circuits, *Proceedings of the 18th ACM Symposium on Theory of Computing*, Association for Computing Machinery, New York, pp. 71–84.

Håstad, J. [1996], Clique is hard to approximate within $n^{1-\epsilon}$, *Proceedings of the 37th IEEE Symposium on Foundations of Computer Science*, IEEE Computer Society Press, Los Angeles, pp. 627–636.

Håstad, J. [1997], Some optimal inapproximability results, *Proceedings of the 28th ACM Symposium on Theory of Computing*, Association for Computing Machinery, New York, pp. 1–10.

Håstad, J., Impagliazzo, R., Levin, L. A. and Luby, M. [1999], A pseudorandom generator from any one-way function, *SIAM J. Comput.* **28**, 1364–1396.

Heller, H. [1984], Relativized polynomial hierarchies extending two levels, *Math. Systems Theory* **17**, 71–84.

Holt, R. C. and Reingold, E. M. [1972], On the time required to detect cycles and connectivity in directed graphs, *Math. Systems Theory* **6**, 103–107.

Homer, S., Kurtz, S. and Royer, J. [1993], On 1-truth-table-hard languages, *Theoret. Comput. Sci.* **115**, 383–389.

Hopcroft, J., Paul, W. and Valiant, L. [1977], On time versus space, *J. Assoc. Comput. Mach.* **24**, 332–337.

Hopcroft, J. and Ullman, J. [1979], *Introduction to Automata Theory, Languages, and Computation*, Addison-Wesley, Reading, MA.

Hunt, H. B. III [1973], The equivalence problem for regular expressions with intersection is not polynomial in tape, Tech. Report TR73-161, Department of Computer Science, Cornell University, Ithaca, NY.

Ibarra, O. H. and Kim, C. E. [1975], Fast approximation algorithms for the knapsack and sum of subset problems, *J. Assoc. Comput. Mach.* **22**, 463–468.

Illies, [1978], *Graph Theory Newsletter*.

Immerman, N. [1988], Nondeterministic space is closed under complement, *Proceedings of the 3rd IEEE Conference on Structure in Complexity Theory*, IEEE Computer Society Press, Los Angeles, pp. 112–115.

Immerman, N. and Landau, S. [1989], The complexity of iterated multiplication, *Proceedings of the 4th IEEE Conference on Structure in Complexity Theory*, IEEE Computer Society Press, Los Angeles, pp. 104–111.

Jerrum, M. R. and Sinclair, A. [1989], Approximating the permanent, *SIAM J. Comput.* **18**, 1149–1178.

Jerrum, M. R., Valiant, L. G. and Vazirani, V. V. [1986], Random generation of combinatorial structures from a uniform distribution, *Theoret. Comput. Sci.* **43**, 169–188.

Jiang, T. and Ravikumar, B. [1993], Minimal NFA problems are hard, *SIAM J. Comput.* **22**, 1117–1141.

Johnson, D. [1988], The NP-completeness column—an ongoing guide, *J. Algorithms* **9**, 426–444.

Johnson, D. [1992], The NP-completeness column—an ongoing guide, *J. Algorithms* **13**, 502–524.

Joseph, D. and Young, P. [1985], Some remarks on witness functions for nonpolynomial and noncomplete sets in NP, *Theoret. Comput. Sci.* **39**, 225–237.

Kahn, J., Saks, M. and Strutevant, D. [1984], A topological approach to evasiveness, *Combinatorica* **4**, 297–306.

Karloff, H. and Zwick, U. [1997], A 7/8-approximation algorithm for MAX 3SAT, *Proceedings of the 38th IEEE Symposium on Foundations of Computer Science*, IEEE Computer Society Press, Los Angeles.

Karmarkar, N. [1984], A new polynomial-time algorithm for linear programming, *Combinatorica* **4**, 373–396.

Karmarkar, N. and Karp, R. [1982], An efficient approximation scheme for the one-dimensional bin-packing problem, *Proceedings of the 23rd IEEE Symposium on Foundations of Computer Science*, IEEE Computer Society Press, Los Angeles, pp. 312–320.

Karp, R. M. [1972], Reducibility among combinatorial problems, in *Complexity of Computer Computations*, Miller, R. and Thatcher, J., eds., Plenum Press, New York, pp. 85–103.

Karp, R. M. and Lipton, R. [1982], Turing machines that take advice, *L'enseigment Mathematique* **28**, 191–209.

Karp, R. M. and Rabin, M. [1987], Efficient randomized pattern-matching algorithms, *IBM J. Res. Develop.*, **31**, 249–260.

Karp, R. M., Upfal, E. and Wigderson, A. [1986], Constructing a maximum matching is in random NC, *Combinatorica* **6**, 35–48.

Khachiyan, L. [1979], A polynomial algorithm in linear programming, *Soviet Math. Dokl.* **20**, 191–194.

Khanna, S, Linial, N. and Safra, S. [1993], On the hardness of approximating the chromatic number, *Proceedings of the 2nd International Symposium on Theoretical Computer Science*, IEEE Computer Science Press, Los Angeles, pp. 250–260.

King, V. [1989], *My Thesis*, Ph.D. thesis, Computer Science Division, University of California, Berkely.

King, V. [1990], A lower bound for the recognition of digraph properties, *Combinatorica*, **10**, 53–59.

Kirkpatrick, D. [1974], Determining graph properties from matrix representations, in *Proceedings of the 6th ACM Symposium on Theory of Computing*, Association for Computing Machinery, New York, pp. 84–90.

Kiwi, M., Lund, C., Russell, A., Spielman, D. and Sundaram, R. [1994], Alternation in interaction, *Proceedings of the 9th IEEE Conference on Structure in Complexity Theory*, IEEE Computer Society Press, Los Angeles, pp. 294–303.

Kleene, S. C. [1979], Origins of recurisve function theory, *Proceedings of the 20th IEEE Symposium on Foundations of Computer Science*, IEEE Computer Society Press, Los Angeles, pp. 371–382.

Kleitman, D. J. and Kwiatkowski, D. J. [1980], Further results on the Aanderaa-Rosenberg Conjecture, *J. Combinatorial Theory (Ser B)*, **28**, 85–95.

Knuth, D. [1981], *The Art of Computer Programming, Vol. 2, Seminumerical Algorithms*, 2nd Ed., Addison-Wesley, Reading, MA.

Ko, K. [1982], Some observations on probabilistic algorithms and NP-hard problems, *Info. Proc. Lett.* **14**, 39–43.

Ko, K. [1983], On self-reducibility and weak P-selectivity, *J. Comput. Systems Sci.* **26**, 209–221.

Ko, K. [1985a], On some natural complete operators, *Theoret. Comput. Sci.* **37**, 1–30.

Ko, K. [1985b], Nonlevelable sets and immune sets in the accepting density hierarchy in *NP*, *Math. Systems Theory* **18**, 189–205.

Ko, K. [1987], On helping by robust oracle machines, *Theoret. Comput. Sci.* **52**, 15–36.

Ko, K. [1989a], Relativized polynomial time hierarchies having exactly k levels, *SIAM J. Comput.* **18**, 392–408.

Ko, K. [1989b], Constructing oracles by lower bound techniques for circuits, in *Combinatorics, Computing and Complexity*, Du, D. and Hu, G., eds., Kluwer Academic Publishers/Science Press, Boston, pp. 30–76.

Ko, K. [1990], Separating and collapsing results on the relativized probabilistic polynomial time hierarchy, *J. Assoc. Comput. Mach.* **37**, 415–438.

Ko, K. [1991a], *Complexity Theory of Real Functions*, Birkhäuser, Boston.

Ko, K. [1991b], Separating the low and high hierarchies by oracles, *Info. Comput.* **90**, 156–177.

Ko, K. [1992], On the computational complexity of integral equations, *Ann. Pure Appl. Logic* **58**, 201–228.

Ko, K. and Friedman, H. [1982], Computational complexity of real functions, *Theoret. Comput. Sci.* **20**, 323–352.

Ko, K. and Lin, C.-L. [1995a], On the complexity of min-max optimization problems and their approximation, in *Minimax and Applications*, Du, D.-Z. and Pardalos, P.M., eds., Kluwer Academic Publishers, Boston.

Ko, K. and Lin, C.-L. [1995b], On the longest circuit in an alterable digraph, *J. Global Opt.* **7**, 279–295.

Ko, K., Long, T. and Du, D. [1986], On one-way functions and polynomial-time isomorphisms, *Theoret. Comput. Sci.* **47**, 263–276.

Ko, K. and Moore, D. [1981], Completeness, approximation and density, *SIAM J. Comput.* **10**, 787–796.

Ko, K. and Tzeng, W.-G. [1991], Three $\Sigma_2^P$-complete problems in computational learning theory, *comput. complex.* **1**, 269–301.

Kozen, D. [1977], Complexity of finitely generated algebra, *Proceedings of the 9th ACM Symposium on Theory of Computing*, Association for Computing Machinery, New York, pp. 164–177.

Kranakis, E. [1986], *Primality and Cryptography*, John Wiley & Sons and Teubner, Stuttgart, Germany.

Kurtz, S. [1983], A relativized failure of the Berman-Hartmanis conjecture, Tech. Report 83-001, University of Chicago, Chicago.

Kurtz, S., Mahaney, S. and Royer, J. [1988], Collapsing degrees, *J. Comput. Systems Sci.* **37**, 247–268.

Kurtz, S., Mahaney, S. and Royer, J. [1989], The isomorphism conjecture fails relative to a random oracle, *Proceedings of the 21st ACM Symposium on Theory of Computing*, Association for Computing Machinery, New York, pp. 157–166.

Kurtz, S., Mahaney, S. and Royer, J. [1990], The structure of complete degrees, in *Complexity Theory Retrospective*, Selman, A., ed., Springer-Verlag, Berlin, pp. 108–146.

Ladner, R. [1975a], On the structure of polynomial-time reducibility, *J. Assoc. Comput. Mach.* **22**, 155–171.

Ladner, R. [1975b], The circuit value problem is log space complete for $P$, *ACM SIGACT News* **7**, 18–20.

Ladner, R. and Lynch, N. [1976], Relativizations of questions about log-space reducibility, *Math. Systems Theory* **10**, 19–32.

Ladner, R., Lynch, N. and Selman, A. [1975], A comparison of polynomial-time reducibilities, *Theoret. Comput. Sci.* **1**, 103–123.

Lautemann, C. [1983], BPP and the polynomial hierarchy, *Info. Proc. Lett.* **14**, 215–217.

Lawler, E. L. [1976], *Combinatorial Optimization: Networks and Matroids*, Holt, Rinehard and Winston, New York.

Lee, C. Y. [1959], Representation of switching functions by binary decision programs, *Bell System Tech. J.* **38**, 985–999.

Levin, L. [1973], Universal search problems, *Problemy Peredaci Informacii* **9**, 115–116; English translation in *Problems of Information Transmission* **9**, 265–266.

Lewis, P. M., Stearns, R. E. and Hartmanis, J. [1965], Memory bounds for recognition of context-free and context-sensitive languages, *Proceedings of the 6th IEEE Symposium on Switching Circuit Theory and Logical Design*, IEEE Computer Society Press, Los Angeles, pp. 191–202.

Li, M. and Vitányi, P. [1997], *An Introduction to Kolmogorov Complexity and Its Applications*, 2nd Ed., Springer-Verlag, Berlin.

Lipton, R. [1991], New directions in testing, in *Distributed Computing and Cryptography*, DIMACS Series in Discrete Mathematics and Theoretical Computer Science, American Mathematical Society, Providence, Rhode Island, pp. 191–202.

Lipton, R. J. and Tarjan, R. E. [1980], Applications of a planar separator theorem, *SIAM J. Comput.* **9**, 615–627.

Liu, C. L. [1968], *Introduction to Combinatorial Mathematics*, McGraw-Hill, New York.

Long, T. [1982], Strong nondeterministic polynomial-time reducibilities, *Theoret. Comput. Sci.* **21**, 1–25.

Long, T. and Selman, A. [1986], Relativizing complexity classes with sparse oracles, *J. Assoc. Comput. Mach.* **33**, 618–627.

Lovász, L. [1973], Coverings and colorings of hypergraphs, *Proceedings of the 4th Southeastern Conference on Combinatorics, Graph Theory, and Computing*, Utilitas Mathematica Publishing, Winnipeg, Manitoba, Canada, pp. 3–12.

Loxton, J. H. (Ed.) [1990], *Number Theory and Cryptography*, London Mathematical Society Lecture Series, Vol. 154, Cambridge University Press, Cambridge, UK.

Lund, C., Fortnow, L., Karloff, H. and Nisan, N. [1992], Algebraic methods for interactive proof systems, *J. Assoc. Comput. Mach.* **39**, 859–868.

Lund, C. and Yannakakis, M. [1994], On the hardness of approximating minimization problems, *J. Assoc. Comput. Mach.* **41**, 960–981.

Lupanov, O. B. [1958], On the synthesis of contact networks, *Dokl. Akad. Nauk. SSSR* **119**, 23–26.

Lynch, N. [1975], On reducibility to complex or sparse sets, *J. Assoc. Comput. Mach.* **22**, 341–345.

Mahaney, S. [1982], Sparse complete sets for NP: Solution of a conjecture of Berman and Hartmanis, *J. Comput. Systems Sci.* **25**, 130–143.

Maier, D. and Storer, J. [1977], A note on the complexity of the superstring problem, Report No. 233, Computer Science Laboratory, Princeton University, Princeton, NJ.

Mayr, E. W. and Subramanian, A. [1989], The complexity of circuit value and network stability, *Proceedings of the 4th IEEE Conference on Structure in Complexity Theory*, IEEE Computer Society Press, Los Angeles, pp. 114–123.

Meyer, A. and Paterson, M. [1979], With what frequency are apparently intractable problems difficult?, Technical Report MIT/LCS/TM-126, MIT, Cambridge, MA.

Meyer, A. and Stockmeyer, L. [1972], The equivalence problem for regular expressions with squaring requires exponential time, *Proceedings of the 13th IEEE Symposium on Switching and Automata Theory*, IEEE Computer Society Press, Los Angeles, pp. 125–129.

Milner, E. C. and Welsh, D. J. A. [1976], On the computational complexity of graph theoretical properties, *Proceedings of the 5th British Combinatorial Conference*, Utilitas Mathematica Publishing, Winnipeg, Manitoba, Canada, pp. 471–487.

Mitchell, J. [1999], Guillotine subdivisions approximate polygonal subdivisions: A simple polynomial-time approximation scheme for geometric TSP, $k$-MST, and related problems, *SIAM J. Comput.* **28**, 1298–1309.

Moran, S. [1987], Generalized lower bounds derived from Hastad's main lemma, *Info. Proc. Lett.* **25**, 383–388.

Motwani, R. and Raghavan, P. [1995], *Randomized Algorithms*, Cambridge University Press, Cambridge, UK.

Mulmuley, K. [1987], A fast parallel algorithm to compute the rank of a matrix over an arbitrary field, *Combinatorica* **7**, 101–104.

Mulmuley, K., Vazirani, U. V. and Vazirani, V. V. [1987], Matching is as easy as matrix inversion, *Combinatorica* **7**, 105–113.

Myhill, J. [1955], Creative sets, *Z. Math. Logik Grundlagen Math.* **1**, 97–108.

Neciporuk, E. I. [1966], A Boolean function, *Boklady Akademii Nauk SSSR* **169** (in Russian); English translation in *Soviet Math. Dokl.* **7**, 999–1000.

Nisan, N. and Wigderson, A. [1994], Hardness and approximation, *J. Comput. Systems Sci.* **49**, 149–167.

Niven, I. and Zuckerman, H. S. [1960], *An Introduction to the Theory of Numbers*, John Wiley & Sons, New York.

Ogihara, M. [1995], Sparse hard sets for P yield space-efficient algorithms, *Proceedings of the 36th IEEE Symposium on Foundations of Computer Science*, IEEE Computer Society Press, Los Angeles, pp. 354–361.

Ogiwara, M. and Watanabe, O. [1991], On polynomial time bounded truth-table reducibility of NP sets to sparse sets, *SIAM J. Comput.* **20**, 471–483.

Oliver, R. [1975], Fixed point sets of group actions on finite cyclic complexes, *Comment. Math. Helv.* **50**, 155–177.

Orponen, P. [1983], Complexity classes of alternating machines with oracles, *Proceedings of the 10th International Colloquium on Automata, Languages and Programming*, Lecture Notes on Computer Science **154**, Springer-Verlag, Berlin, pp. 573–584.

Orponen, P., Russo, D. and Schöning, U. [1986], Optimal approximations and polynomially levelable sets, *SIAM J. Comput.* **15**, 399–408.

Orponen, P. and Schöning, U. [1986], On the density and complexity of polynomial cores for intractable sets, *Info. Contr.* **70**, 54–68.

Papadimirtriou, C. H. and Wolfe, D. [1987], The complexity of facets solved, *J. Comput. Systems Sci.* **37**, 2–13.

Papadimirtriou, C. H. and Yannakakis, M. [1984], The complexity of facets (and some facets of complexity), *J. Comput. Systems Sci.* **28**, 244–259.

Papadimitriou, C. H. and Yannakakis, M. [1991], Optimization, approximation and complexity classes, *J. Comput. Systems Sci.* **18**, 1–11.

Papadimirtriou, C. H. and Zachos, S. [1983], Two remarks on the power of counting, *Proceedings of the 6th GI Conference on Theoretical Computer Science*, Lecture Notes in Computer Science **145**, Springer-Verlag, Berlin, pp. 269–276.

Paul, W. J. [1979], On time hierarchies, *J. Comput. Systems Sci.* **19**, 197–202.

Paul, W., Pippenger, N., Szemeredi, E. and Trotter, W. [1983], On determinism versus nondeterminism and related problems, *Proceedings of the 24th IEEE Symposium on Foundations of Computer Science*, IEEE Computer Society Press, Los Angeles, pp. 429–438.

Pippenger, N. [1979], On simultaneous resource bounds, *Proceedings of the 20th IEEE Symposium on Foundations of Computer Science*, IEEE Computer Society Press, Los Angeles, pp. 307–311.

Plumstead, J. [1982], Inferring a sequence generated by a linear congruence, *Proceedings of the 23rd IEEE Symposium on Foundations of Computer Science*, IEEE Computer Society Press, Los Angeles, pp. 153–159.

Pratt, V. [1975], Every prime has a succinct certificate, *SIAM J. Comput.* **4**, 214–220.

Pratt, V., Rabin, M. and Stockmeyer, L. [1974], A characterisation of the power of vector machines, *Proceedings of the 6th ACM Symposium on Theory of Computing*, Association for Computing Machinery, New York, pp. 122–134.

Rabin, M. [1976], Probabilistic algorithms, in *Algorithms and Complexity: New Directions and Results*, Traub, J. F., ed., Academic Press, San Diego, CA, pp. 21–39.

Rabin, M. [1979], Digital signatures and public-key functions as intractable as factorization, Tech. Report 212, Laboratory for Computer Science, MIT, Cambridge, MA.

Rabin, M. [1980], Probabilistic algorithm for primality testing, *J. Number Theory*, **12**, 128–138.

Rackoff, C. [1982], Relativized questions involving probabilistic algorithms, *J. Assoc. Comput. Mach.* **29**, 261–268.

Raz, R. and Safra, S. [1997], A sub-constant error-probability low-degree test, and a sub-constant error-probability PCP characterization of NP, *Proceedings of the 28th ACM Symposium on Theory of Computing*, Association for Computing Machinery, New York.

Razborov, A. A. [1985a], Lower bounds on the monotone complexity of some Boolean functions, *Doklady Akademii Nauk SSR* **281**, 798–801 (in Russian); English translation in *Soviet Math. Dokl.* **31**, 354–357.

Razborov, A. A. [1985b], A lower bound on the monotone network complexity of the logical permanent, *Matematicheskie Zametki* **37**, 887–900 (in Russian); English translation in *Mathematical Notes of the Academy of Sciences of the USSR* **37**, 485–493.

Rice, H. [1954], Recursive real numbers, *Proc. Amer. Math. Soc.* **5**, 784–791.

Rivest, R., Shamir, A. and Adleman, L. [1978], A method for obtaining digital signatures and public key cryptosystems, *Comm. Assoc. Comput. Mach.* **21**, 120–126.

Rivest, R. and Vuillemin, S. [1975], A generalization and proof of the Aanderaa-Rosenberg conjecture, *Proceedings of the 7th ACM Symposium on Theory of Computing*, Association for Computing Machinery, New York, pp. 6–11.

Rivest, R. and Vuillemin, S. [1976], On recognizing graph properties from adjacency matrices, *Theoret. Comput. Sci.*, **3**, 371–384.

Rogers, H., Jr. [1967], *Theory of Recursive Functions and Effective Computability*, Mc-Graw Hill, New York.

Rogers, J. [1995], The isomorphism conjecture holds and one-way functions exist relative to an oracle, *Proceedings of the 10th IEEE Conference on Structure in Complexity Theory*, IEEE Computer Society Press, Los Angeles, pp. 90–101.

Roman, S. [1992], *Coding and Information Theory*, Springer-Verlag, Berlin.

Rosenberg, A. L. [1973], On the time required to recognize properties of graphs: A problem, *ACM SIGACT News* **5**, 15–16.

Rubinfeld, R. and Sudan, M. [1992], Testing polynomial functions efficiently and over rational domains, *Proceedings of the 3rd ACM-SIAM Symposium on Discrete Algorithms*, SIAM, Philadelphia, pp. 23–43.

Rubinfeld, R. and Sudan, M. [1996], Robust characterizations of polynomials with applications to program testing, *SIAM J. Comput.*, **25**, 252–271.

Ruzzo, W. [1980], Tree-size bounded alternation, *J. Comput. Systems Sci.* **21**, 218–235.

Ruzzo, W., Simon, J. and Tompa, M. [1984], Space-bounded hierarchies and probabilistic computation, *J. Comput. Systems Sci.* **28**, 216–230.

Saks, M. and Zhou, S. [1995], $RSPACE(S) \subseteq DSPACE(S^{3/2})$, *Proceedings of the 36th IEEE Symposium on Foundations of Computer Science*, IEEE Computer Society Press, Los Angeles, pp. 344–353.

Savitch, W. [1970], Relationships between nondeterministic and deterministic tape complexities, *J. Comput. Systems Sci.* **4**, 177–192.

Schaefer, T. J. [1978a], The complexity of satisfiability problems, *Proceedings of the 10th ACM Symposium on Theory of Computing*, Association for Computing Machinery, New York, pp. 216–226.

Schaefer, T. J. [1978b], Complexity of some two-person perfect-information games, *J. Comput. Systems Sci.* **16**, 185–225.

Schöning, U. [1982], A uniform approach to obtain diagonal sets in complexity classes, *Theoret. Comput. Sci.* **18**, 95–103.

Schöning, U. [1983], A low and a high hierarchy within NP, *J. Comput. Systems Sci.* **27**, 14–28.

Schöning, U. [1985], Robust algorithms: A different approach to oracles, *Theoret. Comput. Sci.* **40**, 57–66.

Schöning, U. [1987], Graph isomorphism is in the low hierarchy, *Proceedings of the 4th Symposium on Theoretical Aspects of Computer Science*, Lecture Notes in Computer Science **247**, Springer-Verlag, Berlin, pp. 114–124.

Schwartz, J. [1980], Fast probabilistic algorithms for verification of polynomial identities, *J. Assoc. Comput. Mach.*, **27**, 701–717.

Seiferas, J. I. [1977a], Techniques for separating space complexity classes, *J. Comput. Systems Sci.* **14**, 73–99.

Seiferas, J. I. [1977b], Relating refined complexity classes, *J. Comput. Systems Sci.* **14**, 100–129.

Selman, A. [1979], P-selective sets, tally languages and the behavior of polynomial time reducibilities on *NP*, *Math. Systems Theory* **13**, 55–65.

Selman, A. [1992], A survey of one-way functions in complexity theory, *Math. Systems Theory* **25**, 203–221.

Shamir, A. [1990], $IP = PSPACE$, *Proceedings of the 31st IEEE Symposium on Foundations of Computer Science*, IEEE Computer Society Press, Los Angeles, pp. 11–15.

Shannon, C. E. [1949], The synthesis of two-terminal switching circuits, *Bull. Systems Tech. J.* **28**, 59–98.

Shepherdson, J. C. and Sturgis, H. E. [1963], Computability of recursive functions, *J. Assoc. Comput. Mach.* **10**, 217–255.

Sheu, M. and Long, T. [1992], UP and the low and high hierarchies: A relativized separation, *Proceedings of the 19th International Colloquium on Automata, Languages and Programming*, Lecture Notes in Computer Science **623**, Springer-Verlag, Berlin, pp. 174–185

Shor, P. W. [1997], Polynomial-time algorithms for prime factorization and discrete logarithms on a quantum computer, *SIAM J. Comput.* **26**, 1484–1509.

Simon, J. [1975], *On Some Central Problems in Computational Complexity*, Ph.D. dissertation, Cornell University, Ithaca, NY.

Simon, J. [1981], Space-bounded probabilistic Turing machine complexity classes are closed under complement, *Proceedings of the 13th ACM Symposium on Theory of Computing*, Association for Computing Machinery, New York, pp. 158–167.

Simon, J., Gill, J. and Hunt, J. [1980], Deterministic simulation of tape-bounded probabilistic Turing machine transducers, *Theoret. Comput. Sci.* **12**, 333–338.

Sipser, M. [1983a], Borel sets and circuit complexity, *Proceedings of the 15th ACM Symposium on Theory of Computing*, Association for Computing Machinery, New York, pp. 61–69.

Sipser, M. [1983b], A complexity theoretic approach to randomness, *Proceedings of the 15th ACM Symposium on Theory of Computing*, Association for Computing Machinery, New York, pp. 330–335.

Smolensky, R. [1987], Algebraic methods in the theory of lower bounds for Boolean circuit complexity, *Proceedings of the 19th ACM Symposium on Theory of Computing*, Association for Computing Machinery, New York, pp. 77–82.

Soare, R. I. [1987], *Recursively Enumerable Sets and Degerees*, Springer-Verlag, Berlin.

Solovay, R. and Strassen, V. [1977], A fast Monte-Carlo test for primality, *SIAM J. Comput.* **6**, 84–85.

Stearns, R. E., Hartmanis, J. and Lewis, P. M. [1965], Hierarchies of memory limited computations, *Proceedings of the 6th Symposium on Switching Circuit Theory and Logical Design*, IEEE Computer Society Press, Los Angeles, 179–190.

Stockmeyer, L. [1977], The polynomial time hierarchy, *Theoret. Comput. Sci.* **3**, 1–22.

Stockmeyer, L. [1985], On approximation algorithms for #$P$, *SIAM J. Comput.* **14**, 849–861.

Stockmeyer, L. and Chandra, A. K. [1979], Provably difficult combinatorial games, *SIAM J. Comput.* **8**, 151–174.

Stockmeyer, L. and Meyer, A. [1973], Word problems requiring exponential time, *Proceedings of the 5th ACM Symposium on Theory of Computing*, Association for Computing Machinery, New York, pp. 1–9.

Sudborough, I. H. [1978], On the tape complexity of deterministic context-free languages, *J. Assoc. Comput. Mach.* **27**, 701–717.

Szelepcsényi, R. [1988], The method of forced enumeration for nondeterministic automata, *Acta Info.* **26**, 279–284.

Tardos, E. [1988], The gap between monotone and non-monotone circuit complexity is exponential, *Combinatorica* **8**, 141–142.

Toda, S. [1991], PP is as hard as the polynomial-time hierarchy, *SIAM J. Comput.* **20**, 865–877.

Triesch, E. [1994], Some results on elusive graph properties, *SIAM J. Comput.*, **23**, 247–254.

Turing, A. [1936], On computable numbers, with an application to the Entscheidungs problem, *Proc. London Math. Soc.*, Ser. 2, **42**, 230–265.

Turing, A. [1937], Rectification to 'On computable numbers, with an application to the Entscheidungs problem,' *Proc. London Math. Soc.*, Ser. 2, **43**, 544–546.

Valiant, L. [1976], Relative complexity of checking and evaluating, *Info. Proc. Lett.* **5**, 20–23.

Valiant, L. [1979a], The complexity of computing the permanent, *Theoret. Comput. Sci.* **8**, 189–201.

Valiant, L. [1979b], The complexity of enumeration and reliability problems, *SIAM J. Comput.* **8**, 410–421.

Valiant, L. and Vazirani., V. V. [1986], NP is as easy as detecting unique solutions, *Theoret. Comput. Sci.* **47**, 85–93.

van Emde Boas, P. [1990], Machine models and simulations, in *The Handbook of Theoretical Computer Science, Vol. I: Algorithms and Complexity*, van Leeuwen, ed., MIT Press, Cambridge, MA, pp. 1–61.

van Melkebeek, D. and Ogihara, M. [1997], Sparse hard sets for *P*, in *Advances in Algorithms, Languages, and Complexity*, Du, D. and Ko, K., eds., Kluwer Academic Publishers, Boston, pp. 191–208.

Vollmer, H. and Wagner, K. [1997], Measure one results in computational complexity theory, in *Advances in Algorithms, Languages, and Complexity*, Du, D. and Ko, K., eds., Kluwer Academic Publishers, Boston, pp. 285–312.

Wang, J. [1991], On p-creative sets and p-completely creative sets, *Theoret. Comput. Sci.* **85**, 1–31.

Wang, J. [1992], Polynomial time productivity, approximations, and levelability, *SIAM J. Comput.* **21**, 1100–1111.

Wang, J. [1993], On the E-isomorphism problem, in *DIMACS Series in Discrete Mathematics and Theoretical Computer Science* **13**, pp. 195–209.

Wang, J. [1997], Average-case intractable NP Problems, in *Advances in Algorithms, Languages, and Complexity*, Du, D. and Ko, K., eds., Kluwer Academic Publishers, Boston, pp. 313–378.

Wang, J. and Belanger, J. [1995], On the NP-isomorphism problem with respect to random instances, *J. Comput. Systems Sci.* **50**, 151–164.

Watanabe, O. [1985], On one-one p-equivalence relations, *Theoret. Comput. Sci.* **38**, 157–165.

Watanabe, O. [1987], A comparison of polynomial time completeness notions, *Theoret. Comput. Sci.* **54**, 249–265.

Wegener, I. [1987], *The Complexity of Boolean Functions*, John Wiley & Sons, New York.

Wilson, C. [1988], A measure of relativized space which is faithful with respect to depth, *J. Comput. Systems Sci.* **36**, 303–312.

Wrathall, C. [1977], Complete sets and the polynomial time hierarchy, *Theoret. Comput. Sci.* **3**, 23–33.

Yao, A. [1977], Probabilistic computations: Toward a unified measure of complexity, *Proceedings of the 18th IEEE Symposium on Foundations of Computer Science*, IEEE Computer Society Press, Los Angeles, pp. 222–227.

Yao, A. [1982] Theory and applications of trapdoor functions, *Proceedings of the 23rd IEEE Symposium on Foundations of Computer Science*, IEEE Computer Society Press, Los Angeles, pp. 80–91.

Yao, A. [1985], Separating the polynomial-time hierarchy by oracles, *Proceedings of the 26th IEEE Symposium on Foundations of Computer Science*, IEEE Computer Society Press, Los Angeles, pp. 1–10.

Yao, A. [1986], Monotone bipartite graph properties are evasive, manuscript.

Yao, A. [1987], Lower bounds to randomized algorithms for graph properties, *Proceedings of the 28th IEEE Symposium on Foundations of Computer Science*, IEEE Computer Society Press, Los Angeles, pp. 393–400.

Yao, A. [1994], A lower bound for the monotone depth of connectivity, *Proceedings of the 35th IEEE Symposium on Foundations of Computer Science*, IEEE Computer Society Press, Los Angeles, pp. 302–308.

Zachos, S. [1983], Collapsing probabilistic polynomial hierarchies, *Proceedings of Conference on Computational Complexity Theory*, University of California, Santa Barbara, pp. 75–81.

Zachos, S. [1986], Probabilistic quantifiers, adversaries, and complexity classes, *Proceedings of the Structure in Complexity Theory Conference*, Lecture Notes in Computer Science **223**, Springer-Verlag, Berlin, pp. 383–400.

Zachos, S. and Heller, H. [1986], A decisive characterization of BPP, *Info. Contr.* **69**, 125–135.

# Index

# WILEY-INTERSCIENCE
# SERIES IN DISCRETE MATHEMATICS AND OPTIMIZATION

## ADVISORY EDITORS

### RONALD L. GRAHAM
*AT & T Laboratories, Florham Park, New Jersey, U.S.A.*

### JAN KAREL LENSTRA
*Department of Mathematics and Computer Science,*
*Eindhoven University of Technology, Eindhoven, The Netherlands*

### ROBERT E. TARJAN
*Princeton University, New Jersey, and*
*NEC Research Institute, Princeton, New Jersey, U.S.A.*

PACH AND AGARWAL • Combinatorial Geometry

PLESS • Introduction to the Theory of Error-Correcting Codes, Third Edition

ROOS AND VIAL • Ph. Theory and Algorithms for Linear Optimization: An Interior Point Approach

SCHEINERMAN AND ULLMAN • Fractional Graph Theory: A Rational Approach to the Theory of Graphs

SCHRIJVER • Theory of Linear and Integer Programming

TOMESCU • Problems in Combinatorics and Graph Theory *(Translated by R. A. Melter)*

TUCKER • Applied Combinatorics, Second Edition

WOLSEY • Integer Programming

YE • Interior Point Algorithms: Theory and Analysis